STANDARD LEVEL

Chemistry
for the IB Diploma Programme

**Catrin Brown, Mike Ford,
Oliver Canning, Andreas Economou and Garth Irwin**

Published by Pearson Education Limited, 80 Strand, London, WC2R 0RL.

www.pearson.com/international-schools

Text © Pearson Education Limited 2023
Development edited by Sarah Ryan
Copy edited by Katharine Godfrey Smith and Linnet Bruce
Proofread by Sarah Ryan and Katharine Godfrey-Smith
Indexed by Georgie Bowden
Designed by Pearson Education Limited
Typeset by Tech-Set ltd
Picture research by Integra
Original illustrations © Pearson Education Limited 2023
Cover design © Pearson Education Limited 2023

The rights of Catrin Brown, Mike Ford, Oliver Canning, Andreas Economou, and Garth Irwin to be identified as the authors of this work has been asserted by them in accordance with the Copyright, Designs and Patents Act 1988.

First published 2023

25 24 23
10 9 8 7 6 5 4 3

British Library Cataloguing in Publication Data
A catalogue record for this book is available from the British Library

ISBN 978 1 29242 769 0

Printed in Slovakia by Neografia

Acknowledgements

The author and publisher would like to thank the following individuals and organisations for permission to reproduce photographs, illustrations, and text:

Text extracts relating to the IB syllabus and assessment have been reproduced from IBO documents. Our thanks go to the International Baccalaureate for permission to reproduce its copyright.

The "In cooperation with IB" logo signifies the content in this textbook has been reviewed by the IB to ensure it fully aligns with current IB curriculum and offers high-quality guidance and support for IB teaching and learning.

KEY (t – top, c – center, b – bottom, l – left, r – right)

Images:

123RF: Yhelfman 184, Glevalex 205t, Emzet70 296, Aleksandr Rado 412, Federico Rostagno 448, Aroonsak Thiranuth 590; **Alamy Stock Photo:** Phil Degginger 11, 275, US Navy Photo 12, Wladimir Bulgar/Science Photo Library 21, 553l, The Picture Art Collection 31, Hank Morgan/Rainbow/RGB Ventures/SuperStock 34, The Book Worm 36, David Taylor 48l, 48c, 48r, World History Archive 72t, 587bl, Sciencephotos 83, GOH SEOK THUAN 94t, Anne Gilbert 94b, Photo Researchers/Science History Images 159, 343, Oxford Science Archive/Heritage Images/The Print Collector 231, Artem Evdokimov 274t, Peter Cripps 274b, Steve Dunwell/Agefotostock 283t, GS UK/Greenshoots Communications 313, Kevin Schafer 342, Er Degginger 346b, Jim West 355, 391, North Wind Picture Archives 378, SpaceX 402, Marwan Naamani/dpa/Alamy Live News 447b, Imaginechina Limited 449, Aerial-photos.com 517tl, Jan Eickhoff/Colouria Media 550, PBH Images 552b, GL Archive 587br, Charlie Newham 596; **Catrin Brown:** Catrin Brown 17, 101, 172, 193, 270bl, 473, 507, 613, 623; **Eva Campbell:** Eva Campbell 171; **Fliegende Blätter:** Fliegende Blätter 586t; **Getty Images:** DigitalGlobe/Maxar 379, IgorSPb/iStock 558, Imagenavi 625; **Microsoft Corporation:** Used by the permission fom Microsoft © Microsoft Corporation 324, 409; **Pearson Education Limited:** Jules Selmes 580; **Reuters:** Reuters 219; **Science Photo Library:** Julien Ordan 4, Greg Dimijian 6, Sheila Terry 9, Charles D. Winters 10t, 10b, 72b, 112, 303b, 345tl, 345b, 433, 444, 483t, 487, 494cl, 510, 513, Andrew Lambert Photography 13t, 13c, 16, 56, 79, 85, 114, 139l, 139r, 234t, 245l, 245r, 246, 247t, 247b, 248, 250, 285, 289l, 289r, 290, 348l, 348r, 418, 432, 442, 471c, 486, 492, 509, 517cr, 518, 535t, 536, 570t, 571, 586b, 592r, Giphotostock 13bl, 13br, 95, 588, Martyn F. Chillmaid 18b, 75, 218, 230, 298, 312, 337, 345tr, 349r, 350, 411, 461, 467, 494b, 505, 535b, 589, George Bernard 19, Adam Hart-davis 20, 253t, Ian Cuming/Ikon Images 28, Physics Dept.,Imperial College 30, Patrick Landmann 40, Kenneth Eward/Biografx 47, 326, David A. Hardy 49, Dr Juerg Alean 51, Dept. Of Physics, Imperial College 52, Laguna Design

57, 143, 264, DR MARK J. WINTER 58, 59t, 59b, 60, Mikkel Juul Jensen 99, 526, Karsten Schneider 104, Oxford Science Archive/Heritage Images 105, Edward Kinsman 106, 156, Martin Shields 111l, 111r, Spencer Sutton/Science Source 117, Clive Freeman, The Royal Institution 130, Alfred Pasieka 132, 180l, 190b, 234b Millard h. Sharp 133, Susumu Nishinaga 137t, 178t, Chris Knapton 137b, Science Photo Library 150, 232, 233b, 245c, 349l, 471tr, 472, 494tl, Hagley Museum and Archive 154, Russell Kightley 157, Carlos Clarivan 167, Victor Habbick Visions 181r, Eye of Science 183, European space agency/nasa, c. Gunn 202, Dr Jeremy Burgess 182, 210, Juan Gaertner 221t, Pascal Goetgheluck 222, Steve Gschmeissner 224, 482, Tek Image 228, Dorling Kindersley/Uig 233r, Noaa 253b, Ramon Andrade 3dciencia 270bc, 270br, Prof. K.seddon & Dr. T.evans, Queen's University Belfast 271, Biosym Technologies, Inc. 294, Turtle Rock Scientific/Science Source 303t, John Durham 309, David Hay Jones 333, Ray Ellis 353, Ed Dlugokencky, Noaa/Gml 359l, Simon Fraser/Mauna Loa Observatory 359r, Tony Craddock 361, Nasa/Ssai Edward Winstead 365, Volker Steger 369, Peter Bowater 374, Crown Copyright/Health & Safety Laboratory 419, Paul Kent, National Center for Computational Sciences, Ornl/Jorge Sofo, Penn State University 456, W K Fletcher 458, Simon Fraser 468, Editorial Image 477, Trevor Clifford Photography 480, Bob Gibbons 483b, Monica Schroeder 484t, Gustoimages 484b, Cordelia Molloy 498, Doug Martin 517tr, Steve Horrell 525, Maximilian Stock Ltd 532, Nasa 555bl, Nasa's Goddard Space Flight Center 555br, Stefan Diller 562, Clive Freeman/Biosym Technologies 577, Jean-loup Charmet 583, Library of Congress, Rare Book and Special Collections Division 587tr, Microscopy Australia 592l, Microgen Images 593, Peggy Greb/Us Department of Agriculture 594t, Equinox Graphics 594b, Wladimir Bulgar 595; **Shutterstock:** Production Perig 2, Studio Light and Shade 18t, Pixel-Shot 41, Artem Onoprienko 46, Susan Santa Maria 93, Krakenimages.com 121, GlennV 146, AptTone 178b, Yes058 Montree Nanta 181b, Juancat 190t, Marcin Wos 205b, Studioloco 216l, Vadim Ratnikov 216r, ABCDstock 217, Exopixel 220, Larina Marina 221b, Leonori 282, Tatjana Baibakova 283b, SJ Travel Photo and Video 300tl, DimaBerlin 300tr, Sararwut Jaimassiri 306, Pitsanu suanlim 346t, XXLPhoto 363, Matteo Cozzi 376, Torychemistry 387, Rabbitmindphoto 394, Pattani Studio 395, Mipan 421t, Vetpathologist 421b, Gigra 430, BORDOMARS 431, Evgeniyqw 434, FoodAndPhoto 447t, Albert Russ 511, ANGHI 522, Everett Collection 523, Dashtik 552t, Jason Patrick Ross 553r, Dimbar76 555cr, Love Employee 556, Jitchanamont 570b, Milatas 631, Romix Image xiii; **The Royal Swedish Academy of Sciences:** ©Airi Iliste/The Royal Swedish Academy of Sciences 180r; **Xkcd:** Xkcd 584.

Text:

Amedeo Avogadro: Amedeo Avogadro 383; **American Association of Physics Teachers:** "Richard Feynman (1969) The Physics Teacher Vol. 7, issue 6, 1969, pp. 313–320, © American Association of Physics Teachers" 29, 587; **Eldredge & brother:** Houston, Edwin James. The Elements of Chemistry: For the Use of Schools, Academies, and Colleges. United States: Eldredge & brother, 1883. 98; **Hachette Book Group:** Feynman, R. P., Sands, M., Leighton, R. B. (2011). The Feynman Lectures on Physics, Vol. I: The New Millennium Edition: Mainly Mechanics, Radiation, and Heat. United Kingdom: Basic Books. 297; **HarperCollins:** Heisenberg, W. (1962). Physics and philosophy; the revolution in modern science. United Kingdom: Harper & Row. 589; **Hebrew University of Jerusalem:** Einstein, Albert, 1879-1955. (1950). Out of My Later Years. New York: Philosophical Library. 597; **Henri Poincaré:** Henri Poincaré, 1854–1912. 229; **John Wiley & Sons, Inc:** Cheng, X., Corey, E. J. (1989). The Logic of Chemical Synthesis. United Kingdom: Wiley. 592; **Lindau Nobel Laureate Meetings:** Kroto S W. 2014 May 12. Sir Harold Kroto (2011) - Créativité Sans Frontières <https://www.mediatheque.lindau-nobel.org/videos/31303/creeativitee-sans-frontires-2011/meeting-2011>. Accessed 2022 Aug 2. 582; **Louis Pasteur:** Louis Pasteur 588; **MIT Press:** Feynman, R. P. (1967). The Character of Physical Law. United Kingdom: MIT Press. 588; **NASA:** NASA 359; **Nature Publishing Group:** Molina, M., Rowland, F. Stratospheric sink for chlorofluoromethanes: chlorine atom-catalysed destruction of ozone. Nature 249, 810–812 (1974). https://doi.org/10.1038/249810a0 557; **Oxford University Press:** Warner, D. o. C. J., Anastas, I. C. B. P., Warner, J. C., Anastas, P. T. (1998). Green chemistry : theory and practice. United Kingdom: Oxford University Press. 576; **Penguin Random House:** Bronowski, J. (2011). The Ascent Of Man. United Kingdom: Ebury Publishing. 582; **Philosophical Library:** Planck, M., Laue, M. v. (1949). Scientific Autobiography: And Other Papers. United States: Philosophical Library. 587; **Random House:** Bronowski, J. (2011). The ascent of man. Random House. 31; **Richard L. Apodaca:** Richard L. Apodaca 607; **Samuel Taylor Coleridge:** Taylor Coleridge, Samuel., The Rime of the Ancient Mariner, 1834. 145; **The George Washington University:** The George Washington University 607; **Werner Heisenberg:** Quote from Werner Heisenberg 31;

MIX
Paper from responsible sources
FSC
www.fsc.org
FSC™ C128612

Contents

Structure

Structure 1 Models of the particulate nature of matter   4

Structure 2 Models of bonding and structure   130

Structure 3 Classification of matter   229

Reactivity

Reactivity 1 What drives chemical reactions?   296

Reactivity 2 How much, how fast and how far?   374

Reactivity 3 What are the mechanisms of chemical change?   456

Syllabus roadmap

The aim of the syllabus is to integrate concepts, topic content and the Nature of Science through inquiry. Students and teachers are encouraged to personalize their approach to the syllabus according to their circumstances and interests.

Skills in the study of chemistry			
Structure Structure refers to the nature of matter from simple to more complex forms		**Reactivity** Reactivity refers to how and why chemical reactions occur	
Structure determines reactivity, which in turn transforms structure			
Structure 1 **Models of the particulate nature of matter**	Structure 1.1 – Introduction to the particulate nature of matter	**Reactivity 1** **What drives chemical reactions?**	Reactivity 1.1 – Measuring enthalpy changes
	Structure 1.2 – The nuclear atom		Reactivity 1.2 – Energy cycles in reactions
	Structure 1.3 – Electron configurations		Reactivity 1.3 – Energy from fuels
	Structure 1.4 – Counting particles by mass: The mole		
	Structure 1.5 – Ideal gases		
Structure 2 **Models of bonding and structure**	Structure 2.1 – The ionic model	**Reactivity 2** **How much, how fast and how far?**	Reactivity 2.1 – How much? The amount of chemical change
	Structure 2.2 – The covalent model		Reactivity 2.2 – How fast? The rate of chemical change
	Structure 2.3 – The metallic model		Reactivity 2.3 – How far? The extent of chemical change
	Structure 2.4 – From models to materials		Reactivity 3.1 – Proton transfer reactions
			Reactivity 3.2 – Electron transfer reactions
Structure 3 **Classification of matter**	Structure 3.1 – The periodic table: Classification of elements	**Reactivity 3** **What are the mechanisms of chemical change?**	Reactivity 3.3 – Electron sharing reactions
	Structure 3.2 – Functional groups: Classification of organic compounds		Reactivity 3.4 – Electron-pair sharing reactions

Authors' introduction to the third edition

Welcome to your study of IB Diploma Programme chemistry. This is the third edition of Pearson's highly successful Standard Level (SL) chemistry book, first published in 2009. It has been completely rewritten to match the specifications of the new IB chemistry curriculum for first assessments in 2025 and gives thorough coverage of the entire course content. While there is much new and updated material, we have kept and refined the features that made the previous editions so successful. We are delighted to share our enthusiasm for learning chemistry in the IB programme with you!

Content

This book covers the entire SL course. It is divided into the two main themes, **Structure** and **Reactivity**, which in turn are divided into six topics, Structure 1–3 and Reactivity 1–3. Separate chapters cover each sub-topic within each topic. For example, the Structure 1.1 chapter deals with 'Introduction to the particulate nature of matter'.

The syllabus is presented as a sequence of numbered Understandings, which are shown as three-part boxes. We have given the relevant Understanding from the subject guide at the start of each section within a chapter under a brief header. The Table of Contents shows the full list of these Understanding headers, so you can see what is covered in each chapter.

For example:

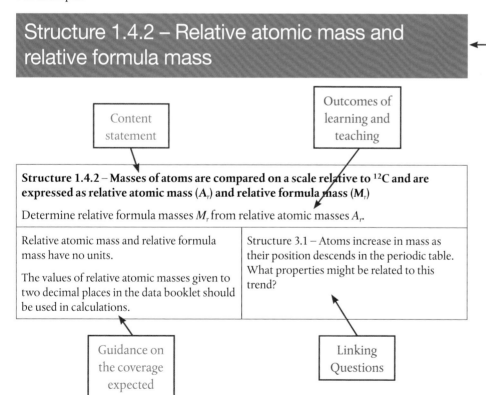

Structure 1.4.2 – Relative atomic mass and relative formula mass

Syllabus header

Content statement

Outcomes of learning and teaching

Structure 1.4.2 – Masses of atoms are compared on a scale relative to ^{12}C and are expressed as relative atomic mass (A_r) and relative formula mass (M_r)

Determine relative formula masses M_r from relative atomic masses A_r.

Relative atomic mass and relative formula mass have no units. The values of relative atomic masses given to two decimal places in the data booklet should be used in calculations.	Structure 3.1 – Atoms increase in mass as their position descends in the periodic table. What properties might be related to this trend?

Guidance on the coverage expected

Linking Questions

The Understandings are presented in the same sequence as in the subject guide.

The text covers the course content using plain language, with all scientific terms explained and shown in bold as they are first introduced. It follows SI notation, and IUPAC nomenclature and definitions throughout. We have been careful also to apply the same terminology you will see in IB examinations in all worked examples and questions.

Conceptual approach

The syllabus emphasizes a conceptual approach, where the two main themes of chemistry, Structure and Reactivity, are shown to be interdependent. There is no suggested sequence for the coverage of the topics, and many different routes through the syllabus are possible. What is important though, is that the relationships between the different topics are recognized, which leads to an increasing depth of understanding.

There are two features, **Guiding Questions** and **Linking Questions**, which are incorporated into each topic in the syllabus. These are designed to help promote the conceptual approach, and we have emphasized them in the text as follows.

Guiding Question

Guiding Question

How do the nuclei of atoms differ?

Each chapter starts with the Guiding Question for the sub-topic from the IB chemistry subject guide. This is followed by a brief consideration of the question itself, which sets the context for the topic and how it relates to your previous knowledge. It is expected that by the end of the chapter you will be able to answer the Guiding Question more fully.

Guiding Question revisited

Guiding Question revisited

How do we quantify matter on the atomic scale?

At the end of each chapter, the Guiding Question is revisited, and can now be answered with significantly more detail and understanding. It is presented as a bulleted list of the material covered, that may help serve as a checklist of your learning at the end of each chapter. The Guiding Question revisited bulleted lists are available as downloadable PDFs from the eBook to help you with revision.

Linking Questions

Reactivity 3.2 – How can oxidation states be used to analyze redox reactions?

Linking Questions are given in many of the Understandings. Linking Questions have a number which indicates a link from the current chapter to another sub-topic, to Tools or to the Nature of Science (NOS). These questions are designed as prompts to help you build a grasp of unifying concepts and to stimulate further learning. Linking Question boxes can be found in the margin next to the content they link to (see example on the left).

By their very nature, the Linking Questions make reference to different parts of the course, some of which you may not have studied yet. As the questions can be asked in either direction, you may choose to consider them as part of the study of either or both of the linked topics. You will again find that you are able to answer the question more fully as your knowledge and understanding increase.

The Linking Questions are designed to lead to a thoughtful response. For this reason, we have given brief answers alongside the questions in the text, and hope these stimulate further consideration of the question.

The Nature of Science

Throughout the course you are encouraged to think about the nature of scientific knowledge and the scientific process as it applies to chemistry. Examples are given of the evolution of chemical theories as new information is gained, the use of models to conceptualize our understanding, and the ways in which experimental work is enhanced by modern technologies. Ethical considerations, environmental impacts, the importance of objectivity, and the responsibilities regarding scientists' code of conduct are also considered here. The emphasis is not on learning any of these examples, but on appreciating the broader conceptual themes in context. We have included several NOS examples in each chapter, and hope you will come up with your own as you keep these ideas at the forefront of your learning.

Key to feature boxes

You will find different coloured feature boxes interspersed throughout each chapter. These are used to enhance your learning and how it applies to real world examples, as explained below.

 Nature of Science

As mentioned this is an overarching theme in the course. Throughout the book you will find NOS themes and questions emerging across different topics. We hope they help you to develop your skills in scientific literacy.

Nature of Science

Dalton's atomic theory was not accepted when it was first proposed. Many scientists considered it as nothing more than a useful fiction which should not be taken too seriously. Over time, as the supporting evidence grew, this changed to its widespread acceptance. These revolutions in understanding, or '**paradigm shifts**', are characteristic of the evolution of scientific thinking.

 Global context

The impact of the study of biology is global, and includes environmental, political and socio-economic considerations. Examples of these are given to help you see the importance of biology in an international context. These examples also illustrate some of the innovative and cutting-edge aspects of research in biology.

 A different unit of concentration is known as **ppm**, parts per million. It denotes one part per 10^6 parts of the whole solution, and is useful in describing very low concentrations..

Precipitation of AgI precipitate from aqueous solutions. Full details of how to carry out this experiment with a worksheet are available in the eBook.

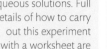

How might developments in scientific knowledge trigger political controversies or controversies in other areas of knowledge?

At equilibrium the rate of the forward reaction is equal to the rate of the backward reaction.

Note the definition of bond enthalpy indicates that all the species have to be in the gaseous state..

Skills in the study of chemistry

These boxes indicate links to the skills section of the course, including ideas for laboratory work and experiments that will support your learning and help you prepare for the Internal Assessment. These link to further resources in the eBook (look out for the grey icon).

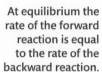 Theory of Knowledge

These questions, which are mostly from the Theory of Knowledge (TOK) guide, stimulate thought and consideration of knowledge issues as they arise in context. The questions are open-ended and will help trigger critical thinking and discussion.

Key fact

Key facts are drawn out of the main text and highlighted in bold. These boxes will help you to identify the core learning points within each section. They also act as a quick summary for review.

Hint for success

These give hints on how to approach questions, and suggest approaches that examiners like to see. They also identify common pitfalls in understanding, and omissions made in answering questions.

Challenge yourself

These boxes contain probing questions that encourage you to think about the topic in more depth, and may take you beyond the syllabus content. They are designed to be challenging and to make you think.

Challenge yourself

1. The components of a mixture can usually be separated by physical means. What might be the challenges of trying to separate the metals from an alloy?

Interesting fact

These give background information that will add to your wider knowledge of the topic and make links with other topics and subjects. Aspects such as historic notes on the life of scientists and origins of names are included here.

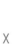 Gilbert Lewis (1875–1946), responsible for the electron-pair theory of the covalent bond, was nominated 41 times for the Nobel Prize in Chemistry without winning. In 1946, he was found dead in his laboratory where he had been working with the toxic compound hydrogen cyanide.

Towards the end of the book, there are chapters on Green Chemistry, TOK as it relates to chemistry, the Internal Assessment, the Extended Essay and Strategies for success in IB chemistry.

Questions

There are three types of question in this book.

1. Worked examples with solutions

Worked examples appear at intervals in the text and are used to illustrate the concepts covered.

They are followed by the solution, which shows the thinking and the steps used in solving the problem.

Worked example

Identify the species with 19 protons, 20 neutrons and 18 electrons.

Solution

- the number of protons tells us the atomic number, $Z = 19$, and so the element is potassium, K
- the mass number $= p + n = 19 + 20 = 39$: $^{39}_{19}K$
- the charge $= p - e = 19 - 18 = +1$ as there is one extra proton: $^{39}_{19}K^+$

2. Exercises

These questions are found throughout the text, usually at the end of each Understanding. They allow you to apply your knowledge and test your understanding of what you have just been reading. The answers to these, together with worked solutions, are accessed via icons on the first page of each chapter. Exercise answers can also be found at the back of the eBook.

Exercise

Q1. Explain why the relative atomic mass of tellurium is greater than the relative atomic mass of iodine, even though iodine has a greater atomic number.

3. Practice questions

These questions are found at the end of each chapter. They are mostly taken from previous years' IB exam papers. The markschemes used by examiners when marking these questions are accessed via icons in the eBook on the first page of each chapter.

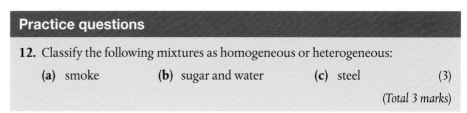

Practice questions

12. Classify the following mixtures as homogeneous or heterogeneous:

    (a) smoke   **(b)** sugar and water   **(c)** steel   (3)

(Total 3 marks)

eBook

In your eBook you will find more information on the Skills section of the course including detailed suggestions for lab work. You will also find links to videos. In addition, there are auto-marked quizzes in the Exercises tab of your eBook account (see screenshot below).

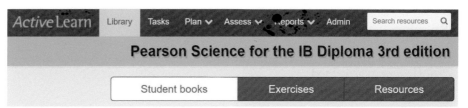

We hope you enjoy your study of IB chemistry from this textbook.

    Catrin Brown, Mike Ford, Oliver Canning, Andreas Economou and Garth Irwin

Introduction to skills in the study of chemistry

"I hear and I forget, I see and I remember, I do and I understand"

Chemistry is an experimental science and its progress continues to be based on the scientific method. The study of chemistry in the IB programme reflects this in the emphasis it places on laboratory work. This approach will help you to understand concepts, learn some practical skills and give you opportunities to explore further through investigations. It is also often the best part about studying chemistry!

The syllabus roadmap on page vi has 'Skills in the study of chemistry' at the top. This unit is a summary of experimental skills and techniques that should be experienced during the course, including the application of technology and mathematics. It is not intended that these 'Tools' are covered as separate content, but should be integrated into the study of all topics. The Skills in the study of chemistry chapter towards the back of this book includes tables that summarize the details of this unit with references to where in the syllabus content it may be suitably included. There are also many links to experimental work and resources available in the ebook.

Nature of Science

The scientific method consists of systematic observation, the formulation and testing of hypotheses, and the collection of data for analysis and evaluation. Central to this process is a consideration of the limits of the usefulness of the data obtained. The interpretation of results must take into account the precision and accuracy of the measurements, the amount and limitations of data collected and the reproducibility of the results. Effective communication of the results must include this information.

The choice of apparatus for measurement in the laboratory determines the precision of the data collected. Here a pipette is used to deliver a single drop of liquid.

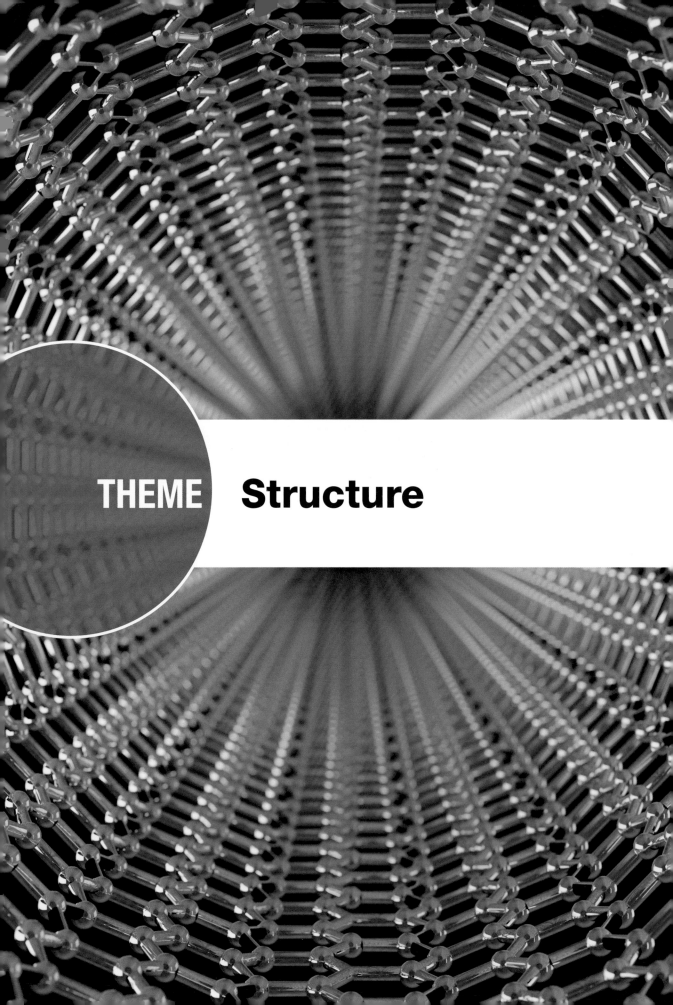

THEME **Structure**

View of graphene with 3D rendering. Graphene is a relatively modern structure, usually composed of a single layer of carbon atoms. It is considered to be the world's thinnest, strongest and most conductive material, of both electricity and heat. Understanding the structure of graphene has helped develop its applications in many fields such as energy generation, sensors, medical equipment and composite materials.

Structure

Structure refers to the nature of matter. Chemists seek to understand the way in which fundamental particles, the building blocks of all chemical structure, combine to form every chemical structure that exists – from single atoms to the most complex compounds. An understanding of structure leads to the ability to explain and predict chemical properties, as studied in our second theme, *Reactivity*.

A major challenge in the study of structure is that the fundamental particles are too small for us to observe directly. Even the most advanced technology gives us only limited information on the nature and behavior of these particles. As a result, since the early days of chemical exploration, scientists have used models to help explain and predict the nature of matter. Over time, these models have developed and changed in the light of increasingly detailed observations and new evidence. As we explore the applications of these models, we must also consider their limitations and how they may continue to evolve.

In Structure 1 we consider evidence for the particulate nature of matter. From an exploration of the properties of sub-atomic particles, we build an understanding of the structure of atoms and how they characterize the unique properties of each element. The problem of scale, how we quantify what we cannot observe directly, is addressed through an introduction to the mole as the unit of amount in chemistry. A detailed study of electron configurations helps us to recognize why atoms of different elements differ in their tendency to attract electrons. This leads to descriptions of the models of different chemical bonds – ionic, covalent and metallic – in Structure 2. The organization of elements in the periodic table, as studied in Structure 3, suggests patterns in elements' properties. This gives predictive power to the types of bonds that they will form. As atoms associate through bond formation, the products have different properties, giving rise to an infinite variety of structures. Chemists have developed clear terminology to communicate about chemical structure, and in Structure 3 we are introduced to the IUPAC (International Union of Pure and Applied Chemistry) system of nomenclature, and learn how compounds are given unambiguous and internationally agreed names.

Our understanding of structure has developed alongside advances in technology and has led to many innovations in materials science. Many modern materials, such as breathable fabrics and biodegradable plastics, and compounds such as therapeutic drugs, are designed for specific functions. You need look no further than your smart phone or clothing to realize how much our lives are influenced by these products. And yet, despite their extraordinary variety and complexity, every chemical structure is based on associations between a relatively small number of different atoms.

Structure determines reactivity, which in turn transforms structure.

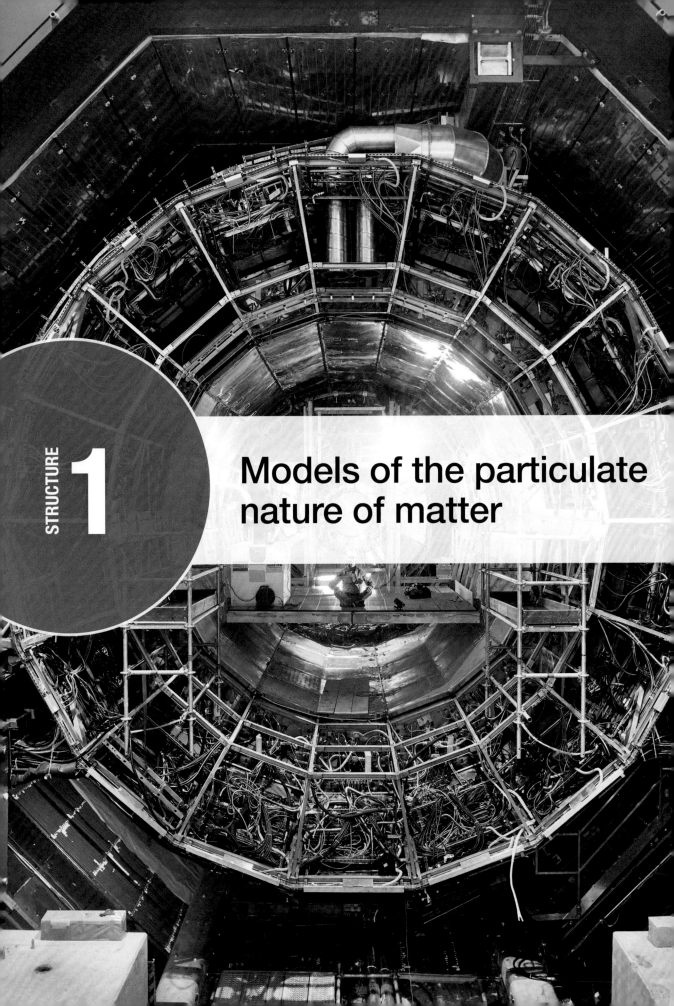

1

Models of the particulate nature of matter

Inside the empty skeleton of the ALICE detector at CERN (the European particle physics laboratory) near Geneva, Switzerland. ALICE (A Large Ion Collider Experiment) is a detector built around the Large Hadron Collider (LHC). The LHC is the world's largest and most powerful particle collider. Beams of ions are accelerated to collide head-on. The collision energy creates new particles that decay into other particles. The LHC energies have allowed the study of exotic materials like quark-gluon plasma, a form of quark matter. CERN announced the Higgs boson discovery on 4 July 2012.

Nature of Science

Progress in science often follows technological developments. The discovery of the Higgs boson was due to the use of particle accelerators, detectors and sophisticated computers. Such technological advances are only possible with international collaboration between scientists. Scientists communicate and collaborate throughout the world. CERN is run by over 20 member states, and many non-European countries are involved in different ways.

One of the earliest questions philosophers asked concerned the divisibility of matter. Could a piece of material be divided again and again continually into smaller and smaller pieces as Aristotle proposed, or would a limit be reached, a single particle, as Democritus argued? The latter idea has generally stood the test of time. This particulate model of matter enables us to explain many aspects of the behavior of matter despite these particles being not directly visible. We have no proof that matter is not infinitely divisible, but we do have evidence that it becomes increasingly more difficult to divide it into smaller pieces. The limiting factor is the energy required to divide a particle. Experimental evidence for this has come from work with particle accelerators, such as at CERN, referenced in the photo at the start of the chapter, where beams of ions undergo high-energy collisions, which produce new particles. We are still learning about how the world is made up of fundamental particles. The Higgs boson, a key elementary particle, was only discovered as recently as 2012 although its existence had been predicted by scientific models in 1964.

Introduction to the particulate nature of matter

Introduction to the particulate nature of matter

◀ Yellowstone National Park, Wyoming, USA, with mist rising on a cold winter morning from waters warmed by thermal springs. Matter can exist in the solid, liquid and gaseous states. The difference in physical properties of the three states is explained by kinetic molecular theory.

Guiding Question

How can we model the particulate nature of matter?

All things are made from atoms. This is one of the most important ideas that the human race has learned about the universe. Atoms are everywhere and they make up everything. You are surrounded by atoms – they make up the foods you eat, the liquids you drink and the fragrances you smell. Atoms make up you! To understand the world and how it changes, you need to understand atoms.

Atoms are the smallest particle of an element and so are fundamental to chemistry. As they are too small ever to be seen directly by a human eye, scientists have developed a series of models to explain the physical and chemical properties of materials. We can understand many aspects of the world simply by considering how these particles move and interact with each other. This particulate model of matter, which includes atoms, molecules and ions, provides a very strong foundation for our chemical understanding and is a good place to start our explorations.

Nature of Science

Scientists construct models as artificial representations of natural phenomena. All models have limitations that need to be considered in their application.

Structure 1.1.1 – Elements, compounds and mixtures

Structure 1.1.1 – Elements are the primary constituents of matter, which cannot be chemically broken down into simpler substances.

Compounds consist of atoms of different elements chemically bonded together in a fixed ratio.

Mixtures contain more than one element or compound in no fixed ratio, which are not chemically bonded and so can be separated by physical methods.

Distinguish between the properties of elements, compounds and mixtures.

Solvation, filtration, recrystallisation, evaporation, distillation and paper chromatography should be covered.	Tool 1 – What factors are considered in choosing a method to separate the components of a mixture?
The differences between homogeneous and heterogeneous mixtures should be understood.	Tool 1 – How can the products of a reaction be purified?
	Structure 2.2 – How do intermolecular forces influence the type of mixture that forms between two substances?
	Structure 2.3 – Why are alloys generally considered to be mixtures, even though they often contain metallic bonding?

Matter can be classified into elements, compounds and mixtures

Classifying matter

Matter is present in an infinite number of different forms. The first step to understanding the chemistry of all these substances is an effective classification system. We start with a basic chemical distinction: some matter is made of pure substances, and some is made from mixtures. The chapter will follow the classification outlined in Figure 1.

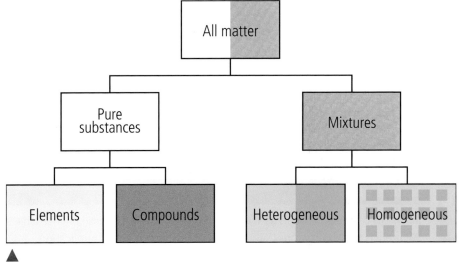

▲ **S1.1 Figure 1** Matter can be classified into pure substances and mixtures. Pure substances can be elements or compounds. Mixtures can be homogeneous or heterogeneous.

Elements are the primary constituents of matter, which cannot be chemically broken down into simpler substances

An element is a pure chemical substance composed of atoms with the same number of protons in the atomic nucleus.

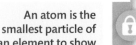

An atom is the smallest particle of an element to show the characteristic properties of that element.

All languages are based on an alphabet of a limited number of characters. The 26 letters of the English alphabet, for example, can be combined in different ways to form the estimated 600 000 words in the language. In a similar way, the substances known as chemical elements can combine to form the material of our universe. Elements are the primary constituents of matter. They cannot be chemically broken down into simpler substances. The near infinite number of different chemical substances in our world are made from only about 100 known elements.

In Structure 2.1 we will learn about atomic structure, and how each element is made up of a particular type of atom. Atoms of the same element all have the same number of protons in the atomic nucleus and the same number of electrons. The distinct make-up of an element's atoms gives each element its individual properties. An atom is the smallest particle of an element to show the characteristic properties of that element.

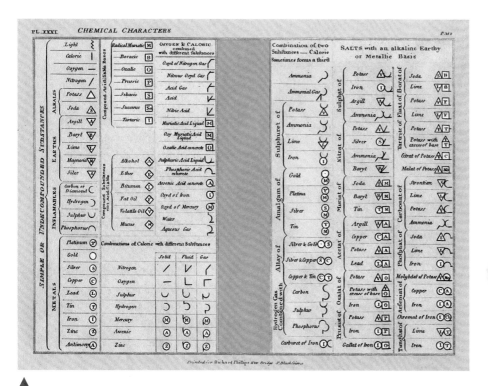

PL. XXXI. CHEMICAL CHARACTERS

Pictographic symbols used at the beginning of the 18th century to represent chemical elements and compounds. They are similar to those of the ancient alchemists. As more elements were discovered during the 18th century, attempts to devise a chemical nomenclature led to the modern alphabetic notational system. This system was devised by the Swedish chemist Berzelius and introduced in 1814.

Each element is denoted by a **chemical symbol** and some examples are given below.

Name of element	Symbol
carbon	C
fluorine	F
potassium	K
calcium	Ca
mercury	Hg
tungsten	W

You will notice that often the letter or letters used to represent the elements are derived from its English name, but in some cases, they derive from other languages. For example, Hg for mercury comes from Latin, whereas W for tungsten has its origin in European dialects. These symbols are all accepted and used internationally, which allows for effective global communication between chemists. A complete list of the names of the elements and their symbols is given in Section 6 of the data booklet.

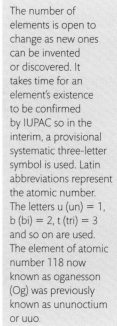

The number of elements is open to change as new ones can be invented or discovered. It takes time for an element's existence to be confirmed by IUPAC so in the interim, a provisional systematic three-letter symbol is used. Latin abbreviations represent the atomic number. The letters u (un) = 1, b (bi) = 2, t (tri) = 3 and so on are used. The element of atomic number 118 now known as oganesson (Og) was previously known as ununoctium or uuo.

Chemistry is an exact subject; it is important to distinguish between upper and lower case letters. Cobalt (Co) is a metallic element whereas carbon monoxide (CO) is a poisonous gas.

Compounds consist of atoms of different elements chemically bonded together in a fixed ratio

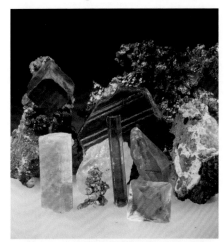

Assorted minerals, including elements such as sulfur and silver, and compounds such as Al_2O_3 (sapphire) and CaF_2 (fluorite). Most minerals are impure and exist as mixtures of different elements and compounds.

Some elements, such as nitrogen and gold, are found in **native** form, uncombined with other elements in nature. But more commonly, elements exist in chemical combinations with other elements, as **chemical compounds**. Compounds consist of atoms of different elements chemically bonded together in a fixed ratio. The physical and chemical properties of compounds are completely different from those of their component elements.

Sodium chloride, for example, is a white crystalline solid that is added to improve the taste of food, whereas sodium is a dangerously reactive metal that reacts violently with water, and chlorine is a toxic gas.

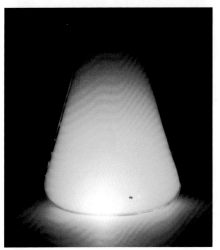

Sodium, Na, reacts violently with chlorine, Cl_2, to produce white crystals of the compound sodium chloride, NaCl.

$$2Na(s) + Cl_2(g) \rightarrow 2NaCl(s)$$

The properties of the compound are completely different from those of its component elements.

The **chemical formula** of a compound uses a combination of chemical symbols of its constituent elements. A subscript is used to show the number of atoms of each element in a unit of the compound. Some examples are given below. (The reasons for the different ratios of elements in compounds will become clearer after we have studied atomic structure and bonding in Structure 2.)

A compound is a substance made by chemically combining two or more elements in a fixed ratio of atoms. The physical and chemical properties of a compound are different from those of its constituent elements.

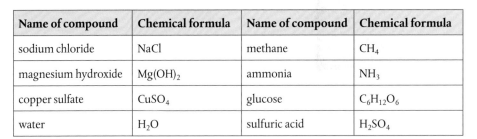

Name of compound	Chemical formula	Name of compound	Chemical formula
sodium chloride	NaCl	methane	CH_4
magnesium hydroxide	$Mg(OH)_2$	ammonia	NH_3
copper sulfate	$CuSO_4$	glucose	$C_6H_{12}O_6$
water	H_2O	sulfuric acid	H_2SO_4

Mixtures contain more than one element or compound in no fixed ratio

Chemistry is primarily concerned with understanding the structure and reactivities of pure substances but dealing with mixtures is a practical reality. Although there are a countable number of pure substances, the elements and compounds, which all have distinctive names, there is an infinite number of mixtures. The composition of any one mixture can vary continuously – the relative amounts of ice cream, milk and flavorings in a milkshake can vary depending on the taste of the person preparing the drink. Names for all the mixtures we could make do not exist.

Air is described as a **mixture** of gases because the separate components – different elements and compounds – are interspersed with each other, but not chemically combined. The gases nitrogen and oxygen when mixed in air, for example, have the same properties as they do when they are in pure samples. Substances burn in air because they react with the oxygen in the same way that they react with pure oxygen. As substances can be mixed in any proportion, mixtures, in contrast to compounds, do not have a fixed composition and cannot be represented by a chemical formula. The composition of air for example varies widely around the world due to the presence of different pollutants.

Air is an example of a **homogeneous mixture**, as it has uniform composition and properties throughout. A solution of salt in water and a metal alloy such as bronze, which is a mixture of copper and tin, are also homogeneous. To form a homogeneous mixture, the inter-particle attraction within the different components must be similar in nature to those between the components in the mixture. Explanations of the different particle interactions are given in Structure 2.2.

Mixture	Component 1	Particle interaction 1	Component 2	Particle interaction 2	Interaction in mixture
air	N_2	dispersion forces	O_2	dispersion forces	dispersion forces
bronze	Cu	metallic bonding	Sn	metallic bonding	metallic bonding
salt water	NaCl	ionic	H_2O	hydrogen bonding	ion-dipole

◀ Making alloys at a steel mill. Alloys are formed by mixing two molten metals that can then form a mixture of uniform composition. There is no chemical reaction between the metallic elements as the metallic bonding present in the individual metals and the alloys is non-directional and not significantly disrupted by the mixing process. Metallic bonding is explored more fully in Structure 2.3.

TOK

Our classification systems are embedded in the language we use. To what extent do the classification systems we use affect the knowledge we obtain? To what extent do the names and labels that we use help or hinder the acquisition of knowledge?

A mixture contains more than one element or compound in no fixed ratio, which are not chemically bonded. The components of a mixture can be separated by physical methods.

Structure 2.3 – Why are alloys generally considered to be mixtures, even though they often contain metallic bonding?

Challenge yourself

1. If you add 10 cm³ of water to 10 cm³ of water, you get 20 cm³ of water. Similarly, if you add 10 cm³ of ethanol to 10 cm³ of ethanol you get 20 cm³ of ethanol. Explain why the volume of the solution formed between 10 cm³ of water and 10 cm³ ethanol is less than 20 cm³.

Structure 2.2 – How do intermolecular forces influence the type of mixture that forms between two substances?

Ocean oil spills can occur when oil is extracted or transported. They cause widespread damage to the environment and can have a major impact on local industries such as fishing and tourism. Efforts to reduce the impact of the spill include the use of dispersants, which break up the oil into smaller droplets, allowing it to mix better with water.

A **heterogeneous mixture**, by contrast, such as water and oil, has a non-uniform composition, and its properties are not the same throughout. The interactions between the components of a heterogeneous mixture are different in nature. The water molecules interact by hydrogen bonding and the oil molecules by dispersion forces, as discussed in Structure 2.2.

It is usually possible to see the separate components in a heterogeneous mixture but not in a homogeneous mixture.

Smoke from a controlled burn of the oil spill caused by the explosion of the Deepwater Horizon oil rig, 20th April 2010. Five thousand barrels of oil a day leaked into the Gulf of Mexico, harming local wildlife and fishing industries.

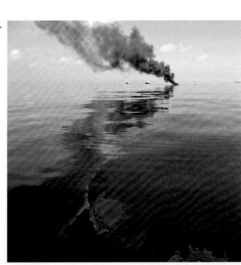

Challenge yourself

2. Does a mixture have the same classification at all scales?

The components in a mixture can be relatively easy to separate if they have a distinct physical property. The technique chosen will depend on this distinct property.

Mixture	Difference in property of components	Technique used
sand and salt	solubility in water	solution and filtration
salt and water	boiling point	distillation
iron and sulfur	magnetism	response to a magnet
pigments in food colouring	adsorption to solid phase	paper chromatography

SKILLS

The required product of a chemical reaction similarly needs to be separated from the reaction mixture, which also contains unreacted reactants and other unwanted products. It is a practical challenge to both maximize the yield and the purity, as some product can be removed with the impurities during the purification process.

Filtration is the process where a solid is separated from a liquid or gas using a membrane. The solid is collected on the membrane as the residue, and the filtrate containing the solute passes through. Sand and salt have different solubilities in water so can be separated by adding water which dissolves the salt – an example of **solvation**. The resulting mixture can be filtered with the insoluble sand removed as the **residue** leaving the salt water solution as the **filtrate**. The water can be separated from the salt by evaporation, which allows the salt crystals to form (**crystallization**).

◄ Filtration apparatus

Tool 1 – What factors are considered in choosing a method to separate the components of a mixture?

Tool 1 – How can the products of a reaction be purified?

Distillation is used to separate a solvent from a solute. The solvent has a lower boiling point than the solute and so is collected as a gas and passes into the condensing tube, which is surrounded by cold flowing water. The gas is condensed into the pure solvent collected in the beaker at the bottom. This method can be used to separate water from sea water, for example.

SKILLS
Preparation of AgI precipitate from aqueous solutions. Full details of how to carry out this experiment with a worksheet are available in the eBook.

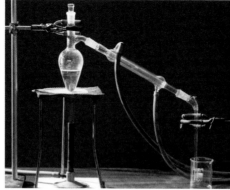

◄ Distillation apparatus

In **paper chromatography**, small spots of solutions containing the samples being tested are placed on the base line. The paper is suspended in a closed container to ensure that the paper is saturated **(a)**. The different components have different affinities for the water in the paper (the solvent) and so separate as the solvent moves up the paper **(b)**. This method can be used to investigate the different pigments in food colouring.

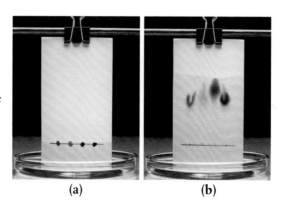

(a)   **(b)**

◄ Paper chromatography apparatus

Exercise

Q1. Identify the homogeneous mixture.

    A water and oil   **B** sand and water

    C salt and water   **D** sand and salt

Q2. Identify the correct descriptions about mixtures.

    I. The components can be elements or compounds.

    II. All components must be in the same state.

    III. The components retain their individual properties.

    A I and II only   **B** I and III only

    C II and III only   **D** I, II and III

Q3. Identify the homogeneous mixtures.

    I. gold

    II. bronze

    III. steel

    A I and II only   **B** I and III only

    C II and III only   **D** I, II and III

Q4. **(a)** River water needs to be purified to make it safe to drink. The water passes through a grid and is then left in a tank where other impurities are removed as they fall to the bottom in a process known as sedimentation. Suggest why pollutants such as fertilizers are not removed in this process.

    (b) Drinking water can be obtained from seawater by distillation.

        Explain the disadvantages of using distillation to obtain large amounts of drinking water.

Q5. A layer of oil paint is left to dry in the air and it hardens. Suggest what causes the oil to harden.

Q6. Metal coins are made from different metals. The composition of a 20-cent and a 50-cent euro coin is given below.

	Mass / g			
Coin	**Copper**	**Aluminium**	**Zinc**	**Tin**
20 cent	5.11	0.287	0.287	0.057
50 cent	6.94	0.390	0.390	0.080

Compare the chemical compositions of the two coins.

Structure 1.1.2 – The kinetic molecular theory

Structure 1.1.2 – The kinetic molecular theory is a model to explain physical properties of matter (solids, liquids and gases) and changes of state.

Distinguish the different states of matter.

Use state symbols (s, l, g and aq) in chemical equations.

Names of the changes of state should be covered: melting, freezing, vaporization (evaporation and boiling), condensation, sublimation and deposition.	Structure 2.4 – Why are some substances solid while others are fluid under standard conditions?
	Structure 2 (all), Reactivity 1.2 – Why are some changes of state endothermic and some exothermic?

The kinetic molecular theory explains physical properties

If you were hit with a bucket of solid water (ice), you could be seriously injured, but the same mass of liquid water would only be annoying. Gaseous water (steam) could also be harmful but for different reasons. These three samples are all made from the same particles: water molecules, H_2O. The difference in physical properties of the three states is explained by **kinetic molecular theory**.

Matter exists in different states as determined by the temperature and the pressure

From our common experience, we know that all matter (elements, compounds, and mixtures) can exist in different forms depending on the temperature and pressure. Liquid water changes into a solid form, such as ice, hail, or snow, as the temperature drops, and it becomes a gas, steam, at higher temperatures. These different forms are known as the **states of matter** and are characterized by the different energies of the particles.

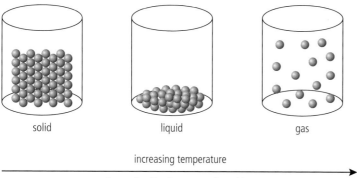

increasing temperature

increasing kinetic energy of particles

◄ **S1.1 Figure 2** Representation of the arrangement of the particles of the same substance in the solid, liquid, and gas states.

i A fourth state of matter, plasma, exists only at conditions of very high temperatures and pressures, such as those commonly found in stars such as our Sun.

Solid	Liquid	Gas
• particles closely packed	• particles more spaced	• particles fully spread out
• inter-particle forces strong, particles vibrate in position	• inter-particle forces weaker, particles can slide over each other	• inter-particle forces negligible, particles move freely
• fixed shape	• no fixed shape	• no fixed shape
• fixed volume	• fixed volume	• no fixed volume

This is known as the kinetic molecular theory of matter as the three states are distinguished by the way the particles move. The temperature of the system is directly related to the average kinetic energy of the particles and the state of matter at a given temperature and pressure is determined by the strength of **inter-particle forces** that exist between the particles relative to this average kinetic energy. If the inter-particle forces are sufficiently strong to keep the particles in position at a given temperature and pressure, the substance will be a solid. If not, it will be a liquid or a gas. The different inter-particle forces are explored more fully in Structure 2.

 Depending on the chemical nature of the substance, matter may exist as atoms such as Ar(g), as molecules such as $H_2O(l)$, or as ions such as Na^+ and Cl^- in aqueous solution NaCl(aq). The term **particle** is therefore used as an inclusive term that is applied in this text to any or all of these entities of matter.

Challenge yourself

3. What evidence, based on simple observations, can you think of that supports the idea that water is made from discrete particles?

Temperature is a measure of the average kinetic energy of the particles of a substance.

Structure 2.4 – Why are some substances solid while others are fluid under standard conditions?

▲
Bromine liquid, $Br_2(l)$, has been placed in the lower gas jar only, and its vapour has diffused to fill both jars. Bromine vaporizes readily at room temperature and its distinctive colour allows the diffusion to be observed.

Worked example

Which of the following has the highest average kinetic energy?

A He at 100 °C   **B** H_2 at 200 °C   **C** O_2 at 300 °C   **D** H_2O at 400 °C

Solution

D   The substance at the highest temperature has the highest average kinetic energy.

Liquids and gases are both referred to as **fluids**, as they have the ability to flow. In the case of liquids, it means that they take the shape of their container. **Diffusion**, the process by which the particles of a substance spread out more evenly, occurs as a result of their random movements. It occurs predominantly in these two fluid states.

Kinetic energy (E_k) refers to the energy associated with movement or motion. It is determined by the mass (m) and speed or velocity (v) of a substance, according to the relationship:

$$E_k = \tfrac{1}{2} mv^2$$

As the average kinetic energy of all particles at the same temperature is the same, there is an inverse relationship between mass and velocity:

$$E_k = \tfrac{1}{2} m_1 v_1{}^2 = \tfrac{1}{2} m_2 v_2{}^2$$

$$\frac{m_1}{m_2} = \frac{v_2{}^2}{v_1{}^2}$$

Particles with smaller mass therefore diffuse more quickly than those with greater mass, at the same temperature.

State symbols are used to show the states of the reactants and products taking part in a reaction. These are abbreviations, which are given in parenthesis after each term in an equation, as shown below.

State	Symbol	Example
solid	(s)	Fe(s)
liquid	(l)	Br_2(l)
gas	(g)	O_2(g)
aqueous	(aq)	H_2SO_4(aq)

For example:

$$2Na(s) + 2H_2O(l) \rightarrow 2NaOH(aq) + H_2(g)$$

Sodium hydroxide, NaOH, is produced during this reaction in **aqueous solution**.

Solutions more generally are mixtures of two components. The less abundant component is the **solute** and the more abundant is the **solvent**. The solute can be solid, liquid or gas but the solvent is generally a liquid. Solutions in water are particularly important and are given the state symbol (aq).

It is good practice to show state symbols in all equations, even when they are not specifically requested.

Matter changes state reversibly

As the movement or kinetic energy of the particles increases with temperature, they will overcome the inter-particle forces and change state. This occurs at a fixed temperature and pressure for each substance. A solid, X(s), changes to a liquid at a defined **melting point** and a liquid, Y(l), changes to a gas at its **boiling point**.

These changes can be reversed when the substance is cooled: a liquid freezes and becomes solid at the same temperature as the solid melts.

A gas or vapour condenses at the same temperature as the liquid boils.

The reversibility of these changes can be represented by reversible arrows as shown in the table.

Change of state		Change of state		Reversible equation
melting	X(s) → X(l)	freezing	X(l) → X(s)	X(s) ⇌ X(l)
boiling	Y(l) → Y(g)	condensing	Y(g) → Y(l)	Y(l) ⇌ Y(g)

Some substances can also change directly between the solid and gaseous states.

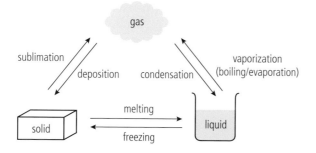

Sublimation is the direct inter-conversion of a solid to a gas without going through the liquid state. It is characteristic of some substances such as iodine, carbon dioxide and ammonium chloride.

Deposition is the reverse of sublimation and occurs when a gas changes directly to a solid. It occurs when snow and frost are formed.

The temperature of a state change can be presented from both perspectives:

melting point = freezing point;

boiling point = condensation point.

S1.1 Figure 3 The different state changes between the solid, liquid and gaseous states.

SKILLS

Sublimation of iron and iodine. Full details of how to carry out this experiment with a worksheet are available in the eBook.

Ice crystals, known as hair ice, formed by deposition on dead wood in a forest on Vancouver Island, Canada.

A liquid can change to gas by two processes: **evaporation** and **boiling**. Boiling occurs at a fixed temperature for a given pressure, when bubbles of gas form throughout the liquid. Evaporation, by contrast, occurs only at the surface and takes place over a range of temperatures below the boiling point.

The heat of the Sun enables all the water to evaporate from the clothes.

More precisely, the boiling point is the temperature at which the vapour pressure is equal to the external pressure. As a liquid is heated, more particles enter the vapour state and the vapour pressure increases. When the vapour pressure reaches the external atmospheric pressure the liquid boils. When the external pressure is lower, the vapour pressure needed to boil is reduced and so boiling occurs at a lower temperature. The relationship between temperature and vapour pressure and the influence of external pressure on the boiling point is demonstrated in Figure 4.

S1.1 Figure 4 Graph showing the boiling point of water at three different pressures.

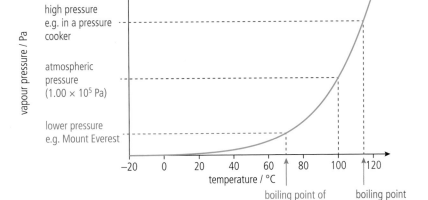

A pressure cooker is a sealed container in which a high pressure can be generated. This raises the boiling point of water, allowing cooking time to decrease. Conversely, at altitude, where the atmospheric pressure is lower, the boiling point of water is reduced so it takes much longer to cook food.

A butane gas camping stove. Butane, C_4H_{10}, is stored as a liquid because the high pressure in the canister raises its boiling point. When the valve is opened the release of pressure causes the butane to boil, releasing a gas that can be burned.

Challenge yourself

4. Propane (C_3H_8) and butane (C_4H_{10}) are both commonly used in portable heating devices. Their boiling points are butane $-1\,°C$ and propane $-42\,°C$.
   Suggest why butane is less suitable for use in very cold climates.

◀ Macrophotograph of freeze-dried instant coffee granules.

Freeze-drying is used to preserve food and some pharmaceuticals. It differs from standard methods of dehydration as the energy needed to evaporate water is produced by the sublimation of ice. The substance to be preserved is first frozen, and then warmed gently at very low pressure which causes the ice to change directly to water vapour. The process is slow but has the significant advantage that the composition of the material, and so also its flavor, are largely conserved. The freeze-dried product is stored in a moisture-free package that excludes oxygen, and can be reconstituted by the addition of water.

Exercise

Q7. A mixture of two gases, X and Y, which both have strong but distinct smells, is released. From across the room, the smell of X is detected more quickly than the smell of Y. What can you deduce about X and Y?

Q8. Use the kinetic theory to explain whether you would expect the rate of diffusion in a liquid to increase or decrease as the temperature is increased.

Q9. Which is the correct descriptor for the movement of particles in the solid, liquid and gaseous states?

	Solid	Liquid	Gas
A	vibrational movement in one dimension	vibrational movement in two dimensions	vibrational movement in three dimensions
B	no movement, fixed in position	only vibrational movement	free movement in all dimensions
C	free movement in all dimensions	free movement in all dimensions	free movement in all dimensions
D	vibrational movement	limited movement in all dimensions	free movement in all dimensions

Q10. A closed flask contains a pure substance, a brown liquid that is at its boiling point.

(a) Explain what you are likely to observe in the flask.

(b) Distinguish between the inter-particle distances and the average speeds of the particles in the two states present.

Q11. A beaker containing solid carbon dioxide is placed in a fume cupboard at room temperature. The carbon dioxide becomes gaseous. Which process describes this change of state?

A boiling   **B** condensation   **C** evaporation   **D** sublimation

Q12. Water exists in three states: ice, liquid water or steam. Which transition can occur over a range of temperatures at constant pressure?

A freezing   **B** melting   **C** evaporation   **D** boiling

At night, as the temperature is lowered, the rate of condensation increases. As the air temperature drops below its saturation point, known as the **dew point**, condensed water called **dew** forms. The temperature of the dew point depends on the atmospheric pressure and the water content of the air – that is, the relative humidity.

Q13. A liquid is placed in an open dish. Which change increases the rate of evaporation?

  I.   increased temperature of the liquid

  II. increased depth of the liquid

  III. increased surface area of the liquid

  A I and II   **B** I and III   **C** II and III   **D** I, II and III

Structure 1.1.3 – Kinetic energy and temperature

Structure 1.1.3 – Temperature (T) in kelvin (K) is a measure of average kinetic energy (E_k) of particles.

Interpret observable changes in physical properties and temperature during changes of state.

Convert between values in the Celsius and Kelvin scales.

The kelvin (K) is the SI unit of temperature and has the same incremental value as the Celsius degree (°C).	Reactivity 2.2 – What is the graphical distribution of kinetic energy values of particles in a sample at a fixed temperature?
	Reactivity 2.2 – What must happen to particles for a chemical reaction to occur?

The heat curve of water. Full details of how to carry out this experiment with a worksheet are available in the eBook.

SKILLS

Temperature is constant during changes of state

Matter changes state when it is heated

Simple experiments can be done to monitor the temperature change while a substance is heated and changes state. Figure 5 shows a typical result.

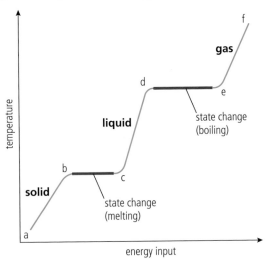

Ice cubes melting in a beaker of water. Heat from the surroundings breaks some inter-particle forces between the water molecules in the ice. In Structure 2.2.8, we will see that the inter-particle forces in ice are hydrogen bonds.

S1.1 Figure 5 Temperature change versus energy input at fixed pressure as a solid substance is heated. The flat regions shown in red are where a change of state is occurring. Heat energy is used to overcome the inter-particle forces.

The graph can be interpreted as follows:

a–b   As the solid is heated, the vibrational energy of its particles increases and so the temperature increases.

b–c   This is the melting point. The vibrations are sufficiently energetic for the particles to move away from their fixed positions and form a liquid. Energy added during this stage is used to break the inter-particle forces, not to raise the kinetic energy, so the temperature remains constant.

c–d   As the liquid is heated, the particles gain kinetic energy and so the temperature increases.

d–e   This is the boiling point. There is now sufficient energy to break all of the inter-particle forces and form a gas. Note that this state change needs more energy than melting, as all the inter-particle forces must be broken. The temperature remains constant as the kinetic energy does not increase during this stage. Bubbles of gas are visible through the volume of the liquid.

▲
Geothermal hot spring, Hveragerdi, Iceland. The geothermal spring provides sufficient energy to separate the water molecules and turn water into steam.

e–f   As the gas is heated under pressure, the kinetic energy of its particles continues to rise, as does the temperature.

Melting (b to c) and boiling (d to e) are endothermic processes as energy is needed to separate the particles. The reverse processes, freezing (c to b) and condensation (e to d), are exothermic. Energy is given out as the inter-particle forces bring the particles closer together. Energy changes are explored in Reactivity 1.2.

Structure 2 (all), Reactivity 1.2 – Why are some changes of state endothermic and some exothermic?

Challenge yourself

5. Which physical properties determine the gradient of the lines for the different states in Figure 5?

Nature of Science

Scientists analyse their observations looking for patterns, trends or discrepancies. You may have noticed that the volume decreases as ice melts and its density increases. This is unusual as density generally decreases as a solid melts. This discrepancy is explained in Structure 2.2.8.

Temperature in kelvin is known as the absolute temperature

The movement or kinetic energy of the particles of a substance depends on the temperature. If the temperature of a substance is decreased, the average kinetic energy of the particles also decreases. Absolute zero (–273 °C) is the lowest possible temperature attainable as this is the temperature at which all movement stops.

The kelvin is the SI unit of temperature. The absolute temperature is directly proportional to the average kinetic energy of its particles. A temperature increase of 1 °C is also an increase of 1 K. This facilitates conversions between the two scales.

The Celsius scale of temperature is defined relative to the boiling and freezing points of water. The original scale, developed by the Swedish astronomer Anders Celsius, made the boiling point of water zero and the freezing point 100. The modern scale reverses this with the boiling point of water 100 °C and the melting point of ice as 0 °C. The original scale may now seem absurd, but the modern scale is just as arbitrary.

An increase of
temperature
$\Delta T = 1\,K = 1\,°C$.
More generally
$T(K) = T(°C) + 273.15$
The kelvin is the SI
unit of temperature.

William Thomson
(1824–1907), who
became known as
Lord Kelvin later in life,
completed most of his
work at the University
of Glasgow, Scotland.
His concept of the
absolute temperature
scale followed from
his recognition of the
relationship between
heat energy and the
ability to do work. The
existence of a minimum
possible temperature
at which no heat can
be extracted from
the system and so no
work done, led him
to the definition of
absolute zero in 1848.
This in turn led to the
formulation of the laws
of thermodynamics.
Kelvin is considered one
of the great scientists of
the 19th century, and
is buried next to Isaac
Newton in London.

Reactivity 2.2 – What
is the graphical
distribution of kinetic
energy values of
particles in a sample at a
fixed temperature?

Reactivity 2.2 – What
must happen to particles
for a chemical reaction
to occur?

Temperature can be converted from Celsius to the Kelvin scale by the relation:

temperature (K) = temperature (°C) + 273.15

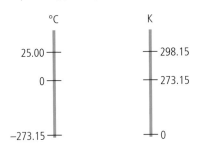

▲
S1.1 Figure 6 The Celsius and Kelvin scales of temperature.

Although we can only relate the average kinetic energy to the temperature, there are many particles in a sample each with a different kinetic energy. The total kinetic energy is conserved in particle collisions, but the kinetic energy of individual particles changes upon each collision. We can however predict the statistical distribution of energies in a sample at a particular temperature, known as the Maxwell–Boltzmann distribution. This distribution will also help us understand the effect of temperature on rates of reaction. To react, particles must collide with sufficient kinetic energy, known as the activation energy. Both these points are explored in Reactivity 2.2.

Exercise

Q14. A solid is heated and the temperature is measured every 20 seconds.

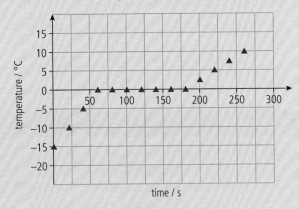

(a) Identify the solid.

(b) Describe how the particles in the solid are moving in the interval 0–60 s.

(c) Describe how the particles are moving in the interval after 180 s.

(d) Explain why the temperature does not change in the interval 60–180 s.

(e) Use the graph to predict the temperature after the material has been heated for 600 s.

(f) Use the graph to predict the temperature after 1200 s. Comment on the validity of your prediction.

Q15. Which of the following shows the correct Celsius temperature for the given kelvin temperature?

	Kelvin scale	Celsius scale
A	0	273
B	50	323
C	283	110
D	323	50

Q16. Which of the following occurs at the freezing point when a liquid is converted to a solid?

    I.   kinetic energy of the particles decreases

    II. separation between the particles decreases

    A I only   **B** II only   **C** I and II   **D** neither I nor II

Q17. Two objects made from different masses of iron have the same temperature. Which of the following is the correct comparison of the average kinetic energy and the total energy of the atoms in the objects?

	Average kinetic energy of the atoms	Total energy of the atoms
A	greater in an object of larger mass	same
B	less in an object of larger mass	same
C	same	greater in an object of larger mass
D	same	less in an object of larger mass

Q18. A substance is heated. The graph shows how the temperature changes with the heat energy added.

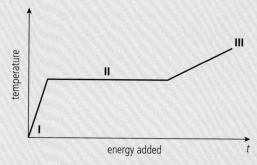

In which region of the graph must the solid state be present?

    A I   **B** II   **C** III   **D** None

Q19. Explain why a burn to the skin caused by boiling water is less harmful than a burn caused by the same amount of steam produced as the water boils.

How can we model the particulate nature of matter?

In this chapter we have used a particulate model to show:

- All matter is made up from small particles.
- An element is a pure chemical substance composed of the same type of atoms.
- Compounds consist of atoms of different elements chemically bonded together in a fixed ratio.
- Mixtures are made from particles of one substance interspersed between particles of at least one other substance. If this mixing is uniform, it is a homogeneous mixture. If it is not uniform, it is heterogeneous.
- The different states of matter are characterized by the movement and arrangement of the particles.
- Changes of state require energy changes as energy is needed to separate particles and is given out when particles come together.

Practice questions

1. Which statements are correct?

    I.   Solids have a fixed shape because the particles can only vibrate about a fixed point.

    II. Liquids can flow because the particles can move freely.

    III. Gases are less dense because the interactions between the particles are very weak.

    A I and II   **B** I and III   **C** II and III   **D** I, II and III

2. The temperature of a gas is reduced. Which of the following statements is true?

    A The molecules collide with the walls of the container less frequently.

    B The molecules collide with each other more frequently.

    C The time of contact between the molecules and the wall is reduced.

    D The time of contact between molecules is increased.

3. Which of the following occurs at the melting point when a solid is converted to a liquid?

    I. kinetic energy of the particles increases

    II. separation between the particles increases

    A I only   **B** II only   **C** I and II   **D** neither I nor II

4. Identify the change(s) that occur at the boiling point when a liquid is converted to a gas.

    I. kinetic energy of the particles increases

    II. separation between the particles increases

    A I only   **B** II only   **C** I and II   **D** neither I nor II

5. A substance is heated at a constant rate.

During which interval does the energy of the substance increase the most?

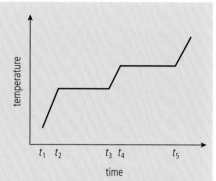

A t_1 to t_2

B t_2 to t_3

C t_3 to t_4

D t_4 to t_5

6. An ice cube of mass 10.0 g with a temperature of 0 °C is placed into a glass containing 90 cm³ of water at 10 °C.

After 5 minutes some of the ice has melted. What is the temperature of the ice remaining?

A 0 °C   **B** 0.5 °C   **C** 2.0 °C   **D** 5 °C

7. Identify the pair of substances that can form a homogeneous mixture.

A olive oil and vinegar   **B** sand and water

C carbon dioxide and water   **D** salt and pepper

8. Identify the equation that represents sublimation.

A $I_2(g) \rightarrow 2I(g)$ **B** $I_2(s) \rightarrow I_2(g)$

C $I_2(s) + I^-(aq) \rightarrow I_3^-(aq)$ **D** $2Al(s) + 3I_2(s) \rightarrow 2AlI_3(s)$

9. Which descriptions are consistent with the diagram?

Diagram	Description
I. ![diagram I]	A compound in the gaseous state.
II. ![diagram II]	A compound in the liquid state.
III. ![diagram III]	A heterogeneous mixture in the solid state.

A I and II only   **B** I and III only   **C** II and III only   **D** I, II and III

10. A mixture of ethanol and methanol can be separated by fractional distillation. This method of separation depends on the difference of which property of the two alcohols?

    A melting point   **B** boiling point   **C** solubility   **D** density

11. The solubilities of four different substances in water and ethanol are given.

	Solubility in water	Solubility in ethanol
P	insoluble	insoluble
Q	insoluble	soluble
R	soluble	insoluble
S	soluble	soluble

    Some mixtures can be separated by solvation and filtration using water or ethanol as solvents.

    Method I uses water as the solvent.

    Method II uses ethanol as the solvent.

    Which substances can be separated from each other using both methods?

    A P and S   **B** Q and S   **C** R and S   **D** Q and P

12. Classify the following mixtures as homogeneous or heterogeneous:

    (a) smoke   **(b)** sugar and water   **(c)** steel   (3)

    (Total 3 marks)

13. During very cold weather, snow often gradually disappears without melting. Explain how this is possible.

14. **(a)** Describe **two** differences, in terms of particle structure, between a gas and a liquid.   (2)

    (b) The temperature of a gas is a measure of the average kinetic energy of the gas particles. Suggest why the **average** kinetic energy is specified. (2)

    (c) A sample of water was heated. The graph shows how the temperature of the water varied with time.

    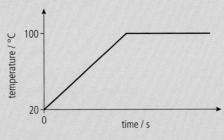

    (i) State why the temperature initially increases.   (1)

    (ii) State why the temperature remains constant at 100 °C.   (2)

    (Total 7 marks)

15. (a) The physical properties of some halogens are shown in the table.

Element	Molecular formula	Melting point / °C	Boiling point / °C
fluorine	F_2	−220	−188.0
chlorine	Cl_2	−101	−35.0
bromine	Br_2		58.8
iodine	I_2	114	184.0

Predict the melting point of bromine. (1)

(b) Describe the trend in the boiling points of the halogens down the group. (1)

(c) Predict the physical state of iodine at 200 °C. (1)

(d) A simplified diagram of the structure for bromine is shown.

(i) Suggest the state of bromine with this structure and justify your answer. (2)

(ii) Describe the changes that occur to this arrangement when bromine is heated. (2)

(iii) Describe the changes that occur when bromine reacts. (1)

(iv) Use the kinetic theory to explain the effect of an increase in temperature on the rate of diffusion of bromine. (1)

(Total 9 marks)

The nuclear atom

◄ The atomic nucleus is at the centre of the atom. The number of protons in the nucleus gives the atom its identity.

Guiding Question

How do the nuclei of atoms differ?

We saw in Structure 1.1 that all matter is built from only about 100 elements. The smallest amount of an element that can exist is an atom, which is the smallest particle of an element to show the element's characteristic properties. It is amazing that particles as small as the atom can have such a huge impact on the universe. Almost all explanations in chemistry refer to the atom, either individually or in the groups we call molecules.

In this chapter we will explore the structure of atoms in more detail. Although the word 'atom' means uncuttable, the atom *can* be split into smaller subatomic particles. All atoms are made from the same basic ingredients: protons, neutrons and electrons, but differ in their composition. If you look at the periodic table, you will see that each element has an atomic number. This was originally used to give the relative position of the element in the table. For example, the first element, hydrogen, has an atomic number of 1. We now know that this number is more fundamental than just a ranking. The atomic number is the defining property of an element as it is the number of protons in the nucleus of the atom.

Although the nucleus is only a small part of the atom, you can think of it as the atom's control centre. As an atom has a neutral charge, the number of protons also determines the number of electrons. The number of protons gives an atom its identity and the number of electrons determines its chemical properties. The mass of an element's atoms can vary due to a difference in the number of neutrons in the nucleus, but this has little impact on the chemistry of the atom. The number of neutrons does, however, affect some physical properties, including the nuclear stability.

Structure 1.2.1 – The atomic model

Structure 1.2.1 – Atoms contain a positively charged, dense nucleus composed of protons and neutrons (nucleons). Negatively charged electrons occupy the space outside the nucleus.

Use the nuclear symbol ($_{Z}^{A}X$) to deduce the number of protons, neutrons and electrons in atoms and ions.

Relative masses and charges of the subatomic particles should be known; actual values are given in the data booklet. The mass of the electron can be considered negligible.	Structure 1.3 – What determines the different chemical properties of atoms?
	Structure 3.1 – How does the atomic number relate to the position of an element in the periodic table?

TOK

Richard Feynman: '... if all of scientific knowledge were to be destroyed, and only one sentence passed on to the next generation of creatures, what statement would contain the most information in the fewest words? I believe it is the atomic hypothesis... that *all things are made of atoms*.'

Are the models and theories that scientists create accurate descriptions of the natural world, or are they primarily useful interpretations for prediction, explanation, and control of the natural world?

Atoms contain a positively charged, dense nucleus

Dalton's model of the atom

One of the first great achievements of chemistry was to show that all matter is built from about 100 **elements.** As mentioned in Structure 1.1, the elements are substances which cannot be broken down into simpler components by chemical reactions. Different elements have different chemical properties but gold foil, for example, reacts in essentially the same way as a single piece of gold dust. Indeed, if the gold dust is cut into smaller and smaller pieces, the chemical properties would remain the same until we reached an **atom**. The atom is the smallest unit of an element. There are only 92 elements which occur naturally on Earth, and they are made up of only 92 different types of atom. (This statement will be qualified when isotopes are discussed later in the chapter.)

The word 'atom' comes from the Greek words for 'not able to be cut'.

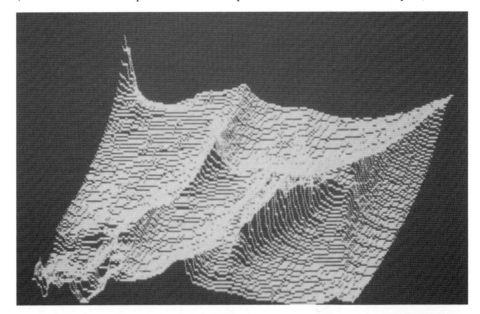

Scanning tunneling microscope (STM) image of the surface of pure gold. STM imaging records the surface structure at the level of the individual atoms. Gold exists in many forms, but all the forms contain the same type of atoms. The 'rolling hills' structure seen here is the result of changes in the surface energy as the gold cooled from its molten state.

Nature of Science

The idea that matter is made up of elements and atoms dates back to the 4th century BCE. This was speculative as there was little evidence to support the idea. Scientists adopt a skeptical attitude to claims and evaluate them using evidence. A significant development for chemistry came with the publication of Robert Boyle's *Sceptical Chymist* in 1661 which emphasized the need for scientific knowledge to be justified by evidence from practical investigations. Boyle was the first person to propose the modern concept of an element as a substance which cannot be changed into anything simpler.

The modern idea of the atom dates from the beginning of the 19th century. John Dalton noticed that the elements hydrogen and oxygen always combined in fixed proportions. To explain this observation, he proposed that:

- all matter is composed of tiny indivisible particles called atoms
- atoms cannot be created or destroyed
- atoms of the same element are alike in every way
- atoms of different elements are different
- atoms can combine together in small numbers to form **molecules**.

Using this model, we can understand how elements react together to make **compounds**. The compound water, for example, is formed when two hydrogen atoms combine with one oxygen atom to produce one water molecule. If we repeat the reaction on a larger scale with $2 \times 6.02 \times 10^{23}$ atoms of hydrogen and 6.02×10^{23} atoms of oxygen, then 6.02×10^{23} molecules of water will be formed. This leads to the conclusion (see Structure 1.4) that 2.02 g of hydrogen will react with 16.00 g of oxygen to form 18.02 g of water. This is one of the observations Dalton was trying to explain.

Nature of Science

Dalton was a man of regular habits. 'For fifty-seven years... he measured the rainfall, the temperature... Of all that mass of data, nothing whatever came. But of the one searching, almost childlike, question about the weights that enter the construction of simple molecules – out of that came modern atomic theory. That is the essence of science: ask an impertinent question: and you are on the way to the pertinent answer.' (J. Bronowski).

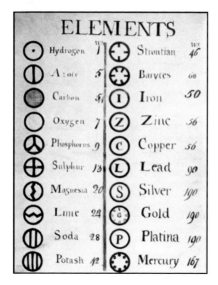

TOK

'What we observe is not nature itself but nature exposed to our mode of questioning.' (Werner Heisenberg).

How does the knowledge we gain about the natural world depend on the questions we ask and the experiments we perform?

John Dalton's symbols for the elements.

Dalton was the first person to assign chemical symbols to the different elements.

Challenge yourself

1. It is now known that some of these substances are not elements but compounds. Lime, for example, is a compound of calcium and oxygen. Find any other examples of compounds in this list, and explain why the component elements had not been extracted at this time.

Following Dalton's example, you can write the formation of water using modern notation:

$$2H + O \rightarrow H_2O$$

But what are atoms really like? It can be useful to think of them as hard spheres (Figure 1), but this tells us little about how the atoms of different elements differ. To understand this, it is necessary to probe deeper.

S1.2 Figure 1 A model of a water molecule made from two hydrogen atoms and one oxygen atom. Dalton's picture of the atom as a hard ball is the basis behind the molecular models we use today.

One of the barriers to the acceptance of Dalton's model of the atom was opposition from leading scientists such as Kelvin. In what ways can influential individuals help or hinder the development of scientific knowledge?

TOK

Atoms contain electrons

The first indication that atoms were destructible came at the end of the 19th century when the British scientist J. J. Thomson discovered that different metals produce a stream of negatively charged particles when a high voltage is applied across two electrodes. As these particles, which we now know as **electrons**, were the same regardless of the metal, he suggested that they are part of the make-up of all atoms.

Nature of Science

The properties of electrons, or cathode rays as they were first called, could only be investigated once powerful vacuum pumps had been invented – and once advances had been made in the use and understanding of electricity and magnetism. Improved instrumentation and new technology are often the drivers for new discoveries.

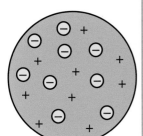

S1.2 Figure 2 Thomson's 'plum pudding' model of the atom. The negatively charged electrons (yellow) are positioned in a positively charged sponge-like substance (pink).

As Thomson knew that the atom had no net charge, he pictured the atom as a 'plum pudding', with the negatively charged electrons positioned in a positively charged sponge-like substance (Figure 2).

Atoms contain a nucleus

Ernest Rutherford (1871–1937) and his research team working at Manchester University in England, tested Thomson's model by firing alpha particles at a piece of gold foil. If Thomson's model was correct, the alpha particles should either pass straight through or get stuck in the positive 'sponge'. Most of the alpha particles did indeed pass straight through, but a very small number were repelled.

When Rutherford's team reported that they had seen a small number of alpha particles deflected by small angles, he asked them to see if any of the alpha particles had bounced back. This was a very unusual suggestion to make at the time, with little logical justification. What is the role of intuition in the pursuit of scientific knowledge?

TOK

The large number of undeflected particles led Rutherford to the conclusion that the atom is mainly empty space. Large deflections occur when the positively charged alpha particles collide with, and are repelled by, a positively charged nucleus. The fact that only a small number of alpha particles bounce back suggests that the nucleus is very small.

Nature of Science

Our knowledge of the nuclear atom came from Rutherford's experiments with the relatively newly discovered alpha particles. Progress in science often follows technological developments that allow new experimental techniques.

Nature of Science

Rutherford used his model of a positively charged nucleus to derive an equation for the scattering pattern of the alpha particles. His derivation assumed that only electrostatic repulsive forces acted between the positive gold nucleus and the positive alpha particles. All models have limitations which need to be considered in their application.

Challenge yourself

2. The derivation of the Rutherford formula is based on the assumption that only electrostatic forces need to be considered during the scattering process. Suggest why the experimental results deviate from this model for high-energy alpha particles.

Subatomic particles

A hundred years or so after Dalton first proposed his model, further experiments showed that the nucleus of an atom is made up of **protons** and **neutrons,** collectively called **nucleons.** Protons and neutrons have almost the same mass and together account for most of the mass of the atom. Electrons, which have a charge equal and opposite to that of the proton, have generally negligible mass and occupy the space in the atom outside of the nucleus.

The absolute masses and charges of these fundamental particles are given in Section 2 of the data booklet and this table gives their relative masses and charges. Note that relative quantities are ratios and so have no units.

Subatomic particle	Relative mass	Relative charge
proton	1	+1
electron	0.0005	−1
neutron	1	0

You should know the relative masses and charges of the subatomic particles. Actual values are given in the data booklet. The mass of the electron can generally be considered to be negligible.

Nature of Science

The description of subatomic particles given here is sufficient to understand chemistry but it is incomplete. Although the electron is indeed a fundamental particle, we now know that protons and neutrons are both themselves made up of more fundamental particles called quarks. We also know that all particles have anti-particles. The positron is the anti-particle of an electron; it has the same mass but has an equal and opposite positive charge. When particles and anti-particles collide, they destroy each other and release energy in the form of high-energy **photons** called gamma rays. Our treatment of subatomic particles is in line with the principle of Occam's razor, which states that theories should be as simple as possible while maximizing explanatory power.

PET (positron-emission tomography) scanners give three-dimensional images of tracer concentration in the body, and can be used to detect cancers. The patient is injected with a tracer compound labelled with a positron-emitting isotope. The positrons collide with electrons after travelling a short distance (≈1 mm) within the body. Both particles are destroyed, and two photons are produced. The photons can be collected by the detectors surrounding the patient, and used to generate an image.

As you are made from atoms, you are also mainly empty space. The particles which make up your mass would occupy the same volume as a flea if they were all squashed together, but a flea with your mass. This gives you an idea of the density of the nucleus.

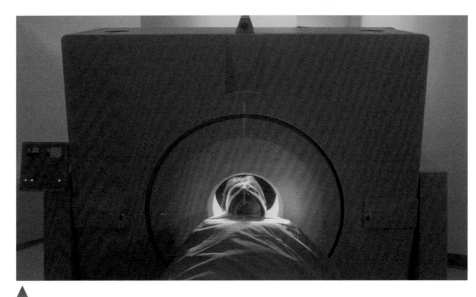

A patient undergoing a positron-emission tomography (PET) brain scan. A radioactive tracer is injected into the patient's bloodstream, which is then absorbed by active tissues of the brain. The PET scanner detects photons emitted by the tracer and produces 'slice' images.

Challenge yourself

3. Construct an equation for the collision of an electron and a positron to give two photons and explain why it is balanced.

Bohr model of the hydrogen atom

The Danish physicist Niels Bohr pictured the hydrogen atom as a small 'solar system', with an electron moving in an orbit (energy level) around the positively charged nucleus of one proton (Figure 3). The electrostatic force of attraction between the oppositely charged subatomic particles prevents the electron from leaving the atom. The nuclear radius is 10^{-15} m and the atomic radius is 10^{-10} m, so most of the volume of the atom is empty space.

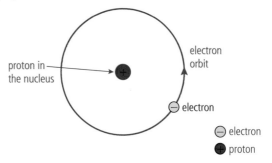

proton in the nucleus

electron orbit

electron

⊖ electron
⊕ proton

S1.2 Figure 3 The Bohr model of the simplest atom. Only one proton and one electron make up the hydrogen atom. Most of the volume of the atom is empty – the only occupant is the single negatively charged electron. It is useful to think of the electron orbiting the nucleus in a similar way to the planets orbiting the Sun. The absence of a neutron is significant – it would be essentially redundant as there is only one proton.

The existence of neutrally charged neutrons is crucial for the stability of nuclei of elements that have more than one proton (Figure 4). Without the neutrons, the positively charged protons would mutually repel each other and the nucleus would fall apart.

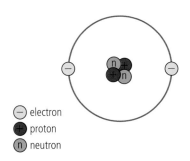

- ⊖ electron
- ⊕ proton
- Ⓝ neutron

◀ **S1.2 Figure 4** The Bohr model of a helium atom. The two neutrons allow the two protons, which repel each other, to stay in the nucleus. A strong nuclear force acts between all the nucleons which is larger than the repulsive electrostatic forces that act between the protons.

Atomic number and mass number

We are now in a position to understand how the atoms of different elements differ. They are all made from the same basic ingredients, the subatomic particles. The only difference is the recipe – how many of each of these subatomic particles are present in the atoms of different elements. If you look at the periodic table, you will see that the elements are each given a number which describes their relative position in the table. This is their **atomic number**. We now know that the atomic number, represented by **Z**, is the defining property of an element as it tells us something about the structure of the atoms of the element. The atomic number is defined as the number of protons in the atom.

As an atom has no overall charge, the positive charge of the protons must be balanced by the negative charge of the electrons. Therefore the atomic number is also equal to the number of electrons in an atom.

The electron has such a very small mass that it is essentially ignored in mass calculations. The mass of an atom depends on the number of protons and neutrons only. The **mass number**, given the symbol **A**, is defined as the number of protons plus the number of neutrons in an atom. An atom is identified in the following way:

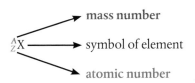

mass number

$^A_Z X$ → symbol of element

atomic number

We can use these numbers to find the composition of any atom.

$$\text{number of protons (p)} = \text{number of electrons} = Z$$

$$\text{number of neutrons (n)} = \text{mass number} - \text{number of protons} = A - Z$$

Consider an atom of aluminium:

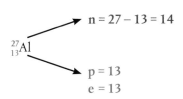

$n = 27 - 13 = 14$

$^{27}_{13}\text{Al}$

$p = 13$
$e = 13$

An aluminium atom has 13 protons and 13 electrons. An atom of gold on the other hand has 79 protons and 79 electrons. Can you find gold in the periodic table? The periodic table arranges the elements in order of their atomic number as discussed in Structure 3.1.

TOK

None of these subatomic particles can be (or ever will be) directly observed.

What assumptions are made when we interpret indirect evidence gained through the use of technology?

The atomic number is defined as the number of protons in the nucleus.

Make sure you have a precise understanding of the terms. The atomic number, for example, is defined in terms of the number of protons, not electrons.

The mass number (*A*) is the number of protons plus the number of neutrons in an atom. It is sometimes called the nucleon number.

SKILLS

Build an atom PhET activity. Full details on how to carry out this activity with a worksheet are available in the eBook.

Structure 3.1 – How does the atomic number relate to the position of an element in the periodic table?

Challenge yourself

4. Explain why the 13 protons in aluminium stay in the nucleus despite their mutual repulsion.

5. Experiments show the nuclear radius R depends on the mass number A according to the expression: $R = 1.2 \times 10^{-15} A^{\frac{1}{3}}$. Deduce an expression for the density of a nucleus and comment on your result.

Worked example

Identify the subatomic particles present in an atom of ^{226}Ra.

Solution

The number identifying the atom is the atomic number. We can find the atomic number from the data booklet (Section 6).

We have $Z = 88$ and $A = 226$

number of protons (p) = 88

number of electrons (e) = 88

number of neutrons (n) = 226 − 88 = 138

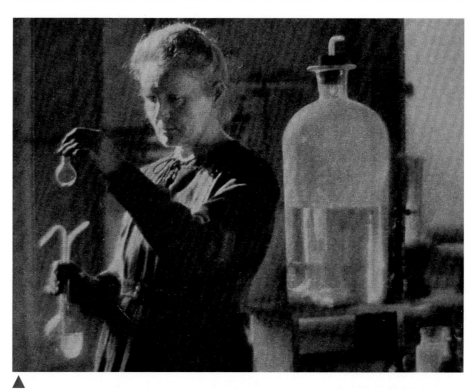

▲
The element radium was first discovered by the Polish–French scientist Marie Curie. She is the only person to win Nobel Prizes in both Physics and Chemistry. The Curies were a remarkable family for scientific honors – Marie shared her first prize with her husband Pierre, and her daughter Irène shared hers with her husband Frédéric. All the Curies' prizes were for work on radioactivity.

Ions

The atomic number is defined in terms of the number of protons because it is a fixed characteristic of the element. The number of protons identifies the element in the same way your fingerprints identify you. The number of protons and neutrons never changes during a chemical reaction. It is the electrons which are responsible for chemical change. Structure 2.1 examines how atoms can lose or gain electrons to form **ions**. When the number of protons in a particle is no longer balanced by the number of electrons, the particle has a non-zero charge. When an atom loses electrons, it forms a positive ion or **cation**, as the number of protons is now greater than the number of electrons. Negative ions or **anions** are formed when atoms gain electrons. The magnitude of the charge depends on the number of electrons lost or gained. The loss or gain of electrons makes a very big difference to the chemical properties. You swallow sodium ions, Na^+, every time you eat table salt, whereas (covered in Structure 3.1) sodium atoms, Na, are dangerously reactive.

An aluminium ion is formed when the atom loses three electrons. There is no change in the atomic or mass numbers of an ion because the number of protons and neutrons remains the same.

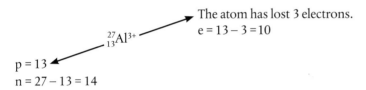

The atom has lost 3 electrons.
e = 13 − 3 = 10

$^{27}_{13}Al^{3+}$

p = 13
n = 27 − 13 = 14

Oxygen forms the oxide ion when its atoms gain two electrons.

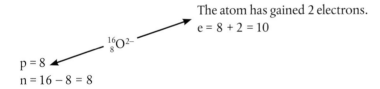

The atom has gained 2 electrons.
e = 8 + 2 = 10

$^{16}_{8}O^{2-}$

p = 8
n = 16 − 8 = 8

Worked example

Most nutrient elements in food are present in the form of ions. The calcium ion $^{40}Ca^{2+}$, for example, is essential for healthy teeth and bones. Identify the subatomic particles present in the ion.

Solution

We can find the atomic number from the data booklet (Section 6). We have Z = 20 and A = 40:

- number of protons (p) = 20

- number of neutrons (n) = 40 − 20 = 20

As the ion has a positive charge of 2+, there are two more protons than electrons:

- number of electrons = 20 − 2 = 18

When an atom loses electrons, a positive ion is formed and when an atom gains electrons, a negative ion is formed. Positive ions are called cations and negative ions are called anions.

Structure 1.3 – What determines the different chemical properties of atoms?

Worked example

Identify the species with 19 protons, 20 neutrons and 18 electrons.

Solution

- the number of protons tells us the atomic number, Z = 19, and so the element is potassium, K
- the mass number = p + n = 19 + 20 = 39: $^{39}_{19}K$
- the charge = p − e = 19 − 18 = +1 as there is one extra proton: $^{39}_{19}K^+$

Challenge yourself

6. We generally make the approximation that the mass of an ion is the same as that of the corresponding atom. To how many significant figures is this approximation valid for the H^+ ion?

Exercise

Q1. Explain why the relative atomic mass of tellurium is greater than the relative atomic mass of iodine, even though iodine has a greater atomic number.

Q2. Use the periodic table to identify the subatomic particles present in the following species.

	Species	No. of protons	No. of neutrons	No. of electrons
(a)	^7Li			
(b)	^1H			
(c)	^{14}C			
(d)	$^{19}F^-$			
(e)	$^{56}Fe^{3+}$			

Q3. Isoelectronic species have the same number of electrons. Identify the following isoelectronic species by giving the correct symbol and charge. You will need a periodic table.

The first one has been done as an example.

	Species	No. of protons	No. of neutrons	No. of electrons
	$^{40}Ca^{2+}$	20	20	18
(a)		18	22	18
(b)		19	20	18
(c)		17	18	18

Q4. What is the difference between two neutral atoms represented by the symbols $^{14}_6C$ and $^{14}_7N$?

I.   the number of neutrons   II.   the number of protons

III. the number of electrons

A   I and II only   B   I and III only   C   II and III only   D   I, II and III

Structure 1.2.2 – Isotopes

Structure 1.2.2 – Isotopes are atoms of the same element with different numbers of neutrons.

Perform calculations involving non-integer relative atomic masses and abundance of isotopes from given data.

Differences in the physical properties of isotopes should be understood. Specific examples of isotopes need not be learned.	Nature of Science, Reactivity 3.4 – How can isotope tracers provide evidence for a reaction mechanism?

Isotopes are atoms of the same element with different numbers of neutrons

Find chlorine in the periodic table. There are two numbers associated with the element, as shown below.

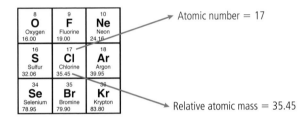

SKILLS

Modeling isotopic abundance. Full details on how to carry out this experiment with a worksheet are available in the eBook.

How can an element have a fractional relative atomic mass if protons and neutrons each have a relative mass of 1? One reason is that atoms of the same element with different mass numbers exist, and the relative atomic mass given in the periodic table is an average value.

To have different mass numbers, the atoms must have different numbers of neutrons – all the atoms have the same number of protons as they are all chlorine atoms. Atoms of the same element with different numbers of neutrons are called **isotopes**.

The isotopes show the same chemical properties, as a difference in the number of neutrons makes no difference to how atoms react and so they occupy the same place in the periodic table.

Isotopes are atoms of the same element with different mass numbers.

Chlorine exists as two isotopes, $^{35}_{17}\text{Cl}$ and $^{37}_{17}\text{Cl}$. The average relative mass of the isotopes is, however, not 36, but 35.45. This value is closer to 35 as there are more $^{35}_{17}\text{Cl}$ atoms in nature – it is the more *abundant* isotope. In a sample of 1000 chlorine atoms, there are 775 atoms of $^{35}_{17}\text{Cl}$ and 225 atoms of the heavier isotope, $^{37}_{17}\text{Cl}$.

The word 'isotope' derives from the Greek for 'same place'. As isotopes are atoms of the same element, they occupy the same place in the periodic table.

To work out the average mass of one atom, we first have to calculate the total mass of the thousand atoms:

$$\text{total mass} = (775 \times 35) + (225 \times 37) = 35\,450$$

$$\text{relative average mass} = \frac{\text{(total mass)}}{\text{(number of atoms)}} = \frac{35\,450}{1000} = 35.45$$

The two isotopes are both atoms of chlorine with 17 protons and 17 electrons.

A common error is to misunderstand the meaning of 'physical property'. A difference in the number of neutrons is not a different physical property. A physical property of a substance can be measured without changing the chemical composition of the substance. Density and boiling point are examples of physical properties.

Uranium consists of two natural isotopes: mostly U-238 and less than 1% U-235. The fuel used in nuclear power stations is U-235; its content needs to be increased to around 3–5% of the overall mixture to make a controlled nuclear fission reaction feasible. The enrichment can be carried out by gas diffusion, by heating solid uranium hexafluoride (UF_6). The UF_6 molecules containing U-235 are lighter and diffuse faster through the pipelines and filters, producing a UF_6 gas that is enriched with U-235.

- $^{35}_{17}Cl$ number of neutrons = $35 - 17 = 18$
- $^{37}_{17}Cl$ number of neutrons = $37 - 17 = 20$

Although both isotopes essentially have the same chemical properties, the difference in mass does lead to different physical properties such as boiling and melting points. As explained in Structure 1.1, heavier particles move more slowly at a given temperature and these differences can be used to separate isotopes.

Gas diffusion machinery in southern France used to enrich uranium for use in nuclear reactors.

Challenge yourself

7. Suggest why isotope nuclear enrichment plants based on gaseous diffusion are so large.

Nature of Science, Reactivity 3.4 – How can isotope tracers provide evidence for a reaction mechanism?

The stability of a nucleus depends on the balance between the number of protons and neutrons. A nucleus which contains either too many or too few neutrons to be stable is radioactive and changes to a more stable nucleus by giving out radiation. As these **radioisotopes** behave chemically in the same way as nonradioactive isotopes, they can be used as tracers to follow the movement of elements or compounds in complex processes such as living systems. The radioisotopes can be located because the radioactivity they emit can be detected. Radioactive isotopes can also provide evidence for the reaction mechanisms explored in Reactivity 3.4.

Human activities have caused increased atmospheric nitrogen pollution, mainly nitrogen oxides and ammonia. This nitrogen is eventually taken up by plants, increasing plant growth.

Evidence for this comes from experiments using tracer isotopes of nitrogen ($^{15}_{7}N$) to follow the nitrogen through the air and plants.

▲ Evidence for increased nitrogen pollution comes from experiments using tracer isotopes of nitrogen ($^{15}_{7}N$) in plants.

Challenge yourself

8. Use the periodic table in the data booklet to identify an element with an atomic number of less than 80 which has no stable isotopes.

The relative atomic mass of an element

The mass of a hydrogen atom is 1.67×10^{-24} g and that of a carbon atom is 1.99×10^{-23} g. As the masses of all elements are in the range 10^{-24} to 10^{-22} g, and as these numbers are beyond our direct experience, it makes more sense to use relative values. The mass needs to be recorded relative to some agreed standard. As carbon is a very common element which is easy to transport and store because it is a solid, its isotope, ^{12}C, was chosen as the standard in 1961. This is discussed in Structure 1.4 and ^{12}C this is given a relative mass of exactly 12, as shown below.

Element	Symbol	Relative atomic mass
carbon	C	12.011
chlorine	Cl	35.453
hydrogen	H	1.008
iron	Fe	55.845
Standard isotope	**Symbol**	**Relative atomic mass**
carbon-12	^{12}C	12.000

The relative atomic mass of an element (A_r) is the average mass of an atom of the element, taking into account all its isotopes and their relative abundance, compared to one atom of carbon-12.

Carbon-12 is the most abundant isotope of carbon but carbon-13 and carbon-14 also exist. This explains why the average value for the element is greater than 12.

Worked example

Deduce the relative atomic mass of the element rubidium from the data given in the table.

Isotope	% Abundance
^{85}Rb	77
^{87}Rb	23

Solution

Consider a sample of 100 atoms.

$$\text{total mass of 100 atoms} = (85 \times 77) + (87 \times 23) = 8546$$

$$\text{relative atomic mass} = \text{average mass of atom} = \frac{\text{total mass}}{\text{number of atoms}} = \frac{8546}{100} = 85.46$$

Worked example

Boron exists as two isotopes, ^{10}B and ^{11}B. ^{10}B is used as a control for nuclear reactors. Use your periodic table to find the abundances of the two isotopes.

Solution

Consider a sample of 100 atoms.

Let x atoms be ^{10}B atoms. The remaining atoms are ^{11}B.

number of ^{11}B atoms $= 100 - x$

total mass of 100 atoms $= [x \times 10] + [(100 - x) \times 11] = 10x + 1100 - 11x = 1100 - x$

$$\text{average mass} = \frac{\text{total mass}}{\text{number of atoms}} = \frac{1100 - x}{100}$$

From the periodic table, the relative atomic mass of boron = 10.81.

$$10.81 = \frac{1100 - x}{100}$$

$$1081 = 1100 - x$$

$$x = 1100 - 1081 = 19$$

The abundances are $^{10}B = 19\%$ and $^{11}B = (100 - 19) = 81\%$

Exercise

Q5. State two physical properties other than boiling point and melting point that would differ for the two isotopes of chlorine.

Q6. Identify the particles which account for the existence of isotopes.
    A electrons   **B** nucleons   **C** neutrons   **D** protons

Q7. Which of the following species contains more electrons than neutrons?
    A $^{2}_{1}H$ **B** $^{11}_{5}B$ **C** $^{16}_{8}O^{2-}$ **D** $^{19}_{9}F^{-}$

Carbon-14 has eight neutrons, which is too many to be stable. It can reduce the neutron-to-proton ratio by radioactive decay. The relative abundance of carbon-14 present in living plants is constant as the carbon atoms are continually replenished from the carbon present in atmospheric carbon dioxide. When organisms die, however, no more carbon-14 is absorbed and the levels of carbon-14 fall as they decay. As this process occurs at a regular rate, it can be used to date carbon-containing materials.

Q8. Which of the following gives the correct composition of the $^{71}Ga^+$ ion?

	Protons	Neutrons	Electrons
A	31	71	30
B	31	40	30
C	31	40	32
D	32	40	31

Q9. Chromium has an atomic number of 24. The mass numbers of its four stable isotopes are 50, 52, 53 and 54.

Identify the correct statements about the isotopes.

I.   All the isotopes have the same chemical properties.

II. All the isotopes have nuclei containing 24 protons.

III. One of the isotopes has 54 neutrons.

A I and II only   **B** I and III only   **C** II and III only   **D** I, II and III

Q10. Identify the atoms which are isotopes.

	Mass number	Atomic number
W	52	24
X	53	25
Y	53	24
Z	52	25

A W and Z   **B** X and Y   **C** W and Y   **D** Y and Z

Q11. The relative atomic mass of silicon is 28.09. Comment on the claim that no atom of silicon exists with this relative mass.

Q12. Deduce the composition of a nucleus of boron-11, $^{11}_{5}B$.

	Protons	Neutrons
A	5	11
B	11	5
C	6	5
D	5	6

Q13. What is the same for an atom of phosphorus-26 and an atom of phosphorus-27?

A atomic number and mass number

B number of protons and electrons

C number of neutrons and electrons

D number of protons and neutrons

Q14. A sample of chromium has the following isotopic composition by mass.

Isotope	^{50}Cr	^{52}Cr	^{53}Cr	^{54}Cr
Relative abundance / %	4.31	83.76	9.55	2.38

Calculate the relative atomic mass of chromium based on this data, giving your answer to two decimal places.

In 1911, a 40 kg meteorite fell in Egypt. Isotopic and chemical analyses of oxygen extracted from this meteorite showed a different relative atomic mass to that of oxygen normally found on Earth. The relative atomic mass value did however match measurements made of the Martian atmosphere by the Viking landing in 1976. This provides strong evidence that the meteorite had originated from Mars.

Q15. Use the periodic table to find the percentage abundance of neon-20, if neon has only one other isotope, neon-22.

Q16. Magnesium has three stable isotopes: ^{24}Mg, ^{25}Mg, and ^{26}Mg. The lightest isotope has an abundance of 78.90%. Calculate the percentage abundance of the other isotopes.

Guiding Question revisited

How do the nuclei of atoms differ?

In this chapter we explored the structure of the atom and how the nuclei of atoms differ.

- All atoms are made up of protons, neutrons and electrons.
- The protons and neutrons, which contribute most of the mass of the atom, are in a small dense nucleus surrounded by electrons which occupy most of the volume of the atom.
- The atomic number gives the atom its identity. This is the number of protons in the nucleus. In a neutral atom this is also the number of electrons.
- The mass number is the number of nucleons: the number of protons and neutrons in the nucleons.
- Evidence shows that most elements have more than one isotope: atoms with the same number of protons but a different number of neutrons.
- The relative atomic mass, which is the average mass of an atom, can be determined from the relative abundance of its isotopes.

Practice questions

1. Which statements about the isotopes of chlorine, $^{35}_{17}Cl$ and $^{37}_{17}Cl$, are correct?

    I.   They have the same chemical properties.

    II. They have the same atomic number.

    III. They have the same physical properties.

    A I and II only   **B** I and III only   **C** II and III only   **D** I, II and III

2. Which statement about the numbers of protons, electrons and neutrons in an atom is always correct?

    A The number of neutrons minus the number of electrons is zero.

    B The number of protons plus the number of neutrons equals the number of electrons.

    C The number of protons equals the number of electrons.

    D The number of neutrons equals the number of protons.

3. Which quantities are the same for all atoms of chlorine?

    I.   number of protons

    II. number of neutrons

    III. number of electrons

    A I and II only   **B** I and III only

    C II and III only   **D** I, II and III

4. How many electrons does the ion $^{31}_{15}P^{3-}$ contain?

   A 12   **B** 15   **C** 16   **D** 18

5. Deduce the number of elementary particles present in the $^{55}_{25}Mn^{2+}$ ion.   (3)

   (Total 3 marks)

6. A sample of iron has the following isotopic composition by mass.

Isotope	^{54}Fe	^{56}Fe	^{57}Fe
Relative abundance / %	5.95	91.88	2.17

   Calculate the relative atomic mass of iron based on this data, giving your answer to two decimal places.   (2)

   (Total 2 marks)

7. **(a)** Explain why the relative atomic mass of cobalt is greater than the relative atomic mass of nickel, even though the atomic number of nickel is greater than the atomic number of cobalt.   (1)

   (b) Deduce the numbers of protons and electrons in the Co^{2+} ion.   (1)

   (Total 2 marks)

8. The table below refers to a sample of silicon.

Mass number of isotope	28	29	30
Relative abundance / %	92.18	4.70	3.12

   (a) Explain why atoms of an element can have different mass numbers.   (1)

   (b) Compare the chemical properties of the four isotopes and justify your answer.   (1)

   (c) Calculate the relative atomic mass of this sample of silicon.   (2)

   (Total 4 marks)

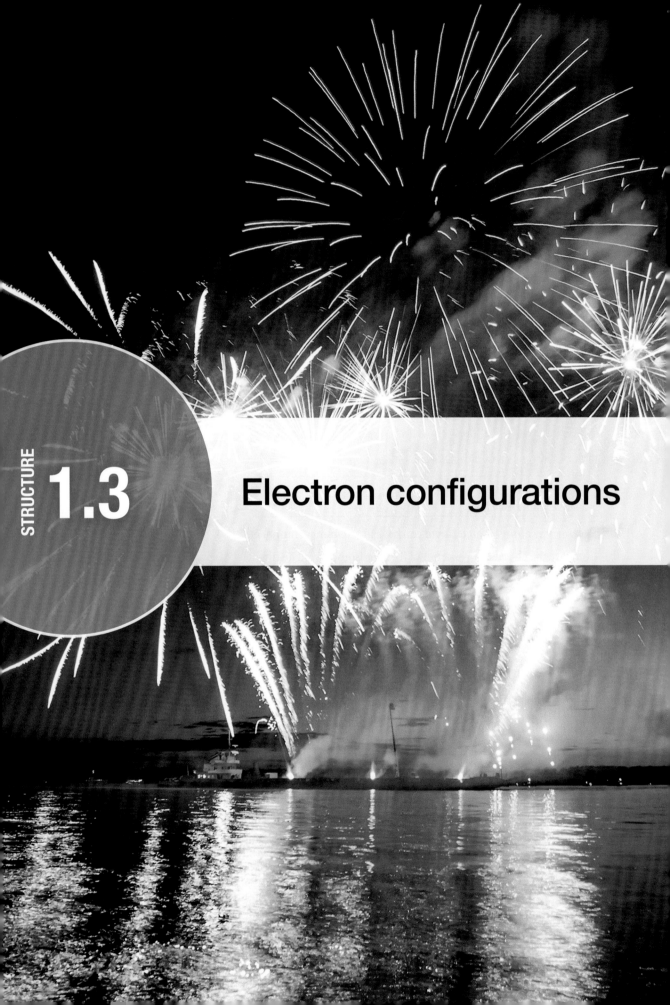

STRUCTURE

1.3

Electron configurations

The dazzling colours observed during a firework display are a result of electrons moving between different energy states.

Guiding Question

How can we model the energy states of electrons in atoms?

The chemical behavior of an atom is determined by its electron configuration. As we cannot look inside the atom directly, we have to look elsewhere for evidence of how the electrons are arranged. The analysis of light emitted from an atom gives us valuable information about the electron configuration within it. It shows that an electron can exist only in certain discrete energy states. This cannot be understood with reference to our everyday experience and demands a new perspective. The notion of particles following fixed trajectories does not apply to the microscopic world of the atom. We can only give a probability description of electron behavior and use quantum theory to adopt a wave description of the electron. The possible positions of an electron are spread out in space in the same way as a wave spreads through space. We will see how the energy states of electrons are best explained in terms of atomic orbitals. These are regions in space where an electron is likely to be found. Within one atom, there are an infinite number of orbitals of different shapes, sizes and energies. These ideas are revolutionary. As Niels Bohr, one of the principal scientists involved in the development of quantum theory said, 'Anyone who is not shocked by quantum theory has not understood it'.

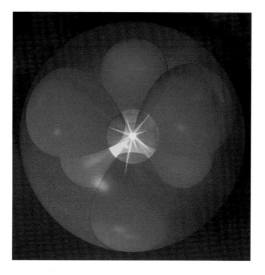

Electrons occupy atomic orbitals of different energy states. The atomic orbitals in an atom of neon, Ne, are represented here. The nucleus is shown by a flash of light and the 1s orbital as a yellow sphere. The 2s orbital is shown as a pink sphere, and the 2p orbitals as blue lobes.

Structure 1.3.1 and 1.3.2 – Emission spectra

Structure 1.3.1 – Emission spectra are produced by atoms emitting photons when electrons in excited states return to lower energy levels.

Qualitatively describe the relationship between colour, wavelength, frequency and energy across the electromagnetic spectrum.

Distinguish between a continuous and a line spectrum.

Details of the electromagnetic spectrum are given in the data booklet.	

> **Structure 1.3.2 – The line emission spectrum of hydrogen provides evidence for the existence of electrons in discrete energy levels, which converge at higher energies.**
>
> Describe the emission spectrum of the hydrogen atom, including the relationships between the lines and energy transitions to the first, second and third energy levels.
>
> | The names of the different series in the hydrogen emission spectrum will not be assessed. | Inquiry 2 – In the study of emission spectra from gaseous elements and from light, what qualitative and quantitative data can be collected from instruments such as gas discharge tubes and prisms? |
> | --- | --- |
> | | Nature of Science, Structure 1.2 – How do emission spectra provide evidence for the existence of different elements? |

Atoms of different elements give out light of distinctive colours

Atoms of different elements give out light of a distinctive colour when an electric discharge is passed through a vapour of the element. Similarly, metals can be identified by the colour of the flame produced when their compounds are heated in a Bunsen burner. Analysis of the light emitted by different atoms gives us insights into the electron configurations within the atom.

Flame tests on the compounds of (a) sodium, (b) potassium and (c) copper.

Flame colours can be used to identify unknown compounds. Full details of how to carry out this experiment with a worksheet are available in the eBook.

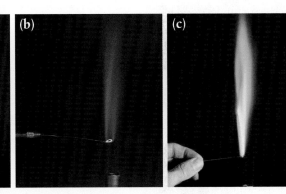

To interpret these results, we must consider the nature of electromagnetic radiation.

Electromagnetic radiation is emitted in different forms of differing energies

Electromagnetic radiation comes in different forms of differing energy. The visible light we need to see the world is only a small part of the full spectrum, which ranges from low-energy radio waves to high-energy gamma (γ) rays. All electromagnetic waves travel at the same **speed (c)** but can be distinguished by their different **wavelengths (λ)** (Figure 1).

S1.3 Figure 1 Snapshot of a wave at a given instant. The distance between successive crests or peaks is called the wavelength (λ).

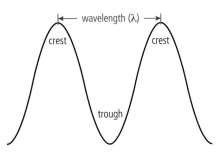

Different colours of visible light have different wavelengths; red light, for example, has a longer wavelength than blue light. The full electromagnetic spectrum is given in Section 5 of the data booklet.

The number of waves that pass a particular point in 1 s is called the **frequency (f)**; the shorter the wavelength, the higher the frequency. Blue light has a higher frequency than red light.

The precise relationship is:

$$c = f\lambda$$

where c is the speed of light.

White light is a mixture of light waves of differing wavelengths or colours. We see this when sunlight passes through a prism to produce a **continuous spectrum** or when light is scattered through water droplets in the air.

All electromagnetic waves travel at the same speed, $c = 3.00 \times 10^8\,\mathrm{m\,s^{-1}}$. This is the cosmic speed limit as, according to Einstein's Theory of Relativity, nothing in the universe can travel faster than this in a vacuum.

The distance between two successive crests (or troughs) is called the wavelength (λ). The frequency (f) of the wave is the number of waves that pass a point in one second. The wavelength and frequency are related by the equation $c = f\lambda$ where c is the speed of light.

◀ A continuous spectrum is produced when white light is passed through a prism. The different **colours** merge smoothly into one another. The two spectra below the illustration of the prism show (top) a continuous spectrum with a series of discrete absorption lines, and (bottom) a line emission spectrum. Details of the absorption spectrum will not be assessed.

As well as visible light, atoms emit infrared (IR) radiation, which has a longer wavelength than red light, and ultraviolet radiation, which has a shorter wavelength than violet light. The complete electromagnetic spectrum is shown in Figure 2.

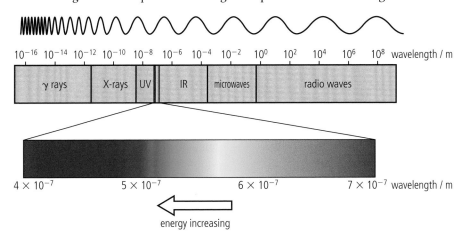

S1.3 Figure 2 The changing wavelength (in m) of electromagnetic radiation through the spectrum is shown by the trace across the top. At the short wavelength end (on the left) of the spectrum are gamma rays, X-rays, and ultraviolet light. In the center of the spectrum are wavelengths that the human eye can see, known as visible light. Visible light comprises light of different wavelengths, energies, and colours. At the longer wavelength end of the spectrum (on the right) are infrared radiation, microwaves, and radio waves. The visible spectrum gives us only a small window to see ◀ the world.

Electromagnetic waves allow energy to be transferred across the universe. They also carry information. Low-energy radio waves are used in radar and television, for example, and higher energy gamma rays are used as medical tracers. The precision with which we view the world is limited by the wavelengths of the colours we can see. This is why we will never be able to see an atom directly; it is too small to interact with the relatively long waves of visible light. What are the implications of this for human knowledge?

TOK

An emission spectrum is produced when an atom moves from a higher to a lower energy level

When electromagnetic radiation is passed through a collection of atoms, some of the radiation is absorbed and used to excite the atoms from a lower energy level to a higher energy level. A spectrometer analyzes the transmitted radiation relative to the incident radiation and an absorption spectrum is produced (Figure 3).

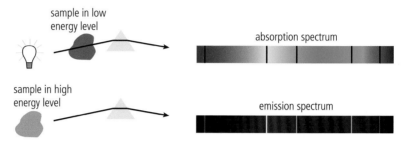

▲
S1.3 Figure 3 The origin of absorption and emission spectra. An absorption spectrum shows the radiation absorbed as atoms move from a lower to a higher energy level. An emission spectrum is produced when an atom moves from a higher to a lower level.

Gases produce a characteristic **emission line spectrum** when they are heated to a high temperature or if a high voltage is applied. Atoms are excited into a higher energy level, which is unstable, so the electron soon falls back to the **ground state**. The energy the electron gives out when it falls into lower levels is in the form of electromagnetic radiation. One packet of energy, a **photon,** is released for each electron transition (Figure 4). Photons of ultraviolet light have more energy than photons of infrared light. The energy of the photon is proportional to the frequency of the radiation.

S1.3 Figure 4 Emission spectra are the result of electrons falling from an excited state E_2 to a lower energy level E_1.

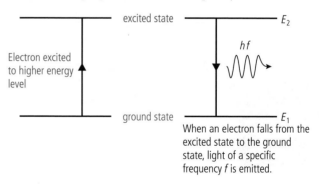

When an electron falls from the excited state to the ground state, light of a specific frequency f is emitted.

- A continuous spectrum shows an unbroken sequence of frequencies, such as the spectrum of visible light.
- A line emission spectrum has only certain frequencies of light because it is produced by excited atoms and ions as they fall back to a lower energy level.

The energy of the photon of light emitted (E_{photon}) is equal to the energy change of the electron in the atom ($\Delta E_{electron}$):

$$\Delta E_{electron} = E_{photon}$$

It is also related to the frequency of the radiation by the Planck equation:

$$E_{photon} = hf$$

This equation and the value of h (Planck's constant) are given in Sections 1 and 2 of the data booklet.

This leads to:

$$\Delta E_{electron} = hf$$

This is a very significant equation because it shows that line spectra allow us to glimpse inside the atom. The atoms emit photons of certain energies, which give lines of certain frequencies because the electron can only occupy certain energy levels. You can think of the energy levels as a staircase. The electron cannot change its energy in a continuous way, in the same way that you cannot stand between steps; it can only change its energy by discrete amounts. This energy of the atom is said to be **quantized**. The line spectrum is crucial evidence for quantization: if the energy were not quantized, the emission spectrum would be continuous.

Nature of Science

The idea that you can think of electromagnetic waves as a stream of photons, or quanta, is one aspect of quantum theory. The theory has implications for human knowledge and technology. The key idea is that energy can only be transferred in discrete amounts or quanta. Quantum theory shows us that our everyday experience cannot be transferred to the microscopic world of the atom. It has led to great technological breakthroughs such as the modern computer. It has been estimated that 30% of the gross national product of the USA depends on the application of quantum theory. Our scientific understanding has led to many technological developments. These new technologies in turn drive developments in science. The implications of quantum theory for the electron are discussed in more detail later (page 54). Note that 'discrete' has a different meaning to 'discreet'.

As different elements have different line spectra, they can be used like barcodes to identify unknown elements. They give us valuable information about the electron configurations of different atoms.

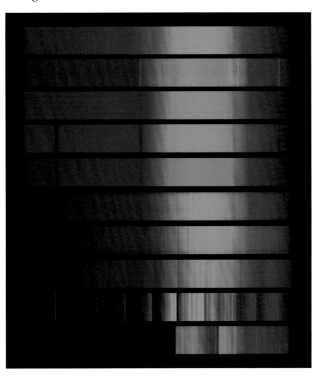

◀ Diagram of the spectra of stars, showing a set of dark absorption lines, which indicate the presence of certain elements, such as hydrogen and helium, in the outer atmosphere of the star.

The words 'discrete' meaning 'separate', and 'discreet' meaning 'unobtrusive' both come from the Latin word 'discretus' for 'to keep separate'.

When asked to distinguish between a line spectrum and a continuous spectrum, references should be made to discrete or continuous energy levels and to specific colours, wavelengths or frequencies.

Elements discovered from their line spectra and named from their flame colours include rubidium (red), caesium (sky blue), thallium (green), and indium (indigo). Emission spectroscopy was a key tool in the discovery of new elements.

Nature of Science, Structure 1.2 – How do emission spectra provide evidence for the existence of different elements?

The elemental composition of stars can be determined by analyzing their spectra. The gases that surround the center of a star absorb some wavelengths of the star's emitted radiation, producing dark absorption lines in the spectrum. These dark bands can be used to identify the elements present. The Sun's spectrum shows dark lines that represent absorbed radiation due to the presence of hydrogen and helium.

The line emission spectrum of hydrogen provides evidence for discrete energy levels

As discussed earlier, a line emission spectrum provides evidence for the electrons in an atom occupying discrete energy levels. A simple picture of the hydrogen atom was considered in Structure 1.2 with the electron orbiting the nucleus in a circular energy level. Niels Bohr proposed that an electron moves into an orbit further from the nucleus (a higher energy level) when an atom absorbs energy. This energy is given out in the form of electromagnetic radiation when the electron falls back from a higher to a lower energy level. In any sample of hydrogen many transitions can occur with each line corresponding to a particular transition. Visible light is produced when the electron falls to the second energy level ($n = 2$; see Figure 5).

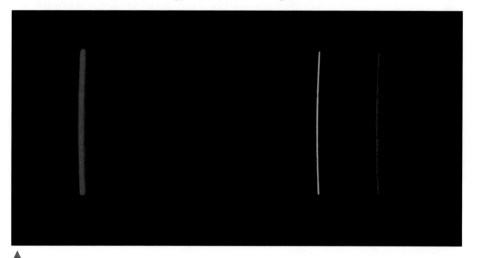

Visible emission spectrum of hydrogen. The energy of lines increases from left to right. They converge at higher energies. Similar series are found in the UV and IR regions.

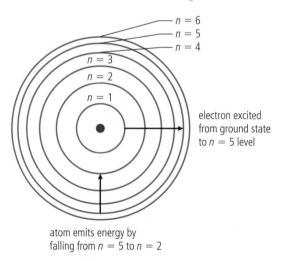

S1.3 Figure 5 An electron is excited from the ground state to a higher energy level. If the unstable electron then falls to a lower $n = 2$ energy level, visible light is emitted.

The transitions to the first energy level ($n = 1$) correspond to the highest energy change and are in the ultraviolet region of the spectrum. Infrared radiation is produced when an electron falls to the third or higher energy levels (Figure 6).

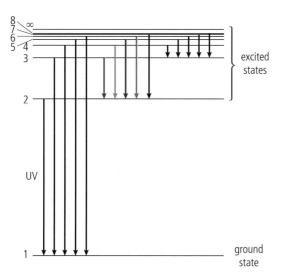

S1.3 Figure 6 Energy levels of the hydrogen atom showing the transitions. The transition $1 \rightarrow \infty$ corresponds to ionization:

$$H(g) \rightarrow H^+(g) + e^-$$

The pattern of the lines in Figure 6 gives us a picture of the energy levels in the atom. The lines converge at higher energies because the energy levels inside the atoms are closer together. When an electron is at the highest energy $n = \infty$, it is no longer in the atom and the atom has been ionized. The energy needed to remove an electron from the ground state of one mole of gaseous atoms, ions, or molecules is called the **ionization energy**. Ionization energies can also be used to support this model of the atom.

SKILLS Emission spectra can be observed using discharge tubes of different gases and a spectroscope. The colours and relative intensities of the lines should be observed. The wavelengths can be measured.

Inquiry 2 – In the study of emission spectra from gaseous elements and from light, what qualitative and quantitative data can be collected from instruments such as gas discharge tubes and prisms?

Exercise

Q1. Emission spectra provide evidence for:

   A   the existence of neutrons

   B   the existence of isotopes

   C   the existence of atomic energy levels

   D   the nuclear model of the atom.

Q2. The diagram shows the lowest five electron energy levels in the hydrogen atom.

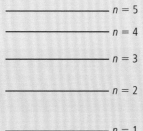

Deduce how many different frequencies in the visible emission spectrum of atomic hydrogen would arise as a result of electron transitions between these levels.

   A   3   **B**   4   **C**   6   **D**   10

Q3. The diagram shows three energy levels of an atom.

(a) Identify the transition that corresponds to the emission of light with the shortest wavelength.

    A   I → II   **B**   II → III   **C**   III → I   **D**   III → II

(b) Identify the emission line spectrum that results from transitions between these energy levels.

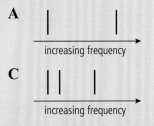

Q4. The visible emission spectrum for hydrogen includes a red line with a wavelength of 657 nm corresponding to the transition 3 → 2. State if the transition from 4 → 2 corresponds to a higher or lower wavelength and justify your answer.

Challenge yourself

1. One of the wavelengths in the emission spectrum of helium occurs at 588 nm.

Some energy levels of the helium atom are shown. The energies of the levels are given in joules. Identify the transition that produces the line at 588 nm.

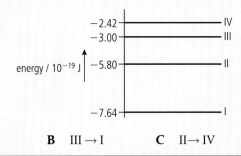

   A   I → III   **B**   III → I   **C**   II → IV   **D**   IV → II

Structure 1.3.3, 1.3.4 and 1.3.5 – Electron configuration

Structure 1.3.3 – The main energy level is given an integer number, n, and can hold a maximum of $2n^2$ electrons.

Deduce the maximum number of electrons that can occupy each energy level.

	Structure 3.1 – How does an element's highest main energy level relate to its period number in the periodic table?

A more sophisticated model is needed for atoms with more than one electron

Wave and particle models of light and the electron

Although the Bohr model of the atom was able to predict the wavelengths of lines in the emission spectrum of hydrogen with great success, it failed to predict the spectral lines of atoms with more than one electron. The model is a simplification. To develop the model of the atom further, we need to reconsider the nature of light and matter.

We saw earlier that light can either be described by its frequency, f, which is a wave property, or by the energy of individual particles, E (called photons, or quanta, of light), which make up a beam of light. The two properties are related by the Planck equation $E = hf$. Both wave and particle models have traditionally been used to explain scientific phenomena and you may be tempted to ask which model gives the 'true' description of light. We now realize that neither model gives a complete explanation of light's properties – both models are needed.

- The diffraction, or spreading out, of light that occurs when light passes through a small slit can only be explained by a wave model.
- The scattering of electrons that occurs when light is incident on a metal surface is best explained using a particle model of light.

TOK

We have outlined the plum pudding and Bohr models of the atom even though we now know they are incorrect. How can a model be useful even if it is obviously false?

In a similar way, quantum theory suggests that it is sometimes preferable to think of an electron (or indeed any particle) as having wave properties. The diffraction pattern produced when a beam of electrons is passed through a thin sheet of graphite demonstrates the wave properties of electrons. To understand the electron configurations of atoms, it is useful to consider a wave description of the electron.

Demonstration of wave–particle duality. An electron gun is fired at a thin sheet of graphite. The electrons pass through the graphite and hit a luminescent screen, producing the pattern of rings associated with diffraction. Diffraction occurs when a wave passes through an aperture similar in size to its wavelength. Quantum theory shows that electrons have wavelengths inversely proportional to their momentum (momentum is the product of their mass and velocity).

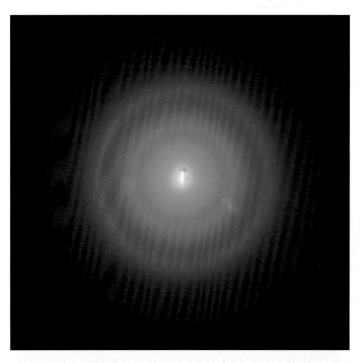

Nature of Science

Scientists use models to explain processes that may not be observable. The models can be simple or complex in nature but must match the experimental evidence if they are to be accepted. The power of the wave and particle models is that they are based on our everyday experience, but this is also their limitation. We should not be too surprised if this way of looking at the world breaks down when applied to the atomic scale, as this is beyond our experience. The model we use depends on the phenomena we are trying to explain.

When differences occur between the theoretical predictions and experimental data, the models must be modified or replaced. Bohr's model of the hydrogen atom was very successful in explaining the line spectra of the hydrogen atom but could not explain the spectra of more complex atoms, or the relative intensities of the lines in the hydrogen spectra. It also suffered from a fundamental weakness in that it was based on postulates, which combined ideas from classical and quantum physics in an ad hoc manner, with little experimental justification. Ideally, models should be consistent with the assumptions and premises of other theories. A modification of Bohr's model could only be achieved at the expense of changing our model of the electron as a particle. Dalton's atomic model and quantum theory are both examples of such radical changes of understanding, often called **paradigm shifts**.

TOK

What role do paradigm shifts play in the progression of scientific knowledge? Do they play a similar role in other areas of knowledge?

The electron's trajectory cannot be precisely described

Another fundamental problem with the Bohr model is that it assumes the electron's trajectory can be precisely described. This is now known to be impossible, as any attempt to measure an electron's position will disturb its motion. The act of focusing radiation to locate the electron sends it hurtling off in a random direction.

According to Heisenberg's **Uncertainty Principle** we cannot know where an electron is at any given moment in time – the best we can hope for is a probability picture of where the electron is *likely* to be. The possible positions of an electron are spread out in space in the same way as a wave is spread across a water surface.

Electrons occupy atomic orbitals

Schrödinger model of the hydrogen atom

We have seen that the electron can be considered to have wave properties and that only a probability description of its location is possible at a given time. Both of these ideas are encapsulated in the Schrödinger model of the hydrogen atom. Erwin Schrödinger (1887–1961) proposed that a wave equation could be used to describe the behavior of an electron in the same way that a wave equation could be used to describe the behavior of light. The equation can be applied to multi-electron systems and its solutions are known as **atomic orbitals**. An atomic orbital is a region around an atomic nucleus in which there is a 90% probability of finding the electron. The shape of the orbitals will depend on the energy of the electron. When an electron is in an orbital of higher energy, it will have a higher probability of being found further from the nucleus.

 TOK

The Uncertainty Principle states that it is impossible to make an exact and simultaneous measurement of both the position and momentum of any given body. It can be thought of as an extreme example of the observer effect discussed on page 587. The significance of the Uncertainty Principle is that it shows the effect cannot be decreased indefinitely by improving the apparatus. There is an inherent uncertainty in our measurements. What are the implications of this for the limits of human knowledge?

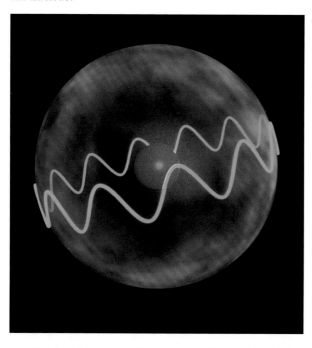

◄ The hydrogen atom shown as a nucleus (a central proton, pink), and an electron orbiting in a wavy path (light blue). It is necessary to consider the wave properties of the electron to understand atomic structure in detail. According to Heisenberg's Uncertainty Principle, the exact position of an electron cannot be defined; atomic orbitals represent regions where there is a high probability of finding an electron.

The progressive nature of scientific knowledge about the atom is illustrated by the Nobel Prizes awarded between 1922 and 1933 to Bohr, Heisenberg and Schrödinger.

Challenge yourself

2. State **two** ways in which the Schrödinger model of the hydrogen atom differs from that of the Bohr model.

Our model of the atom owes a great deal to the work of Niels Bohr and Werner Heisenberg, who worked together in the early years of quantum theory before the Second World War. But they found themselves on different sides when war broke out. The award-winning play and film *Copenhagen* is based on their meeting in that city in 1941 and explores their relationship, the uncertainty of the past, and the moral responsibilities of the scientist.

The Pauli exclusion principle states that no more than two electrons can occupy any one orbital, and if two electrons are in the same orbital, they must spin in opposite directions.

An electron is uniquely characterized by its atomic orbital and spin. If two electrons occupied the same orbital spinning in the same direction, they would be the same electron – which is impossible!

An orbital shows the volume of space in which the electron is likely to be found.

An orbital can hold two electrons of opposite spin

The Schrödinger model of the hydrogen spectrum does not fully explain the fine details of the hydrogen spectrum. The model needs to be further refined: in addition to moving around the nucleus, electrons can also be thought to **spin** on their own axis. They can spin in either a clockwise direction, represented by an upward arrow, or an anti-clockwise direction, represented by a downward arrow. Spin is an important factor in electron–electron interactions. Electrons can occupy the same region of space despite their mutual repulsion if they spin in opposite directions. This leads to the **Pauli exclusion principle**, which states that an orbital can hold only two electrons, of opposite spin.

Atomic orbitals have different shapes and sizes

The first energy level has one 1s orbital

We saw that the electron in hydrogen occupies the first energy level in the ground state. This energy level can hold a maximum of two electrons. To highlight the distinction between this wave description of the electron provided by the Schrödinger model and the circular orbits of the Bohr atom, we say the electron occupies a 1s orbital.

The dots in Figure 7 represent locations where the electron is most likely to be found. The denser the arrangement of dots, the higher the probability that the electron occupies this region of space. The electron can be found anywhere within a spherical space surrounding the nucleus.

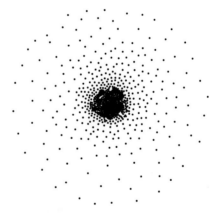

▲ **S1.3 Figure 7** An electron in a 1s atomic orbital. The density of the dots gives a measure of the probability of finding the electron in this region.

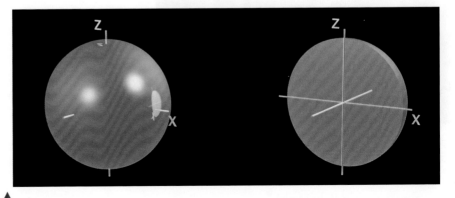

▲ The first energy level consists of a 1s atomic orbital, which is spherical in shape.

The second energy level has a 2s and 2p sublevel

The second energy level of the Bohr model is split into two **sublevels** in the Schrödinger model. The 2s sublevel is one 2s orbital and can hold a maximum of two electrons, and the 2p sublevel is three 2p orbitals and can hold six electrons. The 2s orbital has the same symmetry as a 1s orbital but extends over a larger volume. An electron in a 2s orbital is, on average, further from the nucleus than an electron in a 1s orbital and has higher energy.

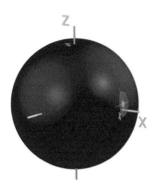 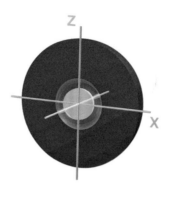

The 2s electron orbital. Just as a water wave can have crests and troughs, an orbital can have positive and negative areas. The blue area shows positive values, and the gold area negative. As it is the magnitude of the wave, not the sign, which determines the probability of finding an electron at particular positions, the sign is often not shown.

The three 2p atomic orbitals in the 2p sublevel have equal energy and are said to be **degenerate**. They all have the same dumbbell shape; the only difference is their orientation in space. They are arranged at right angles to each other with the nucleus at the centre.

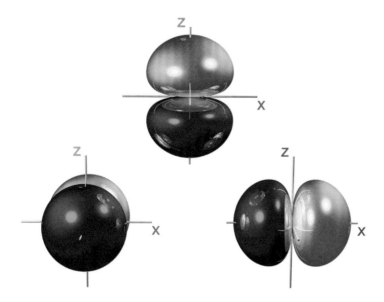

SKILLS

Investigating orbital shapes with modeling clay. Full details of how to carry out this experiment with a worksheet are available in the eBook.

From left to right, the p_y, p_z, and p_x atomic orbitals, localized along the y, z, and x-axes respectively (the y-axis comes out of the page). As they have the same energy, they are said to be degenerate. They form the 2p sublevel.

d and f orbitals

We have seen that the first energy level is made up of one sublevel and the second energy level is made up of two sublevels. This pattern can be generalized; the nth energy level of the atom is divided into n sublevels. The third energy level is made up of three sublevels: the 3s, 3p and 3d. The d sublevel is made up of five d atomic orbitals.

The five electron orbitals found in the 3d sublevel. Four of the orbitals are made up of four lobes, centered on the nucleus.

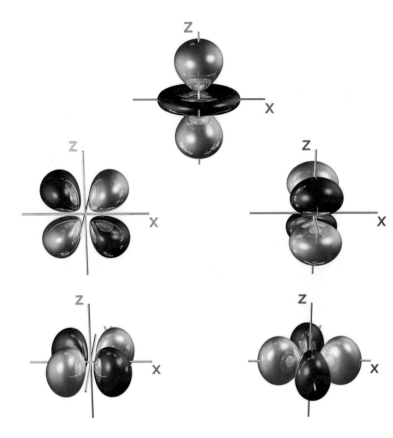

You are expected to know the shapes and names of the s and p atomic orbitals, but not of the d atomic orbitals.

The labels s, p, d and f relate to the nature of the spectral lines the model was attempting to explain. The corresponding spectroscopic terms are *sharp*, *principal*, *diffuse* and *fine*.

The letters **s**, **p**, **d**, and **f** are used to identify different sublevels and the atomic orbitals that comprise them. The fourth level ($n = 4$) is similarly made up from four sublevels. The 4f sublevels are made up of seven f atomic orbitals, but you are not required to know the shapes of these orbitals.

Worked example

Draw the shapes of a 1s orbital and a $2p_x$ orbital.

Solution

The shapes of a 1s orbital and a $2p_x$ orbital. A simple two-dimensional drawing is sufficient.

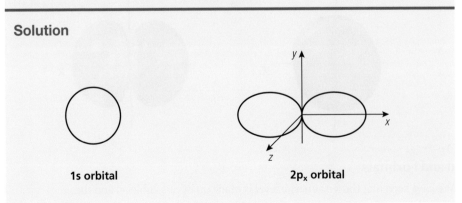

1s orbital $2p_x$ orbital

Each main energy level is divided into sublevels

The atomic orbitals associated with the different energy levels are shown in Figure 8. This diagram is a simplification, as the relative energy of the orbitals depends on the atomic number. The relative energies of the 4s and 3d atomic orbitals are particularly significant and will be discussed in more detail later.

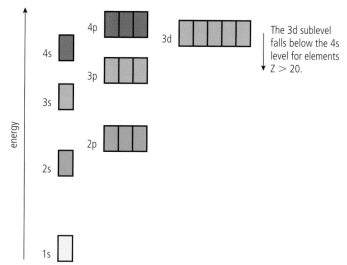

S1.3 Figure 8 The relative energies of the atomic orbitals up to the 4p sublevel.

Degenerate orbitals of the same energy form a sublevel; three p orbitals form a p sublevel, five d orbitals form a d sublevel and seven f orbitals form an f sublevel. A single s orbital makes up an s sublevel.

The number of electrons in the sublevels of the first four energy levels are shown in the table.

Level	Sublevel	Maximum number of electrons in sublevel	Maximum number of electrons in level
$n = 1$	1s	2	2
$n = 2$	2s	2	8
	2p	6	
$n = 3$	3s	2	18
	3p	6	
	3d	10	
$n = 4$	4s	2	32
	4p	6	
	4d	10	
	4f	14	

We can see the following from the table:

- The nth energy level of the atom is divided into n sublevels. For example, the 4th level ($n = 4$) is made up of four sublevels.
- Each main level can hold a maximum of $2n^2$ electrons. For example, the 3rd energy level, can hold a maximum of 18 electrons ($2 \times 3^2 = 18$).
- s sublevels can hold a maximum of 2 electrons.
- p sublevels can hold a maximum of 6 electrons.
- d sublevels can hold a maximum of 10 electrons.
- f sublevels can hold a maximum of 14 electrons.

Models are simplifications of complex systems. Details of the historic atomic models of Dalton, Bohr, Schrödinger and Heisenberg will not be assessed.

The Aufbau principle: constructing arrow-in-box diagrams

Aufbau means 'building up' in German.

The electron configuration of the ground state of an atom of an element can be determined using the **Aufbau principle**, which states that electrons are placed into orbitals of lowest energy first. Boxes can be used to represent the atomic orbitals, with single-headed arrows to represent the spinning electrons. The **electron configurations** of the first five elements are shown in Figure 9. The number of electrons in each sublevel is given as a superscript.

▲ **S1.3 Figure 9** The electron configurations of the first five elements.

The next element in the periodic table is carbon. It has two electrons in the 2p sublevel. These could either pair up, and occupy the same p orbital, or occupy separate p orbitals. Following **Hund's third rule**, we place the two electrons in separate orbitals because this configuration minimizes the mutual repulsion between them. As the 2p orbitals are perpendicular to each other and do not overlap, the two 2p electrons are unlikely to approach each other too closely. The electrons in the different 2p orbitals have parallel spins, as this leads to lower energy. The electron configurations of carbon and nitrogen are shown in Figure 10.

Hund's rule: If more than one orbital in a sublevel is available, electrons occupy different orbitals with parallel spins.

▲ **S1.3 Figure 10** Electron configurations of carbon and nitrogen.

The 2p electrons begin to pair up for oxygen ($1s^2 2s^2 2p_x^2 2p_y^1 2p_z^1$) and fluorine ($1s^2 2s^2 2p_x^2 2p_y^2 2p_z^1$). The 2p sub-shell is completed for neon ($1s^2 2s^2 2p_x^2 2p_y^2 2p_z^2$).

Worked example

Deduce the electron configuration of sulfur.

Solution

Sulfur has an atomic number of 16. Therefore it has 16 electrons.

Two electrons occupy the 1s: $1s^2$

Two electrons occupy the 2s: $2s^2$

Six electrons occupy the 2p: $2p^6$

Two electrons occupy the 3s: $3s^2$

Four electrons occupy the 3p: $3p^4$

So the electron configuration is $1s^2\,2s^2 2p^6\,3s^2\,3p^4$

Exercise

Q5. List the 4d, 4f, 4p, and 4s atomic orbitals in order of increasing energy.

Q6. State the number of 4d, 4f, 4p, and 4s atomic orbitals.

Q7. Apply the *orbital diagram* method to determine the electron configuration of calcium.

Q8. Deduce the number of unpaired electrons present in a phosphorus atom.

Q9. Deduce the number of orbitals in the $n = 4$ level and explain your answer.

Challenge yourself

3. Which of the following provide evidence to support the Bohr model of the hydrogen atom?

I.   The energy of the lines in the emission spectra of atomic hydrogen.

II. The relative intensity of the different spectral lines in the emission spectrum of atomic hydrogen.

     A I only   **B** II only   **C** I and II   **D** Neither I nor II

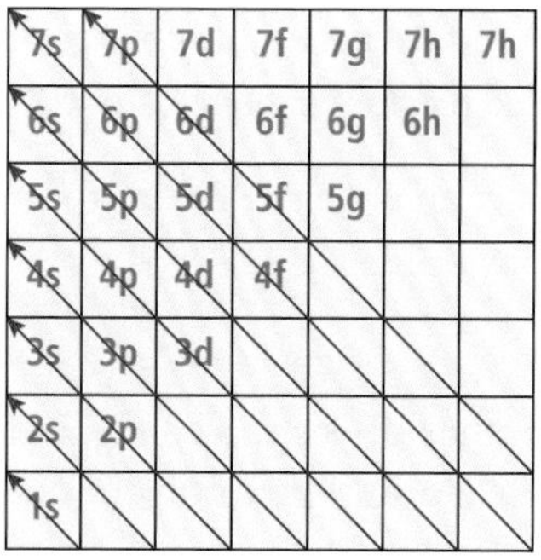

Can you think of a useful mnemonic to help you remember the order of filling orbitals?
Figure 11 shows orbitals filled to sublevel 7s. Follow the arrows to see the order in which the sublevels are filled.

S1.3 Figure 11 Order of filling sublevels:
1s, 2s, 2p, 3s, 3p, 4s, 3d, 4p, 5s, 4d, 5p, 6s, 4f, 5d, 6p, 7s, 5f, 6d, 7p.

The mathematical nature of the orbital description is illustrated by some simple relationships:

- number of sublevels at nth main energy level = n
- number of orbitals at nth energy level = n^2
- number of electrons at nth energy level = $2n^2$
- number of orbitals at lth sublevel = $(2l + 1)$ where n and l are sometimes known as quantum numbers.

Sublevel	s	p	d	f
l	0	1	2	3

The relative energy of the orbitals depends on the atomic number

The energy of an orbital depends on the attraction between the electrons and the nucleus and inter-electron repulsions. As these interactions change with the nuclear charge and the number of electrons – that is, the atomic number – so does the relative energy of the orbitals. All the sublevels in the third energy level (3s, 3p, and 3d), for example, have the same energy for the hydrogen atom and only become separated as extra protons and electrons are added. The situation is particularly complicated when we reach the d block elements. The 3d and 4s levels are very close in energy and their relative separation is very sensitive to inter-electron repulsion. For the elements potassium and calcium, the 4s orbitals are filled before the 3d sublevel. Electrons are, however, first lost from the 4s sublevel when transition metals form their ions, as once the 3d sublevel is occupied the 3d electrons push the 4s electrons to higher energy.

Worked example

State the full electron configuration of vanadium and deduce the number of unpaired electrons.

Solution

The atomic number of vanadium gives the number of electrons: Z = 23.

So the electron configuration is: $1s^2 2s^2 2p^6 3s^2 3p^6 4s^2 3d^3$
Note: the 3d sublevel is filled *after* the 4s sub level.

It is useful, however, to write the electron configuration with the 3d sub-shell before the 4s: $1s^2 2s^2 2p^6 3s^2 3p^6 3d^3 4s^2$ as the 3d sublevel falls below the 4s orbital once the 4s orbital is occupied (i.e. for elements after Ca).

The three 3d orbitals each have an unpaired electron.
Number of unpaired electrons = 3.

The worked example asked for the full electron configuration. Sometimes it is convenient to use an abbreviated form, where only the outer electrons are explicitly shown. The inner electrons are represented as a noble gas core. Using this notation, the electron configuration of vanadium is written [Ar] $3d^3 4s^2$, where [Ar] represents the electron configuration of Ar, which is $1s^2 2s^2 2p^6 3s^2 3p^6$.

The electron configurations of the first 30 elements are shown in the table.

Element	Electron configuration	Element	Electron configuration
$_1$H	$1s^1$	$_{16}$S	$1s^22s^22p^63s^23p^4$
$_2$He	$1s^2$	$_{17}$Cl	$1s^22s^22p^63s^23p^5$
$_3$Li	$1s^22s^1$	$_{18}$Ar	$1s^22s^22p^63s^23p^6$
$_4$Be	$1s^22s^2$	$_{19}$K	$1s^22s^22p^63s^23p^64s^1$
$_5$B	$1s^22s^22p^1$	$_{20}$Ca	$1s^22s^22p^63s^23p^64s^2$
$_6$C	$1s^22s^22p^2$	$_{21}$Sc	[Ar] $3d^14s^2$
$_7$N	$1s^22s^22p^3$	$_{22}$Ti	[Ar] $3d^24s^2$
$_8$O	$1s^22s^22p^4$	$_{23}$V	[Ar] $3d^34s^2$
$_9$F	$1s^22s^22p^5$	$_{24}$Cr	[Ar] $3d^54s^1$
$_{10}$Ne	$1s^22s^22p^6$	$_{25}$Mn	[Ar] $3d^54s^2$
$_{11}$Na	$1s^22s^22p^63s^1$	$_{26}$Fe	[Ar] $3d^64s^2$
$_{12}$Mg	$1s^22s^22p^63s^2$	$_{27}$Co	[Ar] $3d^74s^2$
$_{13}$Al	$1s^22s^22p^63s^23p^1$	$_{28}$Ni	[Ar] $3d^84s^2$
$_{14}$Si	$1s^22s^22p^63s^23p^2$	$_{29}$Cu	[Ar] $3d^{10}4s^1$
$_{15}$P	$1s^22s^22p^63s^23p^3$	$_{30}$Zn	[Ar] $3d^{10}4s^2$

The electrons in the outer energy level are mainly responsible for compound formation and are called **valence electrons**. Lithium has one valence electron in the outer second energy level ($2s^1$), beryllium has two ($2s^2$), boron has three ($2s^2p^1$), and so on. The number of valence electrons follows a periodic pattern, which is discussed fully in Structure 3.1. Atoms can have many other electron configurations when in an excited state. Unless otherwise instructed, assume that you are being asked about ground-state configurations.

For the d block elements, three points should be noted:

- the 3d sublevel is written with the other $n = 3$ sublevels because it falls below the 4s orbital once the 4s orbital is occupied (i.e. for elements after Ca), as discussed earlier
- chromium has the electron configuration [Ar] $3d^54s^1$
- copper has the electron configuration [Ar] $3d^{10}4s^1$.

To understand the electron configurations of copper and chromium, it is helpful to consider the electrons-in-boxes arrangements in Figure 12. As the 4s and 3d orbitals are close in energy, the electron configuration for chromium, with a half-full d sublevel, is relatively stable as it minimizes electrostatic repulsion, with six singly occupied atomic orbitals. This would be the expected configuration using Hund's rule if the 4s and 3d orbitals had exactly the same energy. Half-filled and filled sublevels seem to be particularly stable: the configuration for copper is similarly due to the stability of the full d sublevel.

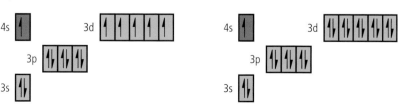

chromium: [Ar]$3d^54s^1$ copper: [Ar]$3d^{10}4s^1$

i The term 'valence' is derived from the Latin word for 'strength'.

◄ **S1.3 Figure 12** The electron configurations of the 3rd and 4th energy levels for chromium and copper.

Exercise

Q10. Identify the sublevel that does not exist.

    A 5d   **B** 4d

    C 3f   **D** 2p

Q11. Which is the correct order of orbital filling according to the Aufbau principle?

    A 4s 4p 4d 4f   **B** 4p 4d 5s 4f

    C 4s 3d 4p 5s   **D** 4d 4f 5s 5p

Q12. State the full ground-state electron configuration of the following elements.

    (a) V   **(b)** K   **(c)** Se   **(d)** Sr

Q13. Determine the total number of electrons in d orbitals in a single iodine atom.

    A 5   **B** 10

    C 15   **D** 20

Q14. Identify the excited state (i.e. not a ground state) in the following electron configurations.

    A [Ne] $3s^23p^3$ **B** [Ne] $3s^23p^34s^1$

    C [Ne] $3s^23p^64s^1$ **D** [Ne] $3s^23p^63d^14s^2$

Q15. Deduce the number of unpaired electrons present in the ground state of a titanium atom.

    A 1   **B** 2

    C 3   **D** 4

Q16. Identify the atom that possesses the most unpaired electrons.

    A O   **B** Mg

    C Ti   **D** Fe

Electron configurations of ions

As discussed earlier, positive ions are formed by the loss of electrons. These electrons are lost from the outer sublevels. For example, the electron configurations of the different aluminium ions formed when electrons are successively removed are as follows:

- Al^+ is $1s^22s^22p^63s^2$
- Al^{2+} is $1s^22s^22p^63s^1$
- Al^{3+} is $1s^22s^22p^6$

When positive ions are formed for transition metals, the outer 4s electrons are removed before the 3d electrons, as discussed earlier.

For example, Cr is [Ar] $3d^54s^1$ and Cr^{3+} is [Ar] $3d^3$

The electron configurations of negative ions are determined by adding the electrons into the next available electron orbital:

S is $1s^22s^22p^63s^23p^4$ and S^{2-} is $1s^22s^22p^63s^23p^6$

Worked example

Deduce the ground-state electron configuration of the Fe^{3+} ion.

Solution

First find the electron configuration of the atom. Fe has 26 electrons:
$1s^2 2s^2 2p^6 3s^2 3p^6 4s^2 3d^6$

As the 3d sublevel is below the 4s level for elements after calcium, we write this as
$1s^2 2s^2 2p^6 3s^2 3p^6 3d^6 4s^2$

Now remove the two electrons from the 4s sublevel and one electron from the 3d sublevel.

Electron configuration of Fe^{3+} is $1s^2 2s^2 2p^6 3s^2 3p^6 3d^5$

> **(!)** Note the abbreviated electron configuration using the noble gas core is not acceptable when asked for the *full* electron configuration.

Exercise

Q17. State the full ground-state electron configuration of the following ions.

    (a) O^{2-} **(b)** Cl^- **(c)** Ti^{3+} **(d)** Cu^{2+}

Q18. State the electron configuration of the following transition metal ions by filling in the boxes below. Use arrows to represent the electron spin.

	Ion	3d					4s
(a)	Ti^{2+}						
(b)	Fe^{2+}						
(c)	Ni^{2+}						
(d)	Zn^{2+}						

Q19. **(a)** State the full electron configuration for neon.

    (b) State the formulas of two oppositely charged ions which have the same electron configuration as neon.

Q20. State the abbreviated electron configuration using the previous noble gas core for:

    (a) Ni^{2+} **(b)** Pb^{2+} **(c)** S^{2-} **(d)** Si^{4+}

Electron configuration and the periodic table

We are now in a position to understand the structure of the periodic table (Figure 13):

- elements whose valence electrons occupy an s sublevel make up the s block
- elements with valence electrons in p orbitals make up the p block
- the d block and the f block are similarly made up of elements with outer electrons in d and f orbitals.

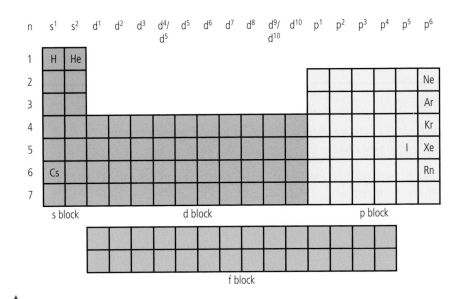

▲

S1.3 Figure 13 The block structure of the periodic table is based on the sublevels of the atom. H and He are difficult elements to classify. Although they have electron configurations that place them in the s block, their chemistry is not typical of group 1 or group 2 elements.

Structure 3.1 – What is the relationship between energy sublevels and the block nature of the periodic table?

The **ns and np sublevels are filled for elements in period n. However the (n – 1)d sublevel is filled for elements in period n.**

Structure 3.1 – How does an element's highest main energy level relate to its period number in the periodic table?

Some versions of the periodic table use the numbering 3–7 for groups 13–17. In this version, group 3 elements have three valence electrons and group 7 elements have seven valence electrons. Although this is simpler, in some respects it can lead to problems. After extensive discussions, the IUPAC concluded that the 1 to 18 numbering provides the clearest and most unambiguous labelling system.

The position of an element in the periodic table is based on the occupied sublevel of highest energy in the ground-state atom. Conversely, the electron configuration of an element can be deduced directly from its position in the periodic table.

Here are some examples.

- Ceasium is in group 1 and period 6 and has the electron configuration: $[Xe]\ 6s^1$.
- Iodine is in group 17 and in period 5 and has the configuration: $[Kr]\ 5s^2 4d^{10} 5p^5$. Placing the 4d sublevel before the 5s gives $[Kr]\ 4d^{10} 5s^2 5p^5$. Iodine has 7 valence electrons, in agreement with the pattern discussed on page 65.

Exercise

Q21. Use the periodic table to find the full ground-state electron configurations of the following elements.

(a) Cl   (b) Nb

(c) Ge   (d) Sb

Q22. Identify the elements that have the following ground-state electron configurations.

(a) $[Ne]\ 3s^2 3p^2$ (b) $[Ar]\ 3d^5 4s^2$

(c) $[Kr]\ 5s^2$ (d) $1s^2 2s^2 2p^6 3s^2 3p^6 3d^1 4s^2$

Q23. State the total number of p orbitals containing one or more electrons in tin.

Q24. How many electrons are there in all the d orbitals in an atom of barium?

Q25. State the electron configuration of the ion Cd^{2+}.

Q26. State the full electron configuration of U^{2+}.

Challenge yourself

4. State the electron configuration of thorium.

5. Only a few atoms of element 109, meitnerium, have ever been made. Isolation of an observable quantity of the element has never been achieved and may well never be. This is because meitnerium decays very rapidly.

(a) Suggest the electron configuration of the ground-state atom of the element.

(b) There is no g block in the periodic table as no elements with outer electrons in g orbitals exist in nature or have been made artificially. Suggest a minimum atomic number for such an element.

6. Consider how the shape of the periodic table is related to the three-dimensional world we live in.

(a) How many 3p and 3d orbitals would there be if only the x and y dimensions existed?

(b) How many groups in the p and d block would there be in such a two-dimensional world?

Nature of Science

We have seen how the model of the atom has changed over time. All these theories are still used today. Dalton's model adequately explains many properties of the states of matter, the Bohr model is used to explain chemical bonding, and the structure of the periodic table is explained by the wave description of the electron. In science, we often follow Occam's razor and use the simplest explanation that can account for the phenomena. As Einstein said 'Explanations should be made as simple as possible, but not simpler'.

TOK

Do atomic orbitals exist or are they primarily useful inventions to aid our understanding? What consequences might questions about the reality of scientific entities have for the public perception and understanding of the subject? If they are only inventions, how is it that they can yield such accurate predictions?

Guiding Question revisited

How can we model the energy states of electrons in atoms?

In this chapter we have developed models of the energy states of electrons in atoms to explain atomic emission line spectra and patterns in successive ionization energies of an element, and first ionizations.

- Electromagnetic radiation can be described using a wave model or a particle model. The speed of the wave (c) is related to the frequency (f) and wavelength (λ) by the expression: $c = f\lambda$

- The existence of lines in an emission spectrum indicates that the electron can only exist in discrete energy levels. The lines in the spectra converge at high energies because the gaps between energy levels in the atom decrease at higher energies.

- In the Bohr model of the hydrogen atom, the electron travels in orbits of discrete radii around the nucleus. This model correctly predicts the frequencies and wavelengths of the line spectra but does not apply to more complex systems with more than one electron.

- According to quantum theory, an electron's trajectory can only be described in terms of probabilities and a wave model of the electron is needed.
- Electrons in the atom occupy atomic orbitals, which are regions in which the electron is most likely to be found. Two electrons of opposite spin can occupy one orbital.
- Atomic orbitals have different shapes, sizes and energies. Orbitals of the same energy form sublevels. The first energy level is made up of one sublevel, the second has two sublevels and so on.
- The ground state configuration of an atom is obtained using the Aufbau principle, with electrons occupying the available orbitals of lowest energy.
- The periodic table reflects the periodicity of the electron configuration. Elements with valence electrons in s orbitals are in the s block, elements with valence electrons in p orbitals are in the p block, and so on.
- The energy of a photon (E_{photon}) depends on the frequency (f) according to Planck's equation: $E_{photon} = hf$.
- When an excited electron in an atom loses energy, the energy is given out as a photon: $\Delta E_{atom} = E_{photon}$

Practice questions

1. What is the electron configuration of the Cr^{2+} ion?

   A   [Ar] $3d^5 4s^1$ **B**   [Ar] $3d^3 4s^1$ **C**   [Ar] $3d^6 4s^1$ **D**   [Ar] $3d^4 4s^0$

2. Which is correct for the following regions of the electromagnetic spectrum?

	Ultraviolet (UV)		Infrared (IR)	
A	high energy	short wavelength	low energy	low frequency
B	high energy	low frequency	low energy	long wavelength
C	high frequency	short wavelength	high energy	long wavelength
D	high frequency	long wavelength	low frequency	low energy

3. An ion has the electron configuration $1s^2 2s^2 2p^6 3s^2 3p^6 3d^{10}$.
   Which ion could it be?

   A Ni^{2+} **B** Cu^+ **C** Cu^{2+} **D** Co^{3+}

4. In the emission spectrum of hydrogen, which electron transition would produce a line in the visible region of the electromagnetic spectrum?

   A $n = 2 \rightarrow n = 1$ **B** $n = 3 \rightarrow n = 2$

   C $n = 2 \rightarrow n = 3$ **D** $n = \infty \rightarrow n = 1$

5. Which is correct for the line emission spectrum for hydrogen?

   A   Line M has a higher energy than line N.

   B   Line N has a lower frequency than line M.

   C   Line M has a longer wavelength than line N.

   D   Lines converge at lower energy.

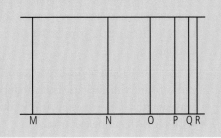

6. Which electron configuration is correct for the selenide ion, Se^{2-}?

A $1s2\ 2s^2\ 2p^6\ 3s^2\ 3p^6\ 4s^2\ 4d^{10}\ 4p^4$

B $1s^2\ 2s^2\ 2p^6\ 3s^2\ 3p^6\ 4s^2\ 4d^{10}\ 4p^6$

C $1s^2\ 2s^2\ 2p^6\ 3s^2\ 3p^6\ 4s^2\ 3d^{10}\ 4p^4$

D $1s^2\ 2s^2\ 2p^6\ 3s^2\ 3p^6\ 4s^2\ 3d^{10}\ 4p^6$

7. The electron configuration of chromium can be expressed as $[Ar]4s^x3d^y$.

(a) Explain what the square brackets around argon, [Ar], represent. (1)

(b) State the values of x and y. (1)

(c) Annotate the diagram below to show the 4s and 3d orbitals for a chromium atom. Use an arrow, ↿ and ⇃, to represent a spinning electron.

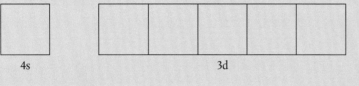

4s   3d
                                                    (1)

(Total 3 marks)

8. (a) List the following types of electromagnetic radiation in order of increasing wavelength (shortest first).

I.   Yellow light   II.   Red light

III. Infrared radiation   IV. Ultraviolet radiation   (1)

(b) Distinguish between a continuous spectrum and a line spectrum. (1)

(Total 2 marks)

9. (a) State the full electron configuration of a bromine atom. (1)

(b) Sketch the orbital diagram of the **valence shell** of a bromine atom (ground state) on the energy axis provided. Use boxes to represent orbitals and arrows to represent electrons.

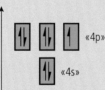

«4p»

«4s»

(1)

(Total 2 marks)

Counting particles by mass: The mole

◀ **(Top)** Entrance to the Bank of England, London, 1872. Samples of precious metals such as gold and silver are being checked for mass on the balance in the centre. **(Bottom)** A modern analytical instrument that measures mass to a high degree of precision, accounting for small factors such as dust and airflow.

Nature of Science

Advances in technology have led to increasingly precise ways of measuring mass, as well as units derived from it, such as concentration. This means that values which were previously too low to be detected can now be reported accurately. This makes possible changes in regulations and laws, such as levels of pollutants in fluids and of illegal drugs in the bloodstream. Authorities and governments often depend on measurements such as these in making judgements and enforcing the law. It is therefore essential that the data collected are reliable and include clearly stated uncertainties.

Guiding Question

How do we quantify matter on the atomic scale?

Chemical change involves interactions between atoms that have fixed mass. Yet the mass of an individual atom is so small that it is not practical to measure it directly in a laboratory. The upper limit of precision of very high quality analytical balances is generally about $1 \times 10^{-8}\,\text{kg}$, whereas, for example, a single atom of carbon weighs $1.99 \times 10^{-26}\,\text{kg}$.

Clearly, we need a way to close the gap between what can be measured and what is happening on the atomic scale. In this chapter we will learn how solving this problem led to the development of the chemical unit of amount, the mole. The mole is a fundamental unit in the SI system and is one of the most widely used tools in chemistry. It allows a form of book-keeping at the atomic level, making that important link between the measurable mass and the number of reacting particles.

Structure 1.4.1 – The mole as the unit of amount

Structure 1.4.1 – The mole (mol) is the SI unit of amount of substance. One mole contains exactly the number of elementary entities given by the Avogadro constant.	
Convert the amount of substance, n, to the number of specified elementary entities.	
An elementary entity may be an atom, a molecule, an ion, an electron, any other particle or a specified group of particles. Avogadro's constant (N_A) is given in the data booklet. It has the units mol^{-1}.	

The Avogadro constant defines the mole as the unit of amount in chemistry

We know that even with the very best experimental apparatus available, we are unable to measure the mass of individual atoms in the laboratory directly. They are simply too small. This is not really a problem, because all we need to do is to weigh an appropriately large number of atoms to give a mass that will be a useful quantity in grams. As atoms do not react individually but in very large numbers, this approach makes sense. So how many atoms shall we lump together in our 'appropriately large number'?

Let us first consider that atoms of different elements have different masses because they contain different numbers of particles, mostly **nucleons** in their nucleus, as discussed in Structure 1.2. This means we can compare their masses with each other in relative terms. For example, an atom of oxygen has a mass approximately 16 times greater than an atom of hydrogen, and an atom of sulfur has a mass about twice that of an atom of oxygen. The good news is that these ratios will stay the same when we increase the number of atoms, *so long as we ensure we have the same number of each type of atom*.

100 atoms of H, O and
S have the same mass
ratio as one atom of each
element, 1 : 16 : 32.

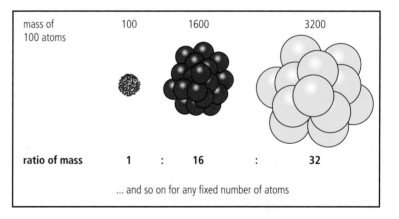

If we could take 6.02×10^{23} atoms of hydrogen, it would have a mass of 1 g. It follows from the ratios above that *the same number of atoms* of oxygen would have a mass of 16 g, while *the same number of atoms* of sulfur would have a mass of 32 g. We now have a quantity of atoms that we can measure in grams (Figure 1).

S1.4 Figure 1 6.02×10^{23} atoms of H, O and S have the same mass ratio as one atom of each element. This number of atoms gives an amount that we can see and measure in grams.

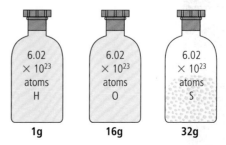

This number, accurately stated as $6.02214129 \times 10^{23}$, is known as the Avogadro number, and it is the basis of the unit of **amount** used in chemistry known as the **mole**. In other words, one mole of a substance contains the Avogadro number of particles. Mole, the unit of amount, is one of the base units in the SI system and has the unit symbol **mol.**

The example in Figure 1 is illustrative only, as in reality hydrogen and oxygen do not occur stably as single atoms, but as diatomic molecules, H_2 and O_2, as explained in Structure 2.2. Other substances exist as particles of different types, so the term **elementary entity** is used to cover the broad range of possible particles. An elementary entity may be an atom, a molecule, an ion, an electron, any other particle or a specified group of particles.

In 2019, the International Union of Pure and Applied Chemistry (IUPAC) published a change to the definition of the mole. The new definition emphasizes that the quantity 'amount of substance' is concerned with counting entities, rather than measuring the mass of a sample.

The SI refers to the metric system of measurement based on seven base units. These are metre (m) for length, kilogram (kg) for mass, second (s) for time, ampere (A) for electric current, kelvin (K) for temperature, candela (cd) for luminous intensity and mole (mol) for amount of substance. All other units are derived from these. The SI system is the world's most widely used system of measurement.

The International Bureau of Weights and Measures (BIPM according to its French initials) is an international standards organization, which aims to ensure uniformity in the application of SI units around the world. The BIPM officially introduced updated definitions to these base units, including the mole, in 2019.

So 'mole' is simply a word that represents a number, just as 'couple' is a word for 2 and 'dozen' is a word for 12. A mole is a very large number, bigger than we can easily imagine or ever count, but it is nonetheless a fixed number. So a mole of any substance contains the Avogadro number, 6.02×10^{23}, of entities. It can refer to atoms, molecules, ions, electrons and so on – it can be applied to any entity because it is just a number. And from this, we can easily calculate the number of particles in any fraction or multiple of a mole of a substance.

Amedeo Avogadro (1776–1856) was an Italian scientist who made several experimental discoveries. He clarified the distinction between atoms and molecules, and used this to propose the relationship between gas volume and number of molecules. His ideas were not accepted in his time, largely due to a lack of consistent experimental evidence. After his death, when his theory was confirmed by fellow Italian Cannizzaro, his name was given in tribute to the famous constant that he helped to establish.

The mole is the SI unit of amount of substance. One mole contains exactly 6.022×10^{23} elementary entities. This number is the fixed numerical value of the Avogadro constant, N_A.

Each sample contains one mole, 6.02×10^{23} particles, of a specific element. Each has a characteristic mass, known as its molar mass. Clockwise from upper left the elements are: carbon (C), sulfur (S), iron (Fe), copper (Cu) and magnesium (Mg).

The Avogadro number, 6.02×10^{23}, is the fixed numerical value of the Avogadro constant, N_A, which has the units mol^{-1}.

Experimental estimation of the Avogadro constant. Full details of how to carry out this experiment with a worksheet are available in the eBook.

SKILLS

Number of particles = number of moles (n) × Avogadro constant (N_A).

The magnitude of the Avogadro constant is beyond the scale of our everyday experience. What is the difference between 'data', 'information' and 'knowledge'?

TOK

The Avogadro constant is so large a number that we cannot comprehend its scale. For example:

- a population of 6.02×10^{23} people would need 75 trillion Earths each with the current population of nearly 8 billion
- 6.02×10^{23} pencil erasers would cover the Earth to a depth of about 500 m
- 6.02×10^{23} drops of water would fill all the oceans of the Earth many times over

Worked example

A tablespoon holds 0.500 moles of water. How many molecules of water are present?

Solution

1.00 mole of water has 6.02×10^{23} molecules of water

So 0.500 moles of water has $0.500\ mol \times 6.02 \times 10^{23}$ molecules

$0.500\ mol = 3.01 \times 10^{23}$ molecules of water

Worked example

A solution of ammonia and water contains 2.10×10^{23} molecules of H_2O and 8.00×10^{21} molecules of NH_3. How many moles of hydrogen atoms are present?

Solution

First total the number of hydrogen atoms.

from water, H_2O: number of H atoms = $2 \times (2.10 \times 10^{23}) = 4.20 \times 10^{23}$

from ammonia, NH_3: number of H atoms = $3 \times (8.00 \times 10^{21}) = 0.240 \times 10^{23}$

so total H atoms = $(4.20 \times 10^{23}) + (0.240 \times 10^{23}) = 4.44 \times 10^{23}$

To convert atoms to moles, divide by the Avogadro constant:

$\dfrac{4.44}{6.02} \times \dfrac{10^{23}}{10^{23}} = 0.738\ mol$ H atoms

divide by Avogadro constant, N_A

number of particles

number of moles, n

multiply by Avogadro constant, N_A

Using the Avogadro constant to calculate the number of particles in a sample has its uses, but it still leaves us with numbers that are beyond our comprehension. What is much more useful, as you have probably realized, is the link between the Avogadro constant and the mass of one mole of a substance, which is based on the relative atomic mass.

Nature of Science

Accurate determinations of the Avogadro constant require the measurement of a single quantity using the same unit on both the atomic and macroscopic scales. This was first done following Millikan's work measuring the charge on a single electron. The charge on a mole of electrons, known as the Faraday constant, was already known through electrolysis experiments. Dividing the charge on a mole of electrons, $96\,485.3383$ C, by the charge on a single electron, $1.60217653 \times 10^{-19}$ C, gives a value for the Avogadro constant of $6.02214154 \times 10^{23}\,\text{mol}^{-1}$. Later work used X-ray crystallography of very pure crystals to measure the spacing between particles and so the volume of one mole. The validity of data in science is often enhanced when different experimental approaches lead to consistent results.

Exercise

Q1. Calculate how many hydrogen atoms are present in:

    (a) 0.020 moles of C_2H_5OH

    (b) 2.50 moles of H_2O

    (c) 0.10 moles of $Ca(HCO_3)_2$

Q2. Propane has the formula C_3H_8. If a sample of propane contains 0.20 moles of C, how many moles of H are present?

Q3. Calculate the amount of sulfuric acid, H_2SO_4, which contains 6.02×10^{23} atoms of oxygen.

> **SKILLS**
>
> When multiplying or dividing, the answer should be given to the same number of significant figures as the data value with the least number of significant figures. When adding or subtracting, the answer should be given to the same number of decimal places as the data value with the least number of decimal places.

Structure 1.4.2 – Relative atomic mass and relative formula mass

> **Structure 1.4.2 – Masses of atoms are compared on a scale relative to ^{12}C and are expressed as relative atomic mass (A_r) and relative formula mass (M_r).**
>
> Determine relative formula masses (M_r) from relative atomic masses (A_r).

Relative atomic mass and relative formula mass have no units.	Structure 3.1 – Atoms increase in mass as their position descends in the periodic table. What properties might be related to this trend?
The values of relative atomic masses given to two decimal places in the data booklet should be used in calculations.	

The isotope carbon–12 is used as the reference point for comparing masses of atoms

Relative atomic mass

On page 74 the whole numbers used to compare the masses of the elements H, O and S are approximate. This is mostly because of the existence of isotopes, atoms of the same element that differ in their mass, as is explained in Structure 1.2. A sample of an element containing billions of atoms will include a mix of these isotopes according to their relative abundance. The mass of an individual atom in the sample is therefore taken as a **weighted average** of these different masses.

The relative scale for comparing the mass of atoms needs a reference point. The international convention for this is to take the specific form of carbon known as the **isotope carbon-12** as the standard, and assign this a value of 12 units. In other words, one twelfth of an atom of carbon-12 has a value of exactly 1.

Putting this together, we can define the **relative atomic mass** as follows:

Relative atomic mass, A_r, is the weighted average mass of one atom of an element relative to $\frac{1}{12}$ the mass of an atom of carbon-12.

$$\text{relative atomic mass, } A_r = \frac{\text{weighted average mass of one atom of the element}}{\frac{1}{12} \text{ mass of one atom of carbon-12}}$$

Values for A_r do not have units as it is a relative term, which simply compares the mass of atoms against the same standard. As they are average values, they are not whole numbers. Section 7 of the data booklet gives A_r values to two decimal places. Some examples are given below.

Element	Relative atomic mass (A_r)
carbon C	12.01
oxygen O	16.00
hydrogen H	1.01
lithium Li	6.94
sodium Na	22.99
potassium K	39.10

A_r values are often rounded to whole numbers for quick calculations, but when using values for more accurate calculations, it is usually best to use the exact value given in section 7 of the data booklet.

You will notice that the A_r of carbon is slightly greater than the mass of the isotope carbon-12 used as the standard, suggesting that carbon has isotopes with mass number greater than 12. In Structure 3.2 we discuss how relative atomic mass is calculated from isotope abundances, using data from mass spectrometry.

The table shows the increase in relative atomic mass of group 1 elements H, Li, Na and K as we descend the group. Successive elements in the group also have an additional energy level of electrons, which increases the atomic radius and therefore the distance of the outermost electrons from the nucleus. The larger atoms tend to lose electrons more easily, increasing their reactivity as metals. This is discussed in greater detail in Structure 3.1.

Structure 3.1 – Atoms increase in mass as their position descends in the periodic table. What properties might be related to this trend?

Challenge yourself

1. Periodic tables usually, but not always, position hydrogen at the top of group 1. What are the arguments for and against different positions for the placement of hydrogen in the periodic table?

Note that the term *relative molecular mass* was previously used, but can accurately be applied only to substances that exist as molecules. The term *relative formula mass* is preferred as it is more inclusive. It can be applied to both ionic and covalently bonded entities.

Relative formula mass

We can extend the concept of relative atomic mass to compounds (and to elements occurring as molecules) to obtain the **relative formula mass**, M_r. This simply involves adding the relative atomic masses of all the atoms or ions present in its formula. Note that M_r, like A_r, is a relative term and so has no units.

One mole of different compounds, each showing the molar mass. The chemical formulas of these ionic compounds are, clockwise from lower left: NaCl, FeCl₃, CuSO₄, KI, Co(NO₃)₂ and KMnO₄.

Within the image:
- Iron (III) chloride 270.3 g
- Copper sulphate 249.7 g
- Potassium iodide 166.0 g
- Potassium manganate (VII) 158.0 g
- Cobalt nitrate 291.0 g
- Sodium chloride 58.5 g

Challenge yourself

2. Three of the compounds in the photograph above are hydrated, containing water of crystallization as described on page 85. Use the formulas given in the caption and the masses marked on the photograph to deduce which compounds are hydrated, and the full formula of each.

Worked example

Use the values for A_r in Section 7 of the data booklet to calculate the M_r of the following:

(a) chlorine Cl_2

(b) ammonium nitrate NH_4NO_3

(c) aluminium sulfate $Al_2(SO_4)_3$

Solution

(a) $M_r = 35.45 \times 2 = 70.90$

(b) $M_r = 14.01 + (1.01 \times 4) + 14.01 + (16.00 \times 3) = 80.06$

(c) $M_r = (26.98 \times 2) + [32.07 + (16.00 \times 4)] \times 3 = 342.17$

Relative formula mass, M_r, is the sum of the weighted average of the atoms of an element in a formula unit relative to $\frac{1}{12}$ of an atom of carbon-12.

Exercise

Q4. The two most common isotopes of chlorine are ^{35}Cl and ^{37}Cl. The A_r of chlorine is 35.45. What can you conclude about the relative abundance of these two isotopes? (no calculation required)

Q5. Calculate the M_r of the following compounds:

(a) magnesium phosphate $Mg_3(PO_4)_2$

(b) ascorbic acid (vitamin C) $C_6H_8O_6$

(c) calcium nitrate $Ca(NO_3)_2$

(d) hydrated sodium thiosulfate $Na_2S_2O_3.5H_2O$

Structure 1.4.3 – Molar mass

Structure 1.4.3 – Molar mass (M) has the units g mol⁻¹.	
Solve problems involving the relationships between the number of particles, the amount of substance in moles and the mass in grams.	
The relationship $n = \dfrac{m}{M}$ is given in the data booklet.	Reactivity 2.1 – How can molar masses be used with chemical equations to determine the masses of the products of a reaction?

Molar mass is the mass of one mole of a substance

The molar mass of a substance is its relative atomic mass, A_r, or its relative formula mass, M_r, expressed in grams. It has the units g mol⁻¹.

The Avogadro constant is defined so that the mass of one mole of a substance is exactly equal to the substance's relative atomic mass or relative formula mass expressed in grams. This is known as the **molar mass** and is given the symbol **M** with the unit g mol⁻¹, which is a derived SI unit. Using examples discussed already in this chapter, we can now deduce the following.

Element or compound	Molar mass (M)
hydrogen H	1.01 g mol⁻¹
oxygen O	16.00 g mol⁻¹
chlorine Cl_2	70.90 g mol⁻¹
ammonium nitrate NH_4NO_3	80.06 g mol⁻¹
aluminium sulfate $Al_2(SO_4)_3$	342.17 g mol⁻¹

Now we are able to use the concept of the mole to make that all-important link between the number of particles and their mass in grams. The key to this is conversions of grams to moles and moles to grams. The following notations are used for these calculations:

- n = number of moles (mol)
- m = mass in grams (g)
- M = molar mass (g mol⁻¹)

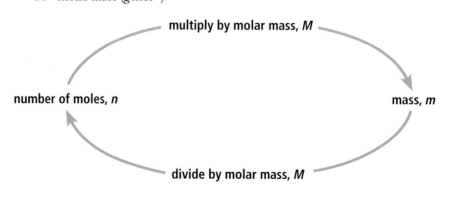

Worked example

What is the mass of the following?

(a) 6.50 moles of NaCl

(b) 0.10 moles of OH^- ions

Solution

In all these questions, we must first calculate the molar mass, M, to know the mass of 1 mole in g mol^{-1}. Multiplying M by the specified number of moles, n, will then give the mass, m, in grams.

(a) $M(NaCl) = 22.99 + 35.45 = 58.44\, g\, mol^{-1}$

$n(NaCl) = 6.50\, mol$

$\therefore m(NaCl) = 58.44\, g\, mol^{-1} \times 6.50\, mol = 380\, g$

(b) OH^- ions carry a charge because electrons have been transferred, but change to the mass is negligible so can be ignored in calculating M.

$M(OH^-) = 16.00 + 1.01 = 17.01\, g\, mol^{-1}$

$n(OH^-) = 0.10\, mol$

$\therefore m(OH^-) = 17.01\, g\, mol^{-1} \times 0.10\, mol = 1.7\, g$

Worked example

What is the amount in moles of the following?

(a) 32.50 g $(NH_4)_2SO_4$

(b) 273.45 g N_2O_5

Solution

Again we calculate the molar mass, M, to know the mass of one mole. Dividing the given mass, m, by the mass of one mole will then give the number of moles, n.

(a) $M((NH_4)_2SO_4) = [14.01 + (1.01 \times 4)] \times 2 + 32.07 + (16.00 \times 4) = 132.17\, g\, mol^{-1}$

$m((NH_4)_2SO_4) = 32.50\, g$

$\therefore n((NH_4)_2SO_4) = \dfrac{32.50\, g}{132.17\, g\, mol^{-1}} = 0.2459\, mol$

(b) $M(N_2O_5) = (14.01 \times 2) + (16.00 \times 5) = 108.02\, g\, mol^{-1}$

$m(N_2O_5) = 273.45\, g$

$\therefore n(N_2O_5) = \dfrac{273.45\, g}{108.02\, g\, mol^{-1}} = 2.532\, mol$

These simple conversions show that:

$$\text{number of moles} = \frac{\text{mass}}{\text{molar mass}} \qquad n(mol) = \frac{m\,(g)}{M\,(g\,mol^{-1})}$$

This is a very useful relationship and it is better for you to understand how it is derived, rather than just memorizing it.

Dimensional analysis, or the factor-label method, is a widely used technique to determine conversion factors on the basis of cancelling the units. This method is not specifically used in the examples here, but the units *are* shown through the calculations. This can be helpful to check that the units on both sides of the equation are balanced, and are appropriate for the answer, which is often a useful check on the steps taken. The units on both sides of the equation cancel to be balanced. As in all cases, there is no one correct way to set out calculations, so long as the steps are clear.

The diagram below summarizes the central role of the number of moles, n, in converting between the number of particles and the mass in grams.

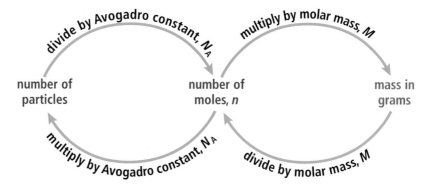

Chemical equations show the simplest ratio of chemical entities reacting together. As molar masses represent a fixed number of entities, these M values can be used directly in chemical equations to calculate the relative masses that react. For example, in the equation for the combustion of hydrogen, we can deduce the following:

	$2H_2(g)$	$+$	$O_2(g)$	$\rightarrow$	$2H_2O(l)$
reacting ratio	2 molecules		1 molecule		2 molecules
moles	2 moles		1 mole		2 moles
grams	$2 \times (1.01 \times 2)$		16.00×2		$2 \times [(1.01 \times 2) + 16.00]$
	$= 4.04\,g$		$= 32.00\,g$		$= 36.04\,g$

This simple example shows that 36.04 g of water can be produced from the combustion of 4.04 g of hydrogen. This type of calculation is the basis of all work on measuring the yield of industrial processes, considerations of atom economy and other aspects of Green Chemistry. These applications are discussed in more detail in Reactivity 2.1.

Calculations involving mass in chemistry always involve converting *grams* to *moles* and *moles* to *grams*. Think of these conversions as fundamental tools for chemists, and so make sure you are fully comfortable with carrying them out effectively.

$$amount = \frac{mass}{molar\ mass}$$

$$n(mol) = \frac{m(g)}{M(g\,mol^{-1})}$$

Reactivity 2.1 – How can molar masses be used with chemical equations to determine the masses of the products of a reaction?

Exercise

Q6. Calcium arsenate $Ca_3(AsO_4)_2$ is a poison that was widely used as an insecticide. What is the mass of 0.475 mol of calcium arsenate?

Q7. How many moles of carbon dioxide are there in 66 g of carbon dioxide, CO_2?

Q8. How many moles of chloride ions, Cl^-, are there in 0.50 g of copper(II) chloride, $CuCl_2$?

Q9. How many carbon atoms are there in 36.55 g of diamond (which is pure carbon)?

Q10. What is the mass in grams of a 0.500 mol sample of sucrose, $C_{12}H_{22}O_{11}$?

Q11. Which contains the greater number of particles, 10.0 g of water, H_2O, or 10.0 g of mercury, Hg?

Q12. Put the following in descending order of mass.

I. 1.0 mol N_2H_4 II. 2.0 mol N_2

III. 3.0 mol NH_3 IV. 25.0 mol H_2

Structure 1.4.4 – Empirical and molecular formulas

Structure 1.4.4 – The empirical formula of a compound gives the simplest ratio of atoms of each element present in that compound. The molecular formula gives the actual number of atoms of each element present in a molecule.

Interconvert the percentage composition by mass and the empirical formula.
Determine the molecular formula of a compound from its empirical formula and molar mass.

	Tool 1 – How can experimental data on mass changes in combustion reactions be used to derive empirical formulas?
	Nature of Science, Tool 3, Structure 3.2 – What is the importance of approximation in the determination of an empirical formula?

The empirical formula of a compound gives the simplest ratio of its atoms

Magnesium burns brightly in air to form a white solid product. If we want to know how many atoms of magnesium combine with how many atoms of oxygen in this reaction, we can use the central role of the mole to relate the number of reacting particles to the measured mass.

The steps in the process are:

- calculate the moles of Mg from the measured mass of Mg used
- calculate the moles of oxygen that reacted from the increase in mass on heating
- express the ratio of moles Mg : moles O in its simplest form
- the ratio of moles is the ratio of atoms, so we can deduce the simplest formula of magnesium oxide.

▲
Magnesium burns with a bright white flame, combining with oxygen from the air to form the white solid magnesium oxide.

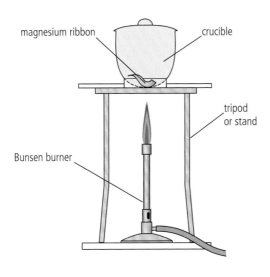

magnesium ribbon

crucible

tripod or stand

Bunsen burner

◀ Apparatus used to measure mass changes on burning magnesium.

Determining the empirical formula of a compound. Full details of how to carry out this experiment with a worksheet are available in the eBook.

Tool 1 – How can experimental data on mass changes in combustion reactions be used to derive empirical formulas?

Nature of Science, Tool 3, Structure 3.2 – What is the importance of approximation in the determination of an empirical formula?

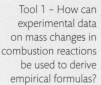

An experiment for determining the empirical formula of magnesium oxide can be found on this page of your eBook (see Skills box on the left). Sample processed data for this experiment are given below.

	Magnesium, Mg	**Oxygen, O**
Mass / g ±0.002	0.043	0.029
M / g mol⁻¹	24.31	16.00
Moles / mol	0.00177	0.00181

ratio moles Mg : moles O = 1 : 1.02

So the ratio atoms Mg : atoms O approximates to 1 : 1.

From the result of this experiment, we conclude that the formula of magnesium oxide is MgO. This is known as an **empirical formula**, which gives the simplest whole-number ratio of the atoms of each element in a compound.

Empirical formulas are often derived from combustion data such as these. As the result must be a whole number ratio, some rounding of the numbers is often required, as we saw above. Sometimes this may involve a further mathematical step.

For example, data from combustion analysis gave:

ratio of atoms Fe : atoms O = 1 : 1.5

multiplying by 2 gives ratio of atoms Fe : atoms O = 2 : 3

so the empirical formula is **Fe_2O_3**

SKILLS Judging when approximation is a valid process in a calculation is an important tool in science. The need for approximation can sometimes indicate experimental errors. Where systematic errors can be identified, modifications to the experiment can lead to increasing the accuracy of the result.

Nature of Science

Scientific investigations based on quantitative measurements are subject to errors, both random and systematic. Analysis of the impact of these errors is inherent in the practice of science. It is good practice in all experimental work to record the sources of errors, consider their effect on the results, and suggest modifications that aim to reduce their impact. Scientists have the responsibility to communicate their results as realistically and honestly as possible, and this must include uncertainties and errors.

The empirical formula is the simplest whole-number ratio of the elements in a compound.

Worked example

Which of the following are empirical formulas?

I. C_6H_6 benzene

II. C_3H_8 propane

III. N_2O_4 dinitrogen tetroxide

IV. $Pb(NO_3)_2$ lead nitrate

Solution

Only II and IV are empirical formulas, as their elements are in the simplest whole-number ratio.

I has the empirical formula CH; III has the empirical formula NO_2.

The formulas of all ionic compounds, made of a metal and a non-metal such as magnesium oxide, are empirical formulas. This is explained in Structure 2.1. But as we see in I and III in the worked example above, the formulas of some covalent compounds are not empirical formulas. We will learn about molecular formulas in the next section.

Worked example

A sample of urea contains 1.210 g N, 0.161 g H, 0.480 g C and 0.640 g O. What is the empirical formula of urea?

Solution

- Convert the mass of each element to moles by dividing by its molar mass, M.
- Divide by the smallest number to give the ratio.
- Approximate to the nearest whole number.

Element	Nitrogen, N	Hydrogen, H	Carbon, C	Oxygen, O
mass / g	1.120	0.161	0.480	0.640
M / g mol^{-1}	14.01	1.01	12.01	16.00
number of moles / mol	0.0799	0.159	0.0400	0.0400
divide by smallest	2.00	3.98	1.00	1.00
nearest whole number ratio	**2**	**4**	**1**	**1**

So the empirical formula of urea is N_2H_4CO, usually written as $CO(NH_2)_2$.

Hydrated copper sulfate, $CuSO_4.5H_2O$, is blue due to the presence of water of crystallization within the molecular structure of the crystals. The anhydrous form is white, as shown in the lower part of the tube. Heating removes the water molecules, leading to the colour being lost. The process is reversible, and the addition of water to the anhydrous crystals restores their blue colour.

A modification of this type of question is to analyze the composition of a **hydrated salt.** These are compounds that contain a fixed ratio of water molecules, known as **water of crystallization**, within the crystalline structure of the compound. The water of crystallization can be driven off by heating, and the change in mass used to calculate the ratio of water molecules to the **anhydrous salt**. The formula of the hydrated salt is shown with a dot before the number of molecules of water, for example $CaCl_2.4H_2O$.

Empirical formulas and percentage composition by mass can be interconverted

Data on the composition of a compound are often given as percentage by mass. Percentage data effectively give us the mass present in a 100 g sample of the compound.

Converting percentage by mass to empirical formulas

We can use percentage by mass data in the same way as in the examples above to determine the ratio of moles of each element in a compound.

- Divide the percentage mass of each element by its molar mass, M, to convert it to moles.
- Divide the moles of each element by the smallest number to give the ratio of moles.
- Approximate to the nearest whole number.

Worked example

The mineral, celestine, consists mostly of a compound of strontium, sulfur and oxygen. It is found by combustion analysis to have the composition 47.70% by mass Sr, 17.46% sulfur and the remainder is oxygen. What is its empirical formula?

Solution

First we must calculate the percentage of oxygen by subtraction of the total given masses from 100.

% O = 100 − (47.70 + 17.46) = 34.84

Element	Strontium, Sr	Sulfur, S	Oxygen, O
% by mass	47.70	17.46	34.84
$M / \text{g mol}^{-1}$	87.62	32.07	16.00
moles	0.5443	0.5444	2.178
divide by the smallest	1.000	1.000	4.001

So the empirical formula of the mineral is $SrSO_4$

When working with percentage figures, always check that they add up to 100. Sometimes an element is omitted from the data and you are expected to deduce its identity and percentage from the information given.

Fertilizers contain nutrients that are added to the soil, usually to replace those used by cultivated plants. The elements needed in the largest quantities, so-called **macronutrients**, include nitrogen, phosphorus and potassium. Fertilizers are often labeled with an N-P-K rating, such as 30–15–30, to show the quantities of each of these three elements. The numbers indicate respectively the percentage by mass, N, percentage by mass diphosphorus pentoxide, P_2O_5, and percentage by mass potassium oxide, K_2O. The percentage data for P_2O_5 and K_2O represent the most oxidized forms of elemental phosphorus and potassium present in the fertilizer. Ammonium salts are the most common source of nitrogen used in fertilizers.

```
        GENERAL PURPOSE 20-10-20
     (For continuous liquid feed programs)

Guaranteed analysis                        F1143
Total nitrogen (N)                          20%
  7.77% ammoniacal nitrogen
  12.23% nitrate nitrogen
Available phosphate (P₂O₅)                  10%
Soluble potash (K₂O)                        20%
Magnesium (Mg)(Total)                       0.05%
  0.05% Water soluble magnesium (Mg)
Boron (B)                                   0.0068%
Copper (Cu)                                 0.0036%
  0.0036% Chelated copper (Cu)
Iron (Fe)                                   0.05%
  0.05% Chelated iron (Fe)
Manganese (Mn)                              0.025%
  0.025% Chelated manganese (Mn)
Molybdenum (Mo)                             0.0009%
Zinc (Zn)                                   0.0025%
  0.0025% Chelated zinc (Zn)

Derived from: ammonium nitrate, potassium phosphate, potassium nitrate,
magnesium sulfate, boric acid, copper EDTA, manganese EDTA, iron EDTA,
zinc EDTA, sodium molybdate. Potential acidity: 487 lbs. calcium carbonate
equivalent per ton.
```

◀ The label on a fertilizer bag shows the percentage by mass of macro- and micronutrients that it contains.

Challenge yourself

3. A fertilizer has an N-P-K rating of 18-51-20. Use the information in the box above to determine the percentage by mass of nitrogen, phosphorus and potassium present.

An understanding of percentage by mass data helps to evaluate information that is commonly given on products such as foods, drinks, pharmaceuticals, household cleaners as well as fertilizers. For example, a common plant fertilizer is labeled as pure sodium tetraborate pentahydrate, $Na_2B_4O_7.5H_2O$, and claims to be 15.2% boron. How accurate is this claim?

Converting empirical formulas to percentage by mass

We can see in the example above that, even though the mineral celestine has only one atom of strontium for every four atoms of oxygen, strontium nonetheless accounts for 47.70% of its mass. This, of course, is because an atom of strontium has significantly greater mass than an atom of oxygen, and the percentage by mass of an element in a compound depends on the *total* contribution of its atoms. We can calculate this as follows.

Worked example

What is the percentage by mass of N, H and O in the compound ammonium nitrate, NH_4NO_3?

Solution

First calculate the molar mass M:

$M (NH_4NO_3) = 14.01 + (1.01 \times 4) + 14.01 + (16.00 \times 3) = 80.06\, g\, mol^{-1}$

Then for each element, total the mass of its atoms, divide by M and multiply by 100:

$\% N = \dfrac{14.01 \times 2}{80.06} \times 100 = 35.00\, \%$ by mass

$\% H = \dfrac{1.01 \times 4}{80.06} \times 100 = 5.05\, \%$ by mass

$\% O = \dfrac{16.00 \times 3}{80.06} \times 100 = 59.96\, \%$ by mass

(or this last term can be calculated by subtraction from 100)

Finally check that the numbers add up to 100%.
Note that rounding here means that the total is 100.01%

The molecular formula of a compound gives the actual number of atoms in a molecule

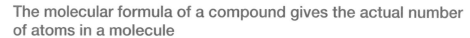

The empirical formula gives us the simplest ratio of atoms present in a compound, but this may not be the full information about the actual *number* of atoms in a molecule. For example, CH_2 is an empirical formula but there is no molecule that exists with just one atom of carbon and two atoms of hydrogen. There are many molecules with multiples of this ratio, such as C_2H_4, C_3H_6 and so on. Formulas that show all the atoms present in a molecule are called **molecular formulas**.

It is possible for different compounds to have the same empirical formula but different molecular formulas. This is particularly the case in organic chemistry.

The molecular formula can be deduced from the empirical formula if the molar mass is known.

$$(\text{mass of empirical formula})_x = M \text{ where } x \text{ is an integer}$$

Worked example

Calomel is a compound once used in the treatment of syphilis. It has the empirical formula HgCl and a molar mass of $472.08\,\text{g}\,\text{mol}^{-1}$. What is its molecular formula?

Solution

First calculate the mass of the empirical formula:

$\text{mass(HgCl)} = 200.59 + 35.45 = 236.04\,\text{g}\,\text{mol}^{-1}$

$(236.04)_x = M = 472.08$

$\therefore x = 2$

molecular formula = Hg_2Cl_2

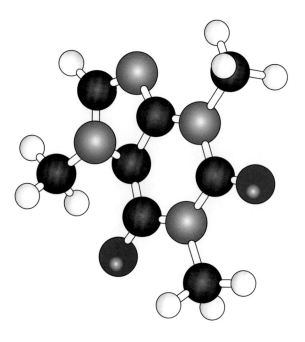

The molecular formula shows all the atoms present in a molecule. It is a multiple of the empirical formula.

▲

A molecular model of the stimulant caffeine. The atoms are color-coded as follows: black = carbon, grey = hydrogen, red = oxygen, blue = nitrogen. Can you deduce the molecular formula, the empirical formula and the molar mass of caffeine?

Combustion analysis usually gives data on the mass of compounds formed

The data presented so far may suggest that combustion analysis directly gives information on the relative masses of individual elements in a compound. In fact this is rarely the case, but instead elements are converted into new compounds, typically their oxides, by reaction with oxygen. So the primary data obtained are the masses of carbon dioxide, water, sulfur dioxide and so on, which are measured by infrared absorption, and are described in Structure 3.2. Processing these data simply involves an extra step.

Worked example

A 0.5438 g sample of a compound known to contain only carbon, hydrogen and oxygen was burned completely in oxygen. The products were 1.0390 g CO_2 and 0.6369 g H_2O. The compound has a molar mass of 46.08 g mol^{-1}. Determine the empirical formula and the molecular formula of the compound.

Solution

First we must convert the mass of each product to moles in the usual way. From the number of moles of CO_2 and H_2O we can deduce the number of moles of C atoms and H atoms.

$$n(CO_2) = \frac{1.0390}{12.01 + (16.00 \times 2)} = 0.02361 \, \text{mol} \, CO_2 \Rightarrow 0.02361 \, \text{mol C atoms}$$

$$n(H_2O) = \frac{0.6369}{(1.01 \times 2) + 16.00} = 0.03534 \, \text{mol} \, H_2O \Rightarrow 0.03534 \times 2$$

$$= 0.07068 \, \text{mol H atoms}$$

In order to know the mass of O in the original sample, we must convert the number of moles of C and H atoms to mass by multiplying by their molar mass, M.

mass C = 0.02361 mol × 12.01 g mol^{-1} = 0.2836 g

mass H = 0.07068 mol × 1.01 g mol^{-1} = 0.07139 g

∴ mass O = 0.5438 − (0.2836 + 0.07139) = 0.1888 g

$$\text{mol O atoms} = \frac{0.1888 \, \text{g}}{16.00 \, \text{g mol}^{-1}} = 0.01180 \, \text{mol}$$

Now we can proceed as with the previous examples, converting mass of O to moles and then comparing the mole ratios.

Element	Carbon, C	Hydrogen, H	Oxygen, O
mass / g			0.1888
moles	0.02361	0.07068	0.0118
divide by smallest	2.00	5.98	1.00
nearest whole number ratio	2	6	1

So the empirical formula is C_2H_6O

Mass of empirical formula = $(12.01 \times 2) + (1.01 \times 6) + 16.00 = 46.08\,g\,mol^{-1}$

(mass of empirical formula)$_x$ = M

$(46.08)_x = 46.08\,g\,mol^{-1} \therefore x = 1$

molecular formula = C_2H_6O

Exercise

Q13. Give the empirical formulas of the following compounds:

    (a) ethyne C_2H_2 **(b)** glucose $C_6H_{12}O_6$

    (c) sucrose $C_{12}H_{22}O_{11}$ **(d)** octane C_8H_{18}

    (e) oct–1–yne C_8H_{14} **(f)** ethanoic acid CH_3COOH

Q14. A sample of a compound contains only the elements sodium, sulfur and oxygen. It is found by analysis to contain 0.979 g Na, 1.365 g S and 1.021 g O. Determine its empirical formula.

Q15. A sample of a hydrated compound was analyzed and found to contain 2.10 g Co, 1.14 g S, 2.28 g O and 4.50 g H_2O. Determine its empirical formula.

Q16. A street drug has the following composition: 83.89% C, 10.35% H, 5.76% N. Determine its empirical formula.

Q17. The following compounds are used in the production of fertilizers. Determine which one has the highest percentage by mass of nitrogen: NH_3, $CO(NH)_2$, $(NH_4)_2SO_4$.

Q18. A compound has the formula M_3N where M is a metal and N is nitrogen. It contains 0.673 g of N per gram of the metal M. Determine the relative atomic mass of M and so its identity.

Q19. Compounds of cadmium are used in the construction of photocells. Show which of the following has the highest percentage by mass of cadmium: CdS, CdSe, CdTe.

Q20. Benzene is a hydrocarbon, a compound of carbon and hydrogen only. It is found to contain 7.74% H by mass. Its molar mass is $78.10\,g\,mol^{-1}$. Determine its empirical and molecular formulas.

Q21. A weak acid has a molar mass of $162\,g\,mol^{-1}$. Analysis of a 0.8821 g sample showed the composition by mass is 0.0220 g H, 0.3374 g P and the remainder oxygen. Determine its empirical and molecular formulas.

Q22. ATP is an important molecule in living cells. A sample with a mass of 0.8138 g was analyzed and found to contain 0.1927 g C, 0.02590 g H, 0.1124 g N, and 0.1491 g P. The remainder was oxygen. Determine the empirical formula of ATP. Its formula mass was found to be $507\,g\,mol^{-1}$. Determine its molecular formula.

Q23. A 0.30 g sample of a compound that contains only carbon, hydrogen and oxygen was burned in excess oxygen. The products were 0.66 g of carbon dioxide and 0.36 g of water. Determine the empirical formula of the compound.

Q24. You are asked to write your name on a suitable surface, using a piece of chalk that is pure calcium carbonate, $CaCO_3$. How could you calculate the number of carbon atoms in your signature?

Structure 1.4.5 – Molar concentration

Structure 1.4.5 – The molar concentration is determined by the amount of solute and the volume of solution.

Solve problems involving the molar concentration, amount of solute and volume of solution.

The use of square brackets to represent molar concentration is required. Units of concentration should include $g\,dm^{-3}$ and $mol\,dm^{-3}$ and conversion between these.	Tool 1 – What are the considerations in the choice of glassware used in preparing a standard solution and a serial dilution? Tool 1, Inquiry 2 – How can a calibration curve be used to determine the concentration of a solution?

The molar concentration of a solution is based on moles of solute and volume of solution

When we are working with liquids, we often focus on measuring their volume. Some liquids in common use are pure substances, such as water (H_2O), bromine (Br_2) and hexane (C_6H_{14}), but more commonly liquids are **solutions** containing two or more components.

A solution is a homogeneous mixture of two or more substances, which may be solids, liquids or gases, or a combination of these. The **solvent** is the component present in the greatest quantity, in which the **solute** is dissolved. Some examples of solutions include:

- solid / solid: metal alloy such as brass (copper and zinc)
- solid / liquid: seawater (salt and water), aqueous copper sulfate (copper sulfate and water)
- liquid / liquid: wine (ethanol and water)
- gas / liquid: fizzy drinks (carbon dioxide and water)

In this section we will be considering solutions made by dissolving a solid solute in a liquid solvent.

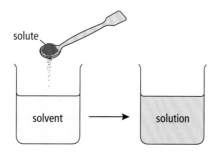

A solution is made by dissolving a solute in a solvent.

For solutions, we express the amount as its **concentration**. The molar concentration of a solution (c) is determined by the amount of solute (n) and the volume of solution (V).

$$\text{molar concentration of solution (mol\,dm}^{-3}) = \frac{\text{amount of solute (mol)}}{\text{volume of solution (dm}^3)} \text{ or } c = \frac{n}{V}$$

$\therefore$ amount of solute (mol) = conc (mol\,dm^{-3}) × volume (dm^3) or $\boldsymbol{n = cV}$

A useful convention in chemistry is to use square brackets to represent 'the molar concentration' of a solution. For example, [HCl] = 1.0 mol\,dm^{-3}.

The molar concentration of a solution refers to the amount of solute per volume of solution. It has the units mol\,dm^{-3} and is often shown using square brackets.

Worked example

A student is supplied with a solution of NaCl(aq) of concentration 0.400 mol\,dm^{-3}. He needs 0.250 mol of NaCl. What volume of solution, in cm^3, should he use?

Solution

Substituting the values given into the equation: $n = cV$

0.250 mol = 0.400 mol\,dm^{-3} × V

$$\therefore V = \frac{0.250 \text{ mol}}{0.400 \text{ mol dm}^{-3}} = 0.625 \text{ dm}^3$$

1 dm^3 = 1000 cm^3 $\Rightarrow$ 0.625 dm^3 = 625 cm^3

Concentration can also be expressed as mass of solute (g) per volume of solution (dm^3).

moles of solute (mol) = concentration of solute (mol\,dm^{-3}) × volume of solution (dm^3)

$n = cV$

$$\text{concentration of solution (g\,dm}^{-3}) = \frac{\text{mass of solute (g)}}{\text{volume of solution (dm}^3)}$$

We can use molar mass to convert grams to moles in order to obtain the molar concentration.

Worked example

What is the molar concentration of a solution of sodium carbonate, Na$_2$CO$_3$, that has a concentration of 4.24 g\,dm^{-3}?

Solution

M(Na$_2$CO$_3$) = (22.99 × 2) + 12.01 + (16.00 × 3) = 105.99 g\,mol^{-1}

Substituting the values given in the equation: $m = nM$

4.24 g = n (mol) × 105.99 g\,mol^{-1}

$$\therefore n = \frac{4.24 \text{ g}}{105.99 \text{ g mol}^{-1}} = 0.0400 \text{ mol}$$

$\therefore$ [Na$_2$CO$_3$] = 0.0400 mol\,dm^{-3}

Note that concentration is specified per volume of final *solution*, not per volume of *solvent* added. This is because volume changes occur on dissolving the solute.

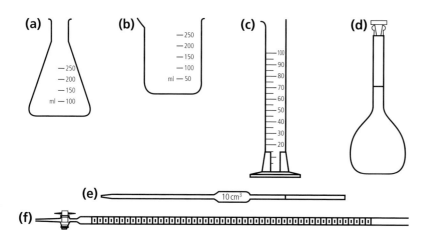

▲
S1.4 Figure 2 Glassware commonly used in the laboratory. (a) conical or Erlenmeyer flask, its shape makes it easy to mix liquids as the flask can be easily swirled; (b) beaker; (c) measuring or graduated cylinder; (d) volumetric flask; (e) pipette; (f) burette.

The term *molarity*, M, has been widely used to express amount concentration, but it is falling out of common usage. It will not be used in IB examination questions, so make sure you are fully familiar with the terms mol dm^{-3} and g dm^{-3}. (Note that M is used specifically to refer to molar mass.)

Challenge yourself

4. When sodium hydroxide pellets (NaOH) dissolve in water, there is a *decrease* in the total volume of the solution. Explain what might cause this.

▲
Different types of laboratory glassware, some of which are identified in Figure 2.

In quantitative work, it is essential to select glassware that measures volume to an appropriate level of precision. Most glassware is marked with a given uncertainty at a specified temperature – the smaller the uncertainty, the more precise the measurement. Beakers and conical flasks have very large uncertainties and are not used for precise volume measurements. Different manufacturers calibrate glassware to different levels of precision, but some typical values for laboratory apparatus are shown below.

Tool 1 – What are the considerations in the choice of glassware in preparing a standard solution and a serial dilution?

Glassware	Volume / cm³	± Uncertainty / cm³	Uncertainty / %
beaker	50	5	10
measuring cylinder	50.0	0.5	1
burette	50.00	0.05	0.1

The accuracy of a measurement is increased by using glassware with the smallest adequate volume. For example, if we need 40 cm³ of liquid, we choose a measuring cylinder of 50 cm³ rather than one of 100 cm³. It is also important to read the volume at the bottom of the meniscus when the glassware is supported on a horizontal surface.

Volumetric flask showing volume of 5000 cm³ and uncertainty of +1.2 cm³ at 20 °C. What percentage uncertainty is this?

Chemists routinely prepare solutions of known concentration, referred to as **standard solutions**. The mass of solute required is accurately measured and then transferred carefully to a volumetric flask, which is precisely calibrated for a specific volume. The solvent is added steadily with swirling to help the solute to dissolve, until the final level reaches the mark on the flask. Note that distilled water, not tap water, must be used as the solvent in the preparation of all aqueous solutions.

Worked example

Explain how you would prepare 100 cm³ of a 0.10 mol dm⁻³ solution of NaCl.

Solution

Ensure that cm³ are converted to dm³ by dividing by 1000.

$n = c\,V$

$n = 0.10\,\text{mol dm}^{-3} \times \dfrac{100}{1000}\,\text{dm}^3 = 0.0100\,\text{mol}$

$M(\text{NaCl}) = 22.99 + 35.45 = 58.44\,\text{g mol}^{-1}$

∴ mass required = 0.0100 mol × 58.44 g mol⁻¹ = 0.584 g

Add 0.584 g NaCl(s) to a 100 cm³ volumetric flask, and make up to the mark with distilled water.

A standard solution is one of accurately known concentration.

The increased popularity, in many countries, of bottled water over tap water for drinking has raised several concerns, including the environmental costs of transport and packaging, and the source of the water and its solute (dissolved mineral) content. Significant differences exist in the regulation of the bottled water industry in different countries. In the USA, the Food and Drug Administration (FDA) requires that mineral water should contain between 500 and 1500 mg dm⁻³ of total dissolved solids. In Europe, mineral water is defined by its origin rather than by content, and the European Union prohibits the treatment of any water bottled from a source. The global cost of bottled water exceeds billions of dollars annually. As the United Nations General Assembly has explicitly recognized that access to safe, clean and affordable drinking water is a human right, there is an urgent need for money and technology to be diverted to improving tap water supplies globally to help make this a reality for all.

A different unit of concentration is known as **ppm**, parts per million. It denotes one part per 10⁶ parts of the whole solution, and is useful in describing very low concentrations. This unit is widely used in reporting levels of pollutants in air, water, soil and food. For example, in the USA the FDA has set a maximum permissible level of 1 part of methylmercury in a million parts of seafood (1 ppm).

Label on water bottle listing the mineral content in milligrams per dm³.

Dilutions of solutions reduce the concentration

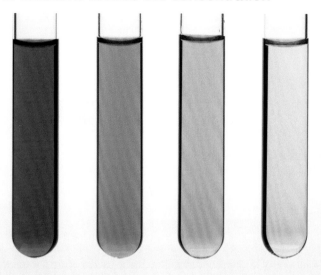

A series of dilutions of cobalt(II) chloride solutions. In coloured solutions such as these, the effect of lowering the concentration of the solution can be observed.

As a solution is diluted, the number of moles of solute remains the same, but because the volume of solution increases, the concentration decreases. In other words, the number of moles n = a constant, and as $n = c\,V \Rightarrow c\,V$ must be constant through dilution.

$\therefore c_1V_1 = c_2V_2$ where c_1V_1 refer to the initial concentration and volume, respectively,

and c_2V_2 refer to the diluted concentration and volume, respectively.

This equation provides an easy way to calculate concentration changes on dilution.

Worked example

Determine the final concentration of a $75\,cm^3$ solution of HCl of concentration $0.40\,mol\,dm^{-3}$, which is diluted to a volume of $300\,cm^3$.

Solution

$c_1V_1 = c_2V_2$

$c_1 = 0.40\,mol\,dm^{-3}; V_1 = 75\,cm^3; V_2 = 300\,cm^3$

$\therefore (0.40\,mol\,dm^{-3}) \times (75\,cm^3) = c_2 \times (300\,cm^3)$

c_2 diluted concentration $= 0.10\,mol\,dm^{-3}$

A quick check shows that the volume has increased four times, so the concentration must have decreased four times.

A **serial dilution** is a series of dilutions of a standard solution, where the concentration is reduced by a fixed amount at each step. It generates a series of solutions of known concentration, e. g. $1.00\,mol\,dm^{-3}, 0.100\,mol\,dm^{-3}, 0.0100\,mol\,dm^{-3}, 0.00100\,mol\,dm^{-3}$. Dilution is carried out into volumetric flasks so that the final volume of the solution is measured, taking account of volume changes that may occur on dilution.

SKILLS

A common practice in laboratory work is to make a **dilution** from a more concentrated starting solution called the **stock solution,** by adding solvent. For all aqueous solutions, distilled water, rather than tap water, must be used.

Note that in the equation $c_1V_1 = c_2V_2$, volume terms appear on both sides of the equation and so their units will cancel. This means that *any* units of volume can be used directly (there is no need to convert them to dm^3), *so long as* they are consistent on both sides of the equation.

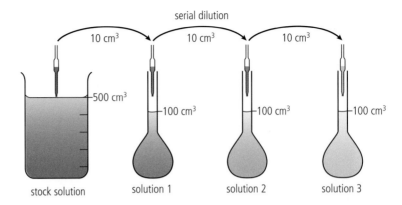

A serial dilution is prepared by using a pipette to transfer 10 cm³ from the stock solution into a 100 cm³ volumetric flask. The solution is then made up to the mark with distilled water. Repeating this process generates a series of solutions, each 10 times less concentrated than the previous one.

This series of solutions of known concentration can be used to form a standard in a technique known as **ultraviolet–visible spectroscopy,** which uses the direct relationship between the concentration of a solution and its absorbance. The absorbance of each solution is measured and the results plotted as a **calibration curve**. This curve is then used to determine the unknown concentration of a sample containing the same solute.

Preparing a calibration curve for spectrophotometric determinations. Full details of how to carry out this experiment with a worksheet are available in the eBook.

SKILLS

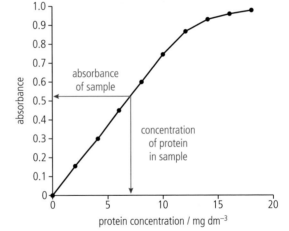

S1.4 Figure 3 A calibration curve can be used to find the protein concentration of a sample.

Tool 1, Inquiry 2 – How can a calibration curve be used to determine the concentration of a solution?

Challenge yourself

5. From the shape of the calibration curve in Figure 3, what can you conclude about the relationship between concentration and absorbance at higher concentration? How might this affect the calculation of the unknown concentration?

Titration is an important technique in volumetric analysis. It is used to determine the concentration of a solution when it reacts exactly with another solution of known concentration. Titrations typically involve reactions between acids and bases, or between oxidizing and reducing agents. Precise glassware including burettes and pipettes is used to obtain the precision required in measurement. These techniques are discussed in Reactivity 3.1 and Reactivity 3.2.

Exercise

Q25. Calculate the mass of potassium hydroxide, KOH, required to prepare 250 cm^3 of a 0.200 mol dm^{-3} solution.

Q26. Calculate the mass of magnesium sulfate heptahydrate, $MgSO_4.7H_2O$, required to prepare 0.100 dm^3 of a 0.200 mol dm^{-3} solution.

Q27. Calculate the number of moles of chloride ions in 0.250 dm^3 of 0.0200 mol dm^{-3} of zinc chloride, $ZnCl_2$, solution.

Q28. 250 cm^3 of a solution contains 5.85 g of sodium chloride. Calculate the concentration of sodium chloride in mol dm^{-3}.

Q29. Concentrated nitric acid, HNO_3, is 16.0 mol dm^{-3}. What volume of concentrated acid would you need to prepare 100 cm^3 of 0.50 mol dm^{-3} HNO_3?

Q30. Sodium sulfate, Na_2SO_4, reacts in aqueous solution with lead nitrate, $Pb(NO_3)_2$, as follows:

$$Na_2SO_4(aq) + Pb(NO_3)_2(aq) \rightarrow PbSO_4(s) + 2NaNO_3(aq)$$

In an experiment, 35.30 cm^3 of a solution of sodium sulfate reacted exactly with 32.50 cm^3 of a solution of lead nitrate. The precipitated lead sulfate was dried and found to have a mass of 1.13 g. Determine the concentrations of the original solutions of lead nitrate and sodium sulfate. State what assumptions are made.

Structure 1.4.6 – Avogadro's law

> **Structure 1.4.6 – Avogadro's law** states that equal volumes of all gases measured under the same conditions of temperature and pressure contain equal numbers of molecules.
>
> Solve problems involving the mole ratio of reactants and/or products and the volume of gases.
>
> | | Structure 1.5 – Avogadro's law applies to ideal gases. Under what conditions might the behaviour of a real gas deviate most from an ideal gas? |
> | --- | --- |

Avogadro's law directly relates gas volumes to moles

Gases, like liquids, are fluids and it is therefore often convenient to focus on their volume as a measure for quantitative work. This means we need to know the relationship between gas volume and the number of moles.

Consider the following demonstration (Figure 4) where two gas jars are each filled with different gases – hydrogen (H_2) in flask A and bromine (Br_2) in flask B. The flasks are at the same temperature and pressure and have equal volumes.

Scientists know, from many experimental measurements on gas volumes, that the number of particles in the two flasks in Figure 4 is the same. At first this might seem surprising – after all, bromine molecules are much larger and heavier than hydrogen molecules. But we need to consider the nature of the gaseous state. We learn on page

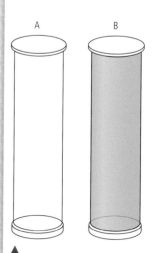

S1.4 Figure 4 Flask A contains hydrogen molecules, flask B contains bromine molecules. The two flasks are under the same conditions of temperature and pressure.

15, the particles in a gas are widely spaced out with negligible forces between them. In simple terms, most of a gas volume is empty space. For this reason, the chemical nature of the gas is irrelevant to its volume. Gas volume is determined only by the number of particles, and by the temperature and pressure.

This understanding is known as Avogadro's law, which states that:

Equal volumes of all gases, when measured at the same temperature and pressure, contain an equal number of particles.

Alternatively, it can be stated that equal numbers of particles of all gases, when measured at the same temperature and pressure, occupy equal volumes.

Using V for volume and n for number of moles, $\boldsymbol{V \propto n}$

This relationship enables us to relate gas volumes of any gas to the number of moles, and so to reacting ratios in equations.

Airbags have become a standard safety fitting in most vehicles. They are designed to act as a cushion or shock absorber by inflating rapidly on sudden impact of the vehicle during a collision. Airbags work on the principle of a chemical reaction triggered by the impact producing a gaseous product that causes a sudden volume change. The key reaction used is the conversion of sodium azide, NaN_3, to nitrogen gas, N_2. To avoid the production of dangerously reactive sodium metal, potassium nitrate, KNO_3 and silicon dioxide, SiO_2, are also included so that harmless silicates are produced instead.

Challenge yourself

6. Use the explanation above to deduce the chemical equations for the reactions taking place in a deployed airbag.

▲
Illustration of an airbag and seat belt in action during a car accident. On impact, the airbag inflates and the seat belt slows the forward force of the body, protecting the driver's head and chest.

Avogadro's law is based on the assumptions outlined above that the particles in a gas occupy negligible volume and have no forces of attraction between them. These are postulates of the **ideal gas model** which is discussed in more detail in Structure 1.5. Real gases will deviate most from this model when conditions bring the gas closer to forming a liquid, that is, lower temperature and higher pressure.

Avogadro's law states that equal volumes of all gases at the same conditions of temperature and pressure contain equal numbers of particles.
$V \propto n$

Structure 1.5 – Avogadro's law applies to ideal gases. Under what conditions might the behaviour of a real gas deviate most from an ideal gas?

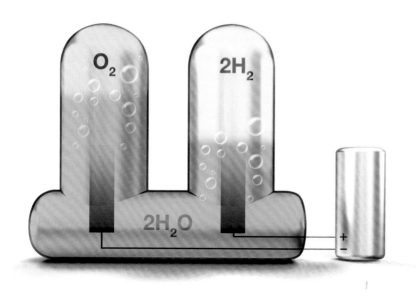

▲

The apparatus shows the use of electrolysis to split water into hydrogen and oxygen in the following reaction: $H_2O(l) \rightarrow 2H_2(g) + O_2(g)$. How can you explain the observed difference in volumes of the two gases produced?

Worked example

40 cm³ of carbon monoxide reacts with excess oxygen in the reaction:

$$2CO(g) + O_2(g) \rightarrow 2CO_2(g)$$

What volume of oxygen will be used and what volume of carbon dioxide is produced?

(Assume all volumes are measured at the same temperature and pressure.)

Solution

Data given in the question are shown in black, deduced data are shown in blue.

First identify the mole ratios in the equation:

$$2CO(g) + O_2(g) \rightarrow 2CO_2(g)$$

2 moles   1 mole   2 moles

The mole ratio is equal to the ratio of reacting gas volumes, so:

$$2CO(g) + O_2(g) \rightarrow 2CO_2(g)$$

40 cm³   20 cm³   40 cm³

Therefore 20 cm³ of oxygen will be used and 40 cm³ of carbon dioxide produced.

Worked example

When 10 cm³ of a gaseous hydrocarbon (a compound containing only carbon and hydrogen) is burned in excess oxygen, the products consist of 30 cm³ of carbon dioxide and 30 cm³ of water vapour, measured under the same conditions of temperature and pressure. Determine the molecular formula of the hydrocarbon.

Solution

'Excess' oxygen indicates that the combustion reaction is complete.

$$C_xH_y + \text{excess } O_2 \rightarrow CO_2 + H_2O$$

Given volumes: 10 cm³   30 cm³ 30 cm³

ratio of volumes / mole ratio: 1   3   3

∴ 1 molecule hydrocarbon → 3 molecules CO_2 + 3 molecules H_2O
                                3 C atoms   6 H atoms

The molecular formula is C_3H_6.

Exercise

Q31. These four balloons are each filled with 1 dm³ of gas. At 25 °C and 100 kPa, they each contain 0.044 mol or 2.65×10^{22} atoms or molecules. Which balloon is the heaviest?

Q32. A balloon contains a certain mass of argon gas. The pressure and temperature are kept constant, and the same mass of neon gas is added to the balloon. What happens?

  A   The volume of the balloon expands by two times.

  B   The volume of the balloon expands by more than two times.

  C   The volume of the balloon expands by less than two times.

  D   None of the above.

Q33. The following reaction takes place in the industrial production of ammonia, NH_3:

$$N_2(g) + 3H_2(g) \rightarrow 2NH_3(g)$$

What volume of hydrogen would be required to produce 30 dm³ of ammonia?

Q34. At a certain temperature and pressure, one mole of hydrogen gas, H_2, occupies a volume of 10 dm³. What would be the volume of one mole of H atoms under the same conditions?

Guiding Question revisited

How do we quantify matter on the atomic scale?

In this chapter we have introduced the mole as the unit of amount in chemistry.

- One mole of any substance contains the Avogadro number of elementary entities.
- Elementary entities can be any form of particle, including atoms, molecules, ions, electrons, or a specified group of particles.
- The Avogadro constant enables us to interconvert the number of moles and the number of entities.
- The molar mass, M, of a substance is its relative atomic mass, A_r, or relative formula mass, M_r, expressed in grams. A_r and M_r are expressed relative to carbon-12, and are a weighted average taking into account the abundance of isotopes.
- The molar mass enables us to interconvert the number of moles and the mass of the substance.
- The empirical formula of a compound can be determined from its percentage composition by mass. It gives the simplest ratio of atoms of elements in the compound.
- The molecular formula can be determined from the empirical formula and the molar mass. It gives the actual number of atoms of each element present in a molecule.
- The molar concentration of a solution is expressed as the moles of solute per volume of solution.
- Equal volumes of all gases at the same temperature and pressure contain the same number of moles.

How many atoms of calcium are in Avogadro's name written in chalk?

Practice questions

1. How many oxygen **atoms** are in 0.100 mol of $CuSO_4.5H_2O$?

   A 5.42×10^{22} **B** 6.02×10^{22} **C** 2.41×10^{23} **D** 5.42×10^{23}

2. Four identical containers under the same conditions are filled with gases as shown below. Which container and contents will have the highest mass?

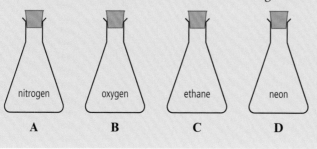

| nitrogen | oxygen | ethane | neon |
| **A** | **B** | **C** | **D** |

3. What is the amount, in moles, of sulfate ions in $100\,cm^3$ of $0.020\,mol\,dm^{-3}$ $FeSO_4(aq)$?

    A 2.0×10^{-3} **B** 2.0×10^{-2} **C** 2.0×10^{-1} **D** 2.0

4. $1.7\,g$ of $NaNO_3$ ($M_r = 85$) is dissolved in water to prepare $0.20\,dm^3$ of solution. What is the concentration of the resulting solution in $mol\,dm^{-3}$?

    A 0.01   **B** 0.1

    C 0.2   **D** 1.0

5. The relative formula mass of a gas is 56 and its empirical formula is CH_2. What is the molecular formula of the gas?

    A CH_2 **B** C_2H_4

    C C_3H_6 **D** C_4H_8

6. What is the total number of hydrogen atoms in $1.0\,mol$ of benzamide, $C_6H_5CONH_2$?

    A 7   **B** 6.0×10^{23}

    C 3.0×10^{24} **D** 4.2×10^{24}

7. What is the concentration of NaCl, in $mol\,dm^{-3}$, when $10.0\,cm^3$ of $0.200\,mol\,dm^{-3}$ NaCl solution is added to $30.0\,cm^3$ of $0.600\,mol\,dm^{-3}$ NaCl solution?

    A 0.450   **B** 0.300

    C 0.500   **D** 0.800

8. On analysis, a compound with molar mass $60\,g\,mol^{-1}$ was found to contain $12\,g$ of carbon, $2\,g$ of hydrogen and $16\,g$ of oxygen. What is the molecular formula of the compound?

    A CH_2O **B** CH_4O

    C C_2H_4O **D** $C_2H_4O_2$

9. $300\,cm^3$ of water is added to a solution of $200\,cm^3$ of $0.5\,mol\,dm^{-3}$ sodium chloride. What is the concentration of sodium chloride in the new solution?

    A $0.05\,mol\,dm^{-3}$ **B** $0.1\,mol\,dm^{-3}$

    C $0.2\,mol\,dm^{-3}$ **D** $0.3\,mol\,dm^{-3}$

10. What is the approximate molar mass, in $g\,mol^{-1}$, of $MgSO_4.7H_2O$?

    A 120   **B** 130

    C 138   **D** 246

11. Which is both an empirical and a molecular formula?

    A C_5H_{12} **B** C_5H_{10}

    C C_4H_8 **D** C_4H_{10}

12. A hydrate of potassium carbonate has the formula $K_2CO_3.xH_2O$. A 10.00 g sample of the hydrated salt is heated and forms 7.93 g of anhydrous salt.

    (a) Calculate the number of moles of water in the hydrated sample.   (1)

    (b) Calculate the number of moles of anhydrous salt that form.   (1)

    (c) Determine the formula of the hydrate.   (1)

    (d) How could you determine when all the hydrated salt has been converted into an anhydrous form?   (1)

    (Total 4 marks)

13. An organic compound contains carbon, hydrogen and oxygen. The percentage by mass of carbon is 62.02% and hydrogen is 10.43%.

    Determine the empirical formula of the compound, showing your working.   (3)

    (Total 3 marks)

14. Phosphorus formed an oxide by reaction with air which contained 43.6% by mass phosphorus.

    (a) Determine the empirical formula of the oxide, showing the steps in your calculation.   (3)

    (b) The molar mass of the oxide is approximately 285 g mol^{-1}. Determine the molecular formula of the oxide.   (1)

    (Total 4 marks)

15. (a) A student needs to prepare 250.00 cm^3 of a solution of sodium hydrogencarbonate with a concentration of 0.500 mol dm^{-3}. Explain the steps in the process starting with the solid salt, and stating the apparatus and glassware used.   (5)

    (b) Explain how, starting with the solution prepared in (a), 100 cm^3 of solutions with concentrations of 0.0500 mol dm^{-3} and 0.00500 mol dm^{-3} could be prepared.   (4)

    (Total 9 marks)

Ideal gases

◀ Computer artwork of the Earth with half of its atmosphere removed. The atmosphere is a layer of gases surrounding the planet. It reaches approximately 100 kilometers above the Earth's surface.

Guiding Question

How does the model of ideal gas behavior help us to predict the behavior of real gases?

Gases surround us in the atmosphere and the first experiments on gases concerned the physical properties of air. Robert Boyle first experimented on the 'springiness of the air' in the 17th century and further investigations on other properties followed in the 18th century when people began to fly in hot air balloons. Simple empirical laws relating pressure, volume, temperature and amount of gases were identified, with the suggestion that all gases behaved in the same way.

The gaseous state is the easiest state to describe and model. In Structure 1.1, we describe a gas as a collection of widely separated particles in chaotic, random motion and it is this randomness that makes the model so simple. A model of an ideal gas, based on kinetic theory, can be defined in terms of a small number of axioms. The volume of the gas particles and the attractive forces between the particles are considered to be negligible and the pressure of a gas is due to particles colliding with the walls of the container. The model, and the application of some simple physics, gives accurate quantitative predictions of how gas pressure changes with temperature and volume and shows that temperature is a measure of the average kinetic energy of the particles.

The initial empirical laws were, however, only approximations and more precise experiments showed that real gases deviate from this ideal behavior at high pressure and low temperature. These deviations highlight some limitations of the initial assumptions of the ideal gas model. The volume of the gas particles and the intermolecular forces cannot be ignored under all conditions.

▲
Joseph Louis Gay-Lussac (1778–1850), French chemist and physicist, in a daring 1804 balloon ascent to investigate the composition of air at high altitude. He, along with another chemical balloonist, Jacques Charles, and Robert Boyle investigated the properties of the gases.

Structure 1.5.1 – The ideal gas model

Structure 1.5.1 – An ideal gas consists of moving particles with negligible volume and no intermolecular forces. All collisions between particles are considered elastic.

Recognize the key assumptions in the ideal gas model.

An ideal gas consists of moving particles with negligible volume and no intermolecular forces

You will be familiar with some behavior of gases through everyday experiences such as blowing up balloons or bicycle tyres. Perhaps you have noticed how inflated balloons shrink in colder temperatures and expand when it is warmer. Experiments show that all gases respond in the same way to changes in volume, pressure and temperature when the mass of gas is fixed. The chemical nature of the gas makes no difference. The gaseous state is the easiest state to describe using kinetic theory.

The theoretical model of an **ideal gas** is assumed to obey the following axioms:

1. The particles in a gas have negligible volume compared with the volume the gas occupies.
2. There are no intermolecular forces between the particles except when the molecules collide.
3. Gas particles have a range of speeds and move randomly. The average kinetic energy of the particles is proportional to the temperature.
4. The collisions of the particles with the walls of the container and with each other are **elastic:** kinetic energy is conserved.

Although no real gas fits this description exactly, the application of Newton's Second Law (force = mass × acceleration) can be used to make accurate quantitative predictions for the physical properties of most gases under typical conditions of temperature and pressure.

The pressure of a gas is due to gas particles colliding with the walls of the container

When a gas particle collides with the wall of a container, it bounces back with the same speed (Figure 1). The collision is elastic, according to the fourth axiom above. The pressure of a gas is a result of a large number of such collisions and is the same on all walls. This is because the particles move randomly with no preferred direction, according to the third axiom.

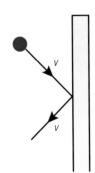

If a gas particle was the size of a tennis ball, it would typically travel the length of a tennis court between collisions.

S1.5 Figure 1 The pressure of a gas is due to the particles colliding with the wall.

The so-called Mentos-soda fountain reaction. Dissolved carbon dioxide in the soda changes quickly into gas in the presence of the mint Mentos. The sudden pressure change causes the soda to be ejected in a fountain of foam.

This simple model allows us to explain the pressure changes that occur when the temperature and volume of a gas changes. The volume of a gas is the total volume of its container as the particles spread out to fill all available space.

- When the volume of a gas increases, the gas particles collide less frequently with the walls as they must travel greater distances between collisions. This decreases the pressure. The model correctly predicts that increasing the volume decreases the pressure (Figure 2).

lower pressure

higher pressure

Smaller volume   Larger volume

S1.5 Figure 2 As the volume of a gas increases, the gas particles collide less frequently with the walls of the container.

- When the temperature of a gas increases, the gas particles have increased kinetic energy according to the third axiom. The collisions with the walls are more energetic and more frequent as the particles are moving faster. Both these factors increase the pressure.

Challenge yourself

1. What would happen to the pressure of a gas if molecular collisions were not perfectly elastic?

Exercise

Q1. The volume of an ideal gas in a container is increased at constant temperature. What happens as a result of this change?

   A The frequency of collisions of the molecules with the container wall decreases.

   B The intermolecular force between the molecules decreases.

   C The speed of the particles increases.

   D The volume of a gas particle increases.

Q2. Identify an assumption of the ideal gas model.

   A Molecules have zero mass.

   B The forces between the molecules of the gas and the container are always zero.

   C Collisions between molecules and the walls of the container are elastic.

   D The kinetic energy of a given molecule of the gas is constant.

Q3. The temperature of a closed gas container is increased. What happens to the pressure and average speed of the particles?

	Pressure	Average speed
A	increases	faster
B	decreases	faster
C	increases	no change in speed
D	decreases	no change in speed

Q4. Which of the following are assumptions of the kinetic model of ideal gases?
   I.   Particles are perfectly elastic spheres.
   II. The interactions between the particles are negligible when not colliding.
   III. The particles are in continuous random motion.
   A I and II   **B** I and III   **C** II and III   **D** I, II and III

Q5. Why does the pressure of an ideal gas change when the temperature is increased?
   A The particles collide with each other more frequently.
   B The time of contact between the molecules and the wall is reduced.
   C The force of attraction between the molecules is reduced.
   D The particles collide with the walls of the container more frequently.

Q6. Identify which of the four assumptions listed on page 106 applies to a monatomic gas.

Structure 1.5.2 – Real gases

Structure 1.5.2 – Real gases deviate from the ideal gas model, particularly at low temperature and high pressure.	
Explain the limitations of the ideal gas model.	
No mathematical coverage is required.	Structure 2.2 – Under comparable conditions, why do some gases deviate more from ideal behavior than others?

Real gases deviate from the ideal gas model

There is no such thing as an ideal gas. All gases, known as **real gases**, deviate to some extent from ideal behavior as the assumptions 1 and 2 made on page 106 do not apply to real gases under all conditions.

1. The volume of the gas particles is not negligible

If the volume of the particles is not negligible, they travel less distance between collisions with the wall (Figure 3). The collisions are more frequent than predicted by the ideal gas model and the pressure is greater.

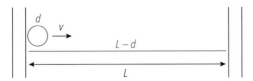

▲
S1.5 Figure 3 An ideal gas particle travels distance 2L between collisions with the same wall. A real gas particle, of diameter d, travels 2(L – d). Real gas particles travel shorter distances and collide more frequently. The difference is most significant when L is small, which occurs at high pressures and low temperatures.

This effect is most noticeable at very high pressures and low temperatures, when the particles are confined in a small volume and the distance between the particles significantly reduced.

2. There are attractive forces between the particles

When a particle is approaching the wall of a container, attractive forces from other particles pull in the opposite direction (Figure 4). This reduces the speed of the colliding particle and leads to a less energetic collision with the wall. The pressure is lower than for an ideal gas.

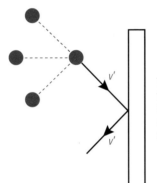

The particle colliding with the wall is slowed down by neighboring particles due to intermolecular attractions

S1.5 Figure 4 Intermolecular forces reduce the speed of particles approaching the wall. They reduce the kinetic energy of the collision and the pressure.

This reduction in speed is most significant when the average speed is relatively low at low temperatures.

Structure 2.2 – Under comparable conditions, why do some gases deviate more from ideal behavior than others?

Exercise

Q7. Gases deviate from ideal gas behavior because their particles:

    A have negligible volume   **B** have forces of attraction between them

    C are polyatomic   **D** are not attracted to one another.

Q8. **(a)** List the main features of the kinetic theory for ideal gases.

    (b) Explain the reason for the difference in behavior between real and ideal gases at low temperature.

Q9. Under what conditions of temperature and pressure are real gases likely to behave like ideal gases?

Structure 1.5.3 and 1.5.4 – The ideal gas laws

Structure 1.5.3 – The molar volume of an ideal gas is constant at a specific temperature and pressure.

Investigate the relationship between temperature, pressure and volume for a fixed mass of an ideal gas and analyze graphs relating these variables.

The names of specific gas laws will not be assessed. The value for the molar volume of an ideal gas under standard temperature and pressure (STP) is given in the data booklet.	Nature of Science, Tools 2 and 3, Reactivity 2.2 – Graphs can be presented as sketches or as accurately-plotted data points. What are the advantages and limitations of each representation?

Structure 1.5.4 – The relationship between the pressure, volume, temperature and amount of an ideal gas is shown in the ideal gas equation $PV = nRT$ and the combined gas law: $\dfrac{P_1V_1}{T_1} = \dfrac{P_2V_2}{T_2}$

Solve problems relating to the ideal gas equation.

Units of volume and pressure should be SI only. The value of the gas constant R, the ideal gas equation, and the combined gas law are given in the data booklet.	Tool 1, Inquiry 2 – How can the ideal gas law be used to calculate the molar mass of a gas from experimental data?

Investigating gases

The state of a fixed mass of gas is described by its pressure, volume and temperature. To discover relationships between these variables, a number of simple experiments can be performed.

The relationship between the pressure and volume

S1.5 Figure 5 As the pressure on a gas is increased, its volume decreases proportionately.

lower pressure

higher pressure

If the temperature of a gas is held constant, it is found that increasing the pressure of a fixed mass of gas decreases its volume (Figure 5).

More precisely, the pressure of a gas is inversely proportional to its volume, and the product of pressure and volume is a constant (Figure 6):

$$P \propto \frac{1}{V}$$

$$PV = \text{a constant}$$

SKILLS

Investigating Boyle's law. Full details of how to carry out this experiment with a worksheet are available in the eBook.

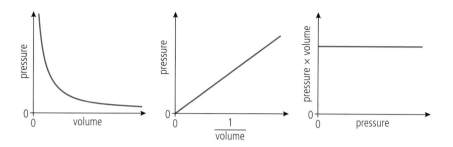

S1.5 Figure 6 Gas pressure is inversely proportional to its volume.

As we discussed at the beginning of the chapter, this relationship can be explained in terms of the decreased frequency of particle collisions with the walls of the container as the gas expands.

Nature of Science

Although many scientific discoveries, such as the gas laws, are named after the key scientists involved, the theories and laws of science stand apart from the individual discoverers.

(a)

(b)

Chip bag on airplane **(a)** at sea level and **(b)** with increased volume because of lower atmospheric pressure at high altitude. Boyle's law has applications in transport and storage.

A note about units of pressure

The SI unit of pressure is the pascal (Pa) which is equivalent to $1\,N\,m^{-2}$. Other commonly used non-SI units are the atmosphere, atm, which is equal to $1.013 \times 10^5\,Pa$ and the bar ($10^5\,Pa$) which it is conveniently close to 1 atm.

A note about units of volume

The SI unit of volume is m^3 but other units such as the dm^3 or cm^3 are commonly used. It is important to be able to interconvert these.

$1\,dm^3 = 10^{-3}\,m^3$ $1000\,dm^3 = 1\,m^3$

$1\,cm^3 = 10^{-3}\,dm^3$ $1000\,cm^3 = 1\,dm^3$

```
              divide by 1000        divide by 1000
   cm³  ───────────────────▶  dm³  ───────────────────▶  m³
        ◀───────────────────      ◀───────────────────
             multiply by 1000       multiply by 1000
```

$1\,m^3 = 1000\,dm^3 = 1\,000\,000\,cm^3$. In the laboratory, volumes are usually measured in cm^3 or dm^3 and often these measurements need to be converted to m^3 in calculations.

Relationship between volume and temperature

If the pressure is held constant, it is found that increasing the temperature of a fixed mass of gas increases its volume. If the temperature is measured in degrees Celsius, it is a **linear** relationship (Figure 7).

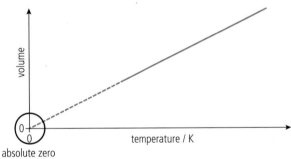

S1.5 Figure 7 The volume of a gas increases linearly with the temperature in degrees Celsius.

 SKILLS

Determining the value of absolute zero. Full details of how to carry out this experiment with a worksheet are available in the eBook.

When the straight line is extended backwards it always crosses the temperature axis, with the volume = 0, at −273 °C. This suggests that there exists a minimum possible temperature, namely −273 °C which is absolute zero or 0 K. If the temperature is measured in kelvin, the relationship can be stated more precisely: the volume of a gas is directly proportional to its absolute temperature measured in kelvin (Figure 8).

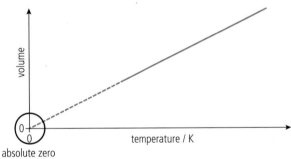

S1.5 Figure 8 Charles' law. Gas volume is proportional to the absolute temperature. Note the dotted line represents an extrapolation, as data at temperatures down to 0 K are not obtainable.

$$V \propto T(K)$$

$$\frac{V}{T(K)} = \text{a constant}$$

Balloons dipped in a mixture of dry ice (solid CO_2) and water rapidly deflate as the air inside contracts at low temperature.

This relationship, often known as Charles' law, was first established by French scientist Jacques Charles in the late 18th century. A demonstration of the relationship is found by immersing dented table tennis balls in warm water. As the air inside the ball reaches the temperature of the water, it expands, pushing the dents out on the surface. As we discussed at the beginning of the chapter, it is explained in terms of the increases in kinetic energy of the gas particles as the temperature increases. As each particle collides with more kinetic energy, the pressure will only remain constant if the frequency of the collisions decreases. This can only happen if the volume of the container increases.

A note about units of temperature

In all calculations involving gases, it is essential to use values for temperature in kelvin (K), not in Celsius (°C). Temperature in kelvin is known as the absolute temperature, and is based on a scale where absolute zero, 0 K, is the point when the gas particles have no kinetic energy. The conversion between the Kelvin and Celsius scales is discussed in Structure 1.1.

Absolute zero, the lowest conceivable temperature, is impossible to reach in practice as it would require an infinite amount of energy to reach it. Even if you could reach it, the uncertainty principle discussed in Structure 1.3 dictates that there would still be some uncertainty about the momentum of the atoms and molecules. The particles would have some energy – the so-called zero-point energy. The closer a substance approaches absolute zero, the stranger its properties become. Liquid helium, for example, turns into a superfluid, a liquid that flows without the resistance of friction. Other materials become superconductors below a critical temperature: they offer no resistance to electric currents.

Relationship between pressure and temperature

If the volume is held constant, increasing the temperature of a fixed mass of gas increases the pressure. If the temperature is measured in kelvin, it is a proportional relationship and the pressure divided by the absolute temperature is a constant (Figure 9).

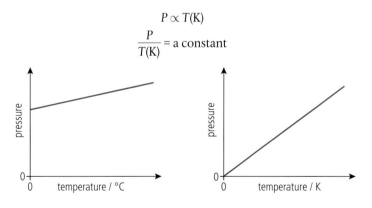

S1.5 Figure 9 Gas pressure is proportional to the absolute temperature.

Again, this relationship can be explained by the ideal gas model. An increase in temperature increases the average kinetic energy of the particles. The particles move faster and collide with the walls of the container with more energy and greater frequency, resulting in an increase in pressure.

$$V \propto T(\text{K})$$
$$\frac{V}{T(\text{K})} = \text{a constant}$$

Without Charles, Boyle and Mariotte the relationships between pressure, volume, and temperature of a gas would still exist. This is very different from the arts. Without Shakespeare, there would have been no *Hamlet*, without Picasso, no *Guernica*. In what ways have influential individuals contributed to the development of the natural sciences as an area of knowledge?

Graphical techniques have widespread applications across the syllabus, and in experimental work. You should be able to recognize the different sketch graphs that illustrate the gas laws and analyze graphs of gas behavior to determine more detailed information from gradients and intercepts.

Nature of Science, Tools 2 and 3, Reactivity 2.2 – Graphs can be presented as sketches or as accurately-plotted data points. What are the advantages and limitations of each representation?

The label on the aerosol can warns of the dangers of exposing the pressurized contents to high temperature. ▶

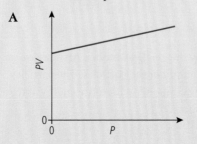

UN No. 1950

Automotive Paint

FLAMMABLE

Keep out of reach of children.
Keep away from sources of ignition –
No smoking.
Do not breathe spray.
Avoid contact with skin and eyes.
Use only in well ventilated areas.
Caution: Pressurised container. Protect from sunlight and do not expose to temperatures exceeding 50°C. Do not pierce or burn even after use. Do not spray on a naked flame or any incandescent material.

TWO STAGE
COLOURS
1. First app
metallic base
colour as n
on can and
dry.
2. Secondl
protective co
ClearCoat
(available on
paint stand
be applied t
your vehicle
original finis
Check in the
Colour Guid
your vehicle
Two Stag

$$P \propto T(K)$$

$$\frac{P}{T(K)} = \text{a constant}$$

🔒

You need to use temperature in kelvin when applying these relationships.

❗

Pressurized cans, such as soda or beer, often carry a warning to be stored in a cool place. The pressure inside the can rises at higher temperatures and could cause it to explode.

Exercise

Q10. Which two values of temperature are equivalent (to the nearest degree)?

	Kelvin scale	Celsius scale
A	25	298
B	50	323
C	283	110
D	323	50

Q11. An ideal gas in a container of fixed volume at a temperature of 27 °C and a pressure of 4.0 atm is heated. Determine the new pressure when the temperature is increased by 300 °C.

    A 0.44 atm   **B** 2.0 atm   **C** 8.0 atm   **D** 44 atm

Q12. The volume V for a fixed mass of an ideal gas was measured at constant temperature and different pressures P. Identify the graph that shows the correct relationship.

A 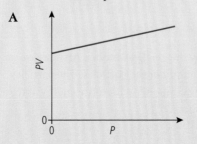 **B**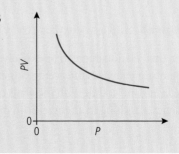

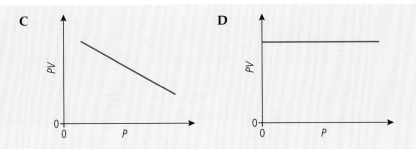

Q13. A 200 cm³ container contains a gas at 15 °C and 1.0×10^5 Pa.
The gas is then heated to 160 °C. Deduce the new pressure of the gas.

  A 1.1×10^5 Pa   **B** 1.5×10^5 Pa   **C** 1.1×10^4 Pa   **D** 1.5×10^4 Pa

Q14. The graph shows the experimental relationship between two properties of a constant amount of a gas.

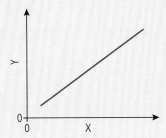

Identify X and Y and the control variable.

	X	Y	Control variable
A	pressure (atm)	volume (dm³)	temperature
B	temperature (°C)	pressure (Pa)	volume
C	temperature (°C)	volume (m³)	pressure
D	1/pressure (Pa⁻¹)	volume (cm³)	temperature

Q15. A student investigated the relationship between temperature and pressure of a gas and produced a graph of their results.

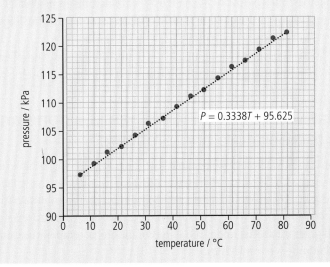

$P = 0.3338T + 95.625$

(a) Describe the relationship between temperature (T) measured in °C and pressure (P).

(b) Use the graph to predict the temperature in °C when the pressure is zero.

(c) Comment on the accuracy of the student's results.

(d) Describe the expected relationship between pressure and temperature when measured in K.

(e) Explain the relationship between pressure and temperature using the ideal gas model. Does the model agree completely with the experimental results?

Q16. A student investigates the pressure and volume of a fixed mass of gas at two temperatures.

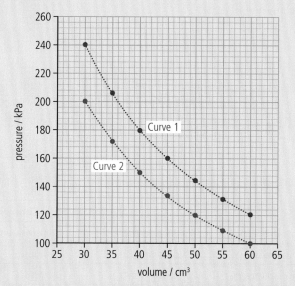

State and explain which curve corresponds to the higher temperature.

Q17. A helium cylinder that inflates a balloon has a volume of $0.0400 \, m^3$ and a pressure of $2.02 \times 10^6 \, N \, m^{-2}$.

(a) Calculate the volume of the balloon assuming there is no change of temperature as the balloon is filled and the pressure is atmospheric pressure ($1.01 \times 10^5 \, N \, m^{-2}$).

(b) The temperature of the air decreases with altitude. State the effect of this temperature change on the helium atoms as the balloon rises.

The molar volume of an ideal gas is a constant at a specific temperature and pressure

The first axiom in the ideal gas model stated that the particles in a gas have negligible volume compared with the volume the gas occupies. This suggests that changing the identity of the gas particles would have a negligible effect on the volume of the gas. This prediction is consistent with Avogadro's law, discussed in Structure 1.4. Equal volumes of all gases measured under the same conditions of temperature and pressure contain equal numbers of molecules.

One consequence of Avogadro's law is that the volume occupied by one mole of any gas must be the same for *all* gases when measured under the same conditions of temperature and pressure. This is known as the **molar volume.**

For reference purposes, the molar volume of a gas given in Section 2 of the data booklet is given under standard conditions of temperature and pressure (**STP**). This shows that when the temperature is 0 °C (273 K) and the pressure is 100 kPa, one mole of a gas has a volume of $2.27 \times 10^{-2}\,\text{m}^3\,\text{mol}^{-1}$ (= $22.7\,\text{dm}^3\,\text{mol}^{-1}$).

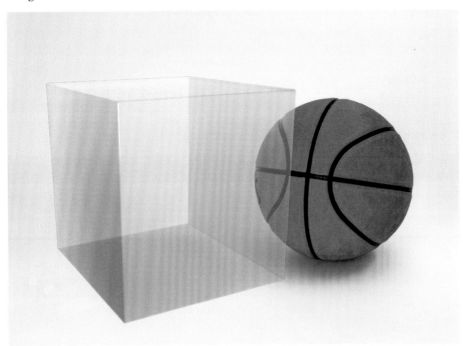

◀ The box contains 1 mole of gas. The molar volume of an ideal gas is 22.7 dm³ at STP. A basketball fits loosely into a box having this volume.

STP refers to a temperature of 273 K and a pressure of 100 kPa.

Note that the conditions of STP (standard temperature and pressure) are not the same as those used to define the 'standard state', used in thermodynamic data and explained in Reactivity 1.1.

Structure 1.4 shows how the molar volume can be used to calculate the amount of gas in a similar way to the use of molar mass. Here though, the calculations are simpler, as all gases have the same molar volume.

$$\text{number of moles of gas } (n) = \frac{\text{volume } (V)}{\text{molar volume}}$$

number of moles of gas (*n*) =

$\dfrac{\text{volume } (V)}{\text{molar volume}}$

Worked example

Calculate the volume occupied by 0.020 g of helium at standard temperature and pressure.

Solution

First convert the mass of helium to moles.

$$n = \frac{m}{M} = \frac{0.020}{4.00} = 0.0050 \text{ mol}$$

Volume = $0.0050\,\text{mol} \times 22.7\,\text{dm}^3\,\text{mol}^{-1} = 0.114\,\text{dm}^3$

The ideal gas equation and the combined gas law

The combined gas law

The three gas laws for a fixed mass of gas:

$$P \propto \frac{1}{V} \text{ at constant temperature}$$

$$V \propto T \text{ at constant pressure}$$

$$P \propto T \text{ at constant volume}$$

can be combined to give one equation:

$$\frac{PV}{T} = \text{a constant}$$

$$\frac{P_1 V_1}{T_1} = \frac{P_2 V_2}{T_2} \text{ where 1 and 2 refer to initial and final conditions respectively.}$$

Application of this **combined gas law** enables gas volume, pressure and temperature to be calculated as conditions change for a fixed mass.

Worked example

What happens to the volume of a fixed mass of gas when its pressure and its absolute temperature are both doubled?

Solution

$$\frac{P_1 V_1}{T_1} = \frac{P_2 V_2}{T_2}$$

$P_2 = 2 \times P_1$ and $T_2 = 2 \times T_1$, so these can be substituted into the equation:

$$\frac{P_1 V_1}{T_1} = \frac{2 P_1 V_2}{2 T_1}$$

We can cancel P_1 and T_1 from both sides, and 2s on the right side, leaving

$$V_1 = V_2$$

The volume does not change.

Worked example

The molar volume of a gas at STP is $22.7 \, dm^3$. Calculate the molar volume at $25.0\,°C$ at the same pressure.

Solution

As the pressure is not changing, we do not need to insert P_1 and P_2 into the combined gas equation. Temperature must be converted from °C to K.

$T_1 = 273 \, K, T_2 = 25.0 + 273 = 298 \, K$

$$\frac{V_1}{T_1} = \frac{V_2}{T_2}$$

$$\frac{22.7 \, dm^3}{273 \, K} = \frac{V_2}{298 \, K}$$

$$V_2 = \frac{298 \times 22.7 \, dm^3}{273} = 24.8 \, dm^3$$

Challenge yourself

2. Consider a sample of gas at pressure P_1, volume V_1, and temperature T_1 (K) which is expanded to a new volume V_2 at constant temperature T_1 with a new pressure P_0 (Figure 10).

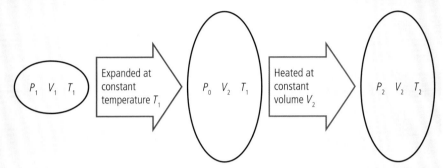

S1.5 Figure 10 A gas undergoes two changes. The volume is increased at constant temperature and then the gas is heated at constant volume.

(a) State the relationship between P_1, V_1, P_0 and V_2 for expansion at constant temperature T_1.

(b) The gas is then heated at constant volume V_2 to new conditions P_2, V_2, and T_2. State the relationship between P_0, T_1, P_2 and T_2 when the gas is heated at constant volume V_2.

(c) Deduce a relationship between the initial conditions P_1, V_1 and T_1 and the final conditions P_2, V_2 and T_2.

Note that temperature must be in kelvin in these calculations. However, the units of volume and pressure are not important, so long as they are consistent on both sides of the equation.

The ideal gas equation is derived from the combined gas equation and Avogadro's law

The combined gas equation and Avogadro's law can be combined to give:

$$\frac{PV}{T} \propto n$$

This can be made into an equation by introducing a constant, **R.** This is known as the **universal gas constant** as it does not depend on the identity of the gas.

$$\therefore \quad \frac{PV}{T} = nR \quad \text{which is usually written as} \quad PV = nRT$$

This equation, known as the **ideal gas equation,** is given in Section 1 of the data booklet. The value of R can be calculated by substituting known values into the equation, such as those for the molar volume of a gas at STP.

Worked example

Determine the gas constant from the molar volume of a gas at STP.

Solution

$P = 10^5\,\text{Pa}\,(\text{N m}^{-2})$, $V = 2.27 \times 10^{-2}\,\text{m}^3$, $T = 273\,\text{K}$, $n = 1$

$$R = \frac{PV}{nT} = \frac{10^5\,\text{N m}^{-2} \times 2.27 \times 10^{-2}\,\text{m}^3}{(1\,\text{mol} \times 273\,\text{K})}$$

$R = 8.31\,\text{N m K}^{-1}\,\text{mol}^{-1}$ or $8.31\,\text{J K}^{-1}\,\text{mol}^{-1}$

Many sources give data for ideal gas law questions in non-SI units such as atmospheres, which require a mathematical conversion. You will not be expected to be familiar with these conversions in IB examinations.

The ideal gas equation can be deduced directly from the assumptions listed on page 106 with the application of simple physics. What is the role of reason, perception, intuition, and imagination in the development of scientific models?

TOK

This value for R, the gas constant, is given in Section 2 in the data booklet and should be used for all calculations involving the ideal gas equation.

Use of the ideal gas equation enables us to calculate how systems respond to changes in pressure, volume and temperature, and to calculate molar mass. Gas density can also be determined by applying the relationship:

$$\text{density} = \frac{\text{mass}}{\text{volume}}$$

These calculations usually involve simply substituting values into the equation, but the use of units needs special attention. The guidelines below, based on the use of SI units only, should help you avoid some of the common mistakes that arise.

- Pressure, P: must be in Pa ($N\,m^{-2}$); if kPa are used, multiply by 10^3
- Volume, V: must be in m^3; if dm^3 are given, divide by 10^3, if cm^3 are given, divide by 10^6
- Number of moles, n: this is often derived by application of $n = \frac{m}{M}$
- Temperature, T: must be in kelvin; if °C is given, add 273.15

Nature of Science

Changes in gas pressure were predicted from the axioms of the kinetic theory of the ideal gas model by considering particle collisions. These collisions are not directly observable and the predictions are of limited value without the evidence provided by the experimental gas laws. The laws are descriptive statements derived from regular patterns in the evidence. The laws supported the theory, and the theory explained the laws. This is typical of how science progresses.

Worked example

A helium party balloon has a volume of $18.0\,dm^3$. At 25 °C the internal pressure is 108 kPa. Calculate the mass of helium in the balloon.

Solution

First ensure that all data are in SI units:

$P = 108\,kPa = 108 \times 10^3\,Pa$

$V = 18.0\,dm^3 = 18.0 \times 10^{-3}\,m^3$

$T = 25\,°C = 298\,K$

As this example shows, if pressure is given in kPa and volume is given in dm^3, the same answer is obtained if these values are used directly in the ideal gas equation.

$$PV = nRT$$

$$108 \times 10^3\,Pa \times 18.0 \times 10^{-3}\,m^3 = n \times 8.31\,J\,K^{-1}\,mol^{-1} \times 298\,K$$

$\therefore\ n(He) = 0.785\,mol$

$\therefore\ \text{mass (He)} = nM = 0.785\,mol \times 4.00\,g\,mol^{-1} = 3.14\,g$

Worked example

A sample of gas has a volume of $445\,cm^3$ and a mass of $1.50\,g$ at a pressure of $95.0\,kPa$ and a temperature of $28.0\,°C$. Calculate its molar mass.

Solution

Substitute $n = \frac{m}{M}$ into the ideal gas equation, and rearrange to solve for M.

$$M = \frac{mRT}{PV}$$

Ensure all data are in SI units:

$P = 95.0\,kPa = 95.0 \times 10^3\,Pa$

$V = 445\,cm^3 = 445 \times 10^{-6}\,m^3$

$T = 28.0\,°C = 301\,K$

$$\therefore M = \frac{1.500\,g \times 8.31\,J\,K^{-1}\,mol^{-1} \times 301\,K}{95 \times 10^3\,Pa \times 445 \times 10^{-6}\,m^3}$$

$$= 88.8\,g\,mol^{-1} \text{ (to 3 significant figures)}$$

Worked example

A gas has a density of $1.65\,g\,dm^{-3}$ at $27\,°C$ and $92.0\,kPa$. Determine its molar mass.

Solution

From the density data we know $1.65\,g$ occupies $1.00\,dm^3$

As before
$$M = \frac{mRT}{PV}$$

$$\therefore M = \frac{1.65\,g \times 8.31\,J\,K^{-1}\,mol^{-1} \times 300\,K}{92 \times 10^3\,Pa \times 1.00 \times 10^{-3}\,m^3} = 44.71\,g\,mol^{-1}$$

Challenge yourself

3. Blowing up a balloon increases its volume as the number of particles increases. What do you think would happen to this inflated balloon on the top of a very high mountain?

SKILLS

The molar mass of carbon dioxide can be experimentally determined by applying the ideal gas equation. Full details with a worksheet are available in the eBook.

Blowing up a balloon increases its volume as the number of particles increases.

Exercise

Q18. The ideal gas equation is $pV = nRT$.

Identify the units of R when the pressure is measured in Pa, and volume is measured in m^3.

A $J K^{-1}$ **B** $Pa\,m^3\,K^{-1}\,mol^{-1}$ **C** $J mol^{-1}$ **D** $J mol\,K^{-1}$

Q19. The volume of a gas was measured with a gas syringe. 0.0673 g of the gas was found to occupy 41.0 cm^3, measured at standard temperature and pressure (25 °C and 1.0×10^5 Pa).

What is the relative molecular mass of the gas?

A 28   **B** 32   **C** 40   **D** 44

Q20. A 2.50 dm^3 container of helium at a pressure of 85 kPa was heated from 25 °C to 75 °C. The volume of the container expanded to 2.75 dm^3. What was the final pressure of the helium?

Q21. After a sample of nitrogen with a volume of 675 cm^3 and a pressure of 1.00×10^5 Pa was compressed to a volume of 350 cm^3 and a pressure of 2×10^5 Pa, its temperature was 27.0 °C. Determine its initial temperature.

Q22. The absolute temperature of 4.0 dm^3 of hydrogen gas is increased by a factor of three and the pressure is increased by a factor of four. Deduce the final volume of the gas.

Q23. To find the volume of a flask, it was first evacuated so that it contained no gas at all. When 4.40 g of carbon dioxide was introduced, it exerted a pressure of 90.0 kPa at 27 °C. Determine the volume of the flask.

Q24. An unknown noble gas has a density of 5.84 g dm^{-3} at STP. Calculate its molar mass, and so identify the gas.

Q25. A 12.1 mg sample of a gas has a volume of 255 cm^3 at a temperature of 25.0 °C and a pressure of 1300 Pa. Determine its molar mass.

Q26. Which has the greater density at STP, hydrogen or helium?

Q27. A cyclist pumps his tyres up very hard before a trip over a mountain pass at high altitude. Near the summit one of his tyres explodes. Suggest why this may have occurred.

Nature of Science

The ideal gas model explains the observable properties of gases in terms of particles which cannot be directly observed. It provides a useful conceptual image of gas behavior, and its predictions are approximately consistent with the results. It has some limitations as we will see on the next page. This does not mean that the model should be discarded – it can still be useful but we should be aware that it is an over-simplification, which does not apply accurately under all conditions.

Real gases do not obey the ideal gas law *PV* = *nRT* under all conditions

An ideal gas is defined as one that obeys the ideal gas law *PV* = *nRT* under all conditions. This means that for one mole of gas, the relationship $\frac{PV}{RT}$ should be equal to 1, and so a graph of $\frac{PV}{RT}$ against *P* for an ideal gas is a horizontal line of intercept 1 (Figure 11).

But, as we noted earlier, there is no such thing as an ideal gas so for real gases *PV* ≠ *nRT* at all conditions, and the value of $\frac{PV}{RT}$ for one mole will vary. An example of the extent of this variation from 1 at different conditions is shown in Figure 12.

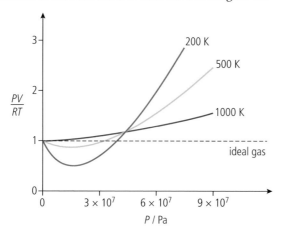

We can draw the following conclusions from the graph:

- The gas behaves most like an ideal gas at low pressure and shows the greatest deviation at high pressure.

- The gas behaves most like an ideal gas at high temperature and shows the greatest deviation at low temperature.

As discussed earlier, this is because assumptions 1 and 2 made on page 106 in defining an ideal gas do not apply under all conditions.

1. If the volume of the gas particles is not negligible, $\frac{PV}{nRT} > 1$

The collisions with the walls of the container are more frequent than predicted by the ideal gas model and the pressure is greater, $\frac{PV}{nRT} > 1$

2. There are attractive forces between the particles, $\frac{PV}{nRT} < 1$

Attractive forces from other particles reduce the speed of the colliding particle and lead to a less energetic collision with the wall. The pressure is lower than for an ideal gas and $\frac{PV}{nRT} < 1$.

Challenge yourself

4. (a) Calculate the volume of 18.0 g of steam at a temperature of 500 K and a pressure of 1.00 × 10⁵ Pa and show that the volume occupied by the particles is negligible.

(b) Show that the volume occupied by the particles is not negligible when the pressure is increased to 5.00 × 10⁷ Pa and the temperature is reduced to 400 K.

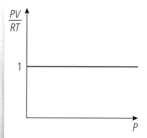

▲

S1.5 Figure 11 For one mole of an ideal gas, the relationship $\frac{PV}{RT}$ is a constant at all pressures.

◀ **S1.5 Figure 12** The deviation from ideal behavior of nitrogen at different temperatures and pressures. Curves go above 1 when the volume of the particles is not negligible and below 1 when the intermolecular forces cannot be ignored.

Overall, we can conclude that real gases deviate from ideal behavior when either or both of the assumptions on the previous page are not valid. This occurs at high pressure and low temperature. It makes sense intuitively that a gas behaves in a less perfect way under these conditions, as they are closest to it changing into a liquid.

Attempts to modify the ideal gas equation to take these factors into account and make it apply accurately to real gases led to the **van der Waals' equation**, which makes corrections for both the volume of the particles and the intermolecular attractions. Happily, for a wide range of conditions under which gases are studied, the ideal gas equation is a sufficiently accurate expression, and has the big advantage that it is a single equation for all gases.

> One form of the van der Waals' equation is:
>
> $$P + \frac{n^2 a}{V^2}(V - nb) = nRT$$
>
> where a is a measure of the attraction between the particles, and b is the volume excluded by 1 mol of particles. The constants a and b are specific to different gases. van der Waals was awarded the 1910 Nobel Physics Prize for his work.

Challenge yourself

5. Show that the van der Waals' equation correctly predicts values of $\frac{PV}{RT} > 1$ for 1 mole of gas at very high pressures.

6. Show that the van der Waals' equation correctly predicts values of $\frac{PV}{RT} < 1$ for 1 mole of gas at very low pressures.

Exercise

Q28. Ammonia, NH_3, forms a relatively strong type of intermolecular attraction known as a hydrogen bond, whereas methane, CH_4, does not. Explain the relative deviation from ideal behavior that each gas is likely to show.

Q29. The behavior of three gases was investigated over a range of pressures.

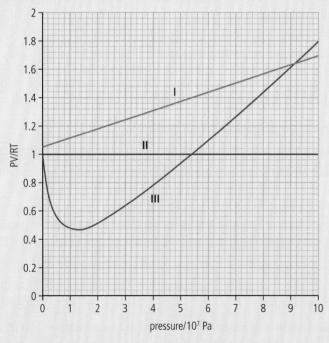

(a) Identify the gas with the strongest intermolecular forces and justify your answer.

(b) Identify the gas with the largest molecular volume and justify your answer.

How does the model of ideal gas behavior help us to predict the behavior of real gases?

We introduced the chapter with the kinetic model of an ideal gas, which can be used to explain the experimental gas laws that relate the pressure, volume and temperature of a real gas under typical conditions. The limitations of the model were also identified with real gases deviating from ideal behavior at high pressure and low temperatures.

- The kinetic theory of gases defines an ideal gas as one in which the particles have negligible volume and move in random chaotic motion with no intermolecular forces.

- The absolute temperature of the gas is proportional to the average kinetic energy of the particles.

- Gases exert pressure due to the impact of their collisions on the walls of the container.

- The gas laws describe the relationship between the pressure, volume and temperature of gases. They are a very good approximation of real gases under most conditions but strictly apply to ideal gases only.

- The gas laws can be summarized by the combined gas law $\dfrac{P_1V_1}{T_1} = \dfrac{P_2V_2}{T_2}$

- The incorporation of Avogadro's hypotheses leads to the ideal gas law: $PV = nRT$

- The ideal gas law can be applied to predict the changes that occur when the conditions of temperature, pressure or volume are changed. The molar mass of a gas can be obtained by measuring the density of a gas:

$$M = \left(\frac{m}{PV}\right)RT$$

- Real gases behave differently from ideal gases especially at high pressure and low temperature. These differences are due to the presence of intermolecular forces which reduce the pressure compared with an ideal gas under the same conditions or the non-negligible volume of the particles which increase the pressure.

Practice questions

1. The volume of a sample of gas is increased and the temperature is kept constant.

   What is the effect of this change on the particles?

	Collisions between particles	Movement of particles
A	occur less often	slower
B	occur with more energy	slower
C	occur less often	no change in speed
D	occur with more energy	no change in speed

2. 1.0 dm³ of an ideal gas at 100 kPa and 25 °C is heated to 50 °C at constant pressure. What is the new volume in dm³?

   A 0.50   **B** 0.90   **C** 1.1   **D** 2.0

3. What is the volume of gas when the pressure on 80 cm³ of gas is changed from 300 kPa to 150 kPa at constant temperature?

   A 40.0 cm³   **B** 80 cm³   **C** 120 cm³   **D** 160 cm³

4. The volume of a sample of gas measured at 27 °C is 5.0 dm³. What is the temperature when the volume is reduced, at the same pressure, to 4.0 dm³?

   A −33.0 °C   **B** −3.0 °C   **C** 21.6 °C   **D** 24.3 °C

5. Identify the graph that shows the relationship between the volume and pressure of a fixed amount of an ideal gas.

   A   **B**

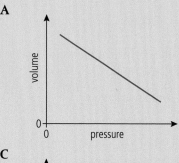

   C   **D**

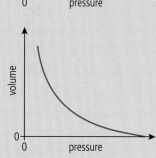

6. Why does nitrogen deviate from the ideal gas law at high pressures?

   A Intermolecular forces increase the volume compared to that for an ideal gas.

   B Increasing the pressure decreases the volume of the molecules.

   C Nitrogen molecules have finite volume.

   D Collisions between molecules occur more frequently at high pressure.

7. Identify the phenomenon that is best explained by the molecular kinetic model.

   A The difference between the Celsius and Kelvin temperature scales

   B The value of the universal gas constant

   C The ideal gases equation

   D The existence of noble gases

8. Identify the conditions when carbon dioxide is best described by the ideal gas model.

   A Low density and low pressure   **B** Low density and high pressure

   C High density and low pressure   **D** High density and high pressure

9. 4 mol of an ideal gas, X, at 25 °C is in a container of fixed volume.

   Which of the following changes would lead to the greatest increase in pressure inside the flask?

   A Increasing the temperature from 25 °C to 300 °C

   B Adding another 2 mol of gas X into the flask at fixed temperature

   C Adding 1 mol of argon gas and increasing the temperature from 25 °C to 200 °C

   D Removing 1 mol of gas X and increasing the temperature from 25 °C to 350 °C

10. Which gas closely approaches ideal behavior at room temperature and pressure?

    A carbon dioxide   **B** chlorine   **C** hydrogen fluoride   **D** neon

11. **(a)** Explain why helium behaves as an ideal gas over a wide range of pressures, volumes and temperatures. (2)

    (b) The volume of a helium gas cylinder is $3.00 \times 10^{-2}\,m^3$. Determine the number of moles of helium in the cylinder if the pressure is $2.00 \times 10^6\,Pa$ and the temperature is 27.0 °C. (2)

    (c) Calculate the number of helium atoms in the cylinder and estimate the average volume occupied by one gas atom in the cylinder. (2)

    (d) Calculate the distance between neighboring helium atoms. (1)

    (Total 7 marks)

12. Airbags are an important safety feature in vehicles. Sodium azide, potassium nitrate, and silicon dioxide have been used in one design of airbag.

    Two students looked at data in a simulated computer-based experiment to determine the volume of nitrogen generated in an airbag.

    Sodium azide, a toxic compound, undergoes the following decomposition reaction under certain conditions.

    $$2NaN_3(s) \rightarrow 2Na(s) + 3N_2(g)$$

    Using the simulation program, the students entered the following data into the computer.

Temperature (T) / °C	Mass of NaN$_3$(s) (m) / kg	Pressure (P) / atm
25.00	0.0650	1.08

    (a) State the number of significant figures for the temperature, mass, and pressure data. (1)

    (b) Calculate the amount, in mol, of sodium azide present. (1)

    (c) Determine the volume of nitrogen gas, in dm^3, produced under these conditions based on this reaction. (4)

    (Total 6 marks)

13. The graph shows the variation in pressure (P) with temperature (T) for a sample of 0.193 g of a gas in a container with a fixed volume of $1.0 \times 10^{-3}\,m^3$.

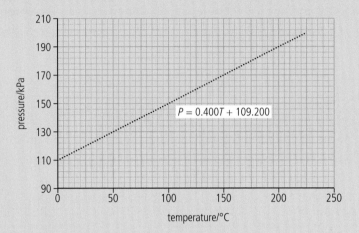

$P = 0.400T + 109.200$

(a) Does the gas behave ideally under these conditions? Justify your answer. (2)

(b) Deduce the identity of the gas in the container. (2)

(Total 4 marks)

14. An alkane gas has a density of 1.94 g dm⁻¹ at STP. Determine the molecular mass of the gas and identify the alkane. (2)

(Total 2 marks)

15. Nitrogen monoxide, NO(g), reacts with oxygen, O_2(g), to form one gaseous product X.

Syringe A contains 50 cm³ of nitrogen monoxide. Syringe B contains 50 cm³ of oxygen gas. In the experiment, 5.0 cm³ portions of oxygen were pushed from syringe B into syringe A. After each addition, the tap was closed. After the gases had returned to their original temperature, the total volume of gases remaining was measured. The results are shown graphically below.

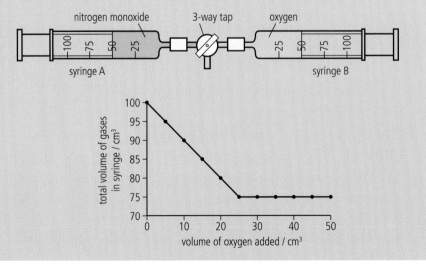

(a) State the total volume of oxygen remaining when the reaction is complete. (1)

(b) Calculate the amount of nitrogen dioxide and oxygen that react. (2)

(c) State the total amount of X produced. (2)

(d) Deduce a balanced equation for the reaction. (2)

(Total 7 marks)

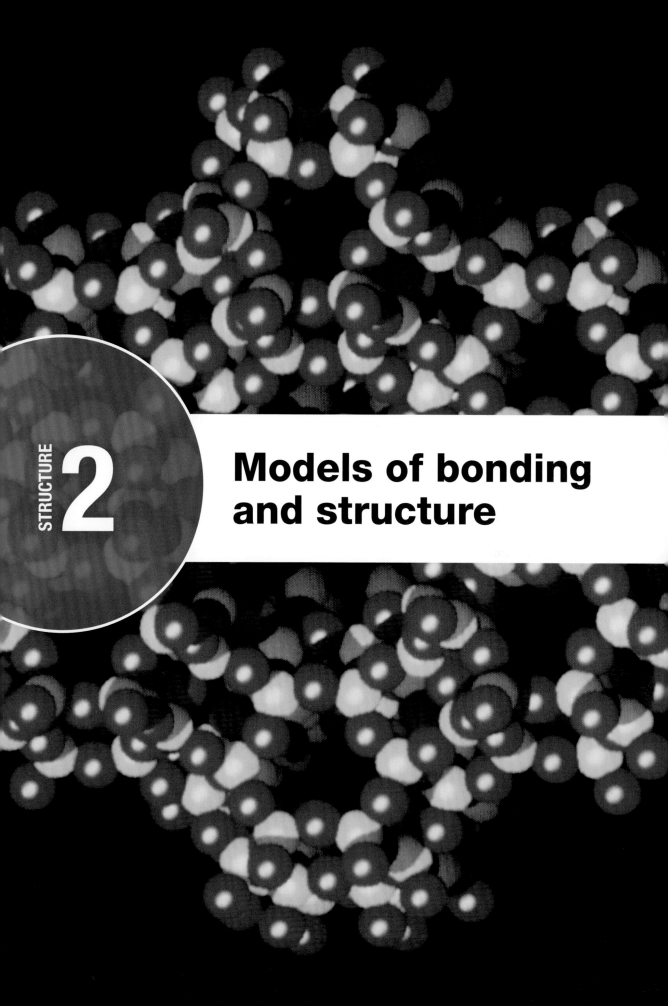

STRUCTURE

2

Models of bonding and structure

We learned in Structure 1.2 that all elements are made of atoms but that there are only just over 100 chemically different types of atom. We know that we live in a world made up of millions of different substances: which somehow must all be formed from combinations of these 100 atomic building blocks. The extraordinary variety of the material world arises from the myriad of different ways these atoms can bond together to form different structures. Atoms come together in small numbers or large, with similar atoms or very different atoms, but a stable association known as a chemical bond is always the result. Atoms linked together by bonds therefore have very different properties from their parent atoms.

All chemical bonds occur as a result of changes in the distribution of electrons, but this can happen in different ways. In this chapter, we will study the ionic bond, where electrons are transferred from one atom to another, the covalent bond where electrons are shared between bonded atoms, and the metallic bond where electrons are delocalized or shared throughout a structure of many atoms. We will also consider intermolecular forces that help to hold covalent substances together.

Our study of the covalent bond at this level will use some of the concepts from quantum theory developed in Structure 1.2 to explain the shapes and properties of molecules in more detail. As electrons are the key to the formation of all these bonds, a solid understanding of electron configurations will help you.

Chemical reactions take place when some bonds break and others re-form. The ability to predict and understand the nature of the bonds within a substance is therefore central to explaining its chemical reactivity.

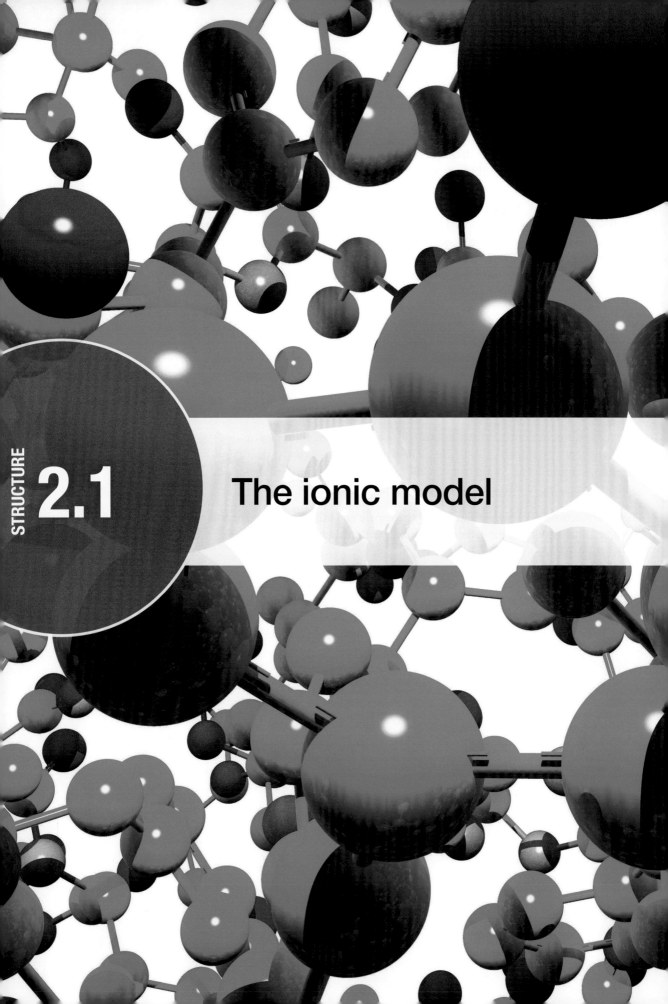

The ionic model

A molecule of insulin, the hormone essential for the regulation of glucose in the body. The ball-and-stick model shows all the atoms and bonds within the protein molecule. Insulin was the first protein to have its entire structure determined

Guiding Question

What determines the ionic nature and properties of a compound?

We saw in Structure 1.2 that all atoms are electrically neutral as the number of protons is equal to the number of electrons. The transfer of an electron from one atom M to another atom X is a possible mechanism to facilitate the increased interaction between the atoms of M and X. The resulting ions have opposite charges and so are electrostatically attracted to each other (Figure 1).

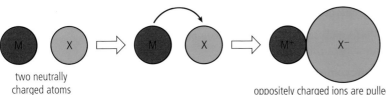

two neutrally charged atoms

oppositely charged ions are pulled together by electrostatic attraction

S2.1 Figure 1 The transfer of an electron between atoms produces ions of opposite charge which are attracted to each other. As we will discover, M is generally a metal and X is a non-metal. Note that the ions have a different size to the atoms from which they are formed. This is discussed in Structure 3.1.

The strength of the ionic bond and the reason it exists is that many ions can arrange themselves in a lattice structure of low energy with oppositely charged ions placed next to each other (Figure 2).

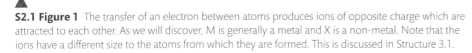
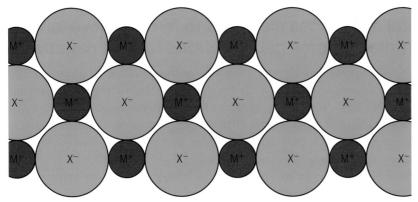

S2.1 Figure 2 Positive and negative ions form a regular structure with oppositely charged ions arranged next to each other.

The resulting regular structure accounts for the crystalline structure of ionic compounds and their physical properties.

The elements M and X play opposite roles in this liaison and it is energetically most feasible when they come from opposite regions of the periodic table. The ionic character of the bond can be related to the distance between the elements' positions in the periodic table.

Crystals of calcium fluoride (CaF_2). Ionic compounds have crystalline structures due to the regular arrangement of the ions in a lattice structure.

133

Structure 2.1.1 and 2.1.2 – The ionic bond

Structure 2.1.1 – When metal atoms lose electrons, they form positive ions called cations.

When non-metal atoms gain electrons, they form negative ions called anions.

Predict the charge of an ion from the electron configuration of the atom.

| The formation of ions with different charges from a transition element should be included. | Structure 3.1 – How does the position of an element in the periodic table relate to the charge of its ion(s)? |

Structure 2.1.2 – The ionic bond is formed by electrostatic attractions between oppositely charged ions.

Deduce the formula and name of an ionic compound from its component ions, including polyatomic ions.

Binary ionic compounds are named with the cation first, followed by the anion. The anion adopts the suffix 'ide'.

Interconvert names and formulas of binary ionic compounds.

| The following polyatomic ions should be known by name and formula: ammonium NH_4^+, hydroxide OH^-, nitrate NO_3^-, hydrogencarbonate HCO_3^-, carbonate CO_3^{2-}, sulfate SO_4^{2-}, phosphate PO_4^{3-}. | Reactivity 3.2 – Why is the formation of an ionic compound from its elements a redox reaction? |

Metal atoms lose electrons to form positive ions and non-metal atoms gain electrons to form negative ions

The atomic number of an element is defined in terms of number of protons. The number of protons and the identity of the element does not change during a chemical reaction. It is the outer electrons, known as the valence electrons, that are involved in chemical reactions. They are furthest from the electrostatic attraction of the nucleus and so more open to external influences. When an atom loses electrons it forms a positive ion or **cation**, as the number of protons is now greater than the number of electrons. Negative ions or **anions** are formed when atoms gain one or more additional negatively charged electrons.

The outer electrons of metal atoms experience a smaller effective nuclear charge than the outer electrons of non-metals

Electron transfers occur if they are energetically feasible. The concept of effective nuclear charge is helpful in explaining when this is likely to be the case. As discussed in Structure 1.2, the **nuclear charge** of the atom is given by the atomic number and so increases by one between successive elements in the periodic table, as a proton is added to the nucleus. The outer electrons which determine the chemical properties of the atom do not, however, experience the full attraction of this charge as they are **shielded** from the nucleus and repelled by the inner electrons.

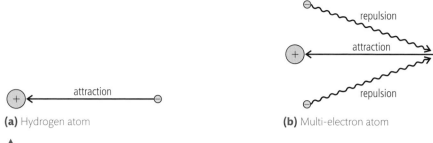

(a) Hydrogen atom   **(b)** Multi-electron atom

S2.1 Figure 3
(a) An electron in the hydrogen atom experiences the full attraction of the nuclear charge.
(b) In a multi-electron atom, the attraction for the nucleus is reduced as the outer electron is shielded from the nucleus and repelled by inner electrons.

The presence of the inner electrons reduces the attraction of the nucleus for the outer electrons (Figure 3(b)) and the **effective charge** experienced by the outer electrons is less than the full nuclear charge.

Consider, for example, a sodium atom with a nuclear charge of +11. The outer electron in the 3s orbital is shielded from this charge by the 10 electrons in the first and second energy levels ($1s^2 2s^2 2p^6$ = [Ne]). If we assume that the 10 electrons of the neon core are completely shielding, they reduce the effective nuclear charge to +1. Similar results for other elements in period 3 are shown in the table.

Element	Na	Mg	Al	Si	P	S	Cl
Nuclear charge	11	12	13	14	15	16	17
Electron configuration	[Ne] $3s^1$	[Ne] $3s^2$	[Ne] $3s^2 3p^1$	[Ne] $3s^2 3p^2$	[Ne] $3s^2 3p^3$	[Ne] $3s^2 3p^4$	[Ne] $3s^2 3p^5$
Effective nuclear charge experienced by valence electrons	$\approx 11-10$ $\approx +1$	$\approx 12-10$ $\approx +2$	$\approx 13-10$ $\approx +3$	$\approx 14-10$ $\approx +4$	$\approx 15-10$ $\approx +5$	$\approx 16-10$ $\approx +6$	$\approx 17-10$ $\approx +7$

As the period is crossed from left to right, one proton is added to the nucleus and one electron is added to the valence electron energy level. The effective charge increases with the nuclear charge as there is no change in the number of inner electrons. The atoms all have a noble gas core of 10 electrons ([Ne] = $1s^2 2s^2 2p^6$).

We can illustrate the changes in effective nuclear charge down a group by considering the elements in group 1.

Element	Nuclear charge	Electron configuration	Effective nuclear charge experienced by valence electrons
Li	3	$1s^2 2s^1$	$\approx 3 - 2 \approx +1$
Na	11	$1s^2 2s^2 p^6 3s^1$	$\approx 11 - 10 \approx +1$
K	19	$1s^2 2s^2 p^6 3s^2 3p^6 4s^1$	$\approx 19 - 10 \approx +1$

As we descend the group, the increase in the nuclear charge is largely offset by the increase in the number of inner electrons; both increase by eight between successive elements in the group. The effective nuclear charge experienced by the outer electrons remains approximately +1 down the group.

Metal atoms form positive ions as they have low ionization energies

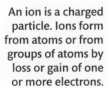

The effective nuclear charge experienced by an atom's outer electrons increases with the group number of the element. It increases across a period but remains approximately the same down a group.

First ionization energies were discussed in Structure 1.3 where we saw that they increase across a period. This can be accounted for by the increase in effective nuclear charge. The increased attraction between the outer electrons and the nucleus makes the valence electrons more difficult to remove. An element M is more likely to lose an electron and form a positive ion if it is a metal on the left of the periodic table where elements have the lowest ionization energies. As ionization energies decrease down a group, metals on the bottom left of the periodic table will have the greatest tendency to lose electrons and form ionic compounds.

Challenge yourself

1. As shown in Structure 1.3, although there is a general increase in ionization energies across a period, there are some regular discontinuities in this trend. The first ionization energy of aluminium, for example, is lower than the first ionization of magnesium. How can the calculation of effective nuclear charge be refined to explain these discontinuities?

Non-metal atoms form negative ions as they have high effective nuclear charges

An ion is a charged particle. Ions form from atoms or from groups of atoms by loss or gain of one or more electrons.

Electron transfer will also be more likely to occur if the non-metal element X attracts the transferred electrons more strongly. This is the case if X is a non-metal at the top right of the periodic table. Such elements have the highest effective nuclear charges and the smallest atomic radii.

Sodium chloride is formed when one electron is transferred from sodium to chlorine

Reactivity 3.2 – Why is the formation of an ionic compound from its elements a redox reaction?

Sodium reacts with chlorine to form the ionic compound sodium chloride. The outer 3s electron from a sodium atom moves to the chlorine atom and acts like a harpoon. Once the transfer is complete, the resulting electrostatic attraction between the oppositely charged Na^+ and Cl^- ions pulls the ions together (Figure 4).

$$Na\cdot \quad + \quad {}^{\times\times}_{\times}\overset{\times\times}{Cl}{}^{\times}_{\times} \longrightarrow [Na]^+ \quad [{}^{\times\times}_{\times}\overset{\times\times}{Cl}{}^{\times}_{\times}]^-$$

| $1s^22s^22p^63s^1$ | $1s^22s^22p^63s^23p^5$ | $1s^22s^22p^6$ | $1s^22s^22p^63s^23p^6$ |
| sodium atom | chlorine atom | sodium ion | chloride ion |

▲

S2.1 Figure 4 An electron is transferred from sodium to chlorine. This is an example of a redox reaction (discussed more fully in Reactivity 3.2). Note that both ion products, Na^+ and Cl^-, have the electron configuration of a noble gas.

Similar reactions occur between other group 1 metals and other group 17 elements. The most vigorous reaction occurs between the elements that are furthest apart in the periodic table. The most reactive alkali metal, caesium, near the bottom of group 1, has the lowest ionization energy and loses electrons most readily. Fluorine is the most reactive halogen, at the top of group 17, as it has the smallest atomic radius and attracts the transferred electron the most strongly.

Coloured scanning electron micrograph of crystals of table salt, sodium chloride, NaCl. The very reactive elements sodium and chlorine have combined to form this stable compound containing Na^+ and Cl^- ions.

The attraction between ions increases with ionic charge but removing multiple electrons has an energy cost

Singly charged ions are formed with the transfer of one electron. More generally, ions of greater charge can be formed if more than one electron is transferred. For example, magnesium oxide is formed when the two valence electrons from a magnesium atom are transferred to an oxygen atom. The resulting ions have a greater charge and there is an increased force of attraction between them.

TOK The electron involved in the bonding of sodium is described as a harpoon. How useful are similes and metaphors in the sciences? Does the language we use have a descriptive or interpretive function?

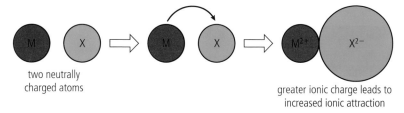

two neutrally charged atoms

greater ionic charge leads to increased ionic attraction

▲ The transfer of two electrons produces doubly charged ions which are attracted to each other more strongly.

The formation of a multiply charged ion does not however come without an energy cost.

Negative ions are more attractive with increased charge

The group 17 elements have one vacancy in their outer energy level and a high effective nuclear charge so they can accept one electron readily into their outer energy level and form the halide X^- ion. Group 16 elements, such as oxygen, have two vacancies but the outer electrons only experience an effective nuclear charge of +6. They can form X^{2-} ions but do not attract electrons as strongly as the halogens. Phosphorus has three vacancies and so can accommodate three additional electrons and form the P^{3-} ion. The outer electrons in phosphorus only experience an effective nuclear charge of +5 so it forms ionic compounds less readily than chlorine or sulfur. The addition of electrons becomes more difficult with increasing negative charge of the ion due to increased electron-electron repulsion. The formation of the silicon Si^{4-} ion is not feasible for this reason.

▲ Sparks fly as an argon welder joins two metals together. The unreactive argon gas provides an inert environment which prevents the hot metals from reacting with the air.

The high ionization energies of the noble gases and their complete energy levels make them unreactive

We saw in Structure 1.3 that the noble gases have very high first ionization energies. The nucleus of a noble gas holds on tightly to its outer electrons making them unavailable for any chemical activity. The noble gases also have complete outer shells with no vacancies to accommodate extra electrons so any added electron would have to occupy an empty outer energy level where they would experience an effective nuclear charge of zero. These factors explain the non-reactivity of the group. The full outer energy level of the noble gases can be thought of as the 'ultimate goal' for all atoms. The noble gases do not react as they have already achieved this goal. Metal atoms lose electrons to achieve the stable electron configuration of the preceding noble gas and non-metals gain electrons to achieve the stable electron configuration of the succeeding noble gas. As we saw earlier, when sodium reacts it loses an electron to achieve the electron configuration of neon and the chloride ion has the electron configuration of argon.

Metals form cations by losing valence electrons. Non-metals form anions by gaining electrons.

It may help you to remember: CATion is PAWsitive and aNion is Negative.

The charge on the ion can be predicted from an element's position in the periodic table

Metals that have a small number of electrons in their outer shells (groups 1, 2, and 13) will lose these electrons and form **cations**. Non-metals that have higher numbers of electrons in their outer shells (groups 15, 16, and 17) will gain electrons and form anions.

We are now able to summarize how the position of an element in the periodic table enables us to predict the ion it forms. The table shows the ions formed by the elements in period 3.

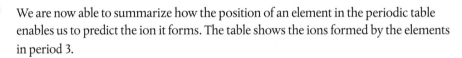

Group number	Element	Electron configuration of atom	Number of valence electrons	Number of electrons transferred	Charge on ion formed	Electron configuration of ion	Type of element
1	Na	[Ne]$3s^1$	1	1 lost	1+	[Ne]	metal
2	Mg	[Ne]$3s^2$	2	2 lost	2+	[Ne]	metal
13	Al	[Ne]$3s^23p^1$	3	3 lost	3+	[Ne]	metal
14	Si	[Ne]$3s^23p^2$	4	–	–	–	metalloid
15	P	[Ne]$3s^23p^3$	5	3 gained	3–	[Ne]$3s^23p^6$ = [Ar]	non-metal
16	S	[Ne]$3s^23p^4$	6	2 gained	2–	[Ne]$3s^23p^6$ = [Ar]	non-metal
17	Cl	[Ne]$3s^23p^5$	7	1 gained	1–	[Ne]$3s^23p^6$ = [Ar]	non-metal

When writing the symbols for ions, note the charge is written as a superscript with the number first and the charge next, e.g., N^{3-}. When an ion X carries a charge of 1^+ or 1^- it is written just as X^+ or X^-.

Silicon and other elements in group 14, with four electrons in their outer shell, do not form ions readily. These elements react to form different, covalent, bonds as discussed in Structure 2.2.

Hydrogen is difficult to place in the periodic table and this is reflected in the ions it forms. It generally loses its 1s electron like the group 1 metals and forms the H^+ ion but can also form the less common H^- (hydride) ion by completing its outer energy level like the halogens.

Structure 3.1 – How does the position of an element in the periodic table relate to the charge of its ion(s)?

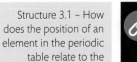

Challenge yourself

2. What evidence based on simple observations supports the existence of ions?

Worked example

Refer to the periodic table to deduce the charge on the ion formed when the following elements react:

(a) lithium   **(b)** oxygen   **(c)** carbon

Solution

(a) lithium is in group 1 so forms Li^+

(b) oxygen is in group 16 so forms O^{2-}

(c) carbon is in group 14 so does not form ions

TOK

Do you *know* that sodium chloride is made up from Na^+ and Cl^- ions or do you simply *believe* this to be the case? What is needed to change a belief into knowledge?

Transition metals form ions of different charge and show a range of oxidation states

One characteristic property of the transition metals is that they from ions with different charges. Notable examples include iron which can form Fe^{2+} and Fe^{3+} ions, and copper which forms $Cu2+$ and $Cu+$ ions. The iron and copper ions of different charge form compounds with different properties including color.

▲ The beaker on the left contains $Fe^{2+}(aq)$ and the beaker on the right contains $Fe^{3+}(aq)$. When iron rusts, it reacts with oxygen to form these different ions.

▲ The tube on the left contains the blue Cu^{2+} ion, the tube on the right contains the red Cu^+ ion. These two ions are observed when Fehling's reagent is warmed with glucose or other 'reducing sugars'. The blue Cu^{2+} is reduced to the red Cu^+.

There are other notable examples of elements forming ions that are not obvious from their position in the periodic table:

- Lead (Pb) and tin (Sn) form stable M^{4+} and M^{2+} ions despite being in group 14.
- Silver, Ag, forms the ion Ag^+.

Oxidation states are discussed more fully in Reactivity 3.2.

Systematic names of compounds use oxidation numbers

The Fe^{2+} ion forms an oxide with the formula FeO and Fe^{3+} ions form Fe_2O_3. These compounds were traditionally distinguished as ferrous oxide and ferric oxide, but IUPAC introduced a nomenclature using **oxidation numbers** to make the names more systematic. A Roman numeral corresponding to the oxidation state is inserted in brackets after the name of the element. So FeO is iron(II) oxide and Fe_2O_3 is iron(III) oxide.

The table shows some common examples of the names of compounds with different oxidation states.

Oxidation state is shown with a '+' or '−' sign and an Arabic numeral, e.g. +2. Oxidation number is shown by inserting a Roman numeral in brackets after the name or symbol of the element.

The terms 'oxidation number' and 'oxidation state' are often used interchangeably, and either term is acceptable.

Formula of compound	Oxidation state	Name using oxidation number
FeO	Fe +2	iron(II) oxide
Fe_2O_3	Fe +3	iron(III) oxide
Cu_2O	Cu +1	copper(I) oxide
CuO	Cu +2	copper(II) oxide
MnO_2	Mn +4	manganese(IV) oxide

Although this nomenclature can be used in the naming of all compounds, it is only necessary when an element has more than one common oxidation state. For example, Na_2O could be called sodium(I) oxide, but as Na only shows this oxidation state $+1$ in compounds, it is simply sodium oxide.

The formula of an ionic compound can be deduced from its component ions

When an ionic compound is formed, there is no net loss or gain of electrons. The ionic compound, like the atoms that formed it, is electrically neutral. Writing the formula for the ionic compound involves *balancing the total number of positive and negative charges*, by taking into account the different charges on each ion.

For example, magnesium oxide is made up of magnesium ions Mg^{2+} and oxide ions O^{2-}. Each magnesium atom has transferred *two* electrons to each oxygen atom and so the compound contains equal numbers of each ion. Its formula, $Mg^{2+}O^{2-}$, is usually written as MgO. When magnesium reacts with fluorine, however, Mg again loses *two* electrons, but *two* F atoms are needed as each F atom can only gain *one* electron. The ionic compound produced has the ratio Mg:F = 1:2 and is written as $Mg^{2+}F^{-}_2$ or MgF_2.

> In many non-metal elements, the ending of the name changes to '-ide' when ionization occurs. For example, *chlorine* (the element) becomes *chloride* (the ion), *oxygen* becomes *oxide*, *nitrogen* becomes *nitride*, etc.

> Note the convention in naming ionic compounds is that the positive ion is written first and the negative ion second.

> Note that the formula of the compound shows the *simplest ratio* of the ions it contains. So, for example, magnesium oxide is not Mg_2O_2 but MgO.

Worked example

Deduce the formula for the compound that forms between aluminium and oxygen.

Solution

1. Check the periodic table for the ions that each element will form.

   aluminium in group 13 forms Al^{3+}

   oxygen in group 16 forms O^{2-}

2. Balance the charges by finding a common multiple. In this case you need six of each charge:

$$2 \times Al^{3+} = 6+ \text{ and } 3 \times O^{2-} = 6-$$

3. State the final formula using subscripts to show the number of ions: Al_2O_3

It is common practice to leave the charges out when showing the final formula.

Important polyatomic ions that should be known

Polyatomic ions are made up of more than one atom which together have lost or gained an electron. Many of these polyatomic ions are found in salts formed from common acids. As we will discuss in Reactivity 3.1, the ions are said to be conjugate bases of these acids as they form when a H^+ ion is removed from the acid. The stronger the acid the more stable the ion.

It will help you to become familiar with the examples in the table below, as you will often use them when writing formulas and equations. (Note that this information is not supplied in the IB data booklet.)

Polyatomic ion name	Formula	Corresponding acid	Example of compound containing this ion	Example of chemical formula
nitrate	NO_3^-	$HNO_3(aq)$	lead(II) nitrate	$Pb(NO_3)_2(s)$
sulfate	SO_4^{2-}	$H_2SO_4(aq)$	copper sulfate	$CuSO_4(s)$
phosphate	PO_4^{3-}	$H_3PO_4(aq)$	calcium phosphate	$Ca_3(PO_4)_2(s)$
hydroxide	OH^-	$H_2O(l)^\Delta$	barium hydroxide	$Ba(OH)_2(s)$
oxide□	O^{2-}	$H_2O(l)^\Delta$	magnesium oxide	$MgO(s)$
hydrogencarbonate	HCO_3^-	$H_2CO_3(aq)*$	potassium hydrogencarbonate	$KHCO_3$
carbonate	CO_3^{2-}	$H_2CO_3(aq)*$	magnesium carbonate	$MgCO_3(s)$
ammonium	NH_4^+	$NH_4Cl(aq)*$	ammonium nitrate	$NH_4NO_3(ag)$

* These are weak acids, as discussed in Reactivity 3.1.

$^\Delta$ Water can act as an acid or a base, as discussed in Reactivity 3.1.

□ The oxide is not polyatomic but is included for reference.

NH_4^+ is unusual in that it is a positive ion formed from non-metal elements.

It should be noted that the bonding within the polyatomic ion is covalent. So in potassium sulfate, for example, the K^+ and SO_4^{2-} ions are attracted together by ionic bonds, but the sulfur and oxygen atoms are held together within the sulfate ion by covalent bonds.

When writing the formula of a compound with more than one polyatomic ion, parentheses are used around the ion before the subscript.

The charge of a polyatomic ion can be deduced from the number of Hs in the formula of the corresponding acid.

The names of compounds give a clue to their composition. The ending '-ate' refers to ions that contain oxygen bonded to another element.

SKILLS

Naming ionic bonds. Full details of how to carry out this activity with a worksheet are available in the eBook.

Worked example

Deduce the formula for ammonium phosphate.

Solution

1. The compound contains two polyatomic ions in the table above and these need to be known: NH_4^+ and PO_4^{3-}

2. Balance the charges by using a common multiple: in this case 3 for each charge:

$$3 \times NH_4^+ = 3+ \text{ and } 1 \times PO_4^{3-} = 3-$$

3. State the formula using subscripts and brackets to show the number of ions:

$$(NH_4)_3PO_4$$

Exercise

Q1. Which is the best description of ionic bonding?

   A The electrostatic attraction between positive ions and non-bonding electrons

   B The electrostatic attraction between positive ions and bonding electrons

   C The electrostatic attraction between positively charged nuclei and an electron pair

   D The electrostatic attraction between oppositely charged ions

Q2. What is the formula for the compound formed by aluminium and sulfur?

    A AlS   **B** AlS_2 **C** AlS_3 **D** Al_2S_3

Q3. What happens when lithium reacts with bromine?

    A lithium atoms gain one electron   bromine atoms gain one electron

    B lithium atoms gain one electron   bromine atoms lose one electron

    C lithium atoms lose one electron   bromine atoms gain one electron

    D lithium atoms lose one electron   bromine atoms lose one electron

Q4. Which of the following are ionic compounds?

    I. NH_4Cl II. NH_4NO_3 III. HCl

    A I and II only   **B** I and III only   **C** II and III only   **D** I, II and III

Q5. Write the formula for each of the following compounds:

    (a) potassium bromide   **(b)** zinc oxide   **(c)** sodium sulfate

    (d) copper(II) bromide   **(e)** chromium(III) sulfate   **(f)** aluminium hydride

Q6. Name the following compounds:

    (a) $Sn_3(PO_4)_2$ **(b)** $Ti(SO_4)_2$ **(c)** $Mn(HCO_3)_2$ **(d)** $BaSO_4$

    (e) Hg_2S **(f)** V_2O_3 **(g)** Cr_2O_3 **(h)** PbO_2

Q7. What is the formula of the compound that forms from element X in group 2 and element Y in group 15?

Q8. Explain what happens to the electron configurations of the elements Mg and Br when they react to form the compound magnesium bromide.

Structure 2.1.3 – Ionic structures and properties

Structure 2.1.3 – Ionic compounds exist as three-dimensional lattice structures, represented by empirical formulas.

Explain the physical properties of ionic compounds to include volatility, electrical conductivity and solubility.

Include lattice enthalpy as a measure of the strength of the ionic bond in different compounds, influenced by ion radius and charge.	Tool 1, Inquiry 2 – What experimental data demonstrates the physical properties of ionic compounds?
	Structure 3.1 – How can lattice enthalpies and the bonding continuum explain the trend in melting points of metal chlorides across period 3?

Ionic compounds have a lattice structure

Once the ions are formed by electron transfer, they are pulled together by the electrostatic attraction that is the **ionic bond**. An ion pair is rarely formed in isolation, however, and many cations and anions arrange themselves in a three-dimensional **lattice** structure held together by ionic bonds between oppositely charged ions. The details of the lattice's geometry vary in different compounds, depending mainly on the relative sizes of the ions, but it always involves a fixed arrangement of ions based on a repeating unit or **unit cell**.

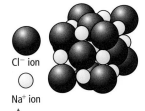

Cl^- ion

Na^+ ion

▲

S2.1 Figure 5 A unit cell of the NaCl lattice is built up from oppositely charged sodium and chloride ions.

A lattice consists of a very large number of ions and can grow indefinitely. Ionic compounds do not have fixed number of ions, so their formulas are *ratios* of ions present. They are an empirical formula and are known as the **formula unit**, as discussed in Structure 1.4.

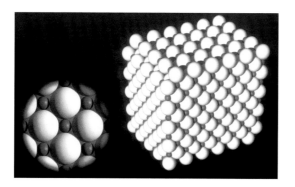

 Make sure that you avoid the term 'molecule' when describing ionic compounds, but instead use the term 'formula unit'.

Computer graphic of crystallized common salt, NaCl. Small spheres represent Na⁺ ions and larger spheres represent Cl⁻. The lattice is arranged so each Na⁺ ion has six oppositely charged nearest neighbours and vice versa.

The physical properties of ionic compounds reflect their lattice structures

Physical properties are those that can be examined without chemically altering the substance. Our knowledge of ionic bonds and lattice structure helps us explain some of these properties of ionic compounds.

 SKILLS Preparing crystals and observing the lattice strcuture. Full details of how to carry out this experiment with a worksheet are available in the eBook.

Lattice enthalpy as a measure of the strength of the ionic bond

Energy is needed to separate particles that are attracted to each other, and energy is released when the particles come together. For example, when one mole of sodium ions forms a lattice with one mole of chloride ions, 790 kJ of energy is given out. As the energy content of the chemical species decreases, this is expressed as a negative enthalpy change (ΔH) as discussed in Reactivity 1.2.

$$Na^+(g) + Cl^-(g) \rightarrow NaCl(s) \qquad (\Delta H = -790 \text{ kJ mol}^{-1})$$

This energy output offsets the energy needed to form the ions and makes the formation of the compound energetically feasible.

The **lattice enthalpy** is defined as the enthalpy change for the reverse process. Energy is needed to separate the ions and the enthalpy change is positive.

$$NaCl(s) \rightarrow Na^+(g) + Cl^-(g) \qquad (\Delta H^\ominus_{lat} = +790 \text{ kJ mol})$$

This enthalpy change can be calculated from the **ionic model**, which assumes the crystal is made up of spherical ions which only interact by electrostatic forces.

The strength of an ionic bond depends on the charge and size of the ions

- An increase in the ionic charge increases the attraction between the ions and increases the lattice enthalpy.
- An increase in the ionic radius of one of the ions decreases the attraction between the ions and decreases the lattice enthalpy.

Ionic compounds have high melting and boiling points and low volatility

The lattice structure results in ionic compounds being crystalline solids at room temperature. They have high melting and boiling points as a large amount of energy is needed to separate the ions in the lattice. This is illustrated in the table below which shows the melting points of four ionic compounds.

Ionic compound	Charge on metal ion	Charge on non-metal ion	Melting point / K
NaF	1+	1−	1266
Na_2O	1+	2−	1548*
MgF_2	2+	1−	1534
MgO	2+	2−	3125

* Na_2O sublimes and turns from a solid directly to a gas.

The melting points are generally higher when the ionic charge is greater, due to the increased strength of the ionic bond. The melting points of magnesium fluoride and sodium oxide are both greater than sodium fluoride because doubling one of the ionic charges increases the attraction between the ions. Magnesium oxide has an even higher melting point because the ionic charge on both ions is double that of the ions in sodium fluoride.

The decrease in melting points of the sodium halides below is explained by the increase in ionic radius of the halide ion down the group. The attraction between the ions decreases with increased ionic radii.

Ionic compound	Melting point / K
NaF	1266
NaCl	1074
NaBr	1020
NaI	934

Challenge yourself

3. We explained that the higher melting point of magnesium fluoride compared to sodium fluoride is due to the increased ionic charge of the Mg^{2+} ion. Explain why this explanation is incomplete.

4. The melting points of aluminium fluoride and oxide are shown in the table. Compare the melting points with the corresponding magnesium compounds and explain any differences.

Ionic compound	Charge on metal ion	Charge on non-metal ion	Melting point / K
AlF_3	3+	1−	1564
Al_2O_3	3+	2−	2345

Ionic compounds have low volatility.

Volatility is a term used to describe the tendency of a substance to vaporize. Ionic compounds also generally have high boiling points and are non-volatile with low volatility.

Ionic compounds are generally soluble in water but not in non-polar liquids

Solubility refers to the maximum amount of solute that can dissolve in a given volume of a solvent to form a solution. Sodium chloride, for example, has a high solubility as a relatively large amount readily dissolves in a given volume of water. As discussed in Structure 2.2, water is a polar molecule. There are small partial positive charges on the hydrogen atoms which attract chloride ions and small negative charges on the oxygen atoms which attract the sodium ions (Figure 6).

$$Cl^- \qquad \begin{matrix} H^{\delta+} \searrow \\ \qquad \quad O \quad \delta- \quad Na^+ \\ H^{\delta+} \nearrow \end{matrix}$$

▲

S2.1 Figure 6 The polar water molecule can attract cations and anions.

At the contact surface of a crystal, the attraction of the ions to the partial charges in the water molecules pulls the ions away from their lattice positions. As these ions separate from the lattice, they become surrounded by water molecules and are said to be **hydrated** and the solid dissolves (Figure 7). This change is represented with the use of state symbols as follows:

$$NaCl(s) \rightarrow NaCl(aq) \text{ or } NaCl(s) \rightarrow Na^+(aq) + Cl^-(aq)$$

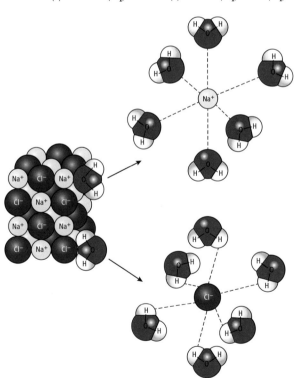

▲

S2.1 Figure 7 Dissolving of NaCl in water involves the attraction of the polar water molecules to the ions in the NaCl lattice. This results in the hydration of the separated ions.

More generally, for solvents other than water, the solute ions are said to be **solvated** and an appropriate state symbol is used to identify the solvent used.

If the liquid is non-polar like oil or hexane, C_6H_{14}, there is no attraction between the liquid molecules and the ions. The ions remain in the lattice and the solid is insoluble.

Ionic liquids are efficient solvents with low volatility and are used increasingly as solvents in Green Chemistry for energy applications and industrial processes. They are usually made of organic cations and anions.

Ionic compounds are generally soluble in ionic or polar solvents but not soluble in non-polar solvents.

'Water, water, every where, Nor any drop to drink.' (From 'The Rime of the Ancient Mariner' by S.T. Coleridge). Nearly 98% of water on Earth is sea water, which contains dissolved ions and so is unfit for drinking and many industrial processes. The solubility of ionic compounds in water has global consequences.

Ionic compounds do not conduct electricity in the solid state but do conduct when molten or in aqueous solution.

The high melting points of ionic compounds is an economic consideration in many industrial processes, such as the electrolysis of molten aluminium oxide used to extract aluminium as discussed in Reactivity 3.2. It is very expensive to maintain the high temperatures needed.

Conductivity of an ionic compound and its aqueous solution. Full details of how to carry out this experiment with a worksheet are available in the eBook.

This leads to the general result that ionic compounds are soluble in polar liquids but insoluble in non-polar liquids. The pattern can be summarized as '*like dissolves like*'. As with many generalizations, you need to be aware of many important exceptions. Calcium carbonate, for example, is an ionic compound that is not soluble in water.

The White Cliffs of Dover, UK, are made of calcium carbonate: an ionic compound which is insoluble in water. The surrounding sea water contains many dissolved ionic compounds such as sodium chloride.

Ionic compounds have high electrical conductivity in the liquid and aqueous states

The ability of an ionic compound to conduct electricity depends on whether the ions are able to move. Ionic compounds are not able to conduct electricity in the solid state because the ions are fixed within the lattice but they can conduct electricity in the liquid or aqueous states where the ions are mobile.

Ionic compounds are generally brittle

Ionic compounds are usually brittle, as the crystal tends to shatter when a shear force is applied. The force displaces some ions within the lattice, which results in ions of the same charge been positioned alongside each other. The repulsive forces between the ions then cause the lattice to split.

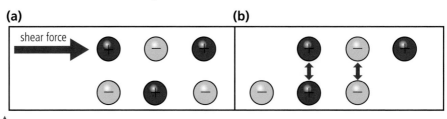

Ionic compounds are brittle. (a) A strong shear force is applied to a layer of ions. (b) Ions of the same charge are forced to be together and the top layer repels the bottom layer and the structure falls apart.

Nature of Science

The ionic model leads to some predictions about the physical properties of ionic compounds, which can be tested by observation and experiment. This is central to the scientific process. The general patterns discussed here on volatility, solubility, and conductivity support the lattice model. There are however some exceptions, which need to be explained. The melting point of aluminium oxide, for example, is not as high as predicted from the ionic model and calcium carbonate is insoluble in water. The model is limited and incomplete. Other types of bonding need to be included to account for these unexpected properties.

Some ionic compounds deviate significantly from the ionic model

When two elements react to form an ionic compound, they show *very different tendencies to lose or gain electrons*. As discussed, this is related to the position of the elements in the periodic table. Metals on the bottom left of the table lose electrons most easily, while non-metals on the top right gain electrons most easily. Caesium and fluorine react together to form the compound with the most ionic character and group 1 halides generally show typical ionic properties. The ionic character of the other halides, however, decreases across a period.

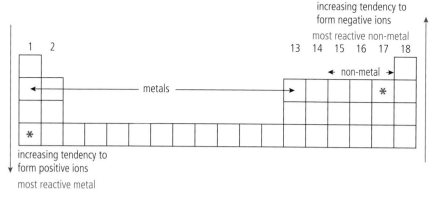

The pairs of elements that react most readily to form ionic compounds are metals on the bottom left of the periodic table and non-metals on the top right, indicated here by asterisks.

Caesium fluoride is used instead of francium fluoride as it has been estimated that at any one time there are only 17 francium atoms on the Earth.

TOK

The periodic table is a diagrammatical representation of general patterns in chemistry. Does it provide an explanation or a description of the ionic character of different compounds? Is the relationship between description and explanations the same in all areas of knowledge?

Period 3 chlorides are less ionic across the period

The melting points and electrical conductivity of the period 3 chlorides are summarized in the table below.

Formula of chloride	NaCl(s)	MgCl$_2$(s)	AlCl$_3$(s)/ Al$_2$Cl$_6$(g)	SiCl$_4$(l)	PCl$_5$(s)	SCl$_2$(l)	Cl$_2$(g)
Melting point / K	1074	987	463	203	435	195	170
Electrical conductivity in molten state	high	high	poor	none	none	none	none

The transition from ionic to covalent character can be seen in the data. Sodium chloride is a typical ionic compound whereas aluminium chloride is generally considered to be covalent, as illustrated by its low conductivity in the liquid state. The move from ionic to covalent is also seen in the decrease in melting points. Aluminium chloride has a layer structure in the solid state but exists as a molecular dimer, Al$_2$Cl$_6$, in the liquid and gaseous states. The chlorides of the later elements are molecular covalent and have very low melting points, as discussed more fully in Structure 3.1.

The difference between theoretical lattice enthalpies based on the ionic model and experimental values can also be used to measure the covalent character of chlorides.

1 H 2.2																	
3 Li 1.0	4 Be 1.6											5 B 2.0	6 C 2.6	7 N 3.0	8 O 3.4	9 F 4.0	
11 Na 0.9	12 Mg 1.3											13 Al 1.6	14 Si 1.9	15 P 2.2	16 S 2.6	17 Cl 3.2	
19 K 0.8	20 Ca 1.0	21 Sc 1.4	22 Ti 1.5	23 V 1.6	24 Cr 1.7	25 Mn 1.6	26 Fe 1.8	27 Co 1.9	28 Ni 1.9	29 Cu 1.9	30 Zn 1.7	31 Ga 1.8	32 Ge 2.0	33 As 2.2	34 Se 2.6	35 Br 3.0	
37 Rb 0.8	38 Sr 1.0	39 Y 1.2	40 Zr 1.3	41 Nb 1.6	42 Mo 2.2	43 Tc 2.1	44 Ru 2.2	45 Rh 2.3	46 Pd 2.2	47 Ag 1.9	48 Cd 1.7	49 In 1.8	50 Sn 2.0	51 Sb 2.1	52 Te 2.1	53 I 2.7	
55 Cs 0.8	56 Ba 0.9	57 La* 1.1	72 Hf 1.3	73 Ta 1.5	74 W 1.7	75 Re 1.9	76 Os 2.2	77 Ir 2.2	78 Pt 2.2	79 Au 2.4	80 Hg 1.9	81 Tl 1.8	82 Pb 1.8	83 Bi 1.9	84 Po 2.0	85 At 2.2	
87 Fr 0.7	88 Ra 0.9	89 Ac* 1.1															

Pauling electronegativity values

- <1.0
- 1.0–1.4
- 1.5–1.9
- 2.0–2.4
- 2.5–2.9
- 3.0–4.0

Periodic trends in electronegativity values show an increase along a period and up a group.

Ionic character and electronegativity

Structure 3.1 – How can lattice enthalpies and the bonding continuum explain the trend in melting points of metal chlorides across period 3?

The tendency of two elements to form an ionic bond can also be approached from the opposite perspective. How does the bonding deviate from the covalent model? How are the bonding pair of electrons shared between the two bonding elements? If they are shared equally, it is a covalent bond, if not, it is a polar covalent bond. An ionic bond between M and X can be considered as an extreme polar bond. The bonding electrons are not shared but both located on the same X atom. As explained in Structure 2.2, electronegativity is a measure of the ability of an atom to attract electrons. This can be quantified using the Pauling scale of values given in Section 9 of the data booklet and summarized here.

CsF is generally considered to be 100% ionic with an electronegativity difference ($\Delta\chi_P$) of 3.2. The % ionic character of other bonds can be approximated by the formula:

$$\% \text{ ionic character} \approx \frac{\Delta\chi_P}{3.2}$$

$\Delta\chi_P$ = **difference in electronegativity**

A difference of electronegativity of 1.6 corresponds to 50% ionic character. A difference of 1.8 is often taken as corresponding to predominantly ionic.

Clearly the ionic model gives an incomplete picture of the bonding in many substances, and we will consider other types of bonding in the next chapters. We will see that the distinction between ionic and covalent bonding is not black and white but is best described as a **bonding continuum** with all intermediate types possible (Figure 7).

ionic compounds	**covalent compounds**
reactive metal + reactive non-metal	two non-metals
electronegativity difference > 1.8	electronegativity difference << 1.8

polar covalent compounds
0 < electronegativity difference < 1.8

S2.1 Figure 7 The bonding continuum.

A more complete method of determining the bonding character of a compound is to use both the difference and average values of the electronegativities. As discussed in Structure 2.4, the nature of the bonding can be identified from the Triangular bonding diagram in section 17 of the IB data booklet.

Worked example

Which pair of elements is most likely to form an ionic bond?

A B and F **B** Si and O **C** N and Cl **D** K and Cl

Solution

Consider the difference in electronegativity of each pair:

A 2.0 and 4.0 **B** 1.9 and 3.4 **C** 3.0 and 3.2 **D** 0.8 and 3.2

$\Delta\chi_P = 2.0$ $\Delta\chi_P = 1.5$ $\Delta\chi_P = 0.2$ $\Delta\chi_P = 2.4$

D has the greatest difference, so the compound KCl is the most ionic.

Exercise

Q9. Describe the bonding in ammonium nitrate.

Q10. Which fluoride is the most ionic?

  A NaF **B** CsF **C** MgF_2 **D** BaF_2

Q11. Which one of the following compounds would be expected to have the strongest ionic bonding?

  A Na_2O **B** MgO **C** CaO **D** KCl

Q12. Element **E** is in group 1 of the periodic table and element **G** is in group 16. E and G are not the symbols for the elements.

  (a) Deduce the formula of the compound formed between E and G.

  A EG **B** EG_6 **C** EG_2 **D** E_2G

  (b) Identify the nature of the bonding in the compound formed between E and G.

  A ionic **B** covalent **C** dative covalent **D** metallic.

  (c) Describe the electrical conductivity of the compound formed between E and G.

  A conducts when solid and liquid
  B conducts when solid but not when liquid
  C conducts when liquid but not when solid
  D does not conduct when solid or liquid

Q13. You are given two white solids and told that only one of them is an ionic compound. Describe three tests you could carry out to determine which is the ionic compound.

The extent of ionic character is determined by the difference in electronegativity between the bonded elements.

Nature of Science

Models are simplifications of complex systems and can be physical representations. All models have limitations that need to be considered in their application. The photo shows a model of the lattice structure of NaCl, with sodium ions shown in green and chloride ions shown in grey.

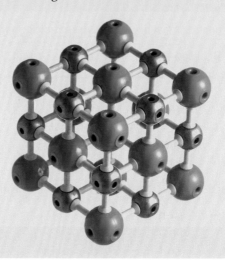

Challenge yourself

5. Discuss the usefulness of this model in describing the real structure of sodium chloride. Compare it with the models pictured earlier in the chapter and consider which you think may be the best model.

Guiding Question revisited

What determines the ionic nature and properties of a compound?

In this chapter we have developed an ionic model to explain the structure and properties of many compounds formed between metals and non-metals. This model is particularly effective in describing the properties of compounds formed between elements at different sides of the periodic table: the group 1 halides. The model does however have some limitations and other compounds show ionic properties to varying degrees as the bonding shows covalent character.

- An ion is a charged particle formed when atoms lose or gain electrons.
- Metals lose electrons to form positive ions (cations); non-metals gain electrons to form negative ions (anions).
- The number of charges on an ion is equal to the number of electrons lost (positive ion) or gained (negative ion) by an atom.
- The charge on an ion can usually be predicted from the group of the element in the periodic table; transition metal elements can form ions with different charges.
- The formula of common molecular ions can be related to the molecular formula of common acids.

- Ionic lattices consist of ions held together by ionic bonds which are electrostatic forces of attraction.
- Ionic compounds are electrically neutral as the total number of positive charges is balanced by the total number of negative charges.
- The formula of the compound is expressed as its simplest ratio, e.g. the ions X^{m+} and Y^{n-} will form the compound X_nY_m. The unit formula of an ionic compound is also the empirical formula.
- Ionic compounds usually have high melting and boiling points, and are more soluble in **polar** solvents such as water than in **non-polar** solvents.
- Positive ions are attracted to the O atom in a water molecule, which have a partial negative charge. Negative ions are attracted to the H atoms, which have a partial positive charge.
- Ions are not free to move in the solid state, so ionic compounds do not conduct electricity in the solid state. They can, however, conduct electricity when molten or in aqueous solution as the ions are free to move.
- The difference in electronegativities of two atoms gives an indication of the ionic character of the compound.
- Most ionic compounds are formed between metals on the bottom left of the periodic table and non-metals on the top right, excluding the noble gases.
- Ionic and covalent bonding are two extremes as many substances show intermediate character. The nature of the bonding can be related to bond polarity and the difference in electronegativity values ($\Delta\chi_P$) of the bonded atoms.

Practice questions

1. Which element does not generally form ions during chemical reactions?

   A aluminium   **B** calcium   **C** carbon   **D** iodine

2. Which statement is true for most ionic compounds?

   I.   They contain metal and non-metal elements.

   II. They conduct electricity in the solid state.

   III. They have high melting and boiling points.

   A I and II only   **B** I and III only   **C** II and III only   **D** I, II and III

3. Which electron transfer describes the reaction between aluminium and oxygen?

   $$2Al(s) + 3O_2(g) \rightarrow Al_2O_3(s)$$

   A Each Al atom loses two electrons.   **B** Each Al atom loses three electrons.

   C Each O atom loses two electrons.   **D** Each O atom loses three electrons.

4. Three substances, **X**, **Y** and **Z** have the physical properties shown.

Substance	X	Y	Z
Melting point / °C	747	1610	2614
Boiling point / °C	1390	2230	2850
Electrical conductivity of solid	high	poor	high

Identify X, Y and Z.

	X	Y	Z
A	CaO	NaBr	SiO_2
B	CaO	SiO_2	NaBr
C	NaBr	CaO	SiO_2
D	NaBr	SiO_2	CaO

5. What is the formula of barium phosphate?

    A Ba_3P_2 **B** $Ba_3(PO_4)_2$ **C** Ba_2P_3 **D** $Ba_2(PO_4)_3$

6. Which substance shows the most ionic character?

	Electrical conductivity in aqueous solution	Solubility in heptane	Volatility
A	low	low	high
B	low	high	low
C	high	low	low
D	high	low	high

7. What are the IUPAC names for CuBr and CuO?

    A copper bromide and copper oxide

    B copper(I) bromide and copper(I) oxide

    C copper(II) bromide and copper(I) oxide

    D copper(I) bromide and copper(II) oxide

8. Which of the following substances show predominantly ionic character?

    I. magnesium chloride   II. iron(II) chloride   III. silicon tetrachloride

    A I and II only   **B** I and III only   **C** II and III only   **D** I, II and III

9. Explain why the melting point of magnesium oxide is higher than that of barium sulfide? (3)

    (Total 3 marks)

10. Iron forms many compounds with oxygen.

    (a) Describe the bonding in Fe_2O_3. (2)

    (b) State the full electron configuration of the bonding species. (2)

    (c) The compound Fe_2O_3 was previously known as ferric oxide. State the IUPAC name for the compound. (1)

    (d) Explain why iron(III) oxide can demonstrate brittleness. (2)

    (e) Magnetite, Fe_3O_4, contains an additional ion of iron. Identify the ion. (1)

    (Total 8 marks)

11. **(a)** Describe the structure and bonding in potassium sulfide at room temperature. (2)

(b) State the full electron configuration of the bonding species. (2)

(c) Compare the size of the bonding species. (2)

(d) Describe and explain the electrical conductivity of solid and molten potassium sulfide. (2)

(Total 8 marks)

12. Successive ionization energies of calcium and a transition element **X** are shown.

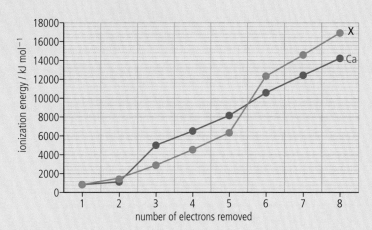

(a) Deduce the maximum oxidation state shown by X and explain your answer. (2)

(b) Deduce the formula of the oxide of X in this oxidation state and explain why it is unlikely to be ionic. (2)

(c) Identify the ions X forms in its ionic compounds. (1)

(Total 5 marks)

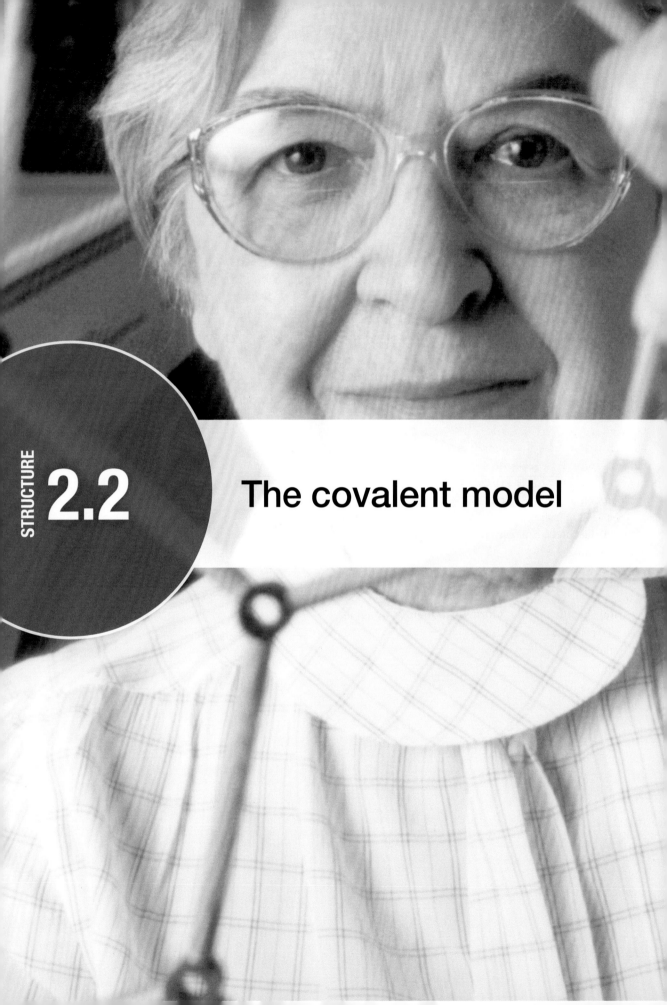

The covalent model

Stephanie Louise Kwolek (1923–2014), US chemist, holding a molecular model. Kwolek spent 40 years working as a research chemist at the DuPont Company. In 1965, she developed the first in a family of liquid crystal polymers that provided the basis for the Kevlar™ brand fibre. Kwolek was inducted into the National Inventors Hall of Fame in 1995, and was awarded the DuPont Company's Lavoisier Medal (1995) and the National Medal of Technology (1996).

Guiding Question

What determines the covalent nature and properties of a substance?

The term 'covalence' relates to the idea that atoms can combine by sharing pairs of valence electrons. This concept dates back to the early 20th century and the work of US chemist Gilbert Lewis, who first proposed the idea of the **covalent bond**. The notation he used to represent the sharing of electrons in a chemical structure is called a **Lewis formula**, in which pairs of electrons are arranged so that each atom gains a full valence shell. The tendency of atoms to gain a full valence shell is known as the **octet rule** because in most instances, eight electrons are required.

Building upon these fundamental understandings, this chapter will explore how Lewis formulas can be used to predict the shapes of molecules by considering the repulsion between regions of electron density, known as **electron domains**, around a central atom. By considering the strength of attraction between individual atoms and shared pairs of electrons, we can also identify when there will be unequal charge distribution, known as a **bond dipole**, in a covalent bond. This, combined with our knowledge of **molecular geometries**, can explain the relative polarity of molecules, which plays a large part in determining the strength of **intermolecular forces** and the resultant physical and chemical properties. For some covalent substances, such as diamond and graphene, we will see distinct sets of properties which can be explained by their **network structure**.

In the context of models of bonding and structure, we should recognize the position of the covalent model on a continuum with other models of bonding to best explain their observed properties. As in all areas of science, it will also be important to understand the limitations of, and exceptions to, the various rules, models and theories that we will apply.

Structure 2.2.1 – Covalent bonding, the octet rule and Lewis formulas

Structure 2.2.1 – A covalent bond is formed by the electrostatic attraction between a shared pair of electrons and the positively charged nuclei.

The octet rule refers to the tendency of atoms to gain a valence shell with a total of eight electrons.

Deduce the Lewis formula of molecules and ions for up to four electron pairs on each atom.

Lewis formulas (also known as electron dot or Lewis structures) show all the valence electrons (bonding and non-bonding pairs) in a covalently bonded species.	Nature of Science – What are some of the limitations of the octet rule?
	Structure 1.3 – Why do noble gases form covalent bonds less readily than other elements?
Electron pairs in a Lewis formula can be shown as dots, crosses and/or dashes.	
Molecules with atoms having fewer than an octet of electrons should be covered.	Structure 2.1 – Why do ionic bonds only form between different elements while covalent bonds can form between atoms of the same element?
Organic and inorganic examples should be used.	

A .22 calibre bullet hitting Kevlar™. The bullet is travelling at 660 feet per second (220 metres per second). This image shows the collision of the bullet and Kevlar™ at $\frac{1}{1\,000\,000}$th of a second. Kevlar™, a synthetic fibre comprised of large covalent molecules, has a high tensile strength-to-weight ratio, five times stronger than steel, leading to applications in fields such as ballistic resistant materials, industrial conveyor belts and automotive components.

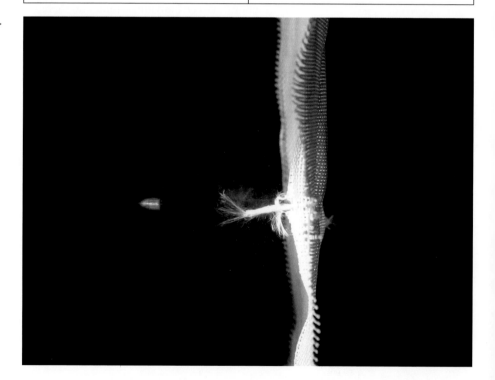

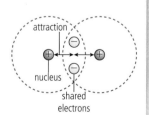

▲
S2.2 Figure 1 In a covalent bond, the nuclei of both atoms are attracted to the shared pair of electrons.

A covalent bond forms by atoms sharing electrons

When atoms of two non-metals react together, each is seeking to gain valence electrons in order to achieve the stable electron structure of a noble gas. By sharing an electron pair, they are effectively each able to achieve this. The shared pair of electrons is concentrated in the region between the two nuclei and is attracted to them both. The electrostatic attraction between the shared pair of electrons and the positively charged nuclei holds the atoms together and is known as a covalent bond (Figure 1).

For example, two hydrogen atoms form a covalent bond as follows.

$$H\overset{\times}{} + \cdot H \longrightarrow H\overset{\times}{\cdot}H$$

two hydrogen   a molecule of
atoms (2H)   hydrogen (H_2)

Note that in the molecule H_2, each hydrogen atom has a share of two electrons so it has gained the stability of the electron arrangement of the noble gas helium, He.

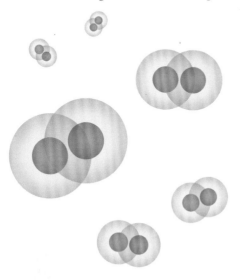

Hydrogen molecules. Each molecule consists of two hydrogen atoms bonded covalently.

The system containing these two hydrogen atoms will be stabilized when the forces of attraction between the nuclei and shared electrons are balanced by the forces of repulsion between the two nuclei. This holds the atoms at a fixed distance apart. Figure 2 shows how the covalent bond forms at this point of lowest energy as two hydrogen atoms approach each other and their electron density shifts.

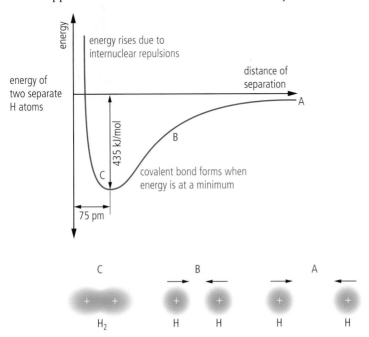

S2.2 Figure 2 When two hydrogen atoms form a covalent bond, the distance between them corresponds to the lowest energy. The letters A, B and C show the correspondence between shifts in electron density and the distance apart of the atoms.

The octet rule can be used to predict stable arrangements of atoms in covalent bonding

Similarly, two chlorine atoms react together to form a chlorine molecule in which both atoms have gained a share of eight electrons in their outer shells (the electron arrangement of argon, Ar). This tendency of atoms to gain a valence shell with a total of eight electrons is referred to as the octet rule. The diagram below only shows the outer electrons, not the inner full shells, as the outer electrons are the only electrons that take part in bonding.

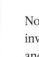

two chlorine atoms (2Cl) → a molecule of chlorine (Cl_2)

bonding pair
non-bonding or lone pairs

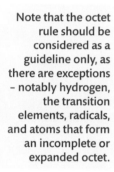

The octet rule refers to the tendency of atoms to gain a valence shell with a total of 8 electrons.

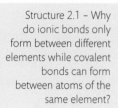

Note that the octet rule should be considered as a guideline only, as there are exceptions – notably hydrogen, the transition elements, radicals, and atoms that form an incomplete or expanded octet.

It is sometimes useful to use different symbols or colours to distinguish between the electrons of different atoms, as you see here. However, in reality all electrons are the same so it is not always necessary to do this. What is crucial is to have the correct number of valence electrons in the atoms.

Note that in the chlorine molecule, each atom has three pairs of electrons that are not involved in forming the bond. These are known as **non-bonding pairs**, or **lone pairs**, and they play an important role in determining the shape of more complex molecules, as we will see later.

Structure 2.1 – Why do ionic bonds only form between different elements while covalent bonds can form between atoms of the same element?

The ability of two identical atoms to form a covalent bond is due to the similar strength with which they attract valence electrons. In contrast, ionic bonds can only form between elements with significant differences in their attraction to valence electrons. This is why we see the formation of ions in a reaction between metal and non-metal atoms instead of the sharing of electrons.

Hydrogen and chlorine also react together to form a molecule of hydrogen chloride, HCl, which has the structure shown below. Note that the formation of the covalent bond enables both H and Cl to gain a stable outer shell.

Structure 1.3 – Why do noble gases form covalent bonds less readily than other elements?

The octet rule can provide a useful insight into the chemical properties of the noble gases. With the exception of helium, the atoms of all elements in group 18 have a complete octet without the need to share electrons. This is why they display low levels of chemical reactivity and do not tend to form covalent bonds with other elements.

Lewis formulas show all of the valence electrons in a covalently bonded species

Gilbert Lewis (1875–1946), responsible for the electron-pair theory of the covalent bond, was nominated 41 times for the Nobel Prize in Chemistry without winning. In 1946, he was found dead in his laboratory where he had been working with the toxic compound hydrogen cyanide.

American chemist Gilbert Lewis (1875–1946).

When describing the structure of covalent molecules, the most convenient method is known as a Lewis formula. This uses a simple notation of dots, crosses or a dash to represent each pair of electrons in the outer energy level, or valence shell, of all the atoms in the molecule. It can be derived as follows.

1. Calculate the total number of valence electrons in the molecule by multiplying the number of valence electrons of each element by the number of atoms of the element in the formula and totalling these.

2. Identify the skeletal structure of the molecule by drawing a pair of crosses, a pair of dots, or a single dash to show one electron pair between the central atom and each of the outer atoms.

3. Beginning with the outer atoms and then the central atom, add electron pairs until each atom has completed the octet rule with 8 valence electrons (except hydrogen, which must have 2 valence electrons, and the other exceptions noted later in the chapter).

4. Check that the total number of electrons in your finished Lewis formula is equal to your calculation in the first step.

Note that the terms 'valence electrons' and 'outer shell electrons' are used interchangeably. They both refer to the electrons that take part in bonding.

Note: If there are not enough electrons to complete the octet on each atom, we can use non-bonding pairs of electrons to form double and triple covalent bonds as described in the next section.

Worked example

Draw the Lewis formula for the molecule CCl_4.

Solution

Note that in Lewis formulas, as in the examples on page 158, only the valence electrons are shown, as these are the only ones that take part in bonding.

1. Total number of valence electrons

   $= (4 \times 1) + (7 \times 4)$

   $= 32$

2. Skeletal structure

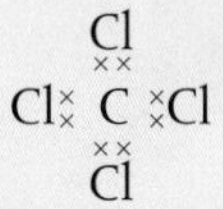

3. Completed Lewis formula 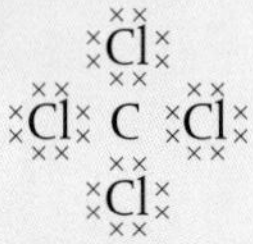 (32 electrons)

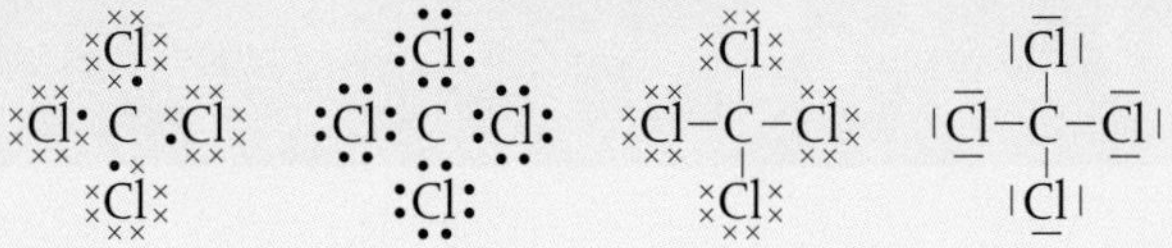

We can represent electron pairs using dots, crosses, a combination of dots and crosses, or a dash. Use whichever notation you prefer, but be prepared to recognize all versions. The important thing is to be clear and consistent, and to avoid vague and scattered dots.

4. Check the total number of electrons.

   The 32 electrons in the Lewis formula correctly match the calculated number in step 1.

   There are several different ways of drawing Lewis formulas, and some alternative acceptable forms are shown below.

This table shows some examples of molecules with their Lewis formulas.

Molecule	Total number of valence electrons	Lewis formula
CH_4	$4 + (1 \times 4) = 8$	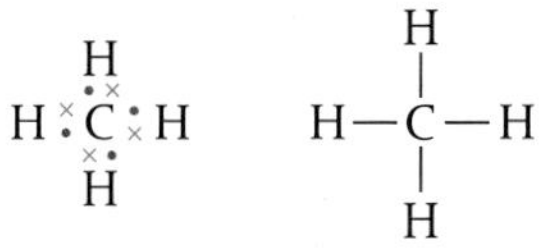
NH_3	$5 + (1 \times 3) = 8$	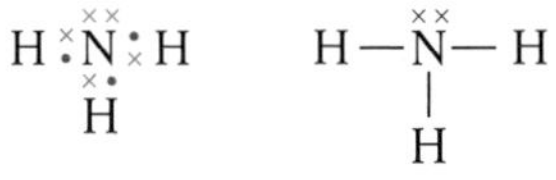
H_2O	$(1 \times 2) + 6 = 8$	

When drawing the Lewis formula of ions, note that you must:

- calculate valence electrons as above and then add one electron for each negative charge and subtract one electron for each positive charge;
- put the Lewis formula in a square bracket with the charge shown outside.

Worked example

Draw the Lewis formulas for:

(a) OH⁻   **(b)** SO₄²⁻

Solution

(a) valence electrons = 6 + 1 + 1 = 8

$$\left[\;{}^{\times\times}_{\times\times}\!O\!\times\!H\;\right]^{-}$$

(b) valence electrons = (6 × 1) + (6 × 4) + 2 = 32

$$\left[\begin{array}{c} {}^{\times\times}O{}^{\times} \\ {}^{\times\times}_{\times\times}O\;{}^{\times}_{\times}S{}^{\times}_{\times}\;O{}^{\times\times}_{\times\times} \\ {}^{\times}O{}^{\times} \end{array}\right]^{2-}$$

When drawing the Lewis formula of an ion, make sure you remember to put a square bracket around the structure with the charge shown clearly outside the bracket.

The octet rule is not always followed

There are a few molecules that are exceptions to the octet rule. Small atoms such as beryllium (Be) and boron (B) form stable molecules in which the central atom has fewer than eight electrons in its valence shell. This is known as an **incomplete octet**. For example, the structures of BeCl₂ and BF₃ are shown below.

Common examples of elements that can form an incomplete octet are hydrogen, beryllium, boron and aluminium.

BeCl₂   valence electrons = 2 + (7 × 2) = 16 $${}^{\times\times}_{\times\times}Cl\!:\!Be\!:\!Cl{}^{\times\times}_{\times\times}$$

BF₃   valence electrons = 3 + (7 × 3) = 24

$${}^{\times}_{\times\times}F\!:\!B\!:\!F{}^{\times}_{\times} \atop {}^{\times}_{\times\times}F{}^{\times}_{\times}$$

These examples demonstrate a limitation of the octet rule when dealing with small central atoms with a limited number of valence electrons. The small size of beryllium and boron, for example, limits the number of atoms that can get close enough to share the required number of electrons for a complete octet.

Nature of Science – What are some of the limitations of the octet rule?

Nature of Science

Scientific *rules* and *principles* are descriptive statements that are generally accurate in predicting the behaviour of matter but will likely have exceptions. For this reason, it is often as important to understand the exceptions as it is to know the rules or principles themselves.

Exercise

Q1. Which is the best definition of a covalent bond?

  A the attraction between positively charged ions and a shared pair of electrons

  B the attraction between pairs of shared electrons

  C the attraction between positively charged nuclei and a shared pair of electrons

  D the attraction between a pair of positively charged nuclei

Q2. Which best describes the octet rule?

  A the tendency of atoms to gain a valence shell with a total of 8 electrons

  B the tendency of atoms to lose 8 electrons

  C the tendency of atoms to gain 8 electrons

  D the tendency of atoms to form bonds with 8 other atoms

Q3. Draw the Lewis formulas of the following molecules:

  (a) HF   **(b)** CF_3Cl **(c)** PCl_3 **(d)** OF_2

  (e) $SeCl_2$ **(f)** C_2H_6 **(g)** N_2H_4

Q4. How many valence electrons are in the following molecules?

  (a) $BeCl_2$ **(b)** BCl_3 **(c)** CCl_4

  (d) PH_3 **(e)** SCl_2 **(f)** NCl_3

Q5. Draw Lewis formulas of the following ions:

  (a) NH_4^+ **(b)** H_3O^+ **(c)** HS^-

Structure 2.2.2 – Single, double and triple covalent bonds

Structure 2.2.2 – Single, double and triple bonds involve one, two and three shared pairs of electrons respectively.

Explain the relationship between the number of bonds, bond length and bond strength.

	Reactivity 2.2 – How does the presence of double and triple bonds in molecules influence their reactivity?

Atoms can share more than one pair of electrons to form multiple bonds

Sometimes it seems there are not enough electrons to achieve octets on all the atoms in a molecule. In these cases, the atoms will have to share more than one electron pair, in other words, form a multiple bond. A **double bond** forms when two electron pairs, a total of four electrons, are shared; a **triple bond** forms when three electron pairs, a total of six electrons, are shared.

Worked example

Draw the Lewis formula for the molecule O_2.

Solution

Following the steps from the flow chart on page 159:

1. Total number of valence electrons
   = (6×2) = 12

2. Skeletal structure

   O ×
   ×
   O

3. Add electron pairs until each atom has completed the octet rule.

   $$\overset{\times\times}{\underset{\times\times}{\times}} O \underset{\times\times}{\overset{\times\times}{\times}} O \longrightarrow O \underset{\times\times}{\overset{\times\times}{\times\times}} O$$

   At this point, you will notice that in order to complete the octet on both atoms, we need to use a non-bonding pair of electrons from the oxygen on the left to form a second shared pair between the atoms. This is a double bond.

4. Check the total number of electrons.

   The 12 electrons in the Lewis formula correctly match the calculated number in step 1.

   There are several different ways of drawing Lewis formulas, and some alternative acceptable forms are shown below.

   $$O \overset{\times\times}{\underset{\times}{\times}} \overset{\bullet\bullet}{\underset{\bullet\bullet}{O}} \quad O = O \quad O = O \quad \text{double bond}$$

Similar to oxygen, the most abundant gas in air, nitrogen, exists as diatomic molecules but requires a triple covalent bond in order to complete the octet of both atoms:

$$\overset{\times}{\underset{\times}{\times}} N \overset{\times}{\underset{\times}{\times}} N \overset{\bullet}{\underset{\bullet}{\times}} \quad \text{or} \quad | N \equiv N | \quad \text{triple bond}$$

The triple bond in nitrogen (N_2) is difficult to break, making the molecule very stable. This is why, although nitrogen makes up about 78% of the atmosphere, it does not readily take part in chemical reactions. For example, although the element is essential for all life forms, N_2 from the atmosphere is only rarely used directly by organisms as their source of nitrogen for synthesis reactions.

Reactivity 2.2 – How does the presence of double and triple bonds in molecules influence their reactivity?

This table shows some examples of Lewis formulas of molecules with multiple bonds.

Molecule	Total number of valence electrons	Lewis formula
CO_2	$4 + (6 \times 2)$ = 16	$:\overset{\bullet}{O}:\overset{\times}{\underset{\times}{C}}:\overset{\bullet}{O}: \quad O = C = O$
HCN	$1 + 4 + 5$ = 10	$H \overset{\times}{\underset{\bullet}{C}} \overset{\bullet}{\underset{\bullet}{N}}: \quad H - C \equiv N$

As shown above, it is often convenient to use a line to represent a shared pair of electrons, two lines to represent a double bond, and three lines to represent a triple bond.

Short bonds are strong bonds

Every covalent bond is characterized by two values.

Data on both these values are given in Sections 10 and 11 of the data booklet.

- **Bond length**: a measure of the distance between the two bonded nuclei;
- **Bond strength**: usually described in terms of bond enthalpy, which is discussed in Reactivity 1.2, and is effectively a measure of the energy required to break the bond.

Since atomic radius increases as we go down a group, we would expect the atoms to form molecules with longer bonds.

$$(F_2 \quad F–F)$$
$$Cl_2 \quad Cl–Cl$$
$$Br_2 \quad Br–Br$$
$$I_2 \quad I–I$$

bond length increases
bond enthalpy decreases

As a result, the shared electron pair is further from the pull of the nuclei in the larger molecules, and so the bond would be expected to be weaker. Data on group 17 elements, which exist as diatomic molecules, mostly confirm these trends.

Challenge yourself

1. Fluorine, F_2, is shown in brackets above because a close look at its bond enthalpy data shows that it is an outlier in the trends described here. What explanation might account for this?

Multiple bonds have a greater number of shared electrons and so have a stronger force of electrostatic attraction to the bonded nuclei. This means there is a greater pulling power between the atoms, bringing them closer together, resulting in bonds that are shorter and stronger than single bonds. We can see this by comparing different bonds between the same atoms, such as in the hydrocarbons.

Hydrocarbon	C_2H_6	C_2H_4	C_2H_2
Structural formula	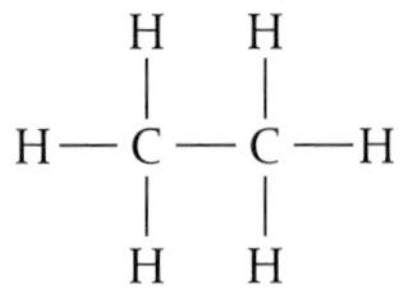 H—C—C—H (with H above and below each C)	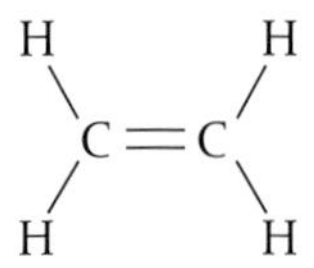 C=C (with H's attached)	 H—C≡C—H
Bond between carbon atoms	single	double	triple
Bond length / pm	154	134	120
Bond enthalpy / kJ mol⁻¹	346	614	839

The data confirm that triple bonds are shorter and stronger than double bonds, which are shorter and stronger than single bonds. Note though that the double bond is not twice as strong as the single bond. The explanation for this is not required in the standard level syllabus content but is related to the nature of the sharing of additional electron pairs found in double and triple bonds.

Also, we can compare two different carbon–oxygen bonds within the molecule of ethanoic acid, CH_3COOH.

$$
\begin{array}{c}
\quad\quad H \quad\quad O \\
\quad\quad | \quad\quad\;\; \diagup\!\diagup \\
H - C - C \\
\quad\quad | \quad\quad\;\; \diagdown \\
\quad\quad H \quad\quad O - H
\end{array}
$$

	C–O	C=O
Bond length / pm	143	122
Bond enthalpy / kJ mol^{-1}	358	804

single bonds   double bonds   triple bonds

———————————————————————————————————————▶

decreasing length, increasing strength

Exercise

Q6. Draw the Lewis formulas of the following molecules:

  (a) C_2H_6 (b) C_2H_4 (c) C_2H_2

Q7. Suggest which of the following molecules would have the shortest carbon–carbon bond length.

  A C_2H_6 B C_2H_4 C C_2H_2

Q8. Draw Lewis formulas of the following ions:

  (a) NO_3^- (b) NO^+ (c) NO_2^-

Q9. Which of the following molecules contains the shortest bond between carbon and oxygen?

  A CO_2 B H_3COCH_3 C CO D CH_3COOH

Q10. Which of the following best describes the general trend in bond length and bond strength?

	Bond strength	Bond length
A	triple > double > single	single > double > triple
B	triple > double > single	triple > double > single
C	single > double > triple	triple > double > single
D	single > double > triple	triple > double > single

Structure 2.2.3 – Coordination bonds

> **Structure 2.2.3 – A coordination bond is a covalent bond in which both the electrons of the shared pair originate from the same atom.**
>
> Identify coordination bonds in compounds.

A coordination bond is a covalent bond in which both shared electrons originate from the same atom

The examples so far involve covalent bonds where each bonded atom contributes one electron to the shared pair. However, sometimes the bond forms by *both* the electrons in the pair originating from the same atom. This means that the other atom accepts and gains a share in a donated electron pair. Such bonds are called **coordination bonds**. An arrow on the head of the bond is sometimes used to show a coordination bond, with the direction indicating the origin of the electrons.

For example:

$$H_3O^+$$

$$NH_4^+$$

$$CO \qquad :C \overset{\times\times}{\underset{\times\times}{:}} O \overset{\times}{} \qquad :C \equiv O \overset{\times}{}$$

A coordination bond is a covalent bond in which both the electrons of the shared pair originate from the same atom.

Note that in CO, the triple bond consists of two bonds that involve sharing an electron from each atom, and the third bond is a coordination bond where both electrons come from the oxygen. The three bonds are nonetheless identical to each other. This illustrates an important point about coordination bonds: once they are formed, they are no different from other covalent bonds.

Exercise

Q11. Which best describes a coordination bond?

    A   A covalent bond containing a shared pair of electrons.

    B   A covalent bond in which each atom shares one electron.

    C   A covalent bond involving a transition metal.

    D   A covalent bond in which both the electrons of the shared pair originate from the same atom.

Q12. Which of the following are features of a coordination bond?

    I.   Contains a shared pair of electrons.

    II.   Both shared electrons originate from the same atom.

    III.   Shorter in length than a normal covalent bond.

    A   I and II   **B**   I and III   **C**   II and III   **D**   I, II and III

Q13. Which of the following species contain a coordination bond?

I. H_3O^+ II. NH_3 III. CO

A I and II **B** I and III **C** II and III **D** I, II and III

Q14. Ammonia, NH_3, can react with boron trifluoride, BF_3, to form ammonia boron trifluoride, NH_3BF_3. Draw a Lewis formula of ammonia boron trifluoride and show the coordination bond using an arrow.

Q15. Aluminum chloride, $AlCl_3$, is a covalent substance that sublimes at 180 °C. In the gaseous state, pairs of $AlCl_3$ molecules combine via two coordination bonds to form a larger molecule with the formula Al_2Cl_6. Draw a Lewis formula for Al_2Cl_6 in which all atoms have a complete octet and show the coordination bonds using arrows.

Structure 2.2.4 – The Valence Shell Electron Pair Repulsion (VSEPR) model

Structure 2.2.4 – The Valence Shell Electron Pair Repulsion (VSEPR) model enables the shapes of molecules to be predicted from the repulsion of electron domains around a central atom.

Predict the electron domain geometry and the molecular geometry for species with up to four electron domains.

Include predicting how non-bonding pairs and multiple bonds affect bond angles.	Nature of Science – How useful is the VSEPR model at predicting molecular geometry?

The shape of a molecule is determined by repulsion between electron domains around a central atom

Once we know the Lewis formula of a molecule, we can predict how the bonds will be orientated with respect to each other in space – in other words, the three-dimensional shape of the molecule. This is often a crucial feature of a substance in determining its reactivity.

Biochemical reactions depend on a precise 'fit' between the enzyme, which controls the rate of the reaction, and the reacting molecule, known as the substrate. Anything that changes the shape of either of these may therefore alter the reaction dramatically. Many drugs work in this way.

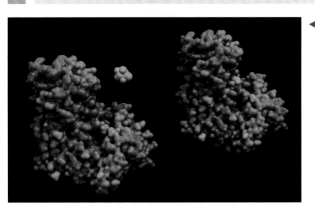

◀ A computer simulation of the hexokinase enzyme (large, orange) binding to glucose (small, blue) to form an enzyme–substrate complex. The function and activity of biochemicals are often dependent on their shape.

TOK

People often find this topic easier to understand when they can build and study models of the molecules in three dimensions. Does this suggest different qualities to the knowledge we acquire in different ways?

SKILLS

Predictions of molecular shape are based on the **Valence Shell Electron Pair Repulsion (VSEPR) model**. As its name suggests, this theory is based on the simple notion that *because electron pairs in the same valence shell carry the same charge, they repel each other and so spread themselves as far apart as possible.*

Before using this model, it is useful to clarify some of the language involved because the term VS**EP**R, evidently referring to **E**lectron **P**airs, is actually an over-simplification. As we have seen, molecules frequently contain multiple pairs of shared electrons, and these behave as a single unit in terms of repulsion because they are orientated together. So a better, more inclusive, term than electron pair is **electron domain**. This includes all electron locations in the valence shell, whether they are occupied by non-bonding pairs, single, double, or triple bonded pairs. What matters in determining shape is the *total number of electron domains*, and this can be determined from the Lewis formula.

Worked example

How many electron domains exist in the central atom of the following molecules whose Lewis formulas are shown?

(a) **(b)** **(c)**

$$H—N—H \qquad O=C=O \qquad O=O—O$$
$$\quad | \qquad\qquad\qquad\qquad\qquad$$
$$\quad H$$

Solution

(a) **(b)** **(c)**

(a) 4 electron domains: 3 bonding and 1 non-bonding

(b) 2 electron domains: 2 bonding and 0 non-bonding

(c) 3 electron domains: 2 bonding and 1 non-bonding

Technically then, the VSEPR model should be called Valence Shell Electron Domain Repulsion. It can be summarized as follows:

- The repulsion applies to electron domains, which can be single, double, or triple bonding electron pairs, or non-bonding pairs of electrons.
- The total number of electron domains around the central atom determines the geometrical arrangement of the electron domains.
- The shape of the molecule is determined by the angles between the bonded atoms.
- Non-bonding pairs (lone pairs) and multiple bonds cause slightly more repulsion than a bonding pair for the following reasons:
  - Non-bonding pairs have a higher concentration of charge than bonding pairs because they are not shared between two atoms.
  - Multiple bonds have a higher concentration of charge because they contain two or three pairs of electrons.

We expect the repulsive force to decrease in the following order:

non-bonding pair > multiple bond > single bond

As a result, molecules with lone pairs or multiple bonds on the central atom have some distortions in their structure that reduce the angle between the bonded atoms.

Note that in the diagrams in this section, non-bonding electrons on the surrounding atoms have been omitted for clarity – these are therefore *not* Lewis formulas.

Species with two electron domains

Molecules with two electron domains will position them at **180°** to each other, giving a **linear** shape to the molecule.

$BeCl_2$	Cl—Be—Cl	
CO_2	O=C=O	linear 180°
C_2H_2	H—C≡C—H	

Species with three electron domains

Molecules with three electron domains will position them at approximately **120°** to each other, giving a **triangular planar** shape to the **electron domain geometry**. If all three electron domains are bonding, the **molecular geometry** will also be triangular planar.

	BH₃	AlCl₃	HCOH
Molecular geometry	Triangular planar H H B)120° | H	Triangular planar Cl Cl Al)120° | Cl	Triangular planar H 118° H C)121° || O
Bond angles	H–B–H is 120°	Cl–Al–Cl is 120°	O=C–H is slightly more than 120° H–C–H is slightly less than 120°

Note that although each of the three molecules has a triangular planar shape, the bond angles in HCOH are slightly distorted due to the increased repulsion from the double bond. The VSEPR model is very useful for predicting molecular geometries but requires additional understanding to explain small differences in bond angles in similar shapes.

However, if one of the electron domains is a lone pair, this will not be 'seen' in the overall shape of the molecule, as it is part of the central atom. The molecular shape is determined by the positions of the atoms fixed by the bonding pairs only. A further consideration is that, as described above, lone pairs and multiple bonds cause slightly more repulsion than bonding pairs, so, in their presence, the angles are slightly altered.

In the example on the next page, the Lewis formula of ozone, O_3, has three electron domains around the central oxygen atom. Therefore the electron domain geometry of the molecule is still triangular planar. But, as only two of the three electron domains contain bonding electrons, the molecular geometry will not be triangular planar. The red outline shows the

You do not need to memorize specific distortions in bond angles but you do need to identify when they might be slightly more or less than the angle predicted by the VSEPR model.

Nature of Science – How useful is the VSEPR model at predicting molecular geometry?

The electron domain geometry is determined by the positions of *all* the electron domains, but the molecular geometry depends only on the positions of the *bonded* atoms.

positions of the bonding electrons that determine the shape. The molecule is described as **bent** or **V-shaped**. The extra repulsion due to the lone pair of electrons and double bond distorts the shape slightly, so the angle is slightly less than 120°; it is approximately 117°.

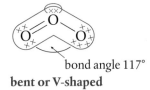

bond angle 117°
bent or V-shaped

Species with four electron domains

Molecules with four electron domains will position them at approximately **109.5°** to each other, giving a **tetrahedral** electron domain geometry. If all four electron domains are bonding, the molecular geometry will also be tetrahedral.

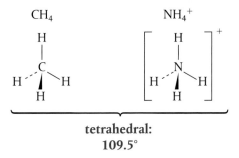

tetrahedral:
109.5°

However, if one or more of the electron domains is a lone pair, we must again focus on the number of bonded pairs to determine the shape of the molecule. The table below compares molecules with 0, 1, and 2 lone pairs respectively.

	CH_4	NH_3	H_2O
Lewis formula	H—C—H with H above and H below	H—N—H with lone pair above and H below	H—O—H with two lone pairs above
Number of electron domains	4	4	4
Electron domain geometry	tetrahedral	tetrahedral	tetrahedral
Number of lone pairs	0	1	2
Number of bonded electron domains	4	3	2
Molecular geometry	tetrahedral	trigonal pyramidal	bent or V-shaped
Bond angles	109.5°	approximately 107°	approximately 104.5°

Note that the presence of two lone pairs in H_2O causes more repulsion than the single lone pair in NH_3. This is why the bond angle is reduced further in H_2O from the 109.5° of the symmetrical tetrahedron.

Models of H_2O, NH_3 and CH_4 showing the different molecular geometries due to the different numbers of non-bonding pairs on the central atom.

Here is a summary of the steps used in determining the shape of a molecule.

1. Draw the Lewis formula, following the steps on page 159.

2. Count the total number of electron domains on the central atom.

3. Determine the electron domain geometry as follows:

        2 electron domains → linear

        3 electron domains → triangular planar

        4 electron domains → tetrahedral

4. Determine the molecular geometry from the number of bonding electron domains.

5. Consider the extra repulsion caused by the lone pairs and adjust the bond angles accordingly.

Always draw the Lewis formula before attempting to predict the shape of a molecule, as you have to know the total number of bonding pairs and lone pairs around the central atom.

Exercise

Q16. Predict the geometry and bond angles of the following molecules:

(a) H_2S (b) CF_4 (c) HCN   (d) NF_3

(e) BCl_3 (f) NH_2Cl (g) OF_2

Q17. Predict the geometry and bond angles of the following ions:

(a) NO_2^+ (b) NO_2^- (c) ClF_2^+ (d) $SnCl_3^-$

Q18. How many electron domains are there around the central atom in molecules that have the following shapes?

(a) tetrahedral   (b) bent   (c) linear

(d) trigonal pyramidal   (e) triangular planar

VSEPR model in nature. Different numbers of snowberries growing together spontaneously adopt the geometry of maximum repulsion. Clockwise from the top: four berries form a tetrahedral shape (bond angle 109.5°), three berries form a triangular planar shape (bond angle 120°), and two berries are linear (bond angle 180°). What other examples of VSEPR shapes can you find in nature?

Structure 2.2.5 – Bond polarity

Structure 2.2.5 – Bond polarity results from the difference in electronegativities of the bonded atoms.

Deduce the polar nature of a covalent bond from electronegativity values.

Bond dipoles can be shown either with partial charges or vectors. Electronegativity values are given in the data booklet.	Structure 2.1 – What properties of ionic compounds might be expected in compounds with polar covalent bonding?

Polar bonds result from unequal sharing of electrons

Electronegativity is a measure of the ability of an atom to attract electrons in a covalent bond. It is described using the Pauling scale from 0–4.

Not all sharing is equal. If electron pairs spend more time with one atom than the other, they are not equally shared. This occurs when there is a difference in the **electronegativities** of the bonded atoms (Figure 3). Electronegativity is a measure of the ability of an atom to attract electrons in a covalent bond, and is described using the Pauling scale of values. These values can be found in the data booklet and the trends are explored in Structure 3.1.3.

Pauling electronegativity values

<1.0	2.0–2.4
1.0–1.4	2.5–2.9
1.5–1.9	3.0–4.0

1 H 2.2																	
3 Li 1.0	4 Be 1.6											5 B 2.0	6 C 2.6	7 N 3.0	8 O 3.4	9 F 4.0	
11 Na 0.9	12 Mg 1.3											13 Al 1.6	14 Si 1.9	15 P 2.2	16 S 2.6	17 Cl 3.2	
19 K 0.8	20 Ca 1.0	21 Sc 1.4	22 Ti 1.5	23 V 1.6	24 Cr 1.7	25 Mn 1.6	26 Fe 1.8	27 Co 1.9	28 Ni 1.9	29 Cu 1.9	30 Zn 1.7	31 Ga 1.8	32 Ge 2.0	33 As 2.2	34 Se 2.6	35 Br 3.0
37 Rb 0.8	38 Sr 1.0	39 Y 1.2	40 Zr 1.3	41 Nb 1.6	42 Mo 2.2	43 Tc 2.1	44 Ru 2.2	45 Rh 2.3	46 Pd 2.2	47 Ag 1.9	48 Cd 1.7	49 In 1.8	50 Sn 2.0	51 Sb 2.1	52 Te 2.1	53 I 2.7
55 Cs 0.8	56 Ba 0.9	57 La* 1.1	72 Hf 1.3	73 Ta 1.5	74 W 1.7	75 Re 1.9	76 Os 2.2	77 Ir 2.2	78 Pt 2.2	79 Au 2.4	80 Hg 1.9	81 Tl 1.8	82 Pb 1.8	83 Bi 1.9	84 Po 2.0	85 At 2.2
87 Fr 0.7	88 Ra 0.9	89 Ac* 1.1														

S2.2 Figure 3 Periodic trends in electronegativity values show an increase along a period and up a group.

Challenge yourself

2. Given the definition of 'electronegativity', why are the group 18 elements often not assigned a Pauling value?

$$H \overset{\times}{\underset{\bullet}{\longrightarrow}} Cl$$

partially partially
positive negative

▲
S2.2 Figure 4 The distribution of electrons in the H–Cl bond is unsymmetrical, with a greater electron density over the Cl atom.

As the more electronegative atom exerts a greater pulling power on the shared electrons, it gains more 'possession' of the electron pair. The bond is now unsymmetrical with respect to electron distribution and is said to be **polar**.

The term **bond dipole** is often used to indicate the fact that this type of bond has two partially separated opposite electric charges. The more electronegative atom with the greater share of the electrons has become partially negative or $\delta-$, and the less electronegative atom has become partially positive or $\delta+$.

For example, in HCl the shared electron pair is pulled more strongly by the Cl atom than by the H atom, resulting in a polar molecule (Figure 4).

Note that the symbol δ (Greek letter, pronounced 'delta') is used to represent a partial charge. This has no fixed value but is always less than the unit charge associated with an ion such as X^+ or X^-.

In water, the electronegativity difference results in polar O–H bonds with the electron density greater on the oxygen. Note that the structure has been drawn to represent the V-shaped geometry of the molecule.

The extent of polarity in a covalent bond varies, depending on how big a difference exists in the electronegativity values of the two bonded atoms. We can estimate this from knowledge of the periodic trends in electronegativity, discussed in Structure 3.1 and summarized here.

electronegativity increases across a period →

electronegativity increases up a group

There are several ways to denote the polar nature of a bond in a structure. One is to write the partial charges $\delta+$ and $\delta-$ over the less electronegative and more electronegative atom respectively. Another is to represent the direction of electron pull as a vector with an arrow on the bond indicating the pull on the electrons by the more electronegative atoms.

▲
General trends in electronegativity.

As fluorine is the most electronegative atom, we can predict it will have the greater electron density whenever it is covalently bonded to another element. Fluorine has a value of 4.0 on the Pauling scale of electronegativity given on page 172. This scale can be used to assess the relative polarity of a bond. For example, the table on the next page compares the polarities of the hydrogen halides.

	Electronegativity values of atoms		Difference in electronegativity
H–F	2.2	4.0	1.8
H–Cl	2.2	3.2	1.0
H–Br	2.2	3.0	0.8
H–I	2.2	2.7	0.5

increasing bond polarity

Challenge yourself

3. Oxygen is a very electronegative element with a value of 3.4 on the Pauling scale. Can you determine the formula of a compound in which oxygen would have a partial positive charge?

Worked example

Use the electronegativity values on page 172 to put the following bonds in order of decreasing polarity.

N–O in NO_2 N–F in NF_3 H–O in H_2O N–H in NH_3

Solution

bond polarity: H–O > N–F > N–H > N–O

difference in electronegativity values: 1.2   1.0   0.8   0.4

The only bonds that are truly non-polar are bonds between the same atoms such as the bonds in F_2, H_2, and O_2, because clearly here the *difference* in electronegativity is zero. These bonds are sometimes referred to as **pure covalent** to express this. All other bonds have some degree of polarity, although it may be very slight. The C–H bond, universal in organic chemistry, is often considered to be largely non-polar, although in fact carbon is slightly more electronegative than hydrogen. As we see in Structure 3.2, the low polarity of the C–H bond is an important factor in determining the properties of organic compounds.

The presence of polar bonds in a molecule has a significant effect on its properties. Effectively, the partial separation of charges introduces some ionic nature into covalent bonds, so the more polar the bond, the more like an ionic compound the molecule behaves. Polar bonds are therefore considered to be intermediates in relation to pure covalent bonds and pure ionic bonds, part of the bonding *continuum* that is discussed in Structure 2.4. This explains why we begin to see overlap in the properties of ionic and covalent substances. Hydrogen chloride, HCl, for example, dissolves in water to form H^+ and Cl^- ions and will therefore be able to conduct electricity in this aqueous state. We now see why the boundaries between ionic and covalent bonds are somewhat 'fuzzy', and why it is often appropriate to describe substances using terms such as 'predominantly' covalent or ionic, or being 'strongly' polar.

The idea of the bonding continuum is summarized on the next page.

In a covalent bond, the greater the difference in the electronegativity values of the atoms, the more polar the bond.

Structure 2.1 – What properties of ionic compounds might be expected in compounds with polar covalent bonding?

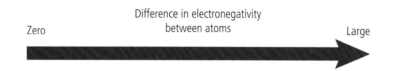

Difference in electronegativity
between atoms

Zero   Large

Pure covalent	**Polar covalent**	**Ionic**
equal sharing of electrons	partial transfer of electrons unequal sharing of electrons	complete transfer of electrons
Cl_2	HF, HCl, HBr, HI	NaCl, LiF, K_2O
discrete molecules	$\longrightarrow$	lattice of oppositely charged ions

Nature of Science

Scientists observe the natural world where variations in structure or properties are often continuous rather than discontinuous. In other words, a spectrum of intermediates exists and this can make classification difficult. At the same time, the properties of such intermediates can strengthen our understanding. For example, we have seen here that the existence of polar covalent molecules in the bonding continuum adds to our interpretation of electron behavior in bond formation.

Exercise

Q19. Which substance contains only ionic bonds?

   A $NaNO_3$ **B** H_3PO_4 **C** NH_4Cl **D** $CaCl_2$

Q20. For each of these molecules, identify any polar bonds and label them using δ+ and δ– appropriately.

   (a) HBr   **(b)** CO_2 **(c)** ClF   **(d)** O_2 **(e)** NH_3

Q21. Use the electronegativity values in Section 9 of the data booklet to predict which bond in each of the following pairs is more polar.

   (a) C–H or C–Cl   **(b)** Si–Li or Si–Cl   **(c)** N–Cl or N–Mg

Structure 2.2.6 – Molecular polarity

Structure 2.2.6 – Molecular polarity depends on both bond polarity and molecular geometry.

Deduce the net dipole of a molecule or ion by considering bond polarity and geometry.

Examples should include species in which bond dipoles do and do not cancel each other.	

Molecular polarity depends on both bond polarity and molecular geometry

We have learned that **bond polarity** depends on the charge separation between its two bonded atoms, which is a result of differences in their electronegativities. The polarity of a *molecule* (**molecular polarity**), however, depends on:

- the polar bonds that it contains;
- the way in which such polar bonds are orientated with respect to each other, in other words, the **molecular geometry**.

If the bonds are of equal polarity (i.e. involving the same elements) *and* are arranged symmetrically with respect to each other, their charge separations will oppose each other and so will effectively cancel each other out. In these cases, the molecule will be non-polar, despite the fact that it contains polar bonds. It is a bit like a game of tug-of-war between players who are equally strong and symmetrically arranged (Figure 5).

S2.2 Figure 5 Equal and opposite pulls cancel each other out.

Molecules with up to four electron domains

The molecules below are all non-polar because the dipoles cancel out. The arrow, shown here in blue, is the notation for a dipole that results from the pull of electrons in the bond towards the more electronegative atom.

Bond dipoles can be thought of as vectors, so the polarity of a molecule is the resultant force. This approach may help you to predict the overall polarity from the sum of the bond dipoles.

CO_2 BF_3 CCl_4

However, if *either* the molecule contains bonds of different polarity, *or* its bonds are not symmetrically arranged, then the dipoles will not cancel out, and the molecule will be polar. Another way of describing this is to say that it has a **net dipole**, which refers to its turning force in an electric field. This is what would happen in a game of tug-of-war if the players were not equally strong or were not pulling in exactly opposite directions (Figure 6).

S2.2 Figure 6 When the pulls are not equal and opposite there is a net pull.

The molecules below are all polar because the dipoles do not cancel out.

CH_3Cl NH_3 H_2O

Exercise

Q22. Predict whether the following will be polar or non-polar molecules.

(a) PH_3 (b) CF_4 (c) HCN (d) $BeCl_2$

(e) C_2H_4 (f) ClF (g) F_2 (h) BF_3

Q23. The molecule $C_2H_2Cl_2$ can exist as two forms known as *cis–trans* isomers, which are shown below.

$$\underset{trans}{\begin{matrix} H & & Cl \\ \diagdown & & \diagup \\ & C=C & \\ \diagup & & \diagdown \\ Cl & & H \end{matrix}} \qquad \underset{cis}{\begin{matrix} H & & H \\ \diagdown & & \diagup \\ & C=C & \\ \diagup & & \diagdown \\ Cl & & Cl \end{matrix}}$$

Determine whether either of these has a net dipole moment.

Q24. Predict the shapes and bond angles of the following molecules and identify which would be expected to be polar:

(a) BeH_2 (b) OF_2 (c) CCl_3F (d) BH_3

Structure 2.2.7 – Covalent network structures

Structure 2.2.7 – Carbon and silicon form covalent network structures.	
Describe the structures and explanation of the properties of silicon, silicon dioxide and carbon's allotropes: diamond, graphite, fullerenes and graphene.	
Allotropes of the same element have different bonding and structural patterns in the same physical state, and so have different chemical and physical properties.	Structure 3.1 – Why are silicon–silicon bonds generally weaker than carbon–carbon bonds?

Some covalent substances form covalent network structures

Most covalent substances exist as discrete molecules with a finite number of atoms. However, there are some substances that have a very different structure, a crystalline lattice in which the atoms are linked together by covalent bonds. Effectively, the crystal is a single molecule with a regular repeating pattern of covalent bonds, so has no finite size. It is referred to as a **covalent network structure**. As we might expect, such crystalline structures have very different properties from other smaller covalent molecules. A few examples are considered here.

Covalent network structures are also commonly known as 'giant molecular' or 'macromolecular' structures.

Allotropes of carbon

Allotropes have different bonding and structural patterns of the same element in the same physical state, and so have different chemical and physical properties. Molecular oxygen (O_2) and ozone (O_3), which both exist as gases, are examples of allotropes of oxygen.

Carbon has several allotropes and these are described and compared in the table on page 179.

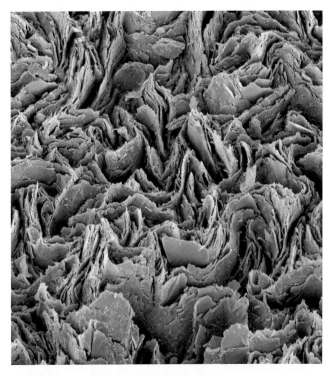

Coloured scanning electron micrograph of layers making up the core of a graphite pencil. Because graphite is a soft form of carbon, the tip of the pencil disintegrates under pressure to leave marks on the paper.

Cut and polished diamond. Diamond is a naturally occurring form of carbon that has crystallized under great pressure. It is the hardest known mineral. Beautiful crystals are found in South Africa, Russia, Brazil, and Sierra Leone.

	Diamond	Graphite	Graphene	C_{60} Fullerene
Structure	Each C atom is sp^3 hybridized and covalently bonded to 4 others, tetrahedrally arranged in a regular repetitive pattern with bond angles of 109.5°.	Each C atom is sp^2 hybridized and covalently bonded to 3 others, forming hexagons (shown in green) in parallel layers with bond angles of 120°. The remaining valence electron on each carbon is delocalized so it can move freely across the layer. The layers are held only by weak London dispersion forces so they can slide over each other.	Each C atom is sp^2 hybridized and covalently bonded to 3 others, as in graphite, forming hexagons with bond angles of 120°. But it is a single layer, so it exists as a two-dimensional material only. It is often described as a honeycomb or chicken wire structure. As with graphite, the remaining valence electron on each carbon is delocalized.	Each C atom is sp^2 hybridized and bonded in a sphere of 60 carbon atoms, consisting of 12 pentagons and 20 hexagons. The structure is a closed spherical cage in which each carbon is bonded to 3 others. (Note it is not a giant molecule as it has a fixed formula.)
Electrical conductivity	Non-conductor of electricity; all electrons are bonded and so non-mobile.	Good electrical conductor; contains one non-bonded, delocalized electron per atom that gives electron mobility.	Very good electrical conductor; one delocalized electron per atom gives electron mobility across the layers.	C_{60} fullerenes are poor conductors of electricity; although individual molecules have delocalized electrons, there is little electron movement between molecules.
Thermal conductivity	Very efficient thermal conductor, better than metals.	Not a good conductor, unless the heat can be forced to conduct in a direction parallel to the crystal layers.	Best thermal conductivity known, even better than diamond.	Very low thermal conductivity.
Appearance	Highly transparent, lustrous crystal.	Non-lustrous, grey crystalline solid.	Almost completely transparent.	Black powder.
Physical and chemical properties	The hardest known natural substance; it cannot be scratched by other materials; brittle; very high melting point.	Soft and slippery due to slippage of layers over each other; brittle; very high melting point; the most stable allotrope of carbon.	The thickness of just one atom so the thinnest material ever to exist; but also the strongest – 100 times stronger than steel; very flexible; very high melting point.	Very light and strong; reacts with K to make superconducting crystalline material; low melting point.
Uses	Polished for jewellery and ornamentation; tools and machinery for grinding and cutting glass.	A dry lubricant; in pencils; electrode rods in electrolysis.	TEM (transmission electron microscopy) grids, photo-voltaic cells, touch screens, high-performance electronic devices; many applications are still being developed.	Lubricants, medical, and industrial devices for binding specific target molecules; related forms are used to make nanotubes and nanobuds used as capacitors in the electronics industry, and catalysts.

Challenge yourself

4. For the most part, good electrical conductors are also good thermal conductors, but this is not the case with diamond. Can you think why this might be, and whether there are other substances that are not electrical conductors but good thermal conductors?

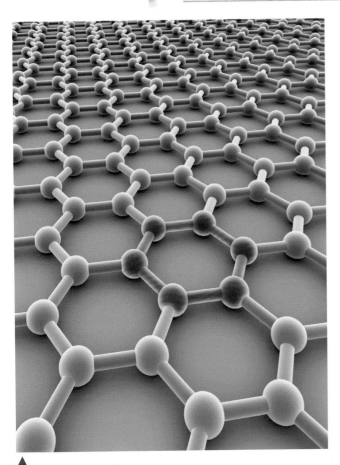

Graphene is a planar sheet of carbon atoms arranged in a hexagonal pattern.

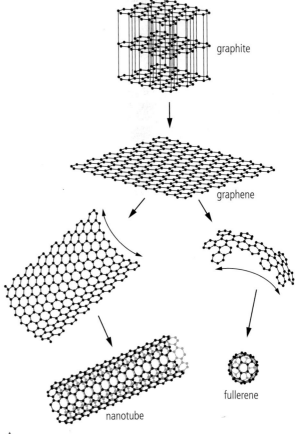

Graphite occurs naturally, and a single separated layer is graphene. A rolled up layer of graphene is a nanotube, and a closed cage is a fullerene.

Graphene is a relatively new form of carbon, isolated for the first time in 2004. Given its hardness, strength, lightness, flexibility, conductivity, and chemical inertness, it has enormous potential applications. Currently though, many of these are only in the development stage as they are dependent on the production of large sheets of pure graphene, which is difficult to achieve economically. But this is an intense field of research and breakthroughs are common.

TOK   Whether graphene will realize its potential and lead to innovative materials and applications remains an open question. What is the role of imagination in helping to direct the research focus of scientists?

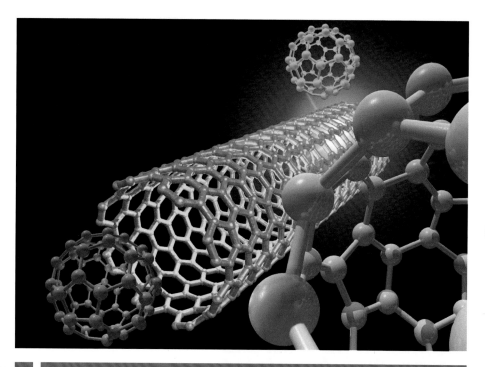

 Computer artwork of spherical and cylindrical fullerenes – buckyballs and carbon nanotubes. These substances are being investigated for a wide range of technical and medical uses. Here the carbon nanotubes have been engineered to have a diameter just large enough to allow buckyballs to pass through them

The name 'fullerene' was given to the newly discovered spheres C_{60} in honor of the American architect R. Buckminster Fuller. He had designed the World Exhibition building in Montreal, Canada, on the same concept of hexagons and a small number of pentagons to create a curved surface known as a geodesic dome. The dome of the Epcot Centre in Disney World, Florida, is similarly designed. Perhaps more familiarly it is also the structure of a European soccer ball. The term 'buckyballs' has slipped into common usage, derived from the full name of the structure 'buckminsterfullerene'.

Nature of Science

The discoveries of the allotropes of carbon – buckminsterfullerene and graphene – illustrate the importance of collaboration between scientists.

The discovery of fullerenes in 1985 was the result of teamwork between scientists with different experience and research objectives. Harold Kroto working at the University of Sussex, UK was interested in red giant stars and how carbon might polymerize near them. Robert Curl and Richard Smalley working in Texas, USA, had developed a technique using a laser beam for evaporating and analyzing different substances. When they worked together and applied this technique to graphite, clusters of stable C_{60} and C_{70} spheres were formed. The three scientists shared the Nobel Prize in Chemistry for 1996.

The isolation of graphene was announced by Kostya Novoselov and Andre Geim in 2004, both working at the University of Manchester, UK. It led to their shared award of the Nobel Prize in Physics in 2010. Novoselov and Geim were research collaborators over many years in different countries, originally as research student and supervisor respectively. Their work involved the use of adhesive tape to rip successively thinner flakes off graphite and analyze the flakes by attachment to a layer of silicon oxide. Many scientists over many years had attempted to isolate graphene, and through their successes and failures indirectly contributed to the later discovery.

The development of structures such as nanotubes, nanobuds, and graphene is part of the growing science of **nanotechnology**, which deals with the atomic scale manipulation of matter. An interesting development is a new material called graphene-hBN, which is made by adding a hexagonal layer of boron and nitrogen atoms onto graphene. The combination of boron nitride, an excellent electrical insulator, and the conductive properties of graphene has some interesting potential for applications in LEDs and quantum computing.

 The structure of fullerene has the same arrangement of hexagons and pentagons as a soccer ball.

181

Challenge yourself

5. Graphite is thermodynamically more stable than diamond. The reaction

$$C(diamond) \rightarrow C(graphite)$$

is accompanied by a loss of energy, $\Delta H = -2 \text{ kJ mol}^{-1}$.

But we commonly hear that 'diamonds are forever'. Are they?

Silicon and silicon dioxide

Silicon is the most abundant element in the Earth's crust after oxygen, occurring as silica (SiO_2) in sand and in silicate minerals. Since silicon is just below carbon in group 14, the possibility of silicon-based life has been proposed. But unlike carbon, silicon is not able to form long chains, multiple bonds, or rings, so cannot compete with the diversity possible in organic chemistry, based on carbon. However, there is some evidence that the first forms of life were forms of clay minerals that were probably based on the silicon atom.

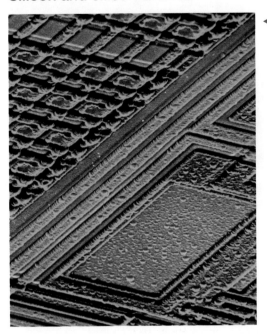

◄ Coloured scanning electron micrograph of the surface of a microprocessor computer chip. Silicon chips are tiny pieces of silicon as small as 1 mm² made to carry minute electrical circuits used in computers and transistors. Silicon is the most widely used semiconductor.

Like carbon, silicon is a group 14 element and so its atoms have four valence shell electrons. In the elemental state, each silicon atom is covalently bonded to four others in a tetrahedral arrangement. This results in a giant lattice structure much like diamond.

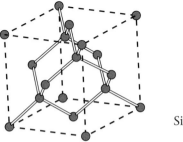

▲ The silicon crystal structure and the arrangement of bonds around a central silicon atom.

Silicon dioxide, SiO_2, commonly known as silica or quartz, also forms a giant covalent structure based on a tetrahedral arrangement (Figure 7). But here the bonds are between silicon and oxygen atoms, where each Si atom is covalently bonded to four O atoms, and each O to two Si atoms. You can think of the oxygen atoms as forming bridges between the tetrahedrally bonded silicon atoms.

Quartz crystals shown in a coloured scanning electron micrograph. Quartz is a form of silica (SiO_2) and the most abundant mineral in the Earth's crust. Quartz is used in optical and scientific instruments and in electronics such as mobile phones and laptops.

Note that here the formula SiO_2 refers to the *ratio* of atoms within the giant molecule – it is an empirical formula and the actual number of atoms present will be a very large multiple of this. As the atoms are strongly held in tetrahedral positions that involve all four silicon valence electrons, the structure has the following properties:

- strong
- insoluble in water
- high melting point
- non-conductor of electricity.

These are all properties we associate with glass and sand – different forms of silica.

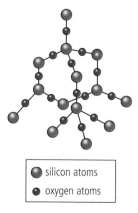

- silicon atoms
- oxygen atoms

S2.2 Figure 7 The structure of quartz, SiO_2.

Structure 3.1 – Why are silicon–silicon bonds generally weaker than carbon-carbon bonds?

A notable difference between the network structures of carbon and silicon is the tendency for carbon to form covalent bonds with other carbon atoms (known as catenation) whereas silicon is more commonly found bonded with oxygen. This is due to the greater strength of a C–C bond, approximately $346 \, kJ \, mol^{-1}$, compared to a Si–Si bond which has a bond enthalpy of $226 \, kJ \, mol^{-1}$. This can be explained by the smaller atomic radius of carbon resulting in a greater electrostatic attraction between nuclei and shared pairs of electrons. In contrast, the Si–O bond found in silica has a bond enthalpy of $466 \, kJ \, mol^{-1}$.

Pure silicon is a poor conductor of electricity at low temperatures. However, as its temperature is increased, the thermal energy provides the necessary energy to free some electrons from the covalent bonds, causing the electrical conductivity to increase dramatically. Each freed electron leaves behind a positively charged hole allowing other electrons to move through the lattice to occupy these holes. This process repeats to allow silicon to conduct electricity. Silicon can also be 'doped' with group 13 and 15 elements such as arsenic and boron in very small concentrations to increase its electrical conductivity. The 'Silicon Valley' region in North California owes its name to the wide range of applications of silicon in the electronics that make these industries possible.

Silicon Valley in Northern California is home to many large high-tech corporations.

Exercise

Q25. Outline the similarities and differences in the structures of diamond and graphite.

Q26. Describe the similarities and differences you would expect in the properties of silicon and diamond.

Q27. Explain why graphite and graphene are good conductors of electricity whereas diamond is not.

Structure 2.2.8 and 2.2.9 – Intermolecular forces

Structure 2.2.8 – The nature of the force that exists between molecules is determined by the size and polarity of the molecules. Intermolecular forces include London (dispersion), dipole-induced dipole, dipole–dipole and hydrogen bonding.	
Deduce the types of intermolecular force present from the structural features of covalent molecules.	
The term 'van der Waals forces' should be used as an inclusive term to include dipole–dipole, dipole-induced dipole, and London (dispersion) forces. Hydrogen bonds occur when hydrogen, being covalently bonded to an electronegative atom, has an attractive interaction with a neighbouring electronegative atom.	Structure 1.5 – To what extent can intermolecular forces explain the deviation of real gases from ideal behaviour? Nature of Science, Structure 1.1, 2.1, 2.3 – How do the terms 'bonds' and 'forces' compare? Nature of Science – How can advances in technology lead to changes in scientific definitions, e.g. the updated International Union of Pure and Applied Chemistry (IUPAC) definition of the hydrogen bond?

Structure 2.2.9 – Given comparable molar mass, the relative strengths of intermolecular forces are generally: London (dispersion) forces < dipole–dipole forces < hydrogen bonding.	
Explain the physical properties of covalent substances to include volatility, electrical conductivity and solubility in terms of their structure.	
	Tool 1, Inquiry 2 – What experimental data demonstrates the physical properties of covalent substances? Structure 3.2 – To what extent does a functional group determine the nature of the intermolecular forces?

Forces of attraction between molecules are called intermolecular forces

Covalent bonds hold atoms together *within* molecules, but of course molecules do not exist in isolation. A gas jar full of chlorine, Cl_2, for example, will contain millions of molecules of chlorine. So what are the attractive forces that exist *between* these molecules, the so-called intermolecular forces? The answer depends on the polarity and size of the molecules involved, and so the intermolecular forces will vary for different molecules. We will consider four types of intermolecular force here, and see how they differ from each other in origin and in strength.

The strength of intermolecular forces determines the physical properties of a substance. Volatility, solubility and conductivity can all be predicted and explained from knowledge of the nature of the forces between molecules.

An understanding of intermolecular forces can also be used to explain one aspect of why the behaviour of *real* gases deviates from that described by the *ideal* gas model. The attractive forces between gaseous atoms and/or molecules become more significant at low temperature and high pressure so limit the predictive accuracy of the ideal gas equation and the combined gas law.

Note that the prefix *intra-* refers to *within*, whereas the prefix *inter-* refers to *between*. For example, the internet is worldwide while an intranet is a computer network within an organization.

intramolecular (covalent bond)

intermolecular

Structure 1.5 – To what extent can intermolecular forces explain the deviation of real gases from ideal behaviour?

Fritz London (1900–1954) was a German physicist who pioneered the interpretation of covalent bond formation and intermolecular forces in the emerging field of quantum mechanics. As he was of Jewish descent, he had to leave his university post in Germany during the Nazi regime before the Second World War, and completed most of his research at Duke University, USA.

London dispersion forces

Non-polar molecules such as chlorine, Cl_2, have no permanent separation of charge within their bonds because the shared electrons are pulled equally by the two chlorine atoms. In other words, they do not have a permanent dipole.

However, because electrons behave somewhat like mobile clouds of negative charge, the density of this cloud may at any one moment be greater over one atom than the other. When this occurs, the bond will have some separation of charge – a weak dipole known as a 'temporary' or 'instantaneous' dipole (Figure 8). This will not last for more than an instant as the electron density is constantly changing, but it may influence the electron distribution in the bond of a neighbouring molecule, causing an **induced dipole**.

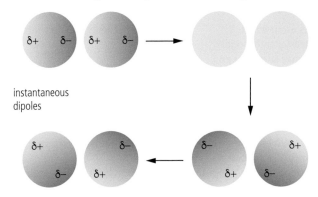

instantaneous dipoles

 S2.2 Figure 8 A series of instantaneous views of electron density in two molecules (each represented as a sphere). Each image is like a snap-shot of a possible electron distribution at a moment in time. A temporary dipole in one molecule can induce a temporary dipole in another. Note that in the top right pair of molecules, there was no temporary dipole in either at that moment in time.

As a result, weak forces of attraction, known as **London dispersion forces**, will occur between opposite ends of these two temporary dipoles in the molecules (Figure 9). These are the weakest form of intermolecular force. Their strength increases with increasing molecular size. This is because a greater number of electrons within a molecule increases the probability and magnitude of temporary dipoles developing.

1. Electron cloud evenly distributed; no dipole.

2. At some instant, more of the electron cloud happens to be at one end of the molecule than the other; the molecule has an instantaneous dipole.

3. Dipole is induced in a neighbouring molecule.

4. This attraction is a London dispersion force.

 S2.2 Figure 9 Instantaneous dipole–induced dipole London dispersion forces between molecules of Cl_2.

London dispersion forces are the *only* forces that exist between non-polar molecules. Such molecules generally have low melting and boiling points, because relatively little energy is required to overcome the weak London dispersion forces and separate the molecules from each other. This is why many non-polar elements and compounds are gases at room temperature, for example, O_2, Cl_2, and CH_4. Boiling point data also indicates how the strength of London dispersion forces increases with increasing molecular size. This is shown in the tables below that compare the boiling points of the halogens (group 17 elements) and of the family of hydrocarbons known as the alkanes.

Element	M_r	Boiling point / °C	State at room temperature
F_2	38	−188	gas
Cl_2	71	−34	gas
Br_2	160	59	liquid
I_2	254	185	solid

boiling point increases with increasing number of electrons

Alkane	M_r	Boiling point / °C
CH_4	16	−164
C_2H_6	30	−89
C_3H_8	44	−42
C_4H_{10}	58	−0.5

boiling point increases with increasing number of electrons

Although London dispersion forces are weak, they are responsible for the fact that non-polar substances can be condensed to form liquids and sometimes solids at low temperatures. For example, nitrogen gas, N_2, liquefies below −196 °C, and even helium gas has been liquefied at the extremely low temperature of −269 °C.

London dispersion forces are also components of the forces between polar molecules, but often get somewhat overlooked because of the presence of stronger forces.

When asked to name forces between molecules, London dispersion forces should be given in all cases, regardless of what additional forces may be present.

Dipole–dipole attraction

Polar molecules such as hydrogen chloride, HCl, have a permanent separation of charge within their bonds as a result of the difference in electronegativity between the bonded atoms. One end of the molecule is electron deficient with a partial positive charge ($\delta+$), while the other end is electron-rich with a partial negative charge ($\delta-$). This is known as a **permanent dipole**. It results in opposite charges on neighbouring molecules attracting each other, generating a force known as a **dipole–dipole attraction**.

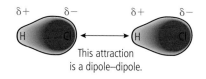

This attraction is a dipole–dipole.

Permanent dipoles cause forces of dipole–dipole attraction between molecules of HCl.

The strength of this intermolecular force will vary depending on the distance and relative orientation of the dipoles. But dipole–dipole forces are always stronger than London dispersion forces, as we might expect from their basis in permanent rather than instantaneous dipoles. These forces cause the melting and boiling points of polar compounds to be higher than those of non-polar substances of comparable molecular mass.

	C_3H_8	CH_3OCH_3
Molar mass / g mol⁻¹	44	46
Intermolecular attraction	London dispersion forces	London dispersion forces and dipole–dipole attraction
Boiling point / K	229	249

Note that dipole–dipole attractions can occur between any combination of polar molecules, such as between PCl_3 and $CHCl_3$, and generally lead to the solubility of polar solutes in polar solvents.

For two substances of similar molecular mass, the more polar substance will have the higher boiling point.

Johannes van der Waals (1837–1923) from the Netherlands established himself as an eminent physicist in his first publication, his PhD thesis. His study of the continuity of the gas and liquid state from which he put forward his 'equation of state' led James Clerk Maxwell to comment, 'there can be no doubt that the name of van der Waals will soon be among the foremost in molecular science'. van der Waals did indeed fulfil this early promise, being awarded the Nobel Prize in Physics in 1910.

Dipole–induced dipole attraction

When a mixture contains both polar and non-polar molecules, the permanent dipole of a polar molecule can cause a temporary separation of charge on a non-polar molecule. The resulting force is known as a **dipole–induced dipole attraction** (Figure 10).

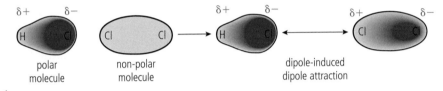

polar molecule   non-polar molecule   dipole-induced dipole attraction

S2.2 Figure 10 Dipole-induced dipole forces between molecules of HCl and Cl_2.

This intermolecular force would act in addition to London dispersion forces between non-polar molecules and dipole-dipole forces between polar molecules.

The term **van der Waals' force** is used to include London dispersion forces, dipole-induced dipole and dipole–dipole attractions. In other words, van der Waals' forces refer to all forces between molecules that do not involve electrostatic attractions between ions or bond formation. We can summarize this as follows:

$$
\left. \begin{array}{ll}
\text{London dispersion force} & Cl_2\text{----}Cl_2 \\
\text{dipole–dipole attraction} & HCl\text{----}HCl \\
\text{dipole–induced dipole} & HCl\text{----}Cl_2
\end{array} \right\} \text{van der Waals, forces}
$$

In some cases, van der Waals' forces may occur within a molecule (intramolecularly) if the different groups can position themselves appropriately. This is an important feature in explaining the structure and behaviour of proteins.

Hydrogen bonding

Hydrogen bonding only occurs between molecules in which hydrogen is bonded directly to fluorine, nitrogen, or oxygen.

When a molecule contains hydrogen covalently bonded to a very electronegative atom (fluorine, nitrogen, or oxygen), these molecules are attracted to each other by a particularly strong type of intermolecular force called a **hydrogen bond**. The hydrogen bond is in essence a particular case of dipole–dipole attraction. The large electronegativity difference between hydrogen and the bonded fluorine, oxygen, or nitrogen results in the electron pair being pulled away from the hydrogen. Given its small size and the fact that it has no other electrons to shield the nucleus, the hydrogen now exerts a strong attractive force on a lone pair in the electronegative atom of a neighbouring molecule (Figure 11). This is the hydrogen bond.

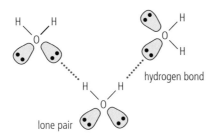

S2.2 Figure 11 Hydrogen bonding between water molecules.

There is ongoing debate about the nature of the hydrogen bond and IUPAC defines six criteria that should be used as evidence for its occurrence. Several of these criteria are dependent on technological advancements, such as the use of infrared and nuclear magnetic resonance spectrometers to analyse the interactions occurring in a substance.

Nature of Science – How can advances in technology lead to changes in scientific definitions, e.g. the updated IUPAC definition of the hydrogen bond?

Hydrogen bonds are the strongest form of intermolecular attraction. Consequently, they cause the boiling points of substances that contain them to be significantly higher than would be predicted from their molar mass. We can see this in Figure 12, in which boiling points of the hydrides from groups 14 to 17 are compared down the periodic table.

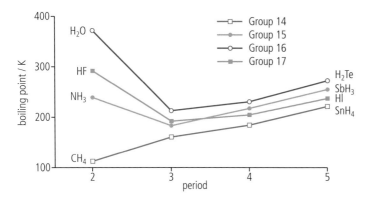

▲
S2.2 Figure 12 Periodic trends in the boiling points of the hydrides of groups 14–17.

In all four groups there is an observable trend that boiling point increases down the group as molar mass increases. The anomalies are NH_3, HF, and H_2O, which all have significantly higher boiling points than would be expected from their molar mass. This can only be explained by the presence of hydrogen bonding in these molecules. If it was not for the fact that it is hydrogen bonded, H_2O would be a gas not a liquid at room temperature.

Likewise, when we compare the boiling points of some organic molecules that have similar or equal values of molar mass, we find a higher value where hydrogen bonding occurs between the molecules. For example, the following table compares two different forms of C_2H_6O, known as isomers, both with a molar mass of 46.

CH_3–O–CH_3	CH_3CH_2–O–H
methoxymethane	ethanol
M = 46 g mol^{-1}	M = 46 g mol^{-1}
does not form hydrogen bonds	forms hydrogen bonds
boiling point = −23 °C	boiling point = +79 °C

Water makes a particularly interesting case for the study of hydrogen bonding. Here, because of the two hydrogen atoms in each molecule and the two lone pairs on the oxygen atom, each H_2O can form up to *four* hydrogen bonds with neighbouring molecules. Liquid water contains fewer than this number, but in the solid form, ice, each H_2O is maximally hydrogen bonded in this way. The result is a tetrahedral arrangement that holds the molecules a fixed distance apart, forming a fairly open structure, which is actually *less* dense than the liquid (Figure 13). This is a remarkable fact – in nearly all other substances the solid form with closer packed particles is *more* dense than its liquid. The fact that ice floats on water is evidence of the power of hydrogen bonds in holding the molecules together in ice. This density change means that water expands on freezing, which can lead to all kinds of problems in cold climates such as burst pipes and cracked machinery. The same force of expansion is at work in the Earth's crust, fragmenting and splitting rocks, ultimately forming sand and soil particles. The humble hydrogen bond is responsible for massive geological changes!

Icebergs float on water. Hydrogen bonds in water create a very open structure in ice, which is less dense than liquid water.

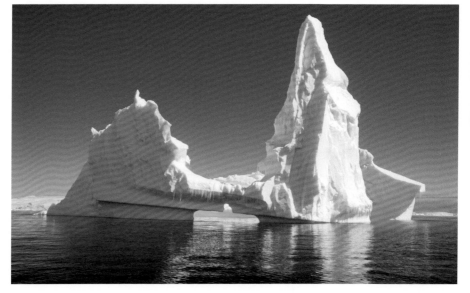

In these examples, we have looked at hydrogen bonds as an intermolecular force, that is, *between* molecules, but like van der Waals' forces, they can also occur *within* large molecules where they play a key role in determining properties. Another fascinating example is DNA (deoxyribonucleic acid) which, as the chemical responsible for storing the genetic information in cells, is able to replicate itself exactly, a feat only possible because of its use of hydrogen bonding.

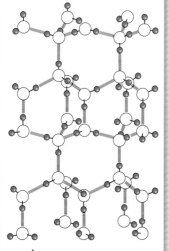

S2.2 Figure 13 The arrangement of water molecules in ice. Each molecule is held by hydrogen bonds to four other molecules in a tetrahedral arrangement.

Computer artwork of a DNA molecule replicating. DNA is composed of two strands, held together by hydrogen bonds and twisted into a double helix. Before replication, the strands separate from each other by breaking the hydrogen bonds and each strand then acts as a template for the synthesis of a new molecule of DNA.

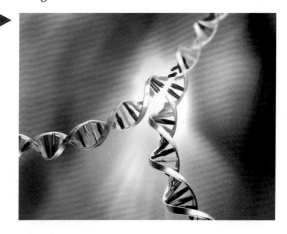

We can now summarize the intermolecular forces we have discussed in this section.

	Non-polar molecules	Polar molecules	Molecules which contain H–O, H–N, or H–F
London dispersion forces	✓	✓	✓
Dipole-induced dipole	requires a mixture of polar and non-polar molecules		
Dipole–dipole attractions		✓	✓
Hydrogen bonding			✓

van der Waals' forces (for London dispersion forces, Dipole-induced dipole, Dipole–dipole attractions)

increasing strength of inter-molecular force

It is important to note that although hydrogen bonds are strong in relation to other types of intermolecular force, they are significantly weaker (about 15–20 times) than covalent bonds and ionic bonds. So although intermolecular forces and intramolecular forces are both categories of electrostatic attraction, there is a significant difference in their relative strengths.

The physical properties of covalent compounds are largely a result of their intermolecular forces

If we know the type of intermolecular force in a substance, we are in a good position to predict its physical properties. Experimental data on melting and boiling points, solubility in polar and non-polar solvents, and electrical conductivity can be used to verify our predictions.

Melting and boiling points

Changing state by melting or boiling both involve separating particles by overcoming the forces between them. So the stronger the interparticle forces, the more energy will be required to do this, and the higher the substance's melting and boiling points will be.

Covalent substances generally have lower melting and boiling points than ionic compounds. The forces to be overcome to separate the molecules are the relatively weak intermolecular forces, which are significantly easier to break than the electrostatic attractions in the ionic lattice. This is why many covalent substances are liquids or gases at room temperature.

As discussed on page 185, the strength of the intermolecular forces increases with increasing molecular size and with an increase in the extent of polarity within the molecule, and so these features help us to predict relative melting and boiling points.

Nature of Science, Structure 1.1, 2.1, 2.3 – How do the terms 'bonds' and 'forces' compare?

Tool 1, Inquiry 2 – What experimental data demonstrate the physical properties of covalent substances?

Make sure you realize that when a covalent substance melts or boils, it is the attractive forces *between* the molecules that are overcome, not the bonds *within* the molecules. When you boil a kettle of water, you do not fill the kitchen with separated oxygen and hydrogen atoms.

Worked example

Put the following molecules in order of increasing boiling point and explain your reasoning:

CH₃CHO   CH₃CH₂OH   CH₃CH₂CH₃

Solution

1. First check the value of M for each molecule:

    CH₃CHO = 44

    CH₃CH₂OH = 46

    CH₃CH₂CH₃ = 44

    These molar masses are nearly identical so we would expect very similar strength London dispersion forces.

2. Now consider the structure of each molecule and predict additional intermolecular forces:

CH₃CHO	**CH₃CH₂OH**	**CH₃CH₂CH₃**
polar molecule ⇒ dipole–dipole forces	O–H bond ⇒ hydrogen bonding	non-polar molecule ⇒ London dispersion forces only

    So the order, starting with the lowest boiling point, will be:

    CH₃CH₂CH₃ < CH₃CHO < CH₃CH₂OH

An understanding of functional groups in organic molecules is extremely helpful to identify the types of intermolecular forces present in these substances. The hydroxyl functional group (R–OH), for example, will lead to the formation of hydrogen bonding in all alcohols. We should recognize, however, that other factors such as the size and orientation of the molecules can also influence the nature of these attractions.

By contrast, macromolecular or giant covalent structures have high melting and boiling points, often at least as high as those of ionic compounds. This is because covalent bonds must be broken in these compounds for the change of state to occur. Diamond, for example, remains solid until about 4000 °C.

Volatility

Volatility is a term used to describe the tendency of a substance to vaporize. As we have seen in the explanation for boiling points, a substance with stronger intermolecular forces will have a lower tendency to vaporize and vice versa. Propan-1-ol, C_3H_7OH, for example, will be considerably less volatile than butane, C_4H_{10}, given the additional hydrogen bonding.

Structure 3.2 – To what extent does a functional group determine the nature of the intermolecular forces?

Solubility

Non-polar substances are generally able to dissolve in non-polar solvents by the formation of London dispersion forces between solute and solvent. For example, the halogens, all of which are non-polar molecules (e.g. Br_2), are readily soluble in the non-polar solvent, paraffin oil. It is another example of *like dissolves like*, which is a useful guiding principle here.

Likewise, polar covalent compounds are generally soluble in water, a highly polar solvent. Here the solute and solvent interact through dipole interactions and hydrogen bonding. Common examples include the aqueous solubility of HCl, glucose ($C_6H_{12}O_6$), and ethanol (C_2H_5OH) (see Figure 14). Biological systems are mostly based on polar covalent molecules in aqueous solution.

The solubility of polar compounds is, however, reduced in larger molecules where the polar bond is only a small part of the total structure. The non-polar parts of the molecule, unable to associate with water, reduce its solubility. For example, while ethanol (C_2H_5OH) is readily soluble in water, the larger alcohol heptanol ($C_7H_{15}OH$) is not. The difference is the size of the non-polar hydrocarbon group.

Predictably, this inability of non-polar groups to associate with water means that non-polar covalent substances do not dissolve well in water. Nitrogen, N_2, for example, has very low solubility in water at normal pressure, as do hydrocarbons such as candle wax and large covalent molecules such as chlorophyll. If you think about the substances that mark clothes and are not easily removed with water – grease, red wine, grass stains, and gravy are good examples – they are all relatively large insoluble non-polar molecules.

Likewise, polar substances have low solubility in non-polar solvents as they will remain held to each other by their dipole–dipole attractions and cannot interact well with the solvent. You can observe this by putting some sugar into cooking oil, and noticing that it does not dissolve.

Giant molecular substances are generally insoluble in all solvents, as too much energy would be required to break the strong covalent bonds in their structure. Beaches made primarily of sand, SiO_2, do not dissolve when the tide comes in and a diamond ring can safely be immersed in water.

Electrical conductivity

Covalent compounds do not contain ions, and so are not able to conduct electricity in the solid or liquid state. Some polar covalent molecules, however, in conditions where they can ionize, will conduct electricity. For example, HCl dissolved in water (hydrochloric acid) is an electrical conductor.

S2.2 Figure 14 Water and ethanol form hydrogen bonds and mix readily, as we know from the homogeneous nature of alcoholic drinks.

The moss plant *Philonotis fontana* has a waxy substance on its surface which repels water. The wax is a large non-polar covalent molecule that is unable to form hydrogen bonds with water.

Dry cleaning is a process where clothes are washed without water. Instead, a less polar organic liquid is used which may be a better solvent for stains caused by large non-polar molecules such as grass and grease.

Non-polar covalent substances dissolve best in non-polar solvents, and polar substances dissolve best in polar solvents.

As discussed on page 179, the giant covalent molecules graphite and graphene are electrical conductors. In addition, fullerene and silicon have semi-conductivity properties. Diamond is a non-conductor of electricity.

Summary of physical properties

We can now summarize the physical properties of ionic and covalent compounds, though it should be recognized that the table below gives only general trends and does not apply universally.

	Ionic compounds	Polar covalent compounds	Non-polar covalent compounds	Giant covalent
Volatility	low	higher	highest	low
Solubility in polar solvent, e.g. water	soluble	solubility increases as polarity increases	non-soluble	non-soluble
Solubility in non-polar solvent, e.g. hexane	non-soluble	solubility increases as polarity decreases	soluble	non-soluble
Electrical conductivity	conduct when molten (l) or dissolved in water (aq)	non-conductors	non-conductors	non-conductors except graphite, graphene and semiconductivity of Si and fullerene

Challenge yourself

6. If you were given two solutions and told one contained a polar substance and the other a non-polar substance, suggest experiments that might enable you to identify which is which.

Exercise

Q28. The physical properties of five solids labelled A, B, C, D and E are summarized below. The substances are: an ionic compound, a non-polar molecular solid, a metal, a polar molecular solid, and a covalent network substance. Classify each correctly.

Sample	Solubility in water	Conductivity of solution	Conductivity of solid	Relative melting point
A	insoluble	–	yes	third to melt
B	insoluble	–	no	highest
C	soluble	no	no	second to melt
D	insoluble	–	no	lowest
E	soluble	yes	no	fourth to melt

Q29. Which substance is the most soluble in water?

    A CH_3OH **B** CH_4 **C** C_2H_6 **D** C_2H_5OH

Q30. State the intermolecular forces that exist between molecules of each of the following:

    (a) dry ice, $CO_2(s)$ **(b)** $NH_3(l)$ **(c)** $N_2(l)$

    (d) $CH_3OCH_3(g)$ **(e)** a mixture of $CH_3OCH_3(l)$ and $C_6H_{14}(l)$

Q31. Which of each pair has the lower boiling point?

    (a) C_2H_6 and C_3H_8 **(b)** H_2O and H_2S

    (c) Cl_2 and Br_2 **(d)** HF and HCl

Structure 2.2.10 – Intermolecular forces and chromatography

Structure 2.2.10 – Chromatography is a technique used to separate the components of a mixture based on their relative attractions involving intermolecular forces to mobile and stationary phases.

Explain, calculate and interpret the retardation factor values, R_f.

The use of locating agents is not required.	Tool 1 – How can a mixture be separated using paper chromatography or thin layer chromatography (TLC)?
The operational details of a gas chromatograph or high-performance liquid chromatograph will not be assessed.	

Chromatography is a technique used to separate the components of a mixture

Chromatography is a useful technique for separating and identifying the components of a mixture. The basic principle is that the components have different affinities for two phases, a **stationary phase** and a **mobile phase**, and so are separated as the mobile phase moves through the stationary phase. The levels of solubility of each component in each of the phases is dependent on the intermolecular forces present.

Paper chromatography

Paper chromatography involves the use of a piece of chromatographic paper and a solvent to separate a mixture. It is used mainly for qualitative analysis. The paper, which contains about 10% water, is the stationary phase. Water is adsorbed by forming hydrogen bonds with the –OH groups in the cellulose of the paper. The solvent is the mobile phase as it rises up the paper by capillary action. As it does so it dissolves the components of the mixture to different extents, so carrying them at different rates. We often have to trial a number of different solvents until we identify one which will separate components of the mixture to a suitable extent.

As shown in Figure 15, the procedure is simple to run. A small sample of the mixture is spotted near the bottom of the chromatographic paper. This position, known as the **origin**, needs to be clearly marked (in pencil so as not to dissolve in the solvent and interfere with

Racing compounds on TLC. Full details of how to carry out this experiment with a worksheet are available in the eBook.

the experiment). The paper is then suspended in a chromatographic tank containing a small volume of solvent, ensuring that the spot is above the level of the solvent.

S2.2 Figure 15 Showing **(a)** the apparatus used to separate a mixture by paper chromatography and **(b)** the resulting separation of components in the mixture after the solvent rises up the chromatography paper.

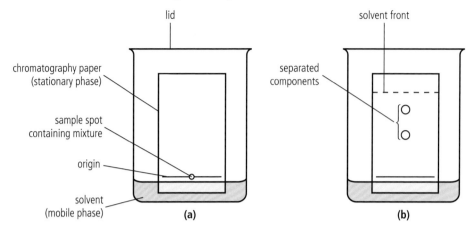

(a)   (b)

As the solvent rises up the paper, it will pass over the sample spot. The components in the spot will distribute themselves between the two phases and so move up the paper at different speeds. As a result, they become spread out according to their different solubilities in the mobile and stationary phases. When the solvent reaches almost to the top of the paper, its final position is marked and is known as the *solvent front*. Each component can be distinguished as a separate isolated spot up the length of the paper. The final result is known as a *chromatogram*.

The position of each component on the chromatogram can be represented as an R_f value (retardation factor), which is calculated as shown in Figure 16.

$$R_f = \frac{\text{distance moved by component}}{\text{distance moved by solvent}}$$

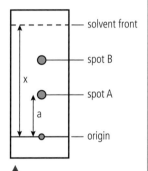

S2.2 Figure 16 Calculation of R_f values in chromatography.

So, for component A, $R_f = \frac{a}{x}$

Specific substances have characteristic R_f values when measured under the same conditions, so can be identified by comparing the values obtained with data tables.

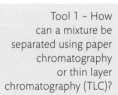

When measuring the distances to calculate R_f values, be careful to measure from the origin and not from the bottom of the paper. Similarly, the solvent moves as far as the solvent front, not to the top of the paper.

In chromatography, the R_f value can be used to identify the components of a mixture.

$$R_f = \frac{\text{distance moved by component}}{\text{distance moved by solvent}}$$

Thin layer chromatography

Another common chromatographic technique is **thin layer chromatography**. It works on the same principles as paper chromatography but the stationary phase is a uniform layer of silica (silicon dioxide) or alumina (aluminium oxide) coated onto a piece of glass, metal or plastic. The surfaces of both silica and alumina contain hydroxyl groups so are considered polar and able to form hydrogen bonding with components in a sample mixture.

Tool 1 – How can a mixture be separated using paper chromatography or thin layer chromatography (TLC)?

Worked example

A sample is known to contain two different substances. After carrying out paper chromatography using a solvent of butan-1-ol, propanone and water, the following chromatogram was obtained.

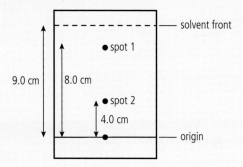

Calculate the R_f values for both spots.

Solution

Use this formula for both spots: $R_f = \dfrac{\text{distance moved by component}}{\text{distance moved by solvent}}$

Spot 1 $R_f = \dfrac{8.0}{9.0} = 0.89$

Spot 2 $R_f = \dfrac{4.0}{9.0} = 0.44$

Exercise

Q32. Which two measurements are required to calculate the retardation factor value for a component spot in chromatography?

	Measurement 1	Measurement 2
A	height of origin	height of spot
B	height of solvent front	height of chromatography paper
C	height of solvent front	height of spot
D	height of origin	height of solvent front

Q33. Calculate the R_f values for the two spots on the following chromatogram.

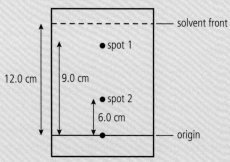

Chromatography is routinely used for the identification of amino acids found in proteins. As amino acids are colourless in solution they are usually treated with a locating reagent such as ninhydrin at the end of the process. Most amino acids will now appear purple and can be distinguished as separate isolated spots up the length of the paper. The R_f of these spots can be compared with known values calculated using the same solvent.

Q34. A mixture of two organic compounds was separated by TLC using a non-polar solvent.

Compound	Distance travelled / mm
X	24
Y	72
Solvent	80

(a) Calculate the R_f values of X and Y.

(b) Outline why compound Y has travelled the greater distance.

Guiding Question revisited

What determines the covalent nature and properties of a substance?

In this chapter we have used the covalent model of bonding to show:

- A covalent bond is formed by the electrostatic attraction between a shared pair of electrons and the positively charged nuclei. Atoms can share one, two or three pairs of electrons to form a single, double or triple bond respectively.

- A covalent bond in which one atom provides both shared electrons is called a coordination bond.

- The octet rule is a useful tool for predicting the arrangement of atoms in covalent molecules and network structures but has many exceptions. An atom with less than four valence pairs of electrons has an incomplete octet.

- Lewis formulas can be used to represent all of the valence electrons in a covalently bonded species.

- The Valence Shell Electron Pair Repulsion (VSEPR) model can be used to accurately predict the shape of molecules by considering the repulsions between electron domains. These can be bonding domains, containing shared pairs of electrons, or non-bonding domains (lone pairs of electrons).

- The electron geometry describes the arrangement of electron domains around a central atom:
  - 2 electron domains → linear
  - 3 electron domains → triangular planar
  - 4 electron domains → tetrahedral

- The molecular geometry describes the position of bonding domains around a central atom. We can predict the bond angles in a molecule knowing its geometry and considering any addition repulsion caused by non-bonding pairs and multiple bonds.

- Bond polarity results from the difference in electronegativities of the bonded atoms. This causes an unequal distribution of electron density, a bond dipole, which can be represented with partial charges ($\delta+$ and $\delta-$) and/or a vector.

- Molecular polarity depends on both bond polarity and molecular geometry. If bond dipoles cancel out on a molecule then it will be non-polar. If there is a net dipole across a molecule, then it will be polar.

- Carbon and silicon are able to form network structures with distinct chemical and physical properties. Examples include silicon, silicon dioxide and the allotropes of carbon.

- Intermolecular forces are determined by the size and polarity of a molecule. In increasing order of strength, they include London dispersion forces, dipole-induced dipole, dipole–dipole and hydrogen bonding. Intermolecular forces can influence physical properties such as melting and boiling point, volatility, electrical conductivity and solubility.

- Paper chromatography and thin layer chromatography are techniques used to separate the components of a mixture based on their relative attractions to the mobile and stationary phases. For a given solvent, the retardation factor value

for a component can be calculated using

$$R_f = \frac{\text{distance moved by component}}{\text{distance moved by solvent}}$$

This value is compared to known values in order to identify the compound.

Practice questions

1. Which bonds are arranged in order of increasing polarity?

   A   H–F < H–Cl < H–Br < H–I   B   H–I < H–Br < H–F < H–Cl

   C   H–I < H–Br < H–Cl < H–F   D   H–Br < H–I < H–Cl < H–F

2. Which row correctly describes the bonding type and melting point of carbon and carbon dioxide?

	Carbon		Carbon dioxide	
A	covalent bonding	high melting point	covalent bonding	low melting point
B	ionic bonding	low melting point	ionic bonding	high melting point
C	ionic bonding	high melting point	ionic bonding	low melting point
D	covalent bonding	low melting point	covalent bonding	high melting point

3. What is the correct order of increasing boiling points?

   A CH_3CH_3 < CH_3CH_2Cl < CH_3CH_2Br < CH_3CH_2I

   B CH_3CH_2Cl < CH_3CH_2Br < CH_3CH_3 < CH_3CH_2I

   C CH_3CH_2I < CH_3CH_2Br < CH_3CH_2Cl < CH_3CH_3

   D CH_3CH_2Br < CH_3CH_2Cl < CH_3CH_2I < CH_3CH_3

4. Which compound forms hydrogen bonds in the liquid state?

   A C_2H_5OH B $CHCl_3$ C CH_3CHO D $(CH_3CH_2)_3N$

5. Which molecule has a non-bonding (lone) pair of electrons around the central atom?

   A BF_3 B SO_2 C $BeCl_2$ D SiF_4

6. Which species contain a coordination bond?

   I.   HCHO   II.   CO   III. H_3O^+

   A   I and II   B   I and III   C   II and III   D   I, II and III

7. Which molecule has an a triangular pyramidal shape?

   A H_2O **B** CH_4 **C** AlF_3 **D** SF_6

8. Which substance can form intermolecular hydrogen bonds in the liquid state?

   A CH_3OCH_3 **B** CH_3CH_2OH **C** CH_3CHO **D** $CH_3CH_2CH_3$

9. Which calculation would be required to calculate the retardation factor value, R_f, for spot 1 in this chromatogram?

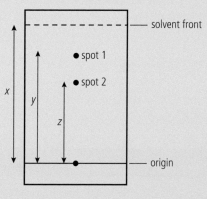

   A $\dfrac{x}{y}$ **B** $\dfrac{y}{x}$ **C** $\dfrac{y}{z}$ **D** $\dfrac{z}{x}$

10. Based on the types of intermolecular force present, explain why butan-1-ol has a higher boiling point than butanal. (2)

    (Total 2 marks)

11. For each of the species BH_3 and PCl_3:

    (a) deduce the Lewis formula (2)

    (b) predict the shape and bond angle (4)

    (c) predict and explain the molecular polarity. (2)

    (Total 8 marks)

12. Carbon and silicon belong to the same group of the periodic table.

    (a) Describe and compare three features of the structure and bonding in the three allotropes of carbon: diamond, graphite and C_{60} fullerene. (6)

    (b) Both silicon and carbon form oxides.

    (i) Describe the structure and bonding in SiO_2. (2)

    (ii) Explain why silicon dioxide is a solid and carbon dioxide is a gas at room temperature. (2)

    (c) Describe the bonding within the carbon monoxide molecule. (2)

    (Total 12 marks)

13. Methoxymethane, CH_3OCH_3, and ethanol, C_2H_5OH, have the same relative molecular mass. Explain why methoxymethane has a much lower boiling point than ethanol. (3)

    (Total 3 marks)

14. Draw the Lewis formulas, state the shape, and predict the bond angles for the following species.

(a) NH_3 (3)

(b) NH_2^- (3)

(c) NH_4^+ (3)

(Total 9 marks)

15. (a) State the electron geometry and the molecular geometry, and predict the bond angles for the following species.

(i) Cl_2O (3)

(ii) NO_2^+ (3)

(b) Explain, using diagrams, why NO_2 is a polar molecule but CO_2 is a non-polar molecule. (3)

(c) Describe the structure and bonding in silicon dioxide. (2)

(Total 11 marks)

16. Ethyne, C_2H_2, and ethene, C_2H_4, are both examples of hydrocarbons.

(a) Draw the Lewis formulas of C_2H_2 and C_2H_4. (2)

(b) Compare, giving a reason, the length of the bond between the carbon atoms in ethyne and ethene. (1)

(c) Identify the type of interaction that must be overcome when liquid ethyne vaporizes. (1)

(Total 4 marks)

17. Paper chromatography is a simple method used to separate and identify the components in a mixture.

(a) State the meaning of the term 'retardation factor value R_f' for a component spot on a chromatogram. (1)

(b) Paper chromatography was used to analyze a sample containing two substances. The following data was collected from the chromatogram.

Compound	Distance travelled / mm
A	32
B	75
solvent	96

Calculate the R_f value of compound A. (1)

(c) Outline how the R_f value of a spot can be used to identify a compound. (1)

(d) Explain why the R_f value for the same component can be very different if different solvents are used for the mobile phase. (2)

(e) State one other type of chromatography that can be used to separate the components of a mixture.

(1)

(Total 6 marks)

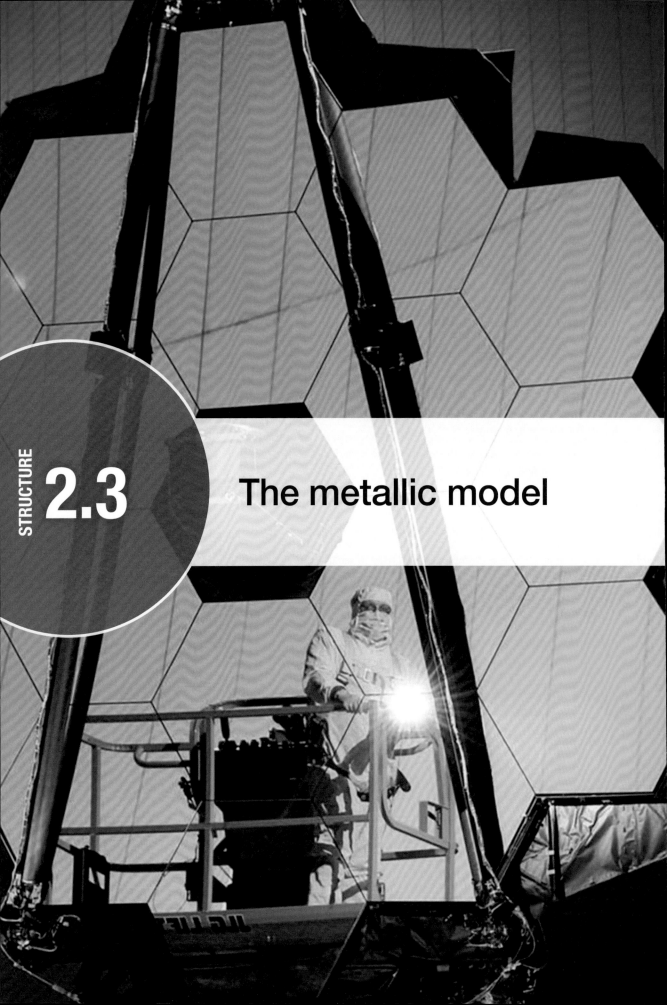

The metallic model

Technicians inspect 18 gold-plated beryllium hexagons on the $10 billion James Webb Space Telescope. These hexagons will act as mirrors to help collect infrared radiation from the furthest reaches of space as the telescope orbits the Sun at a distance of 1.5 million kilometres from Earth. A range of other metals are used in the structural and electrical components of this telescope giving a sense of their value in the development of contemporary technologies.

Nature of Science

As the scope for the application of metals continues to grow alongside the development of technology, scientists have a responsibility to consider the impact of their use. The implications of metal extraction and processing raise ethical questions that must be considered in social, political, economic and environmental domains.

Guiding Question

What determines the metallic nature and properties of an element?

The names of historical periods such as the Bronze Age and the Iron Age hint towards the significant role that metals have played in the transformation of human societies. Gold, for example, has seen its uses develop from art, jewellery and trade in ancient times, to applications as wide-ranging as dentistry, electronics and aerospace in contemporary society.

Given their distinct properties in comparison with both ionic and covalent substances, metals require a separate explanatory model of bonding. This model describes the structure of a metallic element as a lattice of cations surrounded by delocalized electrons and explains the characteristic physical and chemical properties observed in experimentation. This chapter will explore some of the applications of metals that stem from their characteristic properties and consider explanations for trends in their behaviour by considering the charge and radius of metal cations.

Structure 2.3.1 and 2.3.2 – The metallic bond

Structure 2.3.1 – A metallic bond is the electrostatic attraction between a lattice of cations and delocalized electrons.

Explain the electrical conductivity, thermal conductivity and malleability of metals.

Relate characteristic properties of metals to their uses.	Tool 1, Inquiry 2, Structure 3.1 – What experimental data demonstrate the physical properties of metals, and trends in these properties in the periodic table?
	Reactivity 3.2 – What trends in reactivity of metals can be predicted from the periodic table?

Metallic bonding occurs between a lattice of cations and delocalized electrons

Metals are found on the left side of the periodic table and have a small number of electrons in their outer shell. These are atoms with low ionization energies, and so typically react with other elements by losing their valence electrons and forming positive ions. This loss of control over their outer shell electrons is what we are referring to when describing the **metallic character** of these elements.

	1	2	3	4	5	6	7	8	9	10	11	12	13	14	15	16	17	18
1	1 H 1.01						Atomic number Element Relative atomic mass											2 He 4.00
2	3 Li 6.94	4 Be 9.01											5 B 10.81	6 C 12.01	7 N 14.01	8 O 16.00	9 F 19.00	10 Ne 20.18
3	11 Na 22.99	12 Mg 24.31											13 Al 26.98	14 Si 28.09	15 P 30.97	16 S 32.07	17 Cl 35.45	18 Ar 39.95
4	19 K 39.10	20 Ca 40.08	21 Sc 44.96	22 Ti 47.87	23 V 50.94	24 Cr 52.00	25 Mn 54.94	26 Fe 55.85	27 Co 58.93	28 Ni 58.69	29 Cu 63.55	30 Zn 65.38	31 Ga 69.72	32 Ge 72.63	33 As 74.92	34 Se 78.96	35 Br 79.90	36 Kr 83.90
5	37 Rb 85.47	38 Sr 87.62	39 Y 88.91	40 Zr 91.22	41 Nb 92.91	42 Mo 95.96	43 Tc (98)	44 Ru 101.07	45 Rh 102.91	46 Pd 106.42	47 Ag 107.87	48 Cd 112.41	49 In 114.82	50 Sn 118.71	51 Sb 121.76	52 Te 127.60	53 I 126.90	54 Xe 131.29
6	55 Cs 132.91	56 Ba 137.33	57 † La 138.91	72 Hf 178.49	73 Ta 180.95	74 W 183.84	75 Re 186.21	76 Os 190.23	77 Ir 192.22	78 Pt 195.08	79 Au 196.97	80 Hg 200.59	81 Tl 204.38	82 Pb 207.20	83 Bi 208.98	84 Po (209)	85 At (210)	86 Rn (222)
7	87 Fr (223)	88 Ra (226)	89 ‡ Ac (227)	104 Rf (267)	105 Db (268)	106 Sg (269)	107 Bh (270)	108 Hs (269)	109 Mt (278)	110 Ds (281)	111 Rg (281)	112 Cn (285)	113 Uut (286)	114 Uuq (289)	115 Uup (288)	116 Uuh (293)	117 Uus (294)	118 Uuo (294)

†	58 Ce 140.12	59 Pr 140.91	60 Nd 144.24	61 Pm (145)	62 Sm 150.36	63 Eu 151.96	64 Gd 157.25	65 Tb 158.93	66 Dy 162.50	67 Ho 164.93	68 Er 167.26	69 Tm 168.93	70 Yb 173.05	71 Lu 174.97
‡	90 Th 232.04	91 Pa 231.04	92 U 238.03	93 Np (237)	94 Pu (244)	95 Am (243)	96 Cm (247)	97 Bk (247)	98 Cf (251)	99 Es (252)	100 Fm (257)	101 Md (258)	102 No (259)	103 Lr (262)

The periodic table. Metals are shown in yellow, non-metals in red and metalloids in blue.

In the elemental state, when there is no other element present to accept the electrons and form an ionic compound, the outer electrons are held only loosely by the metal atom's nucleus and so tend to 'wander off' or, more correctly, become **delocalized**. As shown in Structure 2.2, delocalized electrons are not fixed in one position. This means that in metals the valence electrons are no longer associated closely with any one metal nucleus but instead can spread themselves through the metal structure. The metal atoms without these electrons have become **cations** (positively charged ions) and form a regular **lattice structure** through which these electrons can move freely.

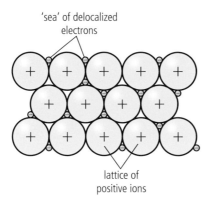

'sea' of delocalized electrons

lattice of positive ions

◄ The metallic model of bonding. It is often described as a lattice of cations surrounded by a sea of electrons.

You can think of it like a close neighbourhood of families where the children do not belong specifically to any one set of parents but are free to wander between the homes. This arrangement causes a close association between the families. Likewise in metals, there is a force of electrostatic attraction between the lattice of cations and the delocalized electrons, and this is known as **metallic bonding**.

🔒 **A metallic bond is the electrostatic attraction between a lattice of cations and delocalized electrons.**

The uses of metals are related to their characteristic properties

Experimental data on melting and boiling points, as well as electrical and thermal conductivity, show that metals share a number of characteristic physical properties. These properties and their trends in the periodic table can be explained using the metallic model of bonding. They are summarized in the table on the next page.

🔗 Tool 1, Inquiry 2, Structure 3.1 – What experimental data demonstrate the physical properties of metals, and trends in these properties in the periodic table?

◄ Pile of ancient Roman copper coins. Due to its relative ease of extraction, copper was one of the first metals used to make tools by humans. The Bronze Age was a historical period named after the alloy *bronze*, made by mixing copper and tin.

◄ Copper wire. Copper is malleable, ductile, and a good conductor of heat and electricity.

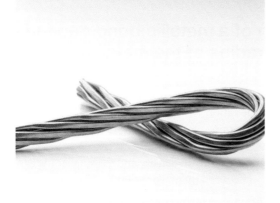

SKILLS

Investigating the properties of metals. Full details of how to carry out this experiment with a worksheet are available in the eBook.

Metallic property	Explanation	Application
good electrical conductivity	delocalized electrons are highly mobile, and so can move through the metal structure in response to an applied voltage	electrical circuits use copper
good thermal conductivity	delocalized electrons and closely packed ions enable efficient transfer of thermal energy	pots and pans used for cooking
malleable, can be shaped under pressure	movement of delocalized electrons is **non-directional** and essentially random through the cation lattice, so the metallic bond remains intact while the conformation changes under applied pressure	moulded into many forms including machinery and structural components of buildings and vehicles
ductile, can be drawn out into threads		electric wires and cables
high melting points	a lot of energy is required to overcome the strong electrostatic forces of attraction in the metallic bond and separate the atoms	high-speed tools and turbine engines
shiny, lustrous appearance	delocalized electrons in metal crystal structure reflect light	ornamental structures and jewellery

Structure 2.4 – What are the features of metallic bonding that make it possible for metals to form alloys?

The non-directional nature of metallic bonding allows us to enhance the properties of a metallic substance by mixing a metal with other metal or non-metal elements in the molten state. The resulting mixture is known as an **alloy**. Steel, for example, is made by the addition of small amounts of carbon to molten iron. Although the carbon atoms sit within the lattice of cations, the cations themselves continue to experience electrostatic attraction to the delocalized electrons in all directions. Steel has a greater tensile strength than pure iron making it a more suitable material for construction. This is discussed in more depth in Structure 2.4.

Nature of Science

Atomic theory enables scientists to explain the distinct properties of metals. This is applied in the development of new materials such as alloys that combine the characteristics of different metals. The ability to explain natural phenomena such as metallic properties through the application of theory is an important feature of science.

The strength of a metallic bond depends on the charge and radius of the metal ions

The strength of a metallic bond depends on the charge and radius of the metal ions.

The strength of the metallic bond is determined by:
- the number of delocalized electrons
- the charge on the cation
- the radius of the cation.

The greater the delocalized electron density and the smaller the cation, the greater the electrostatic attraction between them. For example, if we compare:
- sodium, Na, group 1, electron configuration $1s^2 2s^2 2p^6 3s^1$ and
- magnesium, Mg, group 2, electron configuration $1s^2 2s^2 2p^6 3s^2$

we can deduce that Na will have one delocalized electron per atom and will form the Na^+ ion, whereas Mg will have two delocalized electrons per atom and form the Mg^{2+} ion. In addition, the radius of Mg^{2+} is slightly smaller than that of Na^+. These factors all indicate that Mg should have stronger metallic bonding than Na, which can be confirmed by a comparison of their melting points.

	Na	Mg
Melting point / °C	98	650

The strength of metallic bonding tends to decrease down a group as the size of the cation increases, reducing the attraction between the delocalized electrons and the positively charged protons in the nucleus. Again, melting points confirm this.

	Na	K	Rb
Melting point / °C	98	63	39

Given the tendency for metals to lose electrons during chemical reactions, the trends in melting point across a period and down a group can be used to predict their relative reactivity. These can be summarized as follows:

Left to right across a period

Increasing melting point	⟷	Greater attraction between ions and delocalized electrons	⟷	Lower degree of reactivity

Down a group

Decreasing melting point	⟷	Weaker attraction between ions and delocalized electrons	⟷	Higher degree of reactivity

Specific examples of these predictions are explored in Reactivity 3.2.

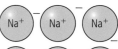

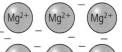
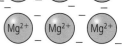

▲
Metallic model of the bonding in sodium and magnesium. Mg has a greater charge on its cations, more delocalized electrons and a smaller ionic radius, leading to stronger metallic bonds and a higher density.

 Reactivity 3.2 – What trends in reactivity of metals can be predicted from the periodic table?

 The extraction of metals from their ores is a major part of industry and development in many countries, but is associated with some complex environmental issues. These include the destruction of landscapes by strip mining, the 'tailings' or waste material that accumulate in spoil tips, and the release of toxic materials that leach out of waste and pollute the environment. An important step to reduce the impact of these effects is to maximize the reuse and recycling of all metal objects, and avoid metal dumping. Despite their apparent abundance, metals are a precious resource, available only at considerable cost.

Challenge yourself

1. The transition elements, found in the d-block of the periodic table, are able to delocalize d-electrons in addition to their valence s-electrons when forming a metallic bond. Can you predict how the melting points, densities and electrical conductivities of the transition metals will compare to those of the group 1 and 2 metals?

Exercise

Q1. Which is the best definition of metallic bonding?

  A the attraction between cations and anions

  B the attraction between cations and delocalized electrons

  C the attraction between nuclei and electron pairs

  D the attraction between nuclei and anions

Q2. Explain, with reference to the features of the metallic model of bonding, the following properties:

(a) electrical conductivity

(b) thermal conductivity

(c) malleability.

Q3. Aluminum is a widely used metal. Which properties make aluminum suitable for the following applications?

(a) baking foil   (b) aircraft bodywork

(c) cooking pans   (d) tent frames

Q4. Explain the difference in the melting points of lithium and beryllium, which melt at 180.5 °C and 1287 °C respectively.

Q5. Use the periodic table to identify the predicted trend in melting points of sodium (Na), potassium (K) and rubidium (Rb).

A Rb > K > Na   **B** K > Na > Rb   **C** Na > K > Rb   **D** Na > Rb > K

Q6. Explain the general trend of decreasing melting point down group 2 metals.

Guiding Question revisited

What determines the metallic nature and properties of an element?

In this chapter we have used the metallic model to show:

- Metallic elements contain delocalized electrons and a lattice of cations. The electrostatic attraction between these electrons and cations is the metallic bond.
- The metallic model helps explain a number of characteristic properties:
  - Electrical conductivity – due to the mobility of delocalized electrons across the metal structure.
  - Thermal conductivity – due to the efficient transfer of heat energy via delocalized electrons and closely packed cations.
  - Malleability – related to the non-directional nature of delocalized electron movement which allows for conformation changes without the breaking of the metallic bond.
- Trends in the properties of metals can be explained by the charge and radius of cations. A stronger metallic bond, caused by ions with smaller radii and greater magnitudes of charge, will have a higher melting point and lower degree of reactivity.

Practice questions

1. Which combination would produce the strongest metallic bond?

	Charge on cations	Ionic radius
A	small	small
B	small	large
C	large	small
D	large	large

2. Which metal has the strongest metallic bonding?

A Na   **B** K

C Mg   **D** Al

3. Which of the following elements is most likely to be metallic?

	Electrical conductivity	
	Molten	**Solid**
A	good	good
B	good	poor
C	poor	good
D	poor	poor

4. Which series shows the correct order of metallic bond strength from strongest to weakest?

A Rb > K > Na > Li   **B** Li > Na> K > Rb

C Rb > Na > K > Li   **D** Li > K > Na > Rb

5. Calcium, Ca, and strontium, Sr, are both group 2 elements that contain metallic bonding.

(a) Describe the metallic model of bonding and explain how it contributes to electrical conductivity. (3)

(b) Explain why calcium has a higher melting point than strontium. (2)

(Total 5 marks)

6. Explain, with reference to the metallic model of bonding, why metals are generally malleable and good thermal conductors. (4)

(Total 4 marks)

From models to materials

◀ False-colour scanning electron micrograph (SEM) of a 'hook-and-loop' material, used as a reversible fastener on clothing and fabric. It is a two-sheet nylon structure with hooks on one surface (right) and loops on the other (left). When the two sheets are pressed together, the hooks attach to the loops, giving a secure grip but one that can be easily undone. Hook-and-loop fasteners are an example of an innovative material that has been designed by application of models of bonding and structure.

Guiding Question

What role do bonding and structure have in the design of materials?

From early history, humans have worked to transform readily available natural resources into more useful materials. Civilizations are sometimes characterized by the technology they have developed to accomplish this. The Bronze Age, for example, marks the time when copper was first produced from smelted ores. The extraction of iron from its ores in the blast furnace is probably one of the most significant developments in the Industrial Revolution of the 18th century. These technological advances, however, were largely based on empirical discovery and usually came without a full understanding of the underlying scientific principles.

As the models we use to explain chemical structure and bonding have advanced, so has our ability to predict the properties of materials. This has enabled us to design structures for specific purposes. Materials that have transformed everyday life, such as superconductors, breathable fabrics and liquid crystals, are only in existence because we have been able to apply advanced bonding theory to their synthesis.

In this chapter, we first develop a more holistic model of bonding from the three discrete models of ionic, covalent and metallic bonding discussed earlier. This gives us a tool, the bonding triangle, which helps to predict the properties of unknown and new substances. We then describe alloys, which are mixtures of metals, as an example of how changing the bonding within a structure can enhance the resulting properties and uses of a material. Finally, we consider the structure of polymers, which include all plastics. These materials have made such a significant impact, both in terms of convenience in our lives and with respect to some of their damaging environmental effects, that it is suggested the current age may become known as the Plastic Age.

As the models we use to explain structure and bonding will continue to evolve, so will our ability to develop materials fit for purpose. This innovation in material science must be accompanied by the application of Green Chemistry principles to control the wider impact of these developments on health and the environment.

Structure 2.4.1 – The bonding triangle

Structure 2.4.1 – Bonding is best described as a continuum between the ionic, covalent and metallic models, and can be represented by a bonding triangle.	
Use bonding models to explain the properties of a material.	
A triangular bonding diagram is provided in the data booklet.	Structure 3.1 – How do the trends in properties of period 3 oxides reflect the trend in their bonding?
	Nature of science, Structure 2.1, 2.2 – What are the limitations of discrete bonding categories?

Chemical bonding is best described as a continuum rather than as discrete bonding types

In Structure 2.1, 2.2 and 2.3 we discussed the three models of bonding – ionic, covalent and metallic – as separate categories. Although there are materials that can be described by one of these models alone, often such a classification is an over-simplification as most materials are an intermediate between these three bonding types, and so show intermediate properties. We have already encountered this concept in Structure 2.2 with the description of polar bonds, which show some ionic character present in covalent bonds.

Therefore, a more accurate picture of bonding is a **continuum** where the different bonding types are present to different degrees. The simplest representation of this is a triangle where each vertex represents one of the three models of bonding. Different materials are then scattered through the triangle, positioned according to the relative amount of each type of bonding they contain.

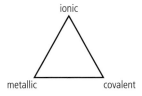

A simple bonding triangle showing the three main types of bonding.

The position of an element or compound in the bonding triangle is determined from its electronegativity values, as these give a measure of the tendency of an atom to gain electrons, and therefore the type of bond it will form. Electronegativity values are given in Section 8 of the data booklet and are shown with the symbol χ. The bonding triangle uses the following axes:

- x-axis: average electronegativity of bonded elements A and B: $\Sigma\Delta = \dfrac{\chi_A + \chi_B}{2}$

- y-axis: electronegativity difference of bonded elements A and B: $\Delta\chi = \chi_A - \chi_B$

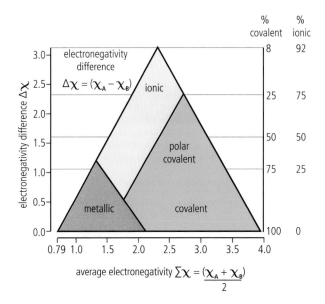

Properties of ionic and covalent compounds. Full details of how to carry out this experiment with a worksheet are available in the eBook.

SKILLS

To what extent do the classification systems we use in the pursuit of knowledge affect the conclusions that we reach?

TOK

The bonding triangle is fully described as the 'van Arkel–Ketelaar triangle of bonding' after the two Dutch chemists credited with its development in the mid-20th century. Anton van Arkel first proposed the triangle and applied it mostly to elements. Jan Ketelaar extended his ideas to more compounds. Others worked in this field and used various parameters to describe the properties of the bonded atoms, but electronegativity values became the established basis for determining positions in the triangle.

The bonding triangle. Calculations of the electronegativity difference and the average electronegativity of the bonded atoms determine the position of a material in the triangle. In turn, this position helps to predict the properties of a material.

Elements have zero electronegativity difference between their atoms, and so are positioned on the *x*-axis, metals on the left (low electronegativity) and non-metals on the right (high electronegativity). Ionic compounds with high electronegativity difference are found at the top centre, while covalent compounds with low electronegativity differences are found in the lower right corner. Polar covalent compounds occupy the intermediate position.

From these positions in the triangle and knowledge of the characteristics of each bonding type, we are able to deduce the expected properties of a material.

Nature of Science

The bonding triangle is a tool that has predictive power for the properties of a substance.

SKILLS

Cement and mortar: investigating the parameters that affect their properties. Full details of how to carry out this experiment with a worksheet are available in the eBook.

Structure 3.1 – How do the trends in properties of period 3 oxides reflect the trend in their bonding?

We can illustrate the importance of electronegativity values in determining the type of bonding by looking at the oxides of the elements in period 3. Oxygen has an electronegativity value of 3.4, and as we move from left to right across period 3, the electronegativity of the elements increases and approaches this value, as shown in Figure 1.

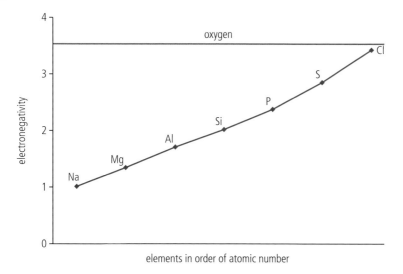

S2.4 Figure 1 Electronegativity values of the elements of period 3 increase across the period, approaching the value of oxygen.

As the *difference* in electronegativity values decreases, so too does the ionic character of the oxides, while the covalent character increases. This trend in bonding explains the trend in physical properties of the period 3 oxides as summarized below.

	Na_2O, MgO and Al_2O_3	SiO_2	P_4O_{10}, SO_2 and Cl_2O
properties	high melting and boiling points, electrical conductors when molten	high melting and boiling points, non-conductor of electricity	low melting and boiling points, non-conductors of electricity
bonding	ionic metallic oxides	giant covalent oxide of the metalloid silicon	covalent non-metallic oxides

This periodic trend could be predicted by considering the positions of the oxides in the bonding triangle, as shown in Figure 2.

213

S2.4 Figure 2 The bonding
triangle showing the positions
of the oxides of period 3
elements. These positions
reflect the trend in their
bonding, and enable us to
predict and explain the trend in
their properties.

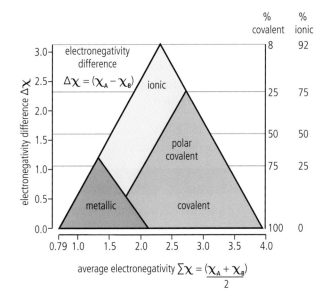

Nature of Science,
Structure 2.1, 2.2 – What
are the limitations
of discrete bonding
categories?

We have described the classification of bonds into ionic, covalent or metallic as an
over-simplification as it can lead to incorrect predictions of physical properties.
For example, aluminium chloride, $AlCl_3$, containing a metal and a non-metal, might be
classified as ionic, which would suggest a high melting point. Yet the melting point of
$AlCl_3$ is significantly lower than that of a similar-sized ionic compound such as NaCl.

- NaCl mp 801 °C
- $AlCl_3$ mp 192 °C

**Electronegativity
values → position
of a material in the
bonding triangle
→ prediction
and explanation of
many properties.**

When we determine the positions of these compounds in the bonding triangle (see
page 212), we find that $AlCl_3$ has significant covalent character, and this explains
its relatively low melting point. The model of a bonding continuum in the bonding
triangle allows us to make more accurate predictions of properties.

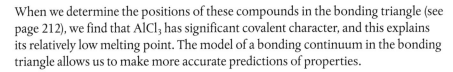

Structure 2.4.2 – Application of the bonding triangle

**Structure 2.4.2 – The position of a compound in the bonding triangle is determined
by the relative contributions of the three bonding types to the overall bond.**

Determine the position of a compound in the bonding triangle from electronegativity data.

Predict the properties of a compound based on its position in the bonding triangle.

To illustrate the relationship between bonding type and properties, include example materials of varying percentage bonding character. Only binary compounds need to be considered. Calculations of percentage ionic character are not required. Electronegativity data are given in the data booklet.	Structure 2.1, 2.2, 2.3 – Why do composites such as reinforced concretes, which are made from ionic and covalently bonded components and steel bars, have unique properties?

Electronegativity data are used to determine the position of a compound in the bonding triangle

We can calculate the position of elements and binary compounds (those containing only two elements) in the bonding triangle from electronegativity values. For example, consider caesium, fluorine and the compound they react together to form, caesium fluoride.

Substance	$\chi_{average}$	$\Delta\chi$	Position in triangle
Cs	0.8	$0.8 - 0.8 = 0.0$	bottom left as 100% metallic; Cs has the lowest absolute electronegativity
F_2	4.0	$4.0 - 4.0 = 0.0$	bottom right corner as 100% molecular covalent
CsF	$\dfrac{4.0 + 0.8}{2} = 2.4$	$4.0 - 0.8 = 3.2$	top of the triangle as 100% ionic compound; made up of the most electropositive metal and most electronegative non-metal

The position of an element or compound in the bonding triangle is determined by the magnitude and difference of the electronegativities of the constituent elements.

Worked example

Locate the position of the following substances in the bonding triangle:

(a) diamond   (b) silicon dioxide

(c) bronze (an alloy of copper and tin)

Solution

Substance	$\chi_{average}$ (x-axis)	$\Delta\chi$ (y-axis)
(a) diamond	2.6	0.0
(b) silicon dioxide	$\dfrac{1.9 + 3.4}{2} = 2.65$	$3.4 - 1.9 = 1.5$
(c) Cu/Sn	$\dfrac{1.9 + 2.0}{2} = 1.95$	$2.0 - 1.9 = 0.1$

Plotting these values on the axes of the bonding triangle gives their positions and so their bonding type.

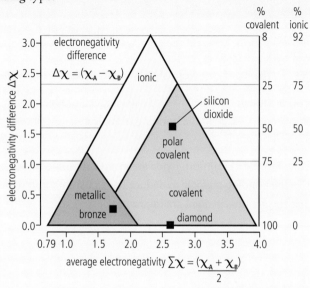

Composite materials have been made by humans for thousands of years. One of the earliest examples is adobe bricks, which are made from a combination of mud and an organic material such as straw or dung. Dried mud alone is brittle, straw is very strong but difficult to hold in place. The combination produces a material which is strong and easy to shape.

In the previous section, we considered the bonding in $AlCl_3$ and $NaCl$. We can determine the positions of these compounds in the bonding triangle, and estimate their percentages of bonding type from the right-hand scale of the triangle.

Substance	$\chi_{average}$	$\Delta\chi$	Position in triangle	Bonding
NaCl	$\dfrac{0.9 + 3.2}{2} = 2.05$	$3.2 - 0.9 = 2.3$	upper centre	ionic compound, approximately 75% ionic
$AlCl_3$	$\dfrac{1.6 + 3.2}{2} = 2.4$	$3.2 - 1.6 = 1.6$	centre, close to the ionic and covalent border	polar covalent, approximately 50% covalent

Composite materials, such as fibreglass and concrete, are heterogeneous mixtures of at least two different materials, which are present as separate phases. As these are mixtures, each component retains its individual properties. The result is a material with enhanced properties designed for specific purposes.

▲ Adobe bricks drying in the sun in Peru.

▲ Construction of a stadium. Reinforcing metal bars (rebar) are fitted before pouring concrete. Concrete is strong in compression but weak in tension. The rebar significantly improves the tensile strength and ductility of concrete structures.

Structure 2.1, 2.2, 2.3 – Why do composites such as reinforced concretes, which are made from ionic and covalently bonded components and steel bars, have unique properties?

Exercise

Q1. Use data from Section 9 and Section 17 of the data booklet to classify the bonding in the following materials:

    (a) Cl_2O **(b)** $PbCl_2$ **(c)** Al_2O_3 **(d)** HBr   **(e)** NaBr

Q2. Copper(II) oxide can be added to give glass a green or blue colour. Deduce the position of copper oxide in the bonding triangle and describe the nature of its structure and bonding.

Q3. Silicon dioxide and aluminium oxide are important minerals in the Earth's crust. Use the bonding triangle to deduce their positions and so determine which has the greater covalent character.

Structure 2.4.3 – Alloys

Structure 2.4.3 – Alloys are mixtures of a metal and other metals or non-metals. They have enhanced properties.	
Explain the properties of alloys in terms of non-directional bonding.	
Illustrate with common examples such as bronze, brass and stainless steel. Specific examples of alloys do not have to be learned.	Structure 1.1 – Why are alloys more correctly described as mixtures rather than as compounds?

Alloys are solutions of metals with enhanced properties

Alloys are produced by adding one metal element to another metal (or carbon) in the liquid state, so that the different atoms can mix. As the mixture solidifies, ions of the different metals are scattered through the lattice forming a structure of uniform composition. Alloys contain metallic bonds as the delocalized electrons bind the lattice. The production of alloys is possible because of the non-directional nature of the delocalized electrons and the fact that the lattice can accommodate ions of different sizes.

SKILLS

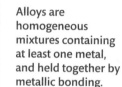

Making polymers. Full details of how to carry out this experiment with a worksheet are available in the eBook.

Alloys are homogeneous mixtures containing at least one metal, and held together by metallic bonding.

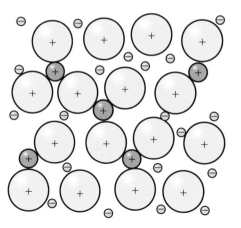

▲ Alloys consist of different metal ions and a sea of delocalized electrons. The smaller cations (blue) are able to fit in the spaces between the larger cations (yellow) in the lattice structure.

Molten metal being poured from a vat at the Magnitogorsk iron and steel works, one of the largest metal-working plants in Russia. The production of alloys depends on mixing metals in the molten state. ▶

Alloys have some properties that are distinct from their component elements due to the different packing of the cations in the lattice. The regular arrangement of atoms in a pure metal is interrupted in the alloy by the presence of different cations, making it more difficult for atoms to slip over each other and so change the shape. The alloy is often stronger, more chemically stable, and more resistant to corrosion than its component elements. For example, steel, which is an alloy of iron with less than 2% carbon and 1% manganese and other trace elements, can be 1000 times stronger than iron.

Changing the regular packing in the lattice gives alloys enhanced properties compared with the pure metal.

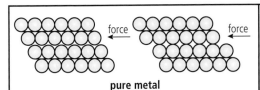 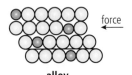

pure metal
The shape of a pure metal can be changed as the atoms can easily slip over each other.

alloy
The presence of atoms of different sizes disrupts the regular structure and prevents the atoms from slipping across each other.

Some examples of common alloys are shown in the table.

You do not need to learn these examples, but you should be able to use examples you choose to explain why the structures of alloys give them enhanced properties compared with their component metals.

Name of alloy	Component metals	Properties and uses
steel	iron with carbon and other elements	high tensile strength but corrodes, used as a structural material
stainless steel	iron with other elements such as nickel and chromium	widely used in domestic and industrial appliances due to strength and corrosion resistance
brass	copper and zinc	variety of plumbing fittings
bronze	copper and tin	coins, medals, tools, heavy gears
pewter	tin, antimony and copper	decorative objects
duralumin	aluminium, copper and manganese	aircraft, boats and machinery due to high strength and resistance to corrosion
nichrome	nickel and copper	heating elements in toasters, electric heaters
solder	lead and tin	joining two metals together, especially in electric circuitry
sterling silver	silver and copper	jewelry, art objects

Nichrome wire, an alloy of nickel and copper, is used when carrying out flame tests in the laboratory. It is a convenient material because it is resistant to oxidation at high temperatures. The lilac colour here indicates that the sample contains potassium.

Alloys, in common with other mixtures, have no fixed composition. For example, 'steel' is the general name for a mixture of iron and carbon and other metals in different proportions. So, unlike compounds, alloys cannot be represented by a chemical formula. Alloys form by mixing liquid metals without chemical reaction, and so the components retain metallic properties such as thermal and electrical conductivity, and magnetism in the case of iron-containing alloys.

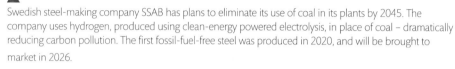

Swedish steel-making company SSAB has plans to eliminate its use of coal in its plants by 2045. The company uses hydrogen, produced using clean-energy powered electrolysis, in place of coal – dramatically reducing carbon pollution. The first fossil-fuel-free steel was produced in 2020, and will be brought to market in 2026.

Structure 1.1 – Why are alloys more correctly described as mixtures rather than as compounds?

Steel manufacture is one of the world's largest industries and is sometimes used as a measure of a country's development and economic progress. Global steel production has tripled in the last 50 years, with China and India becoming the world's top steel-producing nations. Steel is the foundation of buildings, vehicles and industries, including sustainable energy technologies. Plans for sustainability of steel production include moving towards a fossil-fuel-free industry by removing coal from the process. Steel is infinitely recyclable with no loss of quality, and it is estimated that new steel products contain on average 30% recycled steel.

Challenge yourself

1. The components of a mixture can usually be separated by physical means. What might be the challenges of trying to separate the metals from an alloy?

Exercise

Q4. An alloy is a mixture of metal with which of the following?

   A   a second metal or other non-metal material

   B   a non-metal material only

   C   a second metal only

   D   none of the above

Q5. Which of the following is not an alloy?

   A   solder   **B**   silver   **C**   bronze   **D**   brass

Q6. Explain how alloying can modify the structure and properties of metals.

Q7. What is the difference between an alloy and a composite?

Structure 2.4.4 – Polymers

Structure 2.4.4 – Polymers are large molecules, or macromolecules, made from repeating subunits called monomers.

Describe the common properties of plastics in terms of their structure.

Examples of natural and synthetic polymers should be discussed.	Structure 3.2 – What are the structural features of some plastics that make them biodegradable?

219

The polymers described in this chapter are organic compounds, based on the covalent chemistry of carbon and its unique ability to form stable bonds with itself, with hydrogen and with other elements. To understand these structures, you may find it helpful to first study an introduction to organic chemistry in Structure 3.2.

Polymers are large molecules made up of repeating subunits

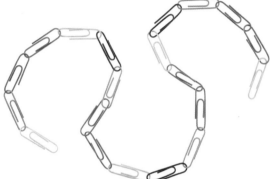

Some small molecules, known as **monomers,** are able to react together to form a linked chain held together by covalent bonds, known as a **polymer**.

Polymers are also known as **macromolecules,** as they are composed of thousands of atoms and so are relatively large compared with other molecules. Polymers vary in the nature of the monomer, and the length and the amount of branching in the chain, and so have a wide range of different properties.

As polymers are such large molecules, we show their structure as a **repeating unit**, which has open bonds at each end. It is put in a bracket with n, the number of monomers in the polymer, as a subscript.

$$n\left(\begin{matrix} X & X \\ | & | \\ C = C \\ | & | \\ X & X \end{matrix}\right) \longrightarrow \left(\begin{matrix} X & X \\ | & | \\ C - C \\ | & | \\ X & X \end{matrix}\right)_n$$

monomer　　　　the repeating unit

▲ We can think of polymer formation as similar to making a chain of paper clips.

This is an example of an addition polymer, described in Structure 2.4.5 on page 222.

Hermann Staudinger was the first polymer chemist to be awarded the Nobel Prize in Chemistry in 1953 'for his discoveries in the field of macromolecular chemistry'.

Does competition between scientists help or hinder the production of knowledge?

Nature of Science

Although plastics were one of the key materials of the 20th century, the notion of very large molecules was not widely accepted by the scientific community until 1929. Hermann Staudinger, a German organic chemist, put forward his idea of macromolecules, which he characterized as polymers. He experienced widespread resistance from his academic peers who thought that the newly invented plastics and natural materials such as rubber, starch, and cellulose were bundles of small molecules held together by unknown intermolecular forces. Staudinger's idea that there was a covalent bond between the units had little support. He was told by one of his colleagues to 'drop the idea of large molecules.' The eventual acceptance of Staudinger's theory was a key step in our understanding that led to the practical development of polymer chemistry. Science is a human activity and it sometimes takes courage to challenge the accepted models of the time.

Polymers are widespread in living things. Molecules such as protein, starch and DNA are examples of **natural polymers**. Understanding the relationship between the structure and function of these molecules has been a major focus of biochemical research over the last 100 years. Many advances in medicine, such as vaccines and drugs, have been developed from knowledge of the detailed structure and interaction of these polymers.

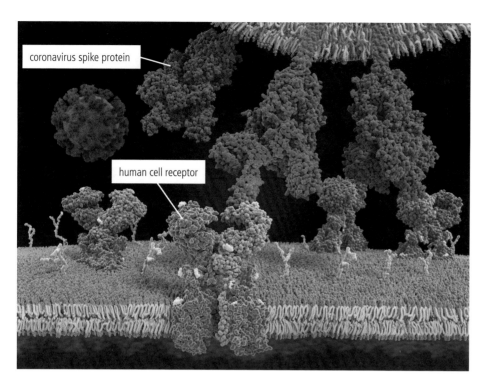

coronavirus spike protein

human cell receptor

◄ Molecular model of a coronavirus spike protein (shown in red) bound to human cell receptors (shown in blue). Binding between the spike protein and the receptor enables the virus to enter the human cell, where it replicates itself and can cause disease. Analysis of the structure of the viral protein, a natural polymer, is critical in enabling researchers to develop vaccines against diseases such as Covid-19.

Human-made polymers, known as **synthetic polymers**, first appeared in the 1930s when raw materials from the petroleum industry became available. This rapidly led to the development of a wide range of new materials known as **plastics**, a trend which continues to this day. Plastics are synthetic polymers that have widespread uses due to their light weight, low reactivity, water resistance and in some cases their strength. We use plastics in almost every aspect of daily life including packaging, construction, clothing, transportation and machinery. As a result, plastics have become widely distributed in both terrestrial and marine environments.

The very features that make plastics so useful, such as their unreactivity and water repellency, mean they are often **non-biodegradable,** that is, they cannot be broken down by natural processes, and so remain in the environment for indefinite periods of time. Accumulation of plastics in natural environments is one of the largest and most pressing environmental concerns of our time.

▲ Plastic waste washed up on a beach in Ireland. Common plastics include polythene, PVC, nylon, polystyrene and Kevlar®.

Global production of plastics has increased exponentially since 1950 and continues to rise. It is suggested that fossilized plastics may form a distinctive stratal layer, possibly marking the start of the proposed geological epoch the 'Anthropocene', the time of significant human impact on Earth's geology and ecosystems. A high proportion of plastics end up in the oceans where they have a widespread impact on the ecosystem. It is estimated that by 2030 there could be 300 million tonnes of accumulated plastic in the oceans. While measures to try to address this include more efficient recycling processes, biodegradable plastics, and plastic-feeding microorganisms, the urgent need is to reduce the source. We can all share some responsibility for decreasing our use of disposable plastics.

Properties of polymers. Full details of how to carry out this experiment with a worksheet are available in the eBook.

Structure 3.2 – What are the structural features of some plastics that make them biodegradable?

Clear plastic pellets made from corn starch. The plastic is biodegradable and can be used for both food and non-food packaging as well as textiles and other materials.

A biodegradable compound can undergo breakdown by microorganisms into end products that are found in nature and therefore not harmful to the environment. Biodegradable plastics may be plant-based, containing starch, lignin or cellulose, and are often produced from corn. The breakdown products are carbon dioxide and water. Swelling of the starch grains in water can help to break down the plastic. Other biodegradable plastics are based on petroleum derivatives and contain catalysts for the breakdown process. **Compostable** plastics break down along with food and garden waste in the specific conditions found in a compost pile.

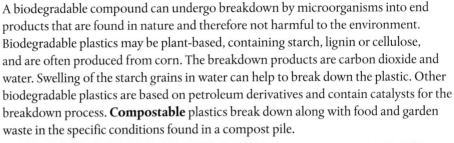

There is growing evidence that some microorganisms are evolving to break down plastic in their habitat. Genetic analysis of bacteria from different environments has found enzymes that can degrade different types of plastic, where the number and type of enzymes match the amount and type of plastic pollution in different locations. For example, in ocean studies, higher levels of degrading enzymes were found at levels of 200–600 m below the surface, matching the higher levels of plastic pollution known to exist at these depths. There is potential to develop these microbial communities for targeted degradation of specific polymers.

Nature of Science

It is hard to assess the scale of the impact that plastics have had on society. The manufacture of diverse plastics has brought innovation and progress to our global food supply, clothing, health care and infrastructure. Yet these innovations have come at a significant cost to the environment, and the long-term consequences are still being determined. While scientific progress is responsible for both the intended and unintended consequences, it is scientists who must respond to the problems.

Structure 2.4.5 – Addition polymers

Structure 2.4.5 – Addition polymers form by the breaking of a double bond in each monomer.	
Represent the repeating unit of an addition polymer from given monomer structures.	
Examples should include polymerization reactions of alkenes. Structures of monomers do not have to be learned but will be provided or will need to be deduced from the polymer.	Structure 3.2 – What functional groups in molecules can enable them to act as monomers in addition reactions? Reactivity 2.1 – Why is the atom economy 100% for an addition polymerization reaction?

Monomers with double bonds can be used to make addition polymers

Addition reactions occur when a multiple bond in a molecule breaks and creates new bonding positions. Alkenes and alkynes are organic compounds that have double and triple carbon–carbon bonds respectively, and so readily undergo addition reactions, as described in Reactivity 3.4. This reactivity also means they can act as monomers in forming **addition polymers**.

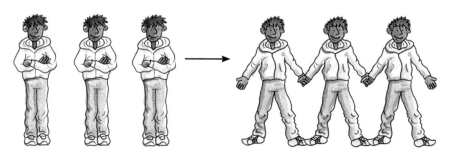

People cannot form a chain until they unfold their arms to release their hands. Similarly, alkenes must break their double bonds in order to join together to form the polymer.

Addition polymerization reactions lead to the synthesis of many common plastics. For example, the reaction of ethene, C_2H_4, undergoing polymerization to form poly(ethene) can be shown as follows.

$$n\left(\begin{array}{c} \text{H} \quad \text{H} \\ | \quad\ | \\ \text{C}=\text{C} \\ | \quad\ | \\ \text{H} \quad \text{H} \end{array}\right) \longrightarrow \left(\begin{array}{c} \text{H} \quad \text{H} \\ | \quad\ | \\ \text{C}-\text{C} \\ | \quad\ | \\ \text{H} \quad \text{H} \end{array}\right)_n$$

monomer   polymer
ethene   polyethene

An addition polymer is formed when the double bonds of monomer molecules break and make new covalent bonds with neighbouring molecules to form a chain. No other product is formed.

Similarly, propene polymerizes to form polypropene, often called polypropylene.

$$n\left(\begin{array}{c} \text{H} \quad \text{H} \\ | \quad\ | \\ \text{C}=\text{C} \\ | \quad\ | \\ \text{H} \quad \text{CH}_3 \end{array}\right) \longrightarrow \left(\begin{array}{c} \text{H} \quad \text{H} \\ | \quad\ | \\ \text{C}-\text{C} \\ | \quad\ | \\ \text{H} \quad \text{CH}_3 \end{array}\right)_n$$

propene   polypropene

Structure 3.2 – What functional groups in molecules can enable them to act as monomers in addition reactions?

Other common addition polymers include polychloroethene, also known as PVC, and PTFE, polytetrafluoroethene, often marketed as Teflon®. Their repeating units are shown below.

$$\left(\begin{array}{c} \text{H} \quad \text{H} \\ | \quad\ | \\ \text{C}-\text{C} \\ | \quad\ | \\ \text{H} \quad \text{Cl} \end{array}\right)_n \qquad \left(\begin{array}{c} \text{F} \quad \text{F} \\ | \quad\ | \\ \text{C}-\text{C} \\ | \quad\ | \\ \text{F} \quad \text{F} \end{array}\right)_n$$

PVC   PTFE

The example with propene shows that it is helpful to draw the structure of the monomer with the double bond in the *middle* and the other groups off at 90°. Then it is easy to see how the monomers link together when the double bond breaks.

Atom economy is a measure of the proportion of reactant that ends up in the desired product, based on the reaction's stoichiometry.

$$\% \text{ atom economy} = \frac{\text{molar mass of desired product}}{\text{molar mass of all reactants}} \times 100$$

Addition polymerization reactions do not generate a by-product and so convert 100% of reactants into product. This does not mean, however, that there is no waste or environmental impact of these reactions. A reaction yield of less than 100% means there will be unreacted monomer at the end of the process. Also, many polymerization reactions use solvents and conditions of high temperature, pressure and/or catalysts – all of which must be considered as part of an environmental assessment of the process.

Reactivity 2.1 – Why is the atom economy 100% for an addition polymerization reaction?

Microplastics are small plastic pieces less than 5 mm long, which harm marine life and enter the food chain. Microplastics come from a variety of sources, including larger plastic debris that degrades into smaller and smaller pieces. **Microbeads** are a type of microplastic, mostly derived from health and beauty products, that pass through water filtration systems.

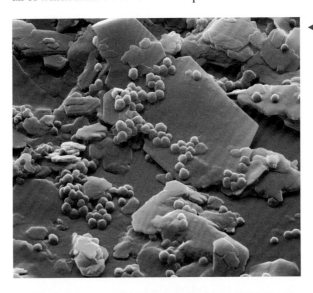

◀ Coloured scanning electron micrograph (SEM) of particles found within a cosmetic facial scrub. A single shower can result in 100 000 plastic particles entering the ocean. Legislation to ban the use of microbeads is currently under discussion in many countries.

Worked example

Deduce the structure of the addition polymer formed from propene. You should include three repeating units in the structure.

Solution

1. Draw three structures with the alkene double bond in the middle:

$$
\begin{array}{cc}
\text{H} & \text{CH}_3 \\
| & | \\
\text{C} & = \text{C} \\
| & | \\
\text{H} & \text{H}
\end{array}
\qquad
\begin{array}{cc}
\text{H} & \text{CH}_3 \\
| & | \\
\text{C} & = \text{C} \\
| & | \\
\text{H} & \text{H}
\end{array}
\qquad
\begin{array}{cc}
\text{H} & \text{CH}_3 \\
| & | \\
\text{C} & = \text{C} \\
| & | \\
\text{H} & \text{H}
\end{array}
$$

2. Open the double bond in each molecule so that single bonds extend in both directions:

$$
\begin{array}{cccccc}
\text{H} & \text{CH}_3 & \text{H} & \text{CH}_3 & \text{H} & \text{CH}_3 \\
| & | & | & | & | & | \\
-\text{C}- & \text{C}- & \text{C}- & \text{C}- & \text{C}- & \text{C}- \\
| & | & | & | & | & | \\
\text{H} & \text{H} & \text{H} & \text{H} & \text{H} & \text{H}
\end{array}
$$

Challenge yourself

2. Draw the repeating unit in polystyrene, given that the formula of the monomer is $C_6H_5CHCH_2$.

Exercise

Q8. What do poly(ethene) and ethene have in common?

    A same molecular formula

    B same empirical formula

    C same physical properties

    D same chemical properties

Q9. The repeating unit of a polymer is shown:

$$\left(\!\!\begin{array}{cccc} H & Cl & H & Cl \\ | & | & | & | \\ -C & -C & -C & -C- \\ | & | & | & | \\ C_2H_5 & H & C_2H_5 & H \end{array}\!\!\right)$$

Which structure shows the monomer used to make this polymer?

A
$$\begin{array}{cc} C_2H_5 & Cl \\ | & | \\ H-C & -C-H \\ | & | \\ H & H \end{array}$$

B
$$\begin{array}{cc} C_2H_5 & H \\ | & | \\ H-C & -C-H \\ | & | \\ Cl & H \end{array}$$

C
$$\begin{array}{cc} C_2H_5 & H \\ | & | \\ C & =C \\ | & | \\ H & Cl \end{array}$$

D
$$\begin{array}{cc} C_2H_5 & H \\ | & | \\ C & =C \\ | & | \\ Cl & H \end{array}$$

Q10. Part of the structure of a polymer is shown:

$$\begin{array}{ccccc} CH_3 & & CH_3 & & CH_3 \\ | & & | & & | \\ -CH-CH_2 & -CH-CH_2 & -CH-CH_2- \end{array}$$

Which is the monomer used to make this polymer?

    A C_3H_6 **B** C_6H_{12} **C** C_3H_8 **D** C_6H_{14}

Q11. Distinguish between the terms recyclable, biodegradable and reusable as applied to plastics. What are the advantages and disadvantages of each of these in helping to reduce the environmental impact of plastics? What other ways can we help to reduce the amount of plastic waste accumulating in the oceans?

Addition polymers are found almost universally in daily life. Poly(ethene) has excellent insulating properties and is commonly used in household containers, water tanks, and piping. Polypropylene is used in the manufacture of clothing, especially thermal wear for outdoor activities. PVC is used in all forms of construction materials, packaging, electrical cable sheathing etc. There is controversy regarding its use, linked to health and environmental concerns, because of its high chlorine content which makes it difficult to recycle. Some countries are banning the use of PVC in packaging. PTFE has a very low surface friction so it is widely used in non-stick pans. It also makes up one of the layers in waterproof, breathable fabrics such as GoreTex®. The resistance of PTFE to van der Waals' forces means that it is the only known surface to which a gecko cannot stick.

Guiding Question revisited

What role do bonding and structure have in the design of materials?

In this chapter we used models of bonding to describe and explain the properties of some important materials.

- Bonding in materials is best described as a continuum rather than as discrete types, and can be represented as a triangle of bonding.
- The position of an element or compound in the bonding triangle is determined from electronegativity values.
- From the position of a substance in the bonding triangle, we can deduce its bonding and predict its properties.
- Alloys are homogeneous mixtures of metals with enhanced properties.
- Metals are able to form alloys because of the non-directional nature of metallic bonding.
- Polymers are macromolecules composed of subunits called monomers held together by covalent bonds.
- Addition polymers form from monomers that possess a double bond which can break to create new bonding positions for the attachment of neighbouring monomers.
- Addition polymerization reactions do not yield a by-product.
- Plastics are polymers with properties that give them widespread uses in almost all aspects of society.
- The distinct properties of plastics also cause them to accumulate in the environment without being broken down.
- Use of biodegradable plastics and recycling programs are important steps to improve the processing of plastic waste, but the urgent need is to reduce the global production of plastic.

Practice questions

1. What are alloys with two components called?
    - **A** binary alloy
    - **B** binary compound
    - **C** ternary alloy
    - **D** metallic compound

2. Which of the following **cannot** be deduced from the position of an unknown substance X in the bonding triangle?
    - **A** the chemical formula of X
    - **B** the type of bonding in X
    - **C** the physical properties of X
    - **D** whether X is an element or a compound

3. Which of the following is not an addition polymer?
    - **A** polyester
    - **B** polystyrene
    - **C** poly(ethene)
    - **D** poly(tetrafluoroethene)

4. In the bonding triangle, substance X is found in the top right, substance Y is found in the bottom left and substance Z is found in the bottom right. Answer the following, giving reasons for your answers:
    - **(a)** Which of these substances is likely to be an element? (2)
    - **(b)** For substances X and Y, give a physical property you would expect them to have in common. (2)
    - **(c)** Which substance will be the most brittle? (2)
    - **(d)** Which substance is likely to have the lowest boiling point? (2)

    (Total 8 marks)

5. Use electronegativity values from Section 9 in the data booklet to complete this table for the five substances given.

Substance	$\chi_{average}$	$\Delta\chi$	Position in triangle and type of bonding	Predicted properties	
Sn					(2)
P_4O_7					(2)
Cd_3Mg					(2)
MgO					(2)
NCl_3					(2)

(Total 10 marks)

6. Justify why, in terms of atom economy, the polymerization reactions of polypropylene could be considered 'Green Chemistry'. (2)

(Total 2 marks)

7. Alloys of aluminium with nickel are used to make engine parts. With reference to the model of metallic bonding, explain why this alloy is used rather than pure aluminium. (3)

(Total 3 marks)

8. Lanthanum, La, and antimony, Sb, form compounds with bromine that have similar formulas, $LaBr_3$ and $SbBr_3$.

   (a) Determine the type of bond present in $SbBr_3$, showing your method. Use Sections 9 and 17 of the IB data booklet. (2)

   (b) Lanthanum has a similar electronegativity to group 2 metals. Explain, in terms of bonding and structure, why crystalline lanthanum bromide is brittle. (2)

(Total 4 marks)

9. Polymers are made up of repeating monomer units which can be manipulated in various ways to give structures with desired properties.

   (a) Deduce the repeating unit of poly(2-methylpropene). (1)

   (b) Deduce the percentage atom economy for polymerization of 2-methylpropene. (1)

   (c) Suggest why incomplete combustion of plastics, such as polyvinyl chloride, is common in industrial and house fires. (1)

(Total 3 marks)

10. (a) Predict the predominant type of bonding for a binary substance AB in which the electronegativity of both atoms is low. Use Section 17 of the data booklet. (1)

   (b) The type of bonding in a compound is sometimes used to classify materials. Outline why this type of classification has limitations by using magnesium diboride, MgB_2, as an example. Refer to Sections 9 and 17 of the data booklet. (2)

(Total 3 marks)

Classification of matter

One of the broadest classifications of materials is to divide them into inorganic and organic substances. Here we have the periodic table, which is the great simplifying concept of inorganic chemistry. We also have the molecular model, which is our guide to organic chemistry. Organic chemistry covers most of the compounds of just one element, carbon, shown as black spheres in the model.

If you have ever visited a large supermarket, you will appreciate the importance of a classification system. Similar products are grouped together to help you find what you want. Classification is an important aspect of all scientific work but is fundamental to chemistry. Chemistry is not the study of a random collection of substances but the trends and patterns in their chemical and physical properties.

The broadest classification is to divide substances into inorganic and organic materials. Inorganic chemistry deals with the chemistry of over 100 elements with different properties, whereas organic chemistry is concerned with the compounds of only one element, carbon. The periodic table is the great simplifying concept of inorganic chemistry because it shows that the elements are related to each other. Elements with similar properties can be put into families or groups. These relations are all ultimately explained by the electron configurations of the atoms as discussed in Structure 1.3. The periodic table has only a limited role, however, in guiding our understanding of organic chemistry. Instead, we use a different classification system based on the groups of atoms or functional groups present in different molecules. Homologous series are the families of organic compounds.

'Science is built of facts the way a house is built of bricks: but an accumulation of facts is no more science than a pile of bricks is a house.' Henri Poincaré, 1854–1912.

Do you agree with this description of science?

Nature of Science

Scientists classify observations and search for patterns and discrepancies as first steps to understanding. The identification of patterns, however, does not explain phenomena. These patterns lead to an underlying mechanism that explains the patterns. We see in Structure 1.3 that, the patterns in the periodic table can be traced back to patterns in electron configuration.

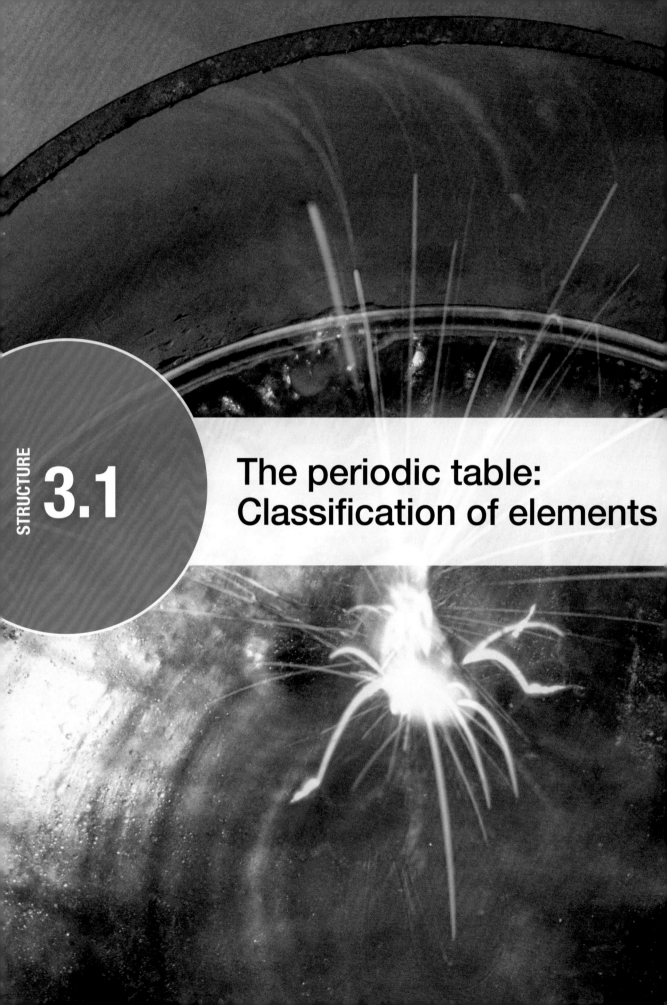

3.1

The periodic table: Classification of elements

Potassium reacts with water to produce potassium hydroxide (soluble) and hydrogen gas. The heat of the reaction ignites the hydrogen gas, which burns and sparks with a pink-lilac flame due to the presence of the potassium. The heat of the reaction also melts the potassium, which forms a molten ball that floats on the water's surface.

Guiding Question

How does the periodic table help us to predict patterns and trends in the properties of elements?

The periodic table has been referred to as the most meaningful and compact system of knowledge that humans have devised. It reveals the organizing principles of matter and without it, inorganic chemistry in particular would be a meaningless compilation of random facts. Instead of studying elements individually, we can class them together in groups. The chemistry of potassium can be accurately predicted from the chemistry of lithium and sodium. Elements at opposite sides of the periodic table show opposing properties: the most reactive metal, francium, is on the bottom left of the table, whereas the most reactive non-metal, fluorine, is found near the top right. The periodic table has guided us to our understanding of the atom. The patterns in the periodic table reflect deeper patterns in electron configuration. The periodic table is the 'map' of chemistry. It suggests new avenues of research for the professional chemist and is a guide for students – because it disentangles a mass of observations and reveals hidden order.

TOK What organizing systems exist in other areas of knowledge? To what extent do the classification systems we use in the pursuit of knowledge affect the conclusions that we reach?

Periodicity is the regular repetition of properties of the elements arising from patterns in their electron configuration.

Structure 3.1.1 – Periods, groups and blocks

Structure 3.1.1 – The periodic table consists of periods, groups and blocks.	
Identify the positions of metals, metalloids and non-metals in the periodic table.	
The four blocks associated with the sublevels s, p, d, f should be recognized.	
A copy of the periodic table is available in the data booklet.	

The development of the periodic table

The periodic law states that after certain regular but varying intervals, the chemical elements show an approximate repetition in their properties. This **periodicity** in the physical and chemical properties was the basis of the periodic table first proposed in 1869 by the Russian chemist Dmitri Mendeleev. Previous attempts had been made to impose order on the then known 62 elements, but Mendeleev had the insight to realize that each element has its allotted place, so he left gaps where no known elements fitted into a position. As a scientific idea it was extremely powerful as it made predictions about the unknown elements that fitted these gaps, predictions which could be tested. When these elements were later discovered, the agreement between the predicted properties and the actual properties was remarkable.

▲
Mendeleev grouped the known elements into families, leaving gaps corresponding to elements that he thought should exist but which had not yet been discovered.

Nature of Science

The periodicity among the elements is not constant like days of the week. Is the periodic 'law' equal in status to other scientific laws?

Mendeleev's periodic table of 1869. The noble gas elements had not been discovered. Reading from top to bottom and left to right, the first four gaps were waiting for scandium (1879), gallium (1875), germanium (1886) and technetium (1937). The later discovery of these elements, with properties that matched their periodicity, helped to confirm Mendeleev's ideas.

		K = 39	Rb = 85	Cs = 133	—	—
		Ca = 40	Sr = 87	Ba = 137	—	—
		—	?Yt = 88?	?Di = 138?	Er = 178?	—
	Ti = 48?	Zr = 90	Ce = 140?	?La = 180?	Tb = 231
	V = 51	Nb = 94	—	Ta = 182	—
	Cr = 52	Mo = 96	—	W = 184	U = 240
	Mn = 55	—	—	—	—
	Fe = 56	Ru = 104	—	Os = 195?	—
	Co = 59	Rh = 104	—	Ir = 197	—
	Ni = 59	Pd = 106	—	Pt = 198?	—

Typische Elemente

H = 1	Li = 7	Na = 23	Cu = 63	Ag = 108	—	Au = 199?	—
	Be = 9,4	Mg = 24	Zn = 65	Cd = 112	—	Hg = 200	—
	B = 11	Al = 27,3	—	In = 113	—	Tl = 204	—
	C = 12	Si = 28	—	Sn = 118	—	Pb = 207	—
	N = 14	P = 31	As = 75	Sb = 122	—	Bi = 208	—
	O = 16	S = 32	Se = 78	Te = 125?	—	—	—
	F = 19	Cl = 35,5	Br = 80	J = 127	—	—	—

Mendeleev had no knowledge of the structure of the atom, which is discussed in Structure 1. With the benefit of hindsight, it is clear that the periodicity of the elements is a direct consequence of the periodicity of the electron configurations within the atom. The position of an element in the periodic table is based on the sublevel of the highest-energy electron in the ground-state atom.

Periods and groups

Mendeleev is said to have made his discovery after a dream. When he awoke he set out his chart in virtually its final form. He enjoyed playing a form of patience (solitaire) and wrote the properties of each element on cards which he arranged into rows and columns.

In the periodic table, the elements are placed in order of increasing atomic number (Z), the number of protons in the nucleus of its atoms. As there are no gaps in the sequence of atomic numbers, we can be confident that the search for new elements in nature is over.

The discovery of the elements was an international endeavor. This is illustrated by some of their names. Some derive from the place where they were found, some from the origins of their discoverers, and some from the geographical origins of the minerals from which they were first isolated. The periodic table hangs in chemistry classrooms and science laboratories throughout the world.

The only way to extend the periodic table is by making elements artificially. In 2022, the International Union of Pure and Applied Chemistry (IUPAC) recognized 118 elements, but more may be discovered in the future.

The columns of the table are called **groups** and the rows are called **periods**.

Challenge yourself

1. Four elements derive their name from a small town called Ytterby, just outside Stockholm, Sweden. Try to find their names.

Metals and non-metals

Metallic elements are found on the left-hand side of the table in the **s block**, in the central **d block**, and in the island of the **f block**. A small number of metals, such as aluminium and lead, are also found on the left of the **p block** (see Figure 1).

	s¹	s²	d¹	d²	d³	d⁵s¹	d⁵	d⁶	d⁷	d⁸	d¹⁰s¹	d¹⁰	p¹	p²	p³	p⁴	p⁵	p⁶
	1	2	3	4	5	6	7	8	9	10	11	12	13	14	15	16	17	18
1	H hydrogen 1																	He helium 2
2	Li lithium 3	Be beryllium 4											B boron 5	C carbon 6	N nitrogen 7	O oxygen 8	F fluorine 9	Ne neon 10
3	Na sodium 11	Mg magnesium 12											Al aluminium 13	Si silicon 14	P phosphorus 15	S sulfur 16	Cl chlorine 17	Ar argon 18
4	K potassium 19	Ca calcium 20	Sc scandium 21	Ti titanium 22	V vanadium 23	Cr chromium 24	Mn manganese 25	Fe iron 26	Co cobalt 27	Ni nickel 28	Cu copper 29	Zn zinc 30	Ga gallium 31	Ge germanium 32	As arsenic 33	Se selenium 34	Br bromine 35	Kr krypton 36
5	Rb rubidium 37	Sr strontium 38	Y yttrium 39	Zr zirconium 40	Nb niobium 41	Mo molybdenum 42	Tc technetium 43	Ru ruthenium 44	Rh rhodium 45	Pd palladium 46	Ag silver 47	Cd 48	In indium 49	Sn tin 50	Sb antimony 51	Te tellurium 52	I iodine 53	Xe xenon 54
6	Cs caesium 55	Ba barium 56	57–71 see below	Hf hafnium 72	Ta tantalum 73	W tungsten 74	Re rhenium 75	Os osmium 76	Ir iridium 77	Pt platinum 78	Au gold 79	Hg mercury 80	Tl thallium 81	Pb lead 82	Bi bismuth 83	Po polonium 84	At astatine 85	Rn radon 86
7	Fr francium 87	Ra radium 88	89–103 see below	Rf rutherfordium 104	Db dubnium 105	Sg seaborgium 106	Bh bohrium 107	Hs hassium 108	Mt meitnerium 109	Ds darmstadtium 110	Rg roentgenium 111	Cp copernicium 112	Uut ununtrium 113	Fl flerovium 114	Uup Ununpentium 115	Lv Livermorium 116	Uus Ununseptium 117	Uuo Ununoctium 118

La lanthanum 57	Ce cerium 58	Pr praseodymium 59	Nd neodymium 60	Pm promethium 61	Sm samarium 62	Eu europium 63	Gd gadolinium 64	Tb terbium 65	Dy dysprosium 66	Ho holmium 67	Er erbium 68	Tm thulium 69	Yb ytterbium 70	Lu Lutetium 71
Ac actinium 89	Th thorium 90	Pa protactinium 91	U uranium 92	Np neptunium 93	Pu plutonium 94	Am americium 95	Cm curium 96	Bk berkelium 97	Cf californium 98	Es einsteinium 99	Fm fermium 100	Md mendelevium 101	No nobelium 102	Lr Lawrencium 103

▲
S3.1 Figure 1 The periodic table. The colours show the blocks formed by the elements with their outer electrons in the same electron sublevel: s block (blue), d block (yellow), p block (red), f block (green).

Challenge yourself

2. State the electron configuration of europium.

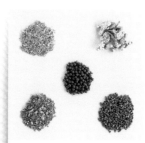

▲
Metals include, clockwise from left, copper, aluminium, zinc, iron and lead.

The element europium is used in the security marking of euro notes and other banknotes. When placed in ultraviolet (UV) radiation, europium compounds fluoresce, making the security markers visible.

▲
A sample of europium (Eu), one of the lanthanides or rare earth elements. Europium is a hard silvery-white metal. Its compounds, such as its oxides, are used in fluorescent light bulbs and television screens.

The non-metals are found on the upper right-hand side of the p block.

SKILLS
Redesigning the periodic table. Full details of how to carry out this activity with a worksheet are available in the eBook.

Some non-metallic elements. Clockwise from top left, sulfur (S), bromine (Br), phosphorus (P), iodine (I) and carbon (C). Non-metals are generally poor conductors of both heat and electricity. Graphite, an allotrope of carbon, is unusual in that it is a non-metal that does conduct electricity.

The metalloid elements have the characteristics of both metals and non-metals. Their physical properties and appearance most resemble the metals, although chemically they have more in common with the non-metals. In the periodic table, the metalloid elements silicon, germanium, arsenic, antimony, tellurium and polonium form a diagonal staircase between the metals and non-metals.

Challenge yourself

3. How many elements in the periodic table are liquids and how many are gases?

4. Distinguish between the terms 'metalloid' and 'semiconductor'.

Silicon is a metalloid used in computers.

Nature of Science

'Theories' and 'laws' are terms that have a special meaning in science and it is important to distinguish these from their everyday use. Scientific laws are descriptive statements derived from observations of regular patterns of behavior. They do not necessarily explain a phenomenon. Mendeleev's periodic law, for example, tells us that there are patterns in the properties of the elements but it does not attempt to explain these patterns. This had to wait for the discovery of the electron by Thomson in 1897 and the work of Rutherford and Bohr at the beginning of the 20th century.

Exercise

Q1. What increases **in equal steps of one** from left to right in the periodic table for the elements sodium to argon?

  A the relative atomic mass

  B the number of occupied electron energy levels

  C the number of neutrons in the most common isotope

  D the number of electrons in the atom

Q2. Which of the following elements is a metalloid?

  A calcium   **B** manganese   **C** germanium   **D** carbon

Q3. Which of the following materials is the best conductor of electricity in the solid state?

  A silicon   **B** graphite   **C** phosphorus   **D** antimony

Q4. Use the periodic table to identify the position of the following elements.

	Element	Period	Group
(a)	helium		
(b)	chlorine		
(c)	barium		
(d)	francium		

Structure 3.1.2 – Periodicity and electron configuration

Structure 3.1.2 – The period number shows the outer energy level that is occupied by electrons.

Elements in a group have a common number of valence electrons.

Deduce the electron configuration of an atom up to Z = 36 from the element's position in the periodic table and vice versa.

Groups are numbered from 1 to 18. The classifications 'alkali metals', 'halogens', 'transition elements' and 'noble gases' should be known.	Nature of Science, Structure 1.2 – How has the organization of elements in the periodic table facilitated the discovery of new elements?

The position of an element is related to the electron configuration of its atoms, as discussed in Structure 1.3. The groups are numbered from 1 to 18. Elements whose valence electrons occupy an s sublevel make up the s block, which includes the **alkali metals** in group 1. Elements with valence electrons in p orbitals make up the p block, which includes the **halogens** in group 17 and the **noble gases** in group 18. The d block and f block are similarly made up of elements with outer electrons in d and f orbitals. The d block includes the **transition metals**.

The alkali metals are in group 1, the halogens are in group 17, the transition elements are in the d block and the noble gases are in group 18.

The period of an element gives the number of occupied energy levels. The element sodium, for example, is in period 3 because it has three occupied principal energy levels, and is in group 1 of the s block because there is one electron in the valence energy level [Ne] $3s^1$. Bromine is in period 4 and in group 17 of the p block because it has 7 electrons in the fourth principal energy level, and 17 more electrons than the previous noble gas, argon: [Ar] $4d^{10}5s^25p^5$.

The rows in the periodic table are called periods. The period number gives the number of occupied electron principal energy levels. The columns in the periodic table are called groups. The group number is related to the number of valence electrons in the outer energy level. All the elements in a group have the same number of valence electrons.

How can the production of artificial elements give us more knowledge of the natural world?

Nature of Science, Structure 1.2 – How has the organization of elements in the periodic table facilitated the discovery of new elements?

Worked example

How many electrons are in the valence energy level of iodine?

Solution

Find the element in the periodic table. It is in period 5, so has the noble gas core of Kr, with electrons added to 5s, then 4d, and then 5p.

As it is in group 17 it has the configuration: [Kr] $5s^24d^{10}5p^5$ or [Kr] $4d^{10}5s^25p^5$ and so has seven electrons in its valence energy level.

Nature of Science

Oganesson (Og), element 118, is one of the most recently added elements to the periodic table. Whereas the discovery of the early elements involved the practical steps of extraction and isolation often performed by one individual, later elements are made by teams of scientists working together. New elements are as much invented as discovered, but their existence provides further knowledge about the natural world. Claims of a new element's existence have to be verified by the IUPAC in a process called peer review.

Exercise

Q5. Phosphorus is in period 3 and group 15 of the periodic table.

  (a) Distinguish between the terms 'period' and 'group'.

  (b) State the electron configuration of phosphorus and relate it to its position in the periodic table.

Q6. How many valence electrons are present in the atoms of the element with atomic number 51?

Q7. Which of the following properties is used to arrange the elements in the modern periodic table?

    A   relative atomic mass

    B   number of valence electrons

    C   atomic number

    D   effective nuclear charge

Q8. The only isotope of oganesson to be synthesized is $^{294}_{118}$Og with a half-life of 700 μs. Which of the following statements is correct?

    I.   The nucleus of the atom has a relative charge of +118.

    II. $^{294}_{118}$Og has the outer electron configuration $7p^6$.

    III. Oganesson is a stable element.

    A   I and II only   **B**   I and III only

    C   II and III only   **D**   I, II and III

Q9. Mendeleev left a space in his table for the element now known as gallium. He was able to predict many properties for the then unknown element.

The densities of some group 13 elements are shown in the table.

Atomic number	Element	Density / g cm^{-3}
5	boron	2.34
13	aluminium	2.70
31	gallium	
49	indium	7.30
81	thallium	11.85

Plot a graph of density against atomic number to predict the density of gallium.

Originally, when gallium was first discovered by the French chemist, Paul-Emile Lecoq, its density did not correspond to the value predicted by Mendeleev. Mendeleev suggested to Lecoq that he needed to prepare a purer sample. When this was done the density matched the value predicted.

Structure 3.1.3 – Periodicity in properties of elements

Structure 3.1.3 – Periodicity refers to trends in properties of elements across a period and down a group.

Explain the periodicity of atomic radius, ionic radius, ionization energy, electron affinity and electronegativity.

Properties

The periodicity of the elements is reflected in their properties. The atomic and ionic radii, electronegativity, ionization energy and electron affinity are of particular interest.

The concept of effective nuclear charge discussed in Structure 2.1 is helpful in explaining trends in both physical and chemical properties. The effective nuclear charge experienced by an atom's outer electrons increases across a period but remains approximately the same down a group.

The effective nuclear charge experienced by an atom's outer electrons increases across a period but remains approximately the same down a group.

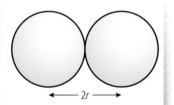

S3.1 Figure 2 The atomic radius *r* is measured as half the distance between neighbouring nuclei in a covalent bond.

Atomic radius increases down a group and decreases across a period

The concept of atomic radius is not as straightforward as you may think. We see in Structure 1.3 that electrons occupy atomic orbitals, which give a probability description of the electrons' locations, but do not have sharp boundaries. The atomic radius *r* is measured as half the distance between neighbouring nuclei (Figure 2). For many purposes, however, it can be considered as the distance from the nucleus to the outermost electrons in the atom.

The data booklet shows that the (covalent) atomic radii increase down a group and decrease across a period. Consider, for example, the group 1 elements – as shown in the table below.

Element	Period	Electron configuration	No. of occupied principal energy levels	Atomic radius / 10^{-12} m
Li	2	$1s^2 2s^1$	2	130
Na	3	$1s^2 2s^2 2p^6 3s^1$	3	160
K	4	$1s^2 2s^2 2p^6 3s^2 3p^6 4s^1$	4	200
Rb	5	$1s^2 2s^2 2p^6 3s^2 3p^6 3d^{10} 4s^2 4p^6 5s^1$ $= [Kr]\, 5s^1$	5	215
Cs	6	$[Rn]\, 6s^1$*	6	238

*The condensed electron configuration is shown for Cs.

The atomic radii increase down a group, as the number of occupied electron levels (given by the period number) increases.

The trend across a period is illustrated by the period 3 elements, as shown below.

Element	Na	Mg	Al	Si	P	S	Cl	Ar
Atomic radius / 10^{-12} m	160	140	124	114	109	104	100	101

All these elements have three occupied principal energy levels. The attraction between the nucleus and the outer electrons increases as the nuclear charge increases so there is a general decrease in atomic radius across the period.

The decrease in radius across a period is significant; a chlorine atom, for example, has a radius that is about 60% that of a sodium atom.

Ionic radius increases down a group and decreases across a period

The atomic and ionic radii of the period 3 elements are shown in the table below.

Element	Na	Mg	Al	Si	P	S	Cl
Atomic radius / 10^{-12} m	160	140	124	114	109	104	100
Ionic radius / 10^{-12} m	102 (Na^+)	72 (Mg^{2+})	54 (Al^{3+})	40 (Si^{4+})	38 (P^{5+})	184 (S^{2-})	181 (Cl^-)
				271 (Si^{4-})	212 (P^{3-})		

Five trends can be identified:

- Positive ions are smaller than their parent atoms. The formation of positive ions involves the loss of the outer energy level.
- Negative ions are larger than their parent atoms. The formation of negative ions involves the addition of electrons into the outer energy level. The increased electron repulsion between the electrons in the outer principal energy level causes the electrons to move further apart and so increases the radius of the outer energy level.
- The ionic radii decrease from groups 1 to 14 for the positive ions due to the increase in nuclear charge with atomic number across the period. The increased attraction between the nucleus and the electrons pulls the outer energy level closer to the nucleus.
- The ionic radii decrease from groups 14 to 17 for the negative ions due to the increase in nuclear charge across the period, as explained above. The positive ions are smaller than the negative ions, as the former have only two occupied electron principal energy levels and the latter have three. This explains the big difference between the ionic radii of the Si^{4+} and Si^{4-} ions and the discontinuity in the middle of the table.
- The ionic radii increase down a group as the number of electron energy levels increases.

Worked example

Describe and explain the trend in radii of the following atoms and ions:

O^{2-}, F^-, Ne, Na^+, and Mg^{2+}.

Solution

The ions and the Ne atom all have 10 electrons and the electron configuration $1s^2 2s^2 2p^6$.

The nuclear charges increase with atomic number:

O: Z = +8   F: Z = +9   Ne: Z = +10   Na: Z = +11   Mg: Z = +12

The increase in nuclear charge results in increased attraction between the nucleus and the outer electrons. The ionic radii decrease as the atomic number increases.

Ionization energies decrease down a group and increase across a period

First ionization energies are a measure of the attraction between the nucleus and the outer electrons. They are defined in Structure 1.3, where they provide evidence for the electron configuration of the atoms of different elements (Figure 3).

The first ionization energy of an element is the energy required to remove one mole of electrons from one mole of gaseous atoms in their ground state.

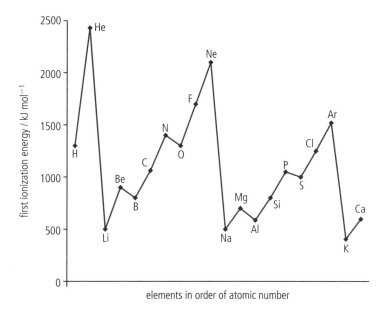

▲
S3.1 Figure 3 First ionization energies of the first 20 elements.

Two general trends can be identified from Figure 3:

- Ionization energies increase across a period. The increase in effective nuclear charge increases the attraction between the outer electrons and the nucleus and makes the electrons more difficult to remove.
- Ionization energies decrease down a group. The electron removed is from the energy level furthest from the nucleus. Although the nuclear charges increase, the effective nuclear charge is about the same, owing to shielding by the inner electrons, and so the increased distance between the electron and the nucleus reduces the attraction between them.

The trend in ionization energy is the reverse of the trend in atomic radii. Both trends are an indication of the attraction between the nucleus and the outer electrons.

Electron affinity decreases down a group and increases across a period

The first **electron affinity** of an element ($\Delta H_{ea}^{\ominus}$) is the energy change that occurs when one mole of electrons is added to one mole of gaseous atoms to form one mole of gaseous ions:

$$X(g) + e^- \rightarrow X^-(g)$$

As the added electron is attracted to the positively charged nucleus, the process generally gives out energy and is exothermic.

Electron affinity values are given in the data booklet. The noble gases do not generally form negatively charged ions so electron affinity values are not available for these elements.

The electron affinities of the first 18 elements are shown on the following page.

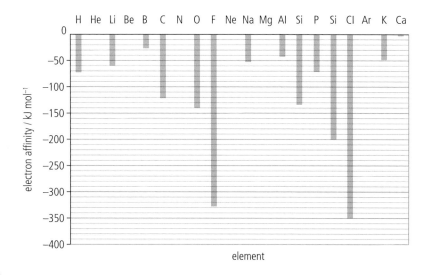

▲ The electron affinities of the first 18 elements. Note there are no values assigned to the noble gases, or to Be, N and Mg.

The group 17 elements have incomplete outer energy levels and a high effective nuclear charge of approximately +7 and so attract electrons the most.

The electron affinities for some group 2 and group 5 elements are not included in the graph. Group 2 elements have an electron configuration ns^2, so the added electron must be placed into a 2p orbital which is further from the nucleus and so experiences reduced electrostatic attraction due to shielding from electrons in the ns orbital.

The value of the electron affinity for beryllium in some sources is positive as there is electrostatic repulsion between the electrons of the Be atom and the added electron. The electrons in the 1s and 2s orbitals of Be shield the added electron from the positively charged nucleus.

A similar effect occurs with nitrogen with the configuration $2s^2 2p_x^1 2p_y^1 2p_z^1$ as the added electron must occupy a p orbital already occupied by a single electron: the attraction between the electron and atom is less than expected as there is increased inter-electron repulsion. Some sources give the electron affinity of nitrogen as zero but there is some uncertainty about its precise value.

The electron affinity of an atom is the energy change that occurs when one mole of electrons is added to one mole of gaseous atoms.

The second and third electron affinities are defined similarly. The second electron affinity for oxygen, for example, corresponds to the change:

$$O^-(g) + e^- \rightarrow O^{2-}(g)$$

This process is endothermic as the added electron is repelled by the negatively charged oxide (O^-) ion, and energy needs to be available for the electron to be added to the negatively charged ion.

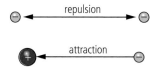

▲ Energy is needed to bring two particles of the same charge closer together because they repel each other. This is an endothermic process. Particles of the opposite charge attract each other. They will spontaneously move closer together. It is an exothermic process.

Electronegativity decreases down a group and increases across a period

The **electronegativity** of an element is a measure of the ability of its atoms to attract electrons in a covalent bond (see Structure 2.2). It is related to ionization energy as it is also a measure of the attraction between the nucleus and its outer electrons – in this case *bonding electrons*. The concept does not apply to the group 18 elements, as they do not generally form covalent bonds.

An element with a high electronegativity has strong electron pulling power and an element with a low electronegativity has weak pulling power. The concept was originally devised by the American chemist Linus Pauling and his values are given in the data booklet. Pauling proposed an electronegativity scale in 1932, which depends on bond energies. The general trends are the same as those for ionization energy.

- Electronegativity increases from left to right across a period owing to the increase in nuclear charge, resulting in an increased attraction between the nucleus and the bond electrons.
- Electronegativity decreases down a group. The bonding electrons are furthest from the nucleus and so there is reduced attraction.

The most electronegative element, fluorine, on the top right of the periodic table, has an electronegativity of 4 and francium on the bottom left has an electronegativity of 0.6.

Although the general trends in ionization energy and electronegativity are the same, they are distinct properties. Ionization energies can be measured directly and are a property of gaseous atoms. Electronegativities are a measure of the atoms' attraction for electrons in a covalent bond.

> Electronegativity is the ability of an atom to attract electrons in a covalent bond.

Linus Pauling has the unique distinction of winning two *unshared* Nobel Prizes – one for Chemistry in 1954 and one for Peace in 1962. His Chemistry Prize was for improving our understanding of the chemical bond and his Peace Prize was for his campaign against nuclear weapons testing.

TOK

The structure of the periodic table was first discovered by observing patterns in the chemical and physical properties of the elements. We now realize that it is determined by the electron configuration of the elements. What roles did inductive and deductive reasoning play in the development of the periodic table? What role does inductive and deductive reasoning have in science in general?

Exercise

Q10. Which properties of the period 3 elements increase from sodium to argon?

    A   nuclear charge and atomic radius

    B   atomic radius and electronegativity

    C   nuclear charge and electronegativity

    D   nuclear charge, atomic radius, and electronegativity

Q11. Which of the following are properties of gaseous atoms?

    I.   ionization energy

    II.   electron affinity

    III.   electronegativity

    A I and II   **B** I and III   **C** II and III   **D** I, II and III

Q12. Which of the following changes is/are endothermic?

    I. $Ca(g) \rightarrow Ca^+(g) + e^-$

    II. $I(g) + e^- \rightarrow I^-(g)$

    III. $O^- + e^- \rightarrow O^{2-}(g)$

    A I and II   **B** I and III   **C** II and III   **D** I, II and III

Q13. Identify the element which is likely to have an electronegativity value most similar to that of sodium.

    A beryllium   **B** calcium   **C** magnesium   **D** hydrogen

Q14. The graph represents the variation of a property of the group 2 elements.

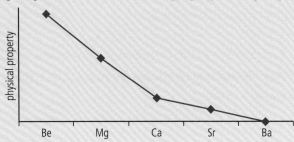

Identify the property.

A	ionic radius	**B**	atomic radius
C	neutron/proton ratio	**D**	first ionization energy

Q15. Which physical property generally increases down a group but decreases from left to right across a period?

A	electron affinity	**B**	electronegativity
C	ionization energy	**D**	atomic radius

Q16. The following graph shows the variation of a physical property, X, of the first 20 elements in the periodic table with the atomic number.

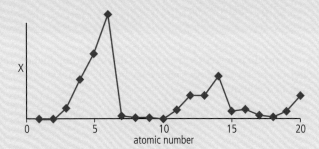

Identify the property X.

A	atomic radius	**B**	first ionization energy
C	ionic radius	**D**	melting point

Q17. Identify the graph which shows the correct ionic radii for the isoelectronic ions Cl^-, K^+, and Ca^{2+}.

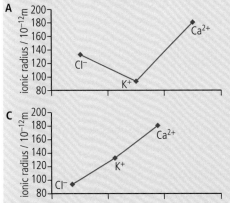

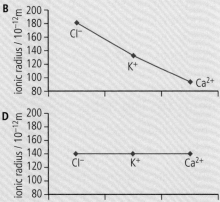

Q18. **(a)** Explain what is meant by the atomic radius of an element.

  (b) The atomic radii of the elements are given in Section 10 of the data booklet.

   (i) Explain why no values for atomic radii are given for the noble gases.

   (ii) Describe and explain the trend in atomic radii across the period 3 elements.

Q19. Si^{4+} has an ionic radius of 4.2×10^{-11} m and Si^{4-} has an ionic radius of 2.71×10^{-10} m.

Explain the large difference in size between the Si^{4+} and Si^{4-} ions.

Q20. Atomic radii and ionic radii are given in the data booklet.

Explain why:

  (a) the potassium ion is much smaller than the potassium atom

  (b) there is a large increase in ionic radius from silicon (Si^{4+}) to phosphorus (P^{3-})

  (c) the ionic radius of Na^+ is less than that of F^-.

Q21. What is the order of decreasing radii for the species Cl, Cl^+, and Cl^-?

Structure 3.1.4 – Periodicity in reactivity

Structure 3.1.4 – Trends in properties of elements down a group include the increasing metallic character of group 1 elements and the decreasing non-metallic character of group 17 elements.

Describe and explain the reactions of group 1 metals with water, and of group 17 elements with halide ions.

	Inquiry 2, Tool 2 – Why are simulations and online reactions often used in exploring the trends in chemical reactivity of group 1 and group 17 elements?

Chemical properties of an element are largely determined by the number of valence electrons in their outer energy level.

Group 18 used to be called the 'inert gases' as it was thought that they were completely unreactive. Xenon, which has the lowest ionization energy, is the most reactive element in the group as reactions involve the withdrawal of electrons from the parent atom. The first compound of xenon was made in 1962.

Trends in chemical properties

The chemical properties of an element are determined by the electron configuration of its atoms. Elements of the same group have similar chemical properties as they have the same number of valence electrons in their outer energy level. The alkali metals in group 1, for example, all have one electron in their outer energy level and the halogens in group 17 all have seven outer electrons.

Group 18: The noble gases

To understand the reactivity of the elements, it is instructive to consider group 18, which contains the least reactive elements – the noble gases.

The reactivity of elements in other groups can be explained by their unstable incomplete electron energy levels. They lose or gain electrons to achieve the electron configuration of their nearest noble gas.

Group 1: The alkali metals

All the group 1 elements are silvery metals and are too reactive to be found in nature. They are usually stored in oil to prevent contact with air and water. The properties of the first three elements are summarized in the table below.

Physical properties	Chemical properties
• They are good conductors of electricity and heat. • They have low densities. • They have shiny grey surfaces when freshly cut with a knife.	• They are very reactive metals. • They form ionic compounds with non-metals.

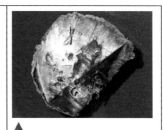

▲ Lithium is a soft reactive metal. When freshly cut, it has a metallic luster. However, it rapidly reacts with oxygen in the air, giving it a dark oxide coat.

▲ A piece of sodium. The shiny surface has been exposed by a knife cutting the soft metal. Sodium is softer and more reactive than lithium.

▲ Freshly cut shiny surface of potassium metal. Potassium is softer and more reactive than sodium.

As discussed in Structure 2.1, group 1 metals form singly charged ions, M^+, with the stable octet of the noble gases when they react. Their low ionization energies give an indication of the ease with which the outer electron is lost. Reactivity increases down the group as the elements with higher atomic numbers have the lowest ionization energies. Their ability to conduct electricity and heat is also due to the mobility of their outer electron.

Reaction with water

The alkali metals react with water to produce hydrogen and the metal hydroxide. When you drop a piece of one of the first three group 1 elements into a small beaker containing distilled water, the following happens.

- Lithium floats and reacts slowly. It releases hydrogen but keeps its shape.
- Sodium reacts with a vigorous release of hydrogen. The heat produced is sufficient to melt the unreacted metal, which forms a small ball that moves around on the water surface.
- Potassium reacts even more vigorously to produce sufficient heat to ignite the hydrogen produced. It produces a lilac flame and moves excitedly on the water surface.

They are called alkali metals because the resulting solution is alkaline owing to the presence of the hydroxide ion formed. For example, for potassium:

$$2K(s) + 2H_2O(l) \rightarrow 2KOH(aq) + H_2(g)$$

As KOH is an ionic compound (Structure 2.1) which dissociates in water (see chapter opener image), it is more appropriate to write the equation as:

$$2K(s) + 2H_2O(l) \rightarrow 2K^+(aq) + 2OH^-(aq) + H_2(g)$$

 SKILLS

Simulations and online reactions are available to explore the trends in the dangerously reactive group 1 elements.

245

The reaction becomes more vigorous as you go down the group. The most reactive element, caesium, has the lowest ionization energy and so forms positive ions most readily.

Group 17: The halogens

The group 17 elements exist as diatomic molecules, X_2. Their physical and chemical properties are summarized in the table below.

Physical properties	Chemical properties
• They are coloured. • They show a gradual change from gases (F_2 and Cl_2), to liquid (Br_2), and solids (I_2 and At_2).	• They are very reactive non-metals. • Reactivity decreases down the group. • They form ionic compounds with metals and covalent compounds with other non-metals.

From left to right: chlorine (Cl_2), bromine (Br_2), and iodine (I_2). These are toxic and reactive non-metals. Chlorine is a green gas at room temperature. Bromine is a dark liquid, although it readily produces a brown vapor. Iodine is a crystalline solid.

The names of diatomic elements all end in *-ine* or *-gen*.

Two halogens are named by their colours: *chloros* means 'yellowish green' and *ioeides* is 'violet' in Greek. One is named by its smell: '*bromos*' is the Greek word for 'stench'.

The word halogen means salt maker.

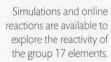

Simulations and online reactions are available to explore the reactivity of the group 17 elements.

Inquiry 2, Tool 2 – Why are simulations and online reactions often used in exploring the trends in chemical reactivity of group 1 and group 17 elements?

The group 17 trend in reactivity can be explained by their readiness to accept electrons, as illustrated by their very exothermic electron affinities discussed earlier. The nuclei have a high effective charge, of approximately +7, and so exert a strong pull on any electron from other atoms. This electron can then occupy the outer energy level of the halogen atom and complete a stable octet. The attraction is greatest for the smallest atom fluorine, which is the most reactive non-metal in the periodic table. Non-metal reactivity decreases down the group as the atomic radius increases and the attraction for outer electrons decreases.

TOK Chlorine was used as a chemical weapon during the First World War. Should scientists be held morally responsible for the applications of their discoveries? Should scientific research be subject to ethical constraints or is the pursuit of all scientific knowledge intrinsically worthwhile?

Displacement reactions

The *relative* reactivity of the group 17 elements can be seen by placing them in direct competition for an extra electron. When chlorine is bubbled through a solution of potassium bromide, the solution changes from colourless to orange owing to the production of bromine:

$$2KBr(aq) + Cl_2(aq) \rightarrow 2KCl(aq) + Br_2(aq)$$

$$2Br^-(aq) + Cl_2(aq) \rightarrow 2Cl^-(aq) + Br_2(aq)$$

A chlorine nucleus has a stronger attraction for an electron than a bromine nucleus does because of its smaller atomic radius and so it takes the electron from the bromide ion. The chlorine has gained an electron and so forms the chloride ion, Cl^-. The bromide ion loses an electron to form bromine.

Other reactions are:

$$2I^-(aq) + Cl_2(aq) \rightarrow 2Cl^-(aq) + I_2(aq)$$

The colour changes from colourless to dark orange/brown owing to the formation of iodine.

$$2I^-(aq) + Br_2(aq) \rightarrow 2Br^-(aq) + I_2(aq)$$

To distinguish between bromine and iodine more effectively, the final solution can be shaken with a hydrocarbon solvent. Iodine forms a violet solution and bromine forms a dark orange solution as shown in the photo below.

When chlorine water is added to the colourless potassium bromide solution, bromine (yellow/orange) is formed. Bromine is displaced from solution by the more reactive chlorine.

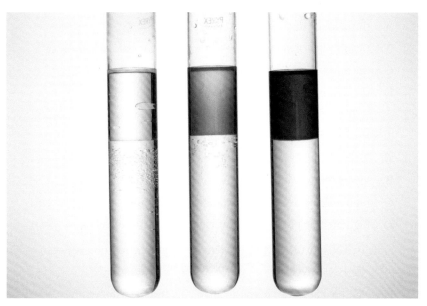

Solutions of chlorine (left), bromine (middle), and iodine (right) in water (lower part) and cyclohexane (upper part). Chlorine dissolves in water, but the halogens are generally more soluble in non-polar solvents such as cyclohexane.

The more reactive halogen displaces the ions of the less reactive halogen from its compounds.

SKILLS

Determining the properties of the halogens. Full details of how to carry out this experiment with a worksheet are available in the eBook.

The halides

The halogens form insoluble salts with silver. Adding a solution containing the halide to a solution containing silver ions produces a **precipitate** that is useful in identifying the halide ion.

$$Ag^+(aq) + X^-(aq) \rightarrow AgX(s)$$

This is shown in the photo on the next page.

Silver halide precipitates formed by reacting silver nitrate ($AgNO_3$) with solutions of the halides. From left to right, these are silver chloride (AgCl), silver bromide (AgBr), and silver iodide (AgI).

The reaction between silver nitrate solution and the different halide solutions can be investigated in the lab.

Exercise

Q22. Which pair of elements reacts most readily?

    A $Li + Br_2$ **B** $Li + Cl_2$ **C** $K + Br_2$ **D** $K + Cl_2$

Q23. Chlorine is a greenish-yellow gas, bromine is a dark red liquid, and iodine is a dark grey solid. Identify the property which most directly causes these differences in volatility.

    A   the halogen–halogen bond energy

    B   the number of neutrons in the nucleus of the halogen atom

    C   the number of outer electrons in the halogen atom

    D   the number of electrons in the halogen molecule

Q24. Caesium is an element in group 1 of the periodic table. Which is the best description of its melting point and reactivity with water?

    A   low melting point and reacts slowly with water

    B   low melting point and reacts vigorously with water

    C   high melting point and reacts slowly with water

    D   high melting point and reacts vigorously with water

Q25. State two observations you could make during the reaction between sodium and water. Give an equation for the reaction.

Q26. How do the reactivities of the alkali metals and the halogens vary down the group?

Q27. Explain the trend in reactivity of the alkali metals in terms of their valence electron configuration.

Q28. Explain the trend in reactivity of the halogens in terms of their valence electron configuration.

Structure 3.1.5 – Metal and non-metal oxides

> **Structure 3.1.5 – Metallic and non-metallic properties show a continuum. This includes the trend from basic metal oxides through amphoteric to acidic non-metal oxides.**
>
> Deduce equations for the reactions with water of the oxides of group 1 and group 2 metals, carbon and sulfur.
>
> | Include acid rain caused by gaseous non-metal oxides, and ocean acidification caused by increasing CO_2 levels. | Structure 2.1, 2.2 – How do differences in bonding explain the differences in the properties of metal and non-metal oxides? |
> | --- | --- |

Bonding of period 3 oxides

The transition from metallic to non-metallic character is illustrated by the bonding of the period 3 oxides. Ionic compounds are generally formed between metal and non-metal elements and so the oxides of elements Na to Al have **giant ionic** structures. Covalent compounds are formed between non-metals, so the oxides of phosphorus, sulfur, and chlorine are **molecular covalent**. The oxide of silicon, which is a metalloid, has a **giant covalent** structure. The different types of structures are explored more fully in Structure 2.

The ionic character of a compound depends on the *difference* in electronegativity between its elements. Oxygen has an electronegativity of 3.4, so the ionic character of the period 3 oxides decreases from left to right, as the electronegativity values of the elements approach this value (Figure 4).

SKILLS

Oxides of period 3. Full details of how to carry out this experiment with a worksheet are available in the eBook.

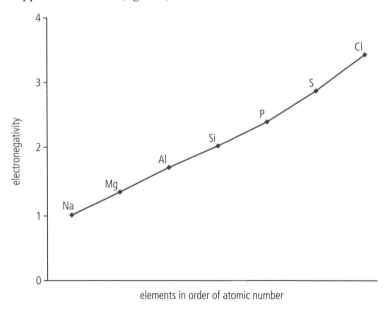

▲
S3.1 Figure 4 Electronegativities increase across the period and approach 3.4, the value for oxygen.

The oxides become more ionic down a group as the electronegativity decreases. The conductivity of the molten oxides gives an experimental measure of their ionic character, as is shown in the table on the next page. Oxides only conduct electricity in the liquid or aqueous state when the ions are free to move.

Note that the maximum oxidation state of a period 3 element is related to the group number. It is +1 for elements in group 1, +2 for elements in group +2, +3 for elements in group 13, +4 for elements in group 14, and so on. Oxidation states are discussed further on page 255.

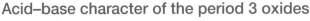

Formula of oxide	$Na_2O(s)$	$MgO(s)$	$Al_2O_3(s)$	$SiO_2(s)$	$P_4O_{10}(s)$ / $P_4O_6(s)$	$SO_3(l)$ / $SO_2(g)$	$Cl_2O_7(l)$ / $Cl_2O(g)$
Oxidation number	+1	+2	+3	+4	+5 / +3	+6 / +4	+7 / +1
Electrical conductivity in molten state	high	high	high	very low	none	none	none
Structure	giant ionic			giant covalent	molecular covalent		

> High melting points are associated with ionic or covalent giant structures, and low melting points with molecular covalent structures.

Acid–base character of the period 3 oxides

The acid–base properties of the oxides are closely linked to their bonding and structure. Metallic elements, which form ionic oxides, are basic; non-metal oxides, which are covalent, are acidic. Aluminium oxide, which can be considered as an ionic oxide with some covalent character, shows amphoteric properties – reacting with both acids and bases. The acid–base properties of period 3 oxides are shown in the table below.

> Oxides of metals are ionic and basic. Oxides of the non-metals are covalent and acidic. Amphoteric oxides show both acidic and basic properties. Aluminium oxide is amphoteric.

Formula of oxide	$Na_2O(s)$	$MgO(s)$	$Al_2O_3(s)$	$SiO_2(s)$	$P_4O_{10}(s)$ / $P_4O_6(s)$	$SO_3(l)$ / $SO_2(g)$	$Cl_2O_7(l)$ / $Cl_2O(g)$
Acid–base character	basic		amphoteric	acidic			

> The pH of aqueous solutions of metal and non-metal oxides can be investigated in the lab.

Aqueous solutions of the oxides of some elements with universal indicator. Sulfur trioxide forms sulfuric acid in water, which is highly acidic. Sodium oxide forms sodium hydroxide, which is a strong alkali. Non-metal oxides have pHs below 7 and are acidic. Metal oxides have high pHs above 7 and are basic.

Basic oxides

Sodium oxide and magnesium oxide dissolve in water to form alkaline solutions owing to the presence of hydroxide ions:

$$Na_2O(s) + H_2O(l) \rightarrow 2NaOH(aq)$$

$$MgO(s) + H_2O(l) \rightarrow Mg(OH)_2(aq)$$

A basic oxide reacts with an acid to form a salt and water. The oxide ion combines with two H^+ ions to form water:

$$O^{2-}(s) + 2H^+(aq) \rightarrow H_2O(l)$$

$$Li_2O(s) + 2HCl(aq) \rightarrow 2LiCl(aq) + H_2O(l)$$

$$MgO(s) + 2HCl(aq) \rightarrow MgCl_2(aq) + H_2O(l)$$

Non-metallic oxides are acidic

The non-metallic oxides react readily with water to produce acidic solutions.

Carbon dioxide dissolves in water to form the weak acid, carbonic acid, H_2CO_3.

$$H_2O(l) + CO_2(g) \rightleftharpoons H_2CO_3(aq)$$

Sulfur trioxide reacts with water to produce sulfuric(VI) acid:

$$SO_3(l) + H_2O(l) \rightarrow H_2SO_4(aq)$$

Sulfur dioxide reacts with water to produce sulfuric(IV) acid:

$$SO_2(g) + H_2O(l) \rightarrow H_2SO_3(aq)$$

Amphoteric oxides

Aluminium oxide does not affect the pH when it is added to water as it is essentially insoluble. It has amphoteric properties, however, as it shows both acid and base behavior. For example, it behaves as a base when it reacts with sulfuric acid:

$$Al_2O_3(s) + 6H^+(aq) \rightarrow 2Al^{3+}(aq) + 3H_2O(l)$$

$$Al_2O_3(s) + 3H_2SO_4(aq) \rightarrow Al_2(SO_4)_3(aq) + 3H_2O(l)$$

and behaves as an acid when it reacts with alkalis such as sodium hydroxide:

$$Al_2O_3(s) + 3H_2O(l) + 2OH^-(aq) \rightarrow 2Al(OH)_4^-(aq)$$

Acid rain and acid deposition are produced by non-metal oxides

All rainwater is naturally acidic due to the presence of dissolved carbon dioxide. The minimum pH of carbonic acid solutions is 5.6. Acid rain refers to solutions with a lower pH owing to the presence of sulfur and nitrogen oxides.

Acid deposition is a broader term than acid rain and includes all processes by which acidic components as precipitates or gases leave the atmosphere.
There are two main types of acid deposition:
- Wet acid deposition: rain, snow, sleet, hail, fog, mist and dew fall to the ground as aqueous precipitates.
- Dry acid deposition: acidifying particles and gases fall to the ground as dust and smoke, and later dissolve in water to form acids.

Alkalis are bases that are soluble in water. They form hydroxide ions in aqueous solution.

Structure 2.1, 2.2 – How do differences in bonding explain the differences in the properties of metal and non-metal oxides?

Although all rainwater is acidic, the term 'acid rain' only applies to some rainwater.
How does language as a way of knowing influence communication in science?

Rainwater is naturally acidic due to dissolved CO_2. Acid rain has a pH < 5.6.

Sulfur oxides produce acid rain

Sulfur dioxide, SO_2, is produced from the burning of fossil fuels, particularly coal and heavy oil in power plants used to generate electricity. It is also released in industrial processes of smelting where metals are extracted from their ores. It is estimated that about 50% of annual global emissions of sulfur dioxide come from coal.

$$S(s) + O_2(g) \rightarrow SO_2(g)$$

Challenge yourself

5. What is the source of sulfur in fossil fuels? Consider why some fuels such as coal contain higher amounts of sulfur.

Sulfur dioxide dissolves in rainwater to form sulfuric(IV) or sulfurous acid, $H_2SO_3(aq)$.

$$H_2O(l) + SO_2(g) \rightarrow H_2SO_3(aq)$$

Sulfur dioxide can be oxidized to sulfur trioxide, SO_3, in the atmosphere, which then dissolves in rainwater to form sulfuric (VI) acid, H_2SO_4.

$$2SO_2(g) + O_2(g) \rightarrow 2SO_3(g)$$

$$H_2O(l) + SO_3(g) \rightarrow H_2SO_4(aq)$$

Nitrogen oxides produce acid rain

Several other oxides of nitrogen exist, and the term NO_x is used somewhat variably. But in atmospheric chemistry it refers specifically to the total of the two oxides of nitrogen, NO and NO_2, present.

Nitrogen monoxide, NO, is produced mainly from internal combustion engines, where the burning of the fuel releases heat energy that causes nitrogen and oxygen from the air to combine.

$$N_2(g) + O_2(g) \rightarrow 2NO(g)$$

A similar reaction gives rise directly to the brown gas, nitrogen dioxide, NO_2.

$$N_2(g) + 2O_2(g) \rightarrow 2NO_2(g)$$

Nitrogen dioxide also forms from the oxidation of nitrogen monoxide.

$$2NO(g) + O_2(g) \rightarrow 2NO_2(g)$$

Nitrogen dioxide dissolves in rainwater to form a mixture of nitric(III) or nitrous acid (HNO_2) and nitric (V) acid (HNO_3).

$$H_2O(l) + 2NO_2(g) \rightarrow HNO_2(aq) + HNO_3(aq)$$

Alternatively, nitrogen dioxide can be oxidized to form nitric(V) acid.

$$2H_2O(l) + 4NO_2(g) + O_2(g) \rightarrow 4HNO_3(aq)$$

One of the most controversial aspects of acid deposition is that its effects often occur far from the source of the pollutants, due to atmospheric weather patterns. In many cases, this means countries are suffering the impact of other countries' industrial processes. Legislation to reduce the impact of acid deposition has therefore been the subject of intense political debate, and several international protocols have developed.

Nature of Science

Intense scientific research into the formation and impacts of acidic rainfall has led to policy responses at the national and local level. The implications of science within environmental, political, and economic domains can be profound and illustrate the importance of local, national and international scientific bodies that engage with the public understanding of science.

Challenge yourself

6. Acid rain damages building materials.

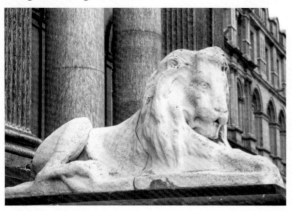

Limestone is a form of calcium carbonate. Deduce an equation for the reaction that accounts for the damage to the stone lion.

Ocean acidification occurs as carbon dioxide dissolves in the oceans

About 50% of the carbon dioxide produced by the combustion of fossil fuels is dissolved by the oceans. The carbon dioxide reacts with the water to form carbonic acid.

$$CO_2(aq) + H_2O(l) \rightleftharpoons H_2CO_3(aq)$$

The carbonic acid then reacts to form hydrogencarbonates or carbonates.

$$H_2CO_3(aq) \rightleftharpoons H^+(aq) + HCO_3^-(aq)$$

$$HCO_3^-(aq) \rightleftharpoons H^+(aq) + CO_3^{2-}(aq)$$

This leads to higher ocean acidity, mainly near the surface, where the carbon dioxide is absorbed. This higher acidity inhibits shell growth in marine animals and is suspected to be a cause of reproductive disorders in some fish.

Ocean acidity dissolving sea snail shell. The shell is being dissolved by increasing ocean acidity due to carbon dioxide released into the atmosphere from human sources. Ocean acidification is predicted to have a disastrous effect on marine ecosystems.

Exercise

Q29. An oxide of a period 3 element is a solid at room temperature and forms a basic solution. Identify the element.

    A Mg   **B** Al   **C** P   **D** S

Q30. Identify the oxide which forms an acidic solution when added to water.

    A $Na_2O(s)$ **B** $MgO(s)$ **C** $SiO_2(s)$ **D** $SO_3(g)$

Q31. Identify the oxides that produce an acidic solution when added to water.

    I. SiO_2 II. N_2O_5 III. SO_2

    A I and II only   **B** I and III only   **C** II and III only   **D** I, II and III

Q32. The melting and boiling points of four oxides are shown.

Oxide	Melting point / K	Boiling point / K
MgO	3125	3873
SiO_2 (quartz)	1883	2503
P_4O_{10}	297	448
SO_2	200	263

(a) Identify the states of the four oxides listed under standard conditions.

(b) Explain the difference in melting points by referring to the bonding and structure in each case.

(c) The oxides are added to separate samples of pure water. State whether each resulting liquid is acidic, neutral, or alkaline. Describe all chemical reactions by giving chemical equations.

(d) Use chemical equations to describe the reactions of aluminium oxide with:

    (i) hydrochloric acid

    (ii) sodium hydroxide.

Q33. Describe the acid–base character of the oxides of the period 3 elements, Na to Ar. For sodium oxide and sulfur trioxide, write balanced equations to illustrate their acid–base character.

Q34. Acid rain has a pH of less than 5.6.

(a) Explain why natural rain has a pH of around 5.6. Give a chemical equation to support your answer.

(b) Acid rain may be 50 times more acidic than natural rain. One of the major acids present in acid rain originates mainly from burning coal. State the name of the acid and give equations to show how it is formed.

(c) The second major acid responsible for acid rain originates mainly from internal combustion engines. State the name of this acid and state two different ways in which its production can be reduced.

Structure 3.1.6 – Oxidation states

Structure 3.1.6 – The oxidation state is a number assigned to an atom to show the number of electrons transferred in forming a bond. It is the charge that the atom would have if the compound were composed of ions.

Deduce the oxidation states of an atom in an ion or a compound.

Explain why the oxidation state of an element is zero.

Oxidation states are shown with a + or − sign followed by the Arabic symbol for the number, e.g. +2, −1. Examples should include hydrogen in metal hydrides (−1) and oxygen in peroxides (−1). The terms 'oxidation number' and 'oxidation state' are often used interchangeably, and either term is acceptable in assessment. Naming conventions for oxyanions use oxidation numbers shown with Roman numerals, but generic names persist and are acceptable. Examples include NO_3^- nitrate, NO_2^- nitrite, SO_4^{2-} sulfate and SO_3^{2-} sulfite.	Reactivity 3.2 – How can oxidation states be used to analyze redox reactions?

The oxidation state is the charge the atom would have if the compound were composed of ions

Oxidation can involve elements other than oxygen

When magnesium reacts with oxygen it is oxidized and forms the ionic compound magnesium oxide.

$$Mg(s) + \tfrac{1}{2}O_2(g) \rightarrow MgO\ (s)$$

If this reaction is considered from the perspective of magnesium, we see that a magnesium atom has been converted into an ion and lost two electrons:

$$Mg \rightarrow Mg^{2+} + 2e^-$$
$$\text{atom} \quad \text{ion}$$

As magnesium also loses electrons when it reacts with other non-metals such as chlorine, it makes sense to extend the term oxidation to other reactions which do not necessarily involve oxygen:

$$Mg(s) + Cl_2(g) \rightarrow MgCl_2(s)$$

As discussed in detail in Reactivity 3.2, the term 'oxidation' is applied to any electron transfer reaction. Magnesium has changed from a neutral atom to an ion with a charge of 2+. We say its oxidation state has increased from 0 to +2. Oxidation has led to an increase in oxidation state.

These points can be summarized as follows.

Oxidation is:

- the addition of oxygen
- the removal of hydrogen
- electron loss
- an increase in oxidation state.

Oxidation states are a measure of how electrons are distributed

An increase in oxidation state means the atom has lost electron control and is oxidized. A decrease in oxidation state means that an atom has gained electron control and is reduced.

Reactivity 3.2 – How can oxidation states be used to analyze redox reactions?

The oxidation state of an element is zero as the atoms have no charge and the electrons are distributed equally between the atoms. In binary ionic compounds such as magnesium oxide, the oxidation states are the same as the ionic charges. The oxidation state of sodium in sodium chloride is +1 as it has lost one electron, and that of chlorine is −1 as it has gained one electron.

In covalent substances, the electron distribution depends on the polarity of the bond and the relative electronegativities of the bonding atoms. If an atom is bonded to a more electronegative atom, it loses some control over its electrons and has a positive oxidation state. When bonded to a less electronegative atom, it gains control over more electrons and has a negative oxidation state.

Rules to assign oxidation states

Here are some rules to follow when assigning the oxidation state of an atom in any species.

	Rule	Example
1.	Atoms in the free (uncombined) element have an oxidation state of zero.	The oxidation states of Mg, O_2, N_2, and Ar are all 0.
2.	In simple ions, the oxidation state is the same as the charge on the ion.	The oxidation state of Mg in Mg^{2+} is +2, oxygen in O^{2-} is −2, and nitrogen in N^{3-} is −3.
3.	The oxidation states of all the atoms in a neutral (uncharged) compound must add up to zero.	The sum of oxidation states in $H_2SO_4 = 0$.
4.	The oxidation states of all the atoms in a polyatomic ion must add up to the charge on the ion.	In SO_4^{2-} the sum of oxidation states = −2.
5.	The usual oxidation state for an element in a compound is the same as the charge on its most common ion.	In compounds: • group 1 metals are +1; • group 2 metals are +2.
6.	F has the oxidation state of −1 in all its compounds as it is the most electronegative element.	
7.	O has the oxidation state of −2 except • when in peroxides such as H_2O_2, where it is −1; • when bonded to fluorine it is positive as it is less electronegative than fluorine.	Oxidation number of O is: • −1 in H_2O_2 • +2 in OF_2
8.	Cl has the oxidation state of −1 except when bonded to the more electronegative F and O.	Oxidation number of Cl is: • +1 in ClF • +1 in OCl_2 • +7 in ClO_4^-
9.	H has the oxidation state of +1 except when bonded to the group 1 and 2 metals when it forms ionic hydrides.	Oxidation number of H is −1 in NaH and in MgH_2.

Worked example

Assign oxidation states to all the elements in **(a)** H_2SO_4 and **(b)** SO_3^{2-}.

Solution

(a) H_2SO_4:

The H atoms can be given a value of $+1$ and the O atoms a value of -2.

As H_2SO_4 is electrically neutral, the sum of all oxidation states $= 0$.

$$H_2SO_4$$

$$2 \times (+1) \ + \ S \ + \ 4 \times (-2) = 0$$

$$\therefore S = +8 - 2 = +6$$

(b) SO_3^{2-}:

We can start by assigning $O = -2$.

Note that here the oxidation states must add up to -2, the charge on the ion.

Therefore, $S + (3 \times -2) = -2$

$S = -2 + 6 = +4$

Worked example

Identify the elements which are oxidized and reduced in the reaction between hydrogen and chlorine to produce hydrogen chloride:

$$H_2(g) + Cl_2(g) \rightarrow 2HCl(g)$$

Solution

Assign oxidation states to the elements in the reactants and products:

$$H_2(g) + Cl_2(g) \rightarrow 2HCl(g)$$

Oxidation states   0   0   (H)$+1$ (Cl)-1

The oxidation state of hydrogen increases: it is oxidized.

The oxidation state of chlorine decreases: it is reduced.

Worked example

Nitrogen dioxide dissolves in rainwater to form a mixture of nitrous acid (HNO_2) and nitric acid (HNO_3).

$$H_2O(l) + 2NO_2(g) \rightarrow HNO_2(aq) + HNO_3(aq)$$

Identify the element that is both oxidized and reduced in the reaction.

Solution

Assign oxidation states to the elements in the reactants and products:

$$H_2O(l) + 2NO_2(g) \rightarrow HNO_2(aq) + HNO_3(aq)$$

Oxidation states	(H) +1	(N) +4	(H) +1	(H) +1
	(O) −2	(O) −2	(N) +3	(N) +5
			(O) −2	(O) −2

The oxidation state of nitrogen increases and decreases. Nitrogen is both oxidized and reduced. This is an example of a disproportionation reaction.

Systematic names of compounds use oxidation numbers

In the first worked example on page 257, we can see that sulfur exhibits different oxidation states in different compounds. We also discussed sulfuric acid ($H_2SO_4(aq)$) and sulfurous acid ($H_2SO_3(aq)$) in acid rain in the last section. Although these traditional names for the acid are acceptable, IUPAC introduced a nomenclature using oxidation numbers to make the names more recognizable and unambiguous. This system uses Roman numerals corresponding to the oxidation state, which are inserted in brackets after the name of the element. As we have seen in the worked example, S is +6 in H_2SO_4, so it is called sulfuric(VI) acid, and the corresponding oxyanion SO_4^{2-} is the sulfate(VI) ion. Using the same rules, H_2SO_3 is sulfuric(IV) acid and the SO_3^{2-} ion is the sulfate(IV) ion.

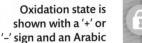

Worked example

Deduce the names of the following compounds using oxidation numbers.

(a) N_2O_5 **(b)** NO_2 **(c)** HNO_2

Solution

1. Deduce the oxidation state of N:

N_2O_5	NO_2	HNO_2
As O = −2	As O = −2	As H = +1 and O = −2
$2N + (5 \times -2) = 0$	$N + (2 \times -2) = 0$	$+1 + N + (2 \times -2) = 0$
$2N = +10$	$N = +4$	$N = +4 - 1$
$N = +5$		$= +3$

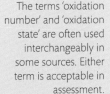

2. Insert the corresponding Roman numeral after the reference to nitrogen in the name. There is no space between the name and the number, and the number is placed in brackets.

N_2O_5	NO_2	HNO_2
nitrogen(V) oxide	nitrogen(IV) oxide	nitric(III) acid

Exercise

Q35. Identify the pair of compounds in which chromium is in the same oxidation state.

   A Cr_2O_3 and CrO_3

   B CrO_3 and $CrCl_3$

   C K_2CrO_4 and $K_2Cr_2O_7$

   D Cr^{2+} and CrO_2

Q36. Deduce the oxidation number of sulfur in $Na_2S_4O_6$.

   A   +6

   B   +3

   C   +2.5

   D   +2

Q37. Assign oxidation states to all elements in the following compounds:

   (a) NH_4^+ **(b)** $SnCl_2$ **(c)** H_2O **(d)** NO_3^-

   (e) PbO_2 **(f)** PO_4^{3-} **(g)** ClO_4^- **(h)** $HOCl$

Q38. Deduce the names of the following compounds, using their oxidation numbers.

   (a) HNO_3 **(b)** HNO_2 **(c)** H_3PO_4 **(d)** PbO_2 **(e)** $PbSO_4$

Q39. Deduce the formulas of the following compounds, using their oxidation numbers.

   (a) copper(I) oxide   **(b)** iron(III) oxide

   (c) tin(IV) oxide   **(d)** sodium bromate(V)

Q40. Use oxidation states to deduce which element is oxidized and which is reduced in the following reactions.

   (a) $Cl_2(aq) + 2NaI(aq) \rightarrow I_2(aq) + 2NaCl(aq)$

   (b) $2H_2O(l) + 2F_2(g) \rightarrow 4HF(aq) + O_2(g)$

   (c) $P_4(s) + 10F_2(g) \rightarrow 4PF_5(g)$

   (d) $H_2(g) + 2Na(s) \rightarrow 2NaH(s)$

Guiding Question revisited

How does the periodic table help us to predict patterns and trends in the properties of elements?

In this chapter, we have seen how the periodic table highlights the patterns and trends in the properties of the elements.

- The periodicity in the physical and chemical properties of elements is due to the periodicity of their electron configuration.
- Metals are on the left of the table, non-metals are on the upper right-hand side and metalloids are in a diagonal area between.
- The columns of the table are called groups and the rows are called periods. Elements in the same group show similar chemical properties as they have the same number of electrons in their valence energy level.
- The group 1 or alkali metals react with water to produce hydroxides and hydrogen. Melting points decrease down the group and metal reactivity increases down the group. They react by losing electrons.
- The group 17 elements or halogens are reactive non-metals. Melting and boiling points increase down the group and non-metal reactivity decreases down the group. They react by gaining electrons.
- Group 1 and group 17 elements are on opposite sides of the table and show opposite trends in reactivities and melting points.
- The group 18 elements or noble gases are unreactive as they have complete outer energy levels.

Practice questions

1. Which property generally decreases across period 3?

   A atomic number

   B electronegativity

   C atomic radius

   D first ionization energy

2. Which statements about period 3 are correct?

   I. The electronegativity of the elements increases across period 3.

   II. The atomic radii of the elements decrease across period 3.

   III. The oxides of the elements change from acidic to basic across period 3.

   A I and II only   **B** I and III only   **C** II and III only   **D** I, II and III

3. Which property decreases down group 17 in the periodic table?

   A melting point   **B** electronegativity

   C atomic radius   **D** ionic radius

4. Which oxides produce an acidic solution when added to water?

   I. P_4O_{10} II. MgO III. SO_3

   A I and II only   **B** I and III only   **C** II and III only   **D** I, II and III

5. The *x*-axis of the graph below represents the atomic number of the elements in Period 3.

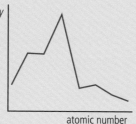

Which variable could represent the *y*-axis?

A melting point   **B** electronegativity

C ionic radius   **D** atomic radius

6. Which is the best definition of *electronegativity*?

A Electronegativity is the energy required for a gaseous atom to gain an electron.

B Electronegativity is the attraction of an atom for a bonding pair of electrons.

C Electronegativity is the attraction between the nucleus and the valence electrons of an atom.

D Electronegativity is the ability of an atom to attract electrons from another atom.

7. What happens when sodium is added to water?

I. A gas is evolved.

II. The temperature of the water increases.

III. A clear, colourless solution is formed.

A I and II only   **B** I and III only   **C** II and III only   **D** I, II and III

8. Which species could be oxidized to form $S_2O_3^{2-}$?

A S   **B** SO_2 **C** H_2SO_3 **D** H_2SO_4

9. Which combination describes the acid–base nature of aluminium and phosphorus oxides?

	Aluminium	Phosphorus
A	Amphoteric oxide	Acidic oxide
B	Basic oxide	Amphoteric oxide
C	Acidic oxide	Amphoteric oxide
D	Amphoteric oxide	Basic oxide

10. Which property shows a general increase from left to right across period 2, Li to F?

A Melting point   **B** Electronegativity

C Ionic radius   **D** Electrical conductivity

11. Which processes could result in acid rain?

    I.   burning coal

    II   internal combustion engines

    III. burning methane

    A   I and II only   **B**   I and III only   **C**   II and III only   **D**   I, II and III

12. (a) Define the term *first ionization energy*.   (2)

    (b) Explain why the first ionization energy of magnesium is higher than
         that of sodium.   (2)

                                                                              (Total 4 marks)

13. Samples of sodium oxide and sulfur trioxide are added to separate beakers of
    water. Deduce the equation for each reaction and identify each oxide as acidic,
    basic, or neutral.   (3)

                                                                              (Total 3 marks)

14. Describe and explain what you will see if chlorine gas is bubbled through a
    solution of:

    (a) potassium iodide   (2)

    (b) potassium fluoride   (1)

                                                                              (Total 3 marks)

15. The periodic table shows the relationship between electron configuration and
    the properties of elements and is a valuable tool for making predictions in
    chemistry.

    (a) Identify the property used to arrange the elements in the periodic table.   (1)

    (b) Outline two reasons why electronegativity increases across period 3
         in the periodic table and one reason why noble gases are not assigned
         electronegativity values.   (3)

                                                                              (Total 4 marks)

16. (a) Outline two reasons why a sodium ion has a smaller radius than a
         sodium atom.   (2)

    (b) Explain why the ionic radius of P^{3-} is greater than the ionic radius of Si^{4+}.
                                                                                              (2)

                                                                              (Total 4 marks)

17. Sodium oxide, Na_2O, is a white solid with a high melting point.

    (a) Explain why solid sodium oxide is a non-conductor of electricity.   (1)

    (b) Molten sodium oxide is a good conductor of electricity. State the half-
         equation for the reaction occurring at the positive electrode during the
         electrolysis of molten sodium oxide.   (1)

    (c) (i)   State the acid/base nature of sodium oxide.   (1)

         (ii) State the equation for the reaction of sodium oxide with water.   (1)

                                                                              (Total 4 marks)

18. The graph of the first ionization energy plotted against atomic number for the first 20 elements shows periodicity.

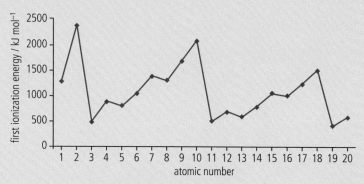

(a) State what is meant by the term *periodicity*. (1)

(b) State the electron arrangement of argon and explain why the noble gases, helium, neon and argon, show the highest first ionization energies for their respective periods. (3)

(c) A graph of atomic radius plotted against atomic number shows that the atomic radius decreases across a period. Explain why chlorine has a smaller atomic radius than sodium. (1)

(d) Explain why a sulfide ion, S^{2-}, is larger than a chloride ion, Cl^-. (1)

(e) Explain why the melting points of the group 1 metals (Li to Cs) decrease down the group whereas the melting points of the group 17 elements (F to I) increase down the group. (3)

(Total 9 marks)

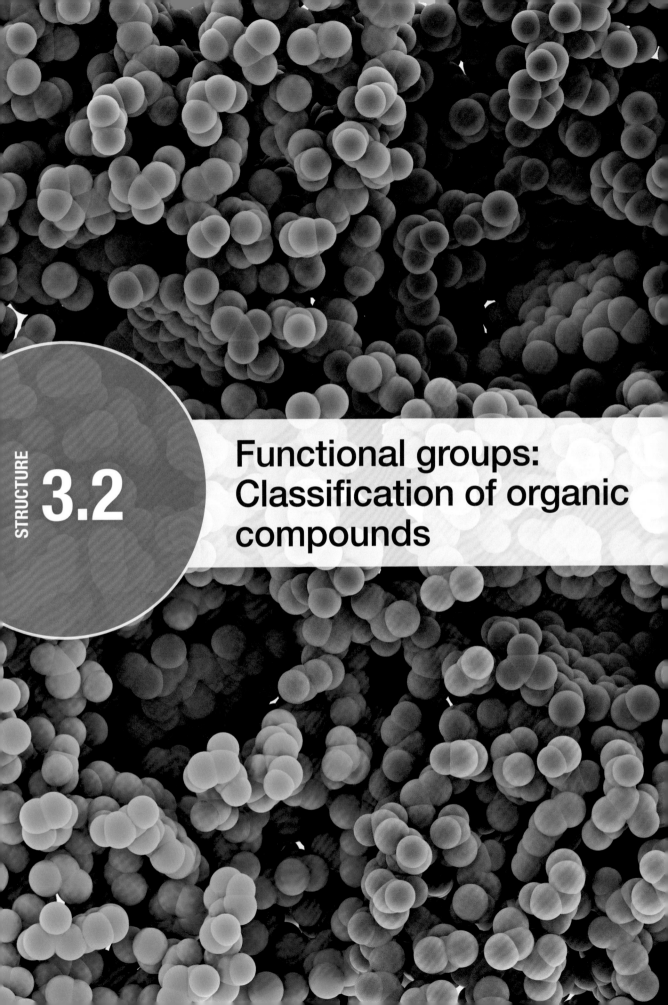

Functional groups: Classification of organic compounds

The development of organic chemistry as a field of study has allowed chemists to investigate increasingly complicated organic compounds, gaining new insights into their physical properties and chemical reactivity. Since the first applications of computerized molecular modelling in the 1960s, the steady growth in computing power has allowed for the modelling of increasingly complex molecules, utilizing current knowledge of chemical bonding and the intramolecular forces associated with different functional groups. With modern computers and software, the 3D structures of extremely complex biological compounds, such as proteins and enzymes, are now able to be modelled, giving new insights into the dynamics of protein folding and enzyme catalysis. The picture shown is a computer-generated model of the human catalase enzyme that shows the shape and folding of the enzyme.

Guiding Question

How does the classification of organic molecules help us to predict their properties?

The total number of organic compounds that exists on Earth is so large that it is impossible to give a definite number. In any case, it is increasing all the time as new materials are synthesized. It is currently estimated that there are over 20 million different organic molecules on the planet and every one of them is unique in its chemical structure and specific properties.

With such a large number of organic compounds, it is necessary to have a system for classifying and naming them. In this chapter, we will learn how the presence of specific **functional groups** allows for the classification of organic compounds into **classes** of compounds with similar properties.

A functional group is usually a small group of atoms attached to a carbon atom in a molecule, and it gives the characteristic properties to the compound. The properties of any individual compound can therefore be predicted and explained by what functional groups are present and the class it belongs to.

Given the incredible diversity of organic compounds, it is possible for compounds with the same molecular formula to have different structures along with different physical and chemical properties. Applying the rules for **IUPAC nomenclature** allows unique names to be obtained for each individual compound, which also assists in their classification.

The international, non-governmental organization IUPAC (International Union of Pure and Applied Chemistry), is best known for its system of nomenclature, and is now recognized as the world authority in this field. IUPAC terminology enables precise communication between scientists across international boundaries.

Structure 3.2.1 – Structural representations of organic compounds

> **Structure 3.2.1 – Organic compounds can be represented by different types of formulas. These include empirical, molecular, structural (full and condensed), stereochemical and skeletal.**
>
> Identify different formulas, and interconvert molecular, skeletal and structural formulas.
>
> Construct 3D models (real or virtual) of organic molecules.

Stereochemical formulas are not expected to be drawn, except where specifically indicated.	Structure 2.2 – What is unique about carbon that enables it to form more compounds than the sum of all the other elements' compounds?
	Nature of Science, Structure 2.2 – What are the advantages and disadvantages of different depictions of an organic compound (e.g. structural formula, stereochemical formula, skeletal formula, 3D models)?

Different types of formulas can provide different information about organic compounds

Structure 1.4 shows that chemical compounds can be represented by empirical and molecular formulas which give the identities of the elements present in the compound. However, for organic compounds, an understanding of *how* the atoms within the compound are linked is often required in order to predict and explain their properties and reactivity. Structural and skeletal formulas that illustrate the connections and spatial arrangements between atoms are commonly used in organic chemistry.

The **empirical formula** of a compound is the simplest whole-number ratio of the atoms it contains (see Structure 1.4). For example, the empirical formula of ethane, C_2H_6, is CH_3. This formula can be derived from percentage composition data obtained from combustion analysis. It is, however, of rather limited use on its own, as it does not tell us the actual number of atoms in the molecule.

The **molecular formula** of a compound is the actual number of atoms of each element present. For example, the molecular formula of ethane is C_2H_6. It is therefore a multiple of the empirical formula, and so can be deduced if we know both the empirical formula and the molar mass M. The relationship can be expressed as:

$$M = (\text{molar mass of empirical formula})_n$$

where n is an integer.

So, for example, if we know that the empirical formula of ethane is CH_3, and its molar mass $M = 30\,\text{g mol}^{-1}$, using the formula above:

$$30\,\text{g mol}^{-1} = (\text{mass of } CH_3) \text{ so } 30 = (12 + (3 \times 1)), \text{ which gives } n = 2$$

Therefore, the molecular formula of ethane is $(CH_3)_2$ or C_2H_6.

However, the molecular formula is also of quite limited value as the properties of a compound are determined not only by the atoms it contains, but also by how those atoms are arranged in relation to each other and in space.

The **structural formula** is a representation of the molecule showing how the atoms are bonded to each other. There are variations in the amount of detail this shows.

- A **full structural formula** (graphic formula or displayed formula) shows every bond and atom. Usually 60°, 90° and 180° angles are used to show the bonds because this is the clearest representation on a two-dimensional page, although it is not the true geometry of the molecule.

- A **condensed structural formula** often omits bonds where they can be assumed, and groups atoms together. It contains the minimum information needed to describe the molecule unambiguously – in other words there is only one possible structure that could be described by this formula.

- The table below shows empirical and structural formulas applied to three compounds.

Compound	Ethane	Ethanoic acid	Glucose
Empirical formula	CH_3	CH_2O	CH_2O
Molecular formula	C_2H_6	$C_2H_4O_2$	$C_6H_{12}O_6$
Full structural formula	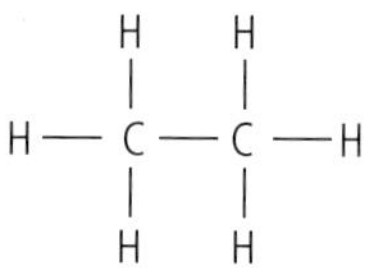	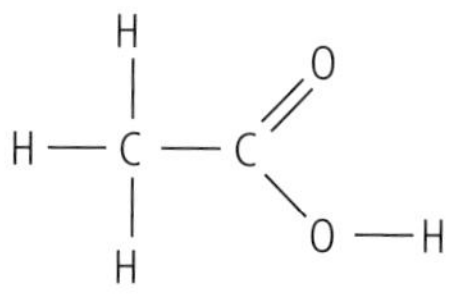	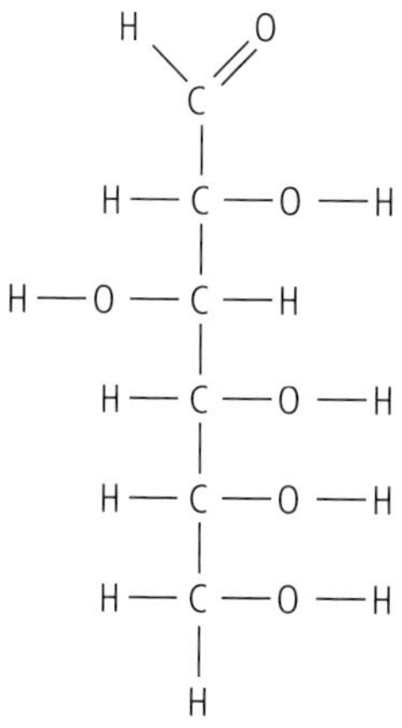
Condensed structural formula	CH_3CH_3	CH_3COOH	$CHO(HCOH)_4CH_2OH$

A **stereochemical formula** attempts to show the relative positions of atoms and groups around a central carbon in three dimensions. The convention is that a bond sticking forwards from the page is shown as a solid wedge, whereas a bond sticking behind the page is shown as a dashed wedge. A bond in the plane of the paper is a solid line.

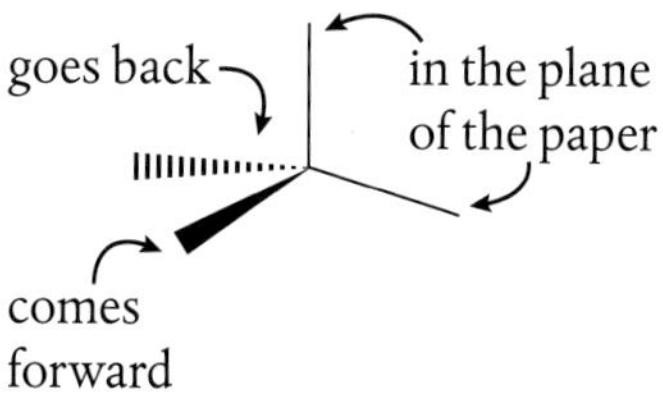

goes back

in the plane of the paper

comes forward

When carbon forms four single bonds, the arrangement is tetrahedral with bond angles of 109.5°. When it forms a double bond, the arrangement is triangular planar with bond angles of 120°.

methanol, CH_3OH ethene, C_2H_4

A **skeletal formula** can be regarded as a shorthand representation of a structural formula. It shows all the bonds present in the molecule, except C–H bonds, and omits the element symbols for carbon and hydrogen atoms. The symbols of all atoms other than C and H are included along with the condensed formula of any functional groups present.

This results in a simpler visualization but it requires the user to become proficient with identifying the positions of carbon atoms, along with the numbers of hydrogens attached to each carbon, as this information is not readily apparent in the skeletal formula.

The skeletal structure of butane, C_4H_{10}, is compared to its full structural formula below:

This point has two bonds attached: it represents a CH_2 This point has one bond attached: it represents a CH_3

C–C bonds in a carbon chain are represented as offset lines in a skeletal formula. The positions and numbers of carbon atoms present are identified by the points where bond lines meet or end.

Each carbon atom in an organic compound has four bonds. The number of hydrogen atoms attached to each carbon can therefore be determined by subtracting the number of bonds drawn to that atom from four. In the example of butane above, the point at the end of the skeletal structure has one bond drawn to it, therefore it has three H attached and represents a CH_3 group.

The table on the next page compares the condensed structural and skeletal formulas for three organic compounds.

Compound	Ethanol	Propanone	Butanoic acid
Molecular formula	C_2H_5OH	C_3H_6O	C_3H_7COOH
Full structural formula	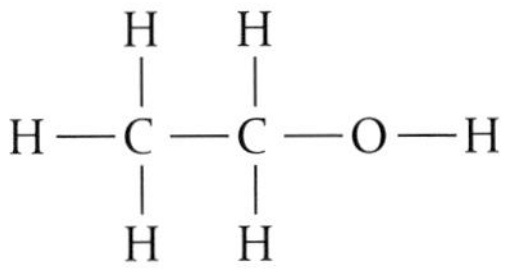	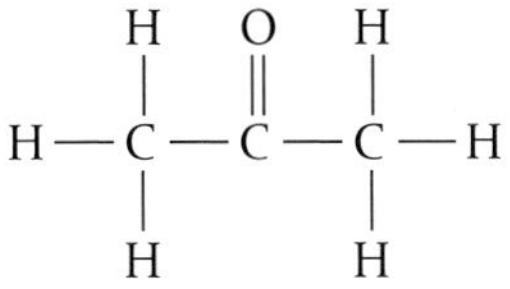	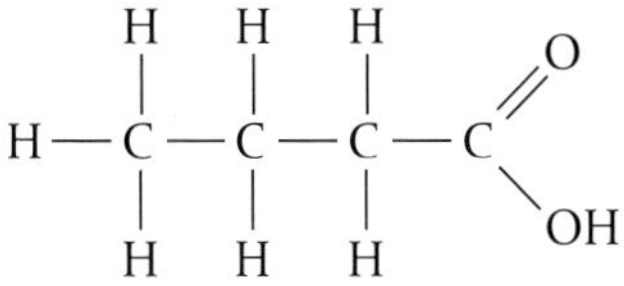
Condensed structural formula	CH_3CH_2OH	CH_3COCH_3	$CH_3CH_2CH_2COOH$
Skeletal formula	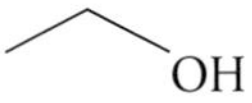	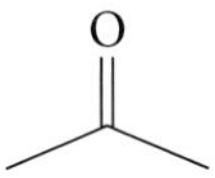	

Sometimes we do not need to show the exact details of the hydrocarbon, or alkyl, part of the molecule in a structure, so we can abbreviate this to R. For molecules which contain a benzene ring, C_6H_6, (page 271) known as **aromatic compounds**, we use  to show the ring.

If you can identify primary, secondary and tertiary carbons (page 288) in a skeletal formula, this should help you to determine the number of hydrogens bonded to each carbon.

Be careful not to confuse structural and skeletal formulas. When a full structural formula is required, the element symbols of all carbon and hydrogen atoms must be included, along with the bonds between them.

Worked example

(a) Convert the following full structural formula to a skeletal formula.

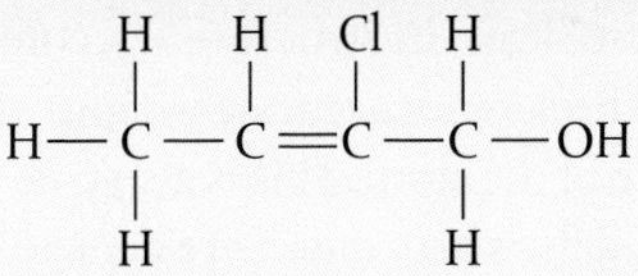

(b) Convert the following skeletal formula to a condensed structural formula.

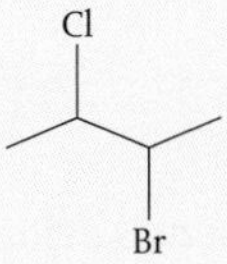

Solution

(a)

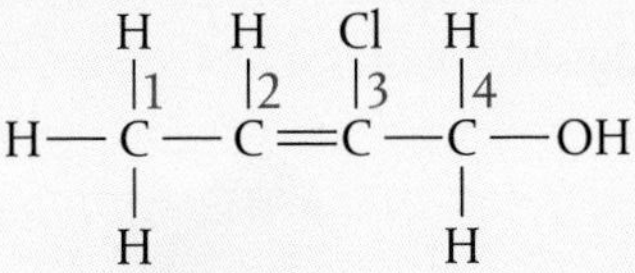

- Carbon 1 has only one bond that is not to a hydrogen atom so it can be represented as a point connected by a line to another point representing carbon 2.
- The double bond between carbons 2 and 3 must be represented as a double line.
- Carbon 3 has a bond to a non-hydrogen atom, Cl, which must be shown.
- Carbon 4 has an OH group attached so this also must be shown.

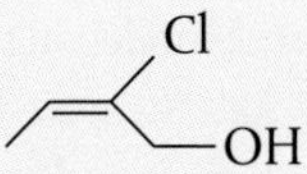

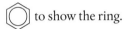

(b)

Cl
1 2 3 4
Br

- The point labelled 1 represents a carbon that has only one C–C bond so it must have three hydrogens attached.
- Point 2 represents a carbon with two C–C bonds and a C–Cl bond so it must have one hydrogen attached.
- Point 3 represents a carbon with two C–C bonds and a C–Br bond so it must also have one hydrogen attached.
- Point 4 is similar to point 1 and also represents a carbon that has one C–C bond and three hydrogens attached.

$$\underset{CH_3CHClCHBrCH_3}{\overset{1\quad2\quad3\quad4}{}}$$

The use of physical and computer models can assist organic chemists in visualizing the three-dimensional shapes of compounds, which can also help explain their physical properties and chemical reactivity.

As we learn in Structure 1.3, common representations of atoms tend to be misleading as they do not accurately depict the true scale of the nucleus compared to the overall size of the atom. This is also true of modelling kits, along with ball-and-stick computer models, where nuclei are depicted as spheres or balls with relatively small distances between them.

SKILLS

Construction of 3D models and working with virtual models online are important experiences that help in gaining an understanding of the structural concepts described in this chapter. Full details of a modeling activity are available in the eBook.

Nature of Science, Structure 2.2 – What are the advantages and disadvantages of different depictions of an organic compound (e.g. structural formula, stereochemical formula, skeletal formula, 3D models)?

Challenge yourself

1. Based on experimental evidence, the bond length of a carbon–carbon single bond is 154 pm (1 pm = 10^{-12} m) and the radius of a carbon nucleus is 2.7×10^{-15} m. If modelling kits were made to scale, how long would the plastic stick representing a single bond need to be if the carbon nucleus was represented as a plastic ball with a 0.5 cm radius?

Students practising their chemistry can find modelling kits very useful in visualizing the shapes of molecules they encounter in class.

The two images above are different computer-generated models of dexamethasone, a drug used to treat arthritis. The *ball-and-stick* model on the left shows the nuclei of the atoms as 'balls' and the covalent bonds as 'sticks'. The *space-filling model* on the right shows the atoms as overlapping spheres whose sizes correspond to the atomic radii of each atom.

In Structure 2.2, we learn that carbon atoms are able to form a variety of different compounds containing single, double or triple bonds. From the examples on previous pages illustrating the different structural representations of organic compounds, it should become apparent that carbon is unique among the elements for its ability to link to itself and form chains of bonded carbon atoms. This ability, known as **catenation**, is one of the main reasons for the vast number of organic compounds that exist. As we continue through this chapter, we will see that the possibility for carbon chains to be branched, and for compounds to contain one or more functional groups or substituents, further increases the incredible variety of organic compounds that are possible.

Structure 2.2 – What is unique about carbon that enables it to form more compounds than the sum of all the other elements' compounds?

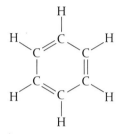

▲
Carbon compounds can form ring structures that are saturated, containing only C-C single bonds, or unsaturated, containing C=C or C≡C bonds. Two common ring structures are cyclohexane, C_6H_{12}, shown as a molecular model on the top, and benzene, C_6H_6, whose structure is shown on the bottom. Benzene is typically represented as a skeletal structure with the three double bonds combined into a circle, ⬡

Exercise

Q1. Convert the following full structural formulas to condensed structural formulas.

(a)
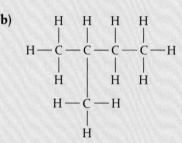

(b)

(c)

(d)

Q2. Convert the following condensed structural formulas to skeletal formulas.

(a) $CH_3CH(CH_3)CH_3$

(b) $CH_2OHCH_2CH_2CH_2Cl$

(c) $CHClCHCH_3$

(d) $CH_3C(CH_3)_2CH_2CH_3$

Q3. Convert the following skeletal formulas to condensed structural formulas.

(a)

(b)

(c)

(d)

Structure 3.2.2 – Functional groups and classes of compounds

Structure 3.2.2 – Functional groups give characteristic physical and chemical properties to a compound. Organic compounds are divided into classes according to the functional groups present in their molecules.

Identify the following functional groups by name and structure: halogeno, hydroxyl, carbonyl, carboxyl, alkoxy, amino, amido, ester, phenyl.

The terms 'saturated' and 'unsaturated' should be included.	Nature of Science, Reactivity 3.2, 3.4 – How can functional group reactivity be used to determine a reaction pathway between compounds, e.g. converting ethane into ethanoic acid?

Different classes of organic compounds contain different functional groups

Functional groups are atoms, or groups of atoms, that are present in organic compounds and are responsible for a compound's physical properties and chemical reactivity. Compounds that contain the same functional group belong to the same **class**. For example, the class of organic compounds called **carboxylic acids** contain the −**COOH** functional group.

Table 1 below gives common classes of organic compounds along with the formula and name of the functional group that defines each class. The table also includes the suffix (or prefix) specific to each class when using IUPAC nomenclature to name a compound as well as the general formula that describes the compounds belonging to each class.

S3.2 Table 1 Important classes of organic compounds along with their associated functional groups, IUPAC suffixes and general formulas. The table is continued on the next page.

For certain classes, the general formula has been given as $C_xH_yO_z$ to show the overall chemical formula and also using R and R′ to represent alkyl chains that are linked by functional groups. This format can help distinguish between classes such as aldehydes and ketones that have the same overall formula.

Class	Functional group	Name of functional group	Suffix in IUPAC name	Example of compound	General formula
alkane			-ane	C_2H_6, ethane	C_nH_{2n+2}
alkene	$\diagdown C = C \diagup$	alkenyl	-ene	$H_2C{=}CH_2$, ethene	C_nH_{2n}
alkyne	$-C \equiv C-$	alkynyl	-yne	$HC{\equiv}CH$, ethyne	C_nH_{2n-2}
alcohol	−OH	hydroxyl	-anol	C_2H_5OH, ethanol	$C_nH_{2n+1}OH$
ether	R−O−R′	alkoxy	-oxyalkane	$H_3C{-}O{-}C_2H_5$, methoxyethane	$C_nH_{2n+2}O$ R−O−R′

continued

Class	Functional group	Name of functional group	Suffix in IUPAC name	Example of compound	General formula
aldehyde	![aldehyde structure]	carbonyl (aldehyde)	-anal	C_2H_5CHO, propanal	$C_nH_{2n}O$ R–CHO
ketone	![ketone structure]	carbonyl (ketone)	-anone	CH_3COCH_3, propanone	$C_nH_{2n}O$ R–CO–R′
carboxylic acid	![carboxylic acid structure]	carboxyl (acid)	-anoic acid	C_2H_5COOH, propanoic acid	$C_nH_{2n+1}COOH$
ester	![ester structure]	carboxyl (ester)	-anoate	$C_2H_5COOCH_3$, methyl propanoate	$C_nH_{2n}O_2$ R–COO–R′
amide	![amide structure]	amido	-anamide	$C_2H_5CONH_2$, propanamide	$C_nH_{2n+1}CONH_2$
amine	$-NH_2$	amino	-anamine	$C_2H_5NH_2$, ethanamine	$C_nH_{2n+1}NH_2$
halogenoalkane*	–F, –Cl, –Br, –I	halogeno	fluoro, chloro, bromo, iodo (prefixes, not suffixes. See below)	C_2H_5Cl, chloroethane $CH_3CH_2BrCH_3$, 2-bromopropane	$C_nH_{2n+1}X$ X = F, Cl, Br, I
arene**	![arene structure] C_6H_5-	phenyl	-benzene	$C_6H_5CH_3$, methylbenzene	

* For halogenoalkane compounds, the halogen atoms are regarded as **substituents** as they have taken the position of a hydrogen atom.

When applying IUPAC nomenclature, substituents such as halogens are typically identified using prefixes: chloroethane, 2-bromopropane etc.

** The syllabus does not require knowledge of arenes as a class of compounds, but you are expected to recognize the phenyl group when it is present in a structure. The naming of compounds containing a phenyl group is also not required knowledge.

Although there will only be one systematic IUPAC name, compounds containing phenyl groups often have alternative names. Methylbenzene (IUPAC name) consists of a methyl group (–CH_3) attached to a phenyl group (–C_6H_5) and it is therefore also known as phenylmethane. Historically, it was known as toluene and this name is still in common use.

Methylbenzene is used as a solvent in many applications, including markers, paint thinners, industrial cleaners and glues.

Quickly sketching the full structural formula can sometimes make it easier to identify the functional groups and structural features of an organic compound.

One important way of classifying organic compounds is by their **saturation**. Compounds that are **saturated** only contain single bonds whereas **unsaturated** compounds contain double or triple bonds. As we see in Reactivity 3.3 and 3.4, saturated and unsaturated compounds undergo different types of reactions.

Food sources rich in unsaturated fats. Fats are biological compounds that contain long carbon chains which, in unsaturated fats, include multiple double bonds. The World Health Organization and other health agencies recommend diets higher in unsaturated fats than saturated fats due to a decreased risk of cardiovascular disease.

Functional groups influence the chemical reactivity of organic compounds

Nature of Science, Reactivity 3.2, 3.4 – How can functional group reactivity be used to determine a reaction pathway between compounds, e.g., converting ethene into ethanoic acid?

In Reactivity 3.2–3.4, we learn in more detail about different chemical reactions that occur for compounds with specific functional groups. Synthetic organic chemists can use their knowledge of these reactions to synthesize target compounds from readily available starting materials. This often requires organizing several reactions in sequence so that the product of one reaction is the reactant of the next. Each step will typically involve a functional group interconversion and the series of discrete steps is known as a **reaction pathway**.

For example, ethene (C_2H_4) can be converted into ethanoic acid (CH_3COOH) via a two-step reaction pathway where ethanol (CH_3CH_2OH) is formed as a product in the first step and then used as a reactant in the second step.

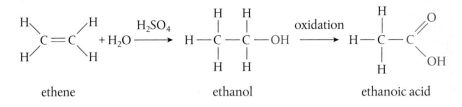

| ethene | ethanol | ethanoic acid |

Amino acids link together when condensation reactions occur between their functional groups

The reaction of two amino acids with each other provides a specific example of how functional groups on different molecules can react via a specific type of reaction to give a new class of product.

Ethanoic acid is an important industrial chemical. One major use involves its conversion to vinyl acetate and the production of polymers, such as polyvinyl acetate (PVA) which is used in many glues. A fun home use of PVA glue is making artificial slime.

Amino acids each contain two functional groups, an amine group ($-NH_2$) and a carboxylic acid group ($-COOH$). As shown on the next page, these two groups are able to react together in a **condensation reaction** in which a molecule of water is eliminated and a new bond is formed between the acid group of one amino acid and the amino group of the other. This bond is a substituted **amide link** known as a **peptide bond**, and two amino acids linked in this way form an amide product known as a **dipeptide**.

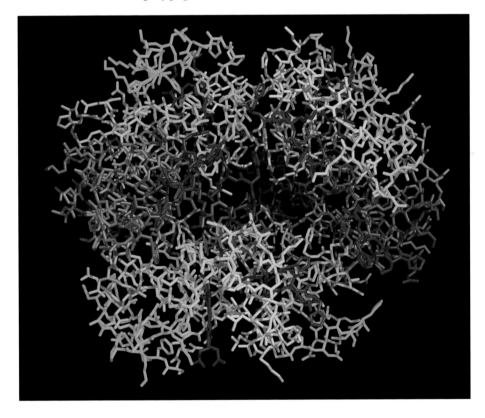

free amino
group (N-terminal)

peptide
bond

free carboxylic acid
group (C-terminal)

We can see that the dipeptide still has a functional group at each end of the molecule, with $-NH_2$ at one end and $-COOH$ at the other. So it can react again by another condensation reaction, forming a **tripeptide** and eventually a chain of many linked amino acids known as a **polypeptide**.

Haemoglobin is a metalloprotein that transports oxygen in the bloodstream. The structure of haemoglobin includes four subunits which are folded polypeptide chains.

Exercise

Q4. For each of the following molecules, give its class and the name of its functional group:

(a) CH_3CH_2COOH

(b) $CH_3CHCHCH_3$

(c) $CH_2OHCH_2CH_2CH_3$

(d) CH_3COOCH_3

(e) CH_3CH_2CHO

(f) $CH_3NH(CH_3)$

Q5. Which of the following is an amide?

    A $CH_3NH(CH_3)$ **B** CH_3CONH_2

    C $CH_3CH_2NH(CH_3)$ **D** $CH_3CH_2COOCH_3$

Q6. Which of the following is an ester?

    A CH_3COCH_3 **B** $CH_3COCH_2NH_2$

    C $CH_3CH(OH)CH_3$ **D** $CH_3COOCH_2CH_3$

Structure 3.2.3 and 3.2.4 – Homologous series

Structure 3.2.3 – A homologous series is a family of compounds in which successive members differ by a common structural unit, typically CH$_2$. Each homologous series can be described by a general formula. Identify the following homologous series: alkanes, alkenes, alkynes, halogenoalkanes, alcohols, aldehydes, ketones, carboxylic acids, ethers, amines, amides and esters.	
	Nature of Science, Tool 2 – How useful are 3D models (real or virtual) to visualize the invisible?

Structure 3.2.4 – Successive members of a homologous series show a trend in physical properties. Describe and explain the trend in melting and boiling points of members of a homologous series.	
	Structure 2.2 – What is the influence of the carbon chain length, branching and the nature of the functional groups on intermolecular forces?

Homologous series differ by a CH$_2$ group

Organic compounds are classified into 'families' of compounds known as **homologous series**. The members of each such series possess certain common features, as described below.

Consider the following compounds of carbon and hydrogen where the carbons are all bonded by single covalent bonds.

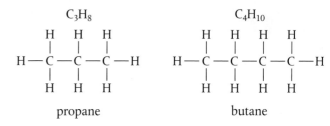

These are members of a homologous series known as the **alkanes**. It can be seen that neighbouring members differ from each other by $-CH_2-$. This same increment applies to successive members of each homologous series, and means that the molecular mass increases by a fixed amount as we go up a series.

Members of a homologous series can be represented by the same general formula

The four alkanes shown on the previous page can all be represented by the general formula C_nH_{2n+2}.

Other homologous series are characterized by the presence of a particular functional group, and this will be shown in the general formula for the series. For example, the homologous series known as the **alcohols** possesses the functional group –OH as shown below.

CH$_3$OH $\qquad\qquad$ C$_2$H$_5$OH $\qquad\qquad$ C$_3$H$_7$OH $\qquad\qquad$ C$_4$H$_9$OH

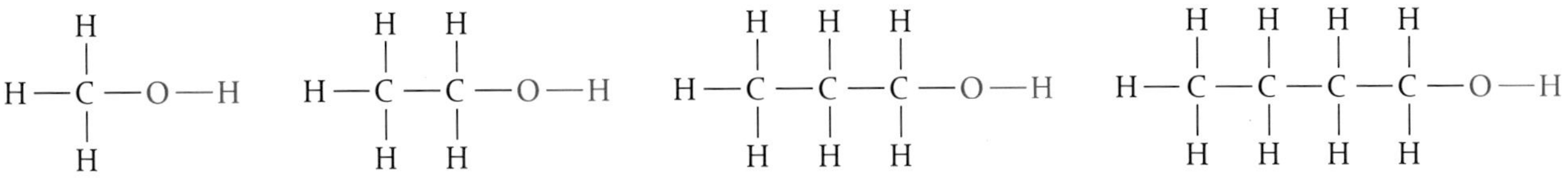

methanol $\qquad\qquad$ ethanol $\qquad\qquad$ propanol $\qquad\qquad$ butanol

The four alcohols shown above can all be represented by the general formula $C_nH_{2n+1}OH$.

General formulas for the homologous series of other classes of organic compounds have been provided in Table 1 on pages 272–273.

Members of a homologous series show a trend in physical properties

As successive members of a homologous series differ by a –CH$_2$– group, they have successively longer carbon chains. This is reflected in a gradual trend in the physical properties of the members of the series. For example, the effect of the length of the carbon chain on the boiling point of three classes of compounds, alkanes, chloroalkanes and alcohols is shown in Table 2.

S3.2 Table 2 Boiling points for the first eight members in the homologous series of alkanes, chloroalkanes and alcohols.

▼

Alkane		Boiling point / °C	Chloroalkane		Boiling point / °C	Alcohol		Boiling point / °C
methane	CH$_4$	−164	chloromethane	CH$_3$Cl	−24	methanol	CH$_3$OH	65
ethane	C$_2$H$_6$	−89	chloroethane	C$_2$H$_5$Cl	12	ethanol	C$_2$H$_5$OH	78
propane	C$_3$H$_8$	−42	1-chloropropane	C$_3$H$_7$Cl	46	propan-1-ol	C$_3$H$_7$OH	97
butane	C$_4$H$_{10}$	−0.5	1-chlorobutane	C$_4$H$_9$Cl	78	butan-1-ol	C$_4$H$_9$OH	118
pentane	C$_5$H$_{12}$	36	1-chloropentane	C$_5$H$_{11}$Cl	108	pentan-1-ol	C$_5$H$_{11}$OH	138
hexane	C$_6$H$_{14}$	69	1-chlorohexane	C$_6$H$_{13}$Cl	135	hexan-1-ol	C$_6$H$_{13}$OH	156
heptane	C$_7$H$_{16}$	98	1-chloroheptane	C$_7$H$_{15}$Cl	159	heptan-1-ol	C$_7$H$_{15}$OH	176
octane	C$_8$H$_{18}$	125	1-chlorooctane	C$_8$H$_{17}$Cl	182	octan-1-ol	C$_8$H$_{17}$OH	194

The compounds shaded in yellow exist as gases at room temperature and those shaded in green exist as liquids at room temperature.

The data for each class of compound in Table 2 show that boiling point increases with increasing carbon number. This is because a longer chain length increases both the instantaneous and induced dipoles occurring, which then results in stronger London dispersion forces, and therefore stronger overall intermolecular forces, between the molecules.

Other physical properties that show this predictable trend with increasing carbon number are melting point, density and viscosity.

Functional groups affect the physical properties of organic compounds

For the alkanes listed in Table 2, the increase in boiling point is not linear, but steeper near the beginning as the influence of the increased chain length is proportionally greater for small molecules. For the chloroalkanes and alcohols, there is a more consistent increase in boiling point with carbon chain length. This is because the strongest intermolecular forces present are associated with their functional groups rather than carbon chain length.

Owing to their different polarities and resulting differences in intermolecular forces, functional groups have a significant effect on physical properties. One such property is volatility and the effect of different functional groups on volatility can be summarized as follows.

> **most volatile**   **least volatile**
>
> **alkane > halogenoalkane > aldehyde > ketone > alcohol > carboxylic acid**
>
> London (dispersion) force → dipole–dipole interaction → hydrogen bonding
>
> increasing strength of intermolecular attraction ⟶
>
> increasing boiling point ⟶

If we look across any row in Table 2, the compounds listed have the same number of carbons but their boiling points increase in the order: alkane < chloroalkane < alcohol. This is consistent with the strength of the intermolecular forces present due to the different functional groups in these compounds: London dispersion forces (alkane) < dipole–dipole (chloroalkane) < hydrogen bonding (alcohol). As Table 2 shows, the hydrogen bonding present in alcohols is so strong that even the smallest alcohol, methanol (CH_3OH), is a liquid at room temperature.

In this section we have seen that both chain length and the functional groups present affect the nature of the intermolecular forces in a compound. Longer chains result in increased London dispersion forces and polar functional groups result in dipole–dipole interactions or hydrogen bonding. In the next section we will see that branching of the carbon chain in a compound also affects intermolecular forces as it results in weaker London dispersion forces.

The trend in boiling points of the alkanes is of great significance in the oil industry, as it makes it possible to separate the many components of crude oil into **fractions** that contain molecules of similar molecular mass on the basis of their boiling points. As the crude oil is heated, the smaller hydrocarbons boil off first while larger molecules distil at progressively higher temperatures in the fractionating column. The different fractions are used as fuels, industrial lubricants, and as starting molecules in the manufacture of synthetic compounds.

Volatility is a term used to describe the tendency of a substance to vaporize. Substances with strong intermolecular forces have a high boiling point and a low volatility.

Structure 2.2 – What is the influence of the carbon chain length, branching and the nature of the functional groups on intermolecular forces?

Exercise

Q7. Which of the following is a member of the same homologous series as butanoic acid, $CH_3CH_2CH_2COOH$?

    A $CH_3CH_2COOCH_3$ **B** $CH_3CH_2CH_3$

    C $CH_3OCH_2CH_3$ **D** CH_3COOH

Q8. Which of the following pairs of compounds belong to the same homologous series?

    I. CH_3CH_2Cl, $CH_2ClCH_2CH_2CH_3$

    II. CH_3CH_2Br, $CH_2ClCH_2CH_3$

    III. CH_3CH_2Br, CH_2BrCH_2Br

    A   I only   **B**   I and III   **C**   III only   **D**   I, II and III

Q9. Which option ranks the compounds in order of increasing boiling point?

    A $CH_3CH_2COOH < CH_3CH_2CH_3 < CH_3CH_2CH_2F$

    B $CH_3CH_3 < CH_3COOH < CH_3CH_2COOH$

    C $CH_3COOH < CH_3COOCH_3 < CH_3COCH_3$

Q10. Rank the following compounds in order of increasing volatility and provide an explanation for your answer.

    bromopropane ($CH_3CH_2CH_2Br$); propanoic acid (CH_3CH_2COOH); propane ($CH_3CH_2CH_3$)

Q11. Predict how the aqueous solubility of the alcohols changes with increasing chain length and provide a reason for your answer.

Structure 3.2.5 – IUPAC nomenclature

Structure 3.2.5 – 'IUPAC nomenclature' refers to a set of rules used by the International Union of Pure and Applied Chemistry to apply systematic names to organic and inorganic compounds.
Apply IUPAC nomenclature to saturated or mono-unsaturated compounds that have up to six carbon atoms in the parent chain and contain one type of the following functional groups: halogeno, hydroxyl, carbonyl, carboxyl.

Include straight and branched-chain isomers.	

Unique names for organic compounds are derived using IUPAC nomenclature

For over a hundred years, chemists have recognized the need for a specific set of rules for the naming of organic compounds. Rules developed by the IUPAC generate names that are logically based on the chemistry of the compounds, and so give information about the functional groups present and the size of the molecules.

Some guidelines for applying the IUPAC nomenclature are discussed below.

Rule 1: Identify the longest straight chain of carbon atoms

The longest chain of carbon atoms gives the **stem** of the name as follows.

Number of carbon atoms in longest chain	Stem in IUPAC name	Example of compound
1	meth-	CH_4, methane
2	eth-	C_2H_6, ethane
3	prop-	C_3H_8, propane
4	but-	C_4H_{10}, butane
5	pent-	C_5H_{12}, pentane
6	hex-	C_6H_{14}, hexane

Note that 'straight chain' refers to continuous or unbranched chains of carbon atoms – it does not mean angles of 180°. Be careful when identifying the longest straight chain not to be confused by the way the molecule may appear on paper, because of the free rotation around carbon–carbon single bonds. For example, in Figure 1 all three structures are the same straight-chain molecule pentane, C_5H_{12}, even though they may look different.

S3.2 Figure 1 Different representations of the same molecule, C_5H_{12}, pentane. These can all be interconverted by rotating the carbon–carbon bonds.

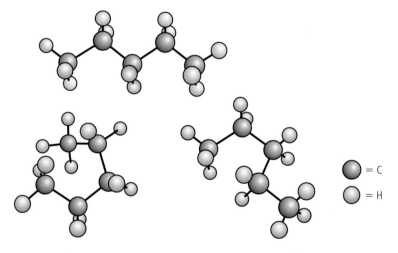

= C
= H

Rule 2: Identify the functional group

The functional group usually determines the specific ending or **suffix** to the name, which replaces the '-ane' ending in the parent alk*ane*. The suffixes used for some common functional groups are shown in Table 1 on pages 272–273. Note the distinction between class, which refers to the type of compound, and functional group, which refers to the site of reactivity in the molecule.

 Remember that the name for the stem is derived from the *longest* carbon chain, which may include the carbon of the functional group. Look at Table 1 on pages 272–273 and see how this applies to carboxylic acids and their derivatives (esters and amides).

The position of a functional group is shown by a number between dashes inserted before the functional group ending. The number refers to the carbon atom to which the functional group is attached. *The chain is always numbered starting at the end that will give the smallest number to the functional group.*

You are expected to be familiar with the nitrogen-containing functional groups at the bottom of Table 1 on page 273 (amine and amide), but nomenclature of these classes of compounds will not be assessed.

For example:

propan-2-ol   but-1-ene

As shown in the structure of but-1-ene above, here we number the carbon chain starting from the right-hand side so that the number of the group will be 1 and not 3.

Sometimes a functional group can only be in one place; in these cases we do not need to give a number to show its position.

For example, the carboxylic acid group must always be at the end of a chain so it does not need a number assigned. A ketone with three carbons has only one possible position for the carbonyl group so it also does not need a number assigned.

butanoic acid   propanone

Rule 3: Identify the side chains or substituent groups

Side chains, or functional groups in addition to the one used as the suffix, are known as **substituents** and are given as the first part or **prefix** of the name. Some common examples are shown in the table below.

Class	Functional group	Name of functional group	Prefix in IUPAC name	Example of compound
alkane			methyl, ethyl, propyl etc.	$CH_3CH(CH_3)C_2H_5$, 2-methylbutane $CH(C_2H_5)_3$, 3-ethylpentane $CH(C_3H_7)_3$, 4-propylheptane
halogenoalkane	$-F, -Cl, -Br, -I$	halogeno	fluoro, chloro, bromo, iodo	C_2H_5Cl, chloroethane $CH_3CH_2BrCH_3$, 2-bromopropane
amine	$-NH_2$	amine	amino	$CH_2(NH_2)COOH$, 2-aminoethanoic acid

You will notice that $-NH_2$ appears as both a possible suffix in Table 1 on pages 272–273 and a prefix in the table above. Usually when it is the only functional group it will take the suffix, but if there are two or more functional groups in the molecule it will be a prefix, as for example in *amino* acids.

As shown in the table on the previous page, the position of the substituent groups is given by a number followed by a dash in front of its name showing the carbon atom to which it is attached, again numbering the chain to give the smallest number to the group.

For example:

2-methylbutane   2-methylpentane

If there is more than one substituent group of the same type, we use commas between the numbers and the prefixes di-, tri-, or tetra- before the name. Substituents are given in order of the number of the carbon atom to which they are attached; if there is more than one group on the same atom they are put in alphabetical order.

For example:

1,2-dichloropropane   2-bromo-2-chloropropane

1-chloro-2-methylpropane

Naming esters and ethers

Esters

Esters form when the alkyl group of an alcohol replaces the hydrogen of a carboxylic acid in a condensation reaction:

$$R-COOH + R'OH \rightarrow R-COO-R' + H_2O$$

Esters are named in a similar way to salts, which form when a metal has replaced the hydrogen of a carboxylic acid. Salts take the stem of the name from the parent acid and the prefix is the cation. For example, CH_3COONa is sodium ethanoate.

In esters, the stem still comes from the parent acid but the alkyl group of the alcohol is the prefix. The ester made from the reaction of ethanol with ethanoic acid will be ethyl ethanoate, $CH_3COOCH_2CH_3$.

ethyl ethanoate

When carboxylic acids lose H⁺ they become **alkanoate ions.** In naming esters this also applies to the parent acid the ester is made from: methanoic acid (HCOOH) becomes methanoate (HCOO⁻) in the ester; propanoic acid (CH_3CH_2COOH) becomes propanoate ($CH_3CH_2COO^-$) etc.

▲
A selection of fresh fruits. The distinctive tastes and smells of fruits and berries are typically caused by ester compounds.

Ethers

Ethers consist of two alkyl chains linked by an oxygen atom (R–O–R′).

When naming ethers:

- the longer chain will be the stem and retains its alkane name;
- the shorter chain is regarded as a substituent and is given the prefix alkoxy.

methoxy**propane**

When the two chains are the same length, then one is assigned as the alkane stem and the other becomes the substituent.

ethoxyethane

A selection of foods rich in carbohydrates. Carbohydrates are complex sugars with individual sugar units (saccharides) joined by ether linkages.

Ethers are used commercially as solvents and in the production of polymers. They are also used for medicinal purposes such as antiseptics and some of the first anaesthetics were ethers.

The ether monument in Boston, USA. It commemorates the first use of ethoxyethane (diethyl ether) as an anaesthetic, as well as the many lives saved through surgical procedures only possible with the use of anaesthesia.

Nature of Science

Classification is an important aspect of many disciplines as it brings some order to large amounts of data, and helps in the recognition of patterns and trends. In organic chemistry, classification is particularly helpful given the almost infinite number of compounds. The IUPAC nomenclature system described earlier is broadly accepted by scientists because it provides effective communication. Yet, in common with most systems of classification, there are limits to its usefulness. One of these limits is in biochemistry, where IUPAC nomenclature can result in seemingly complex and extended names for relatively simple molecules. For example, sucrose, $C_{12}H_{22}O_{11}$, becomes (2R,3R,4S,5S,6R)-2-[(2S,3S,4S,5R)-3,4-dihydroxy-2,5-bis(hydroxymethyl) oxolan-2-yl]oxy-6-(hydroxymethyl)oxane-3,4,5-triol. Although there is a place for such terminology, it is not widely used. Classification and nomenclature are artificial constructs, and when not helpful need not be used. Throughout this book we will commonly use non-IUPAC names for biological molecules for this reason.

Nature of Science

As the study of organic chemistry has grown to include increasingly large and complex molecules, IUPAC nomenclature has had to adapt accordingly. New rules have been introduced with the identification of complex structural features that could not be named using earlier rules.

Summary of IUPAC nomenclature

In summary, IUPAC nomenclature has three possible parts, which are usually written together as a single word.

Prefix	··········	**Stem**	··········	**Suffix**
position, number and name of substituents		number of carbon atoms in longest chain		class of compound determined by functional group

For example:

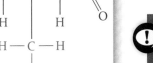

3,3-dimethylbutanal

In the upcoming section on structural isomers you will encounter many examples of different organic compounds that have the same molecular formula (isomers). You should use these examples as an opportunity to confirm your ability to apply IUPAC rules and determine the correct systematic name for each isomer.

Exercise

Q12. For each of the following molecules, give its class and its IUPAC name.

(a) $CH_3CH_2CH(CH_3)CH_3$

(b) $CH_3CH(CH_3)COOH$

(c) $CH_3COCH_2CH_3$

(d) $CH_3COOCH_2CH_2CH_3$

(e) $CH_3CH_2CH_2OH$

(f) $CH_3CH_2CH_2CH_2CHO$

Q13. Give the structural formulas for the following molecules (condensed form is acceptable).

    (a) methanoic acid   **(b)** hex-3-yne

    (c) pentanal   **(d)** butanone

    (e) ethoxypropane   **(f)** 2-bromopropan-1-ol

Q14. A student draws the structure of what they believe is 2-ethylpentane. Draw the structure that is consistent with this name and then determine the correct IUPAC name for this compound.

Structure 3.2.6 – Structural isomers

Structure 3.2.6 – Structural isomers are molecules that have the same molecular formula but different connectivities.	
Recognize isomers, including branched, straight-chain, position and functional group isomers.	
Primary, secondary and tertiary alcohols, halogenoalkanes and amines should be included.	

Organic compounds with the same molecular formula can have different structures

The molecular formula of a compound shows the atoms that are present in the molecule, but gives no information on how they are arranged. Consider for example the formula C_4H_{10}. There are two possible arrangements for these atoms that correspond to different molecules with different properties.

butane,
boiling point –0.5 °C

2-methylpropane,
boiling point –11.7 °C

Such molecules, having the same molecular formula but different arrangements of the atoms, are known as **structural isomers**.

Structural isomers are molecules that have the same molecular formula but different arrangements of the atoms. Because of their different structures, they also have different physical and chemical properties.

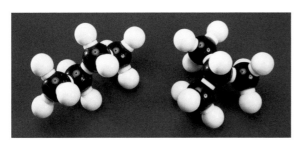

Molecular models of the two isomers of C_4H_{10}. On the left is butane (sometimes called *n*-butane to denote a straight chain), and on the right is 2-methylpropane.

The number of isomers that exist for a molecular formula increases as the molecular size increases. In fact, the increase is exponential; there are 75 possible isomers of $C_{10}H_{22}$ and 366 319 isomers of $C_{20}H_{42}$!

Each isomer is a distinct compound, having unique physical and chemical properties. As we will see, the number of isomers that exist for a molecular formula increases with the size of the molecule. This is one of the reasons for the vast number of compounds that can be formed from carbon.

Alkanes can have straight chain or branched structural isomers

As we saw on the previous page, butane, C_4H_{10}, has two structural isomers, one that forms a straight chain and another that has a branched structure. Pentane, C_5H_{12}, has three structural isomers as shown below.

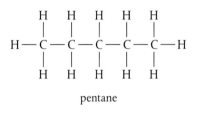

pentane	2-methylbutane	2,2-dimethylpropane
straight chain	one branch	two branches
boiling point 36.1 °C	boiling point 27.8 °C	boiling point 9.5 °C

From the examples of C_4H_{10} and C_5H_{12} we can see a trend in the boiling points of their isomers: *the more branching that is present in an isomer, the lower its boiling point.*

Branching of the hydrocarbon chain influences the strength of the London dispersion forces occurring between neighbouring molecules because it reduces the amount of surface contact. Think, for example, how tree logs with lots of branches sticking out cannot stack together as closely as a pile of logs which have no branches. In a similar way, branched-chain isomers have less contact with each other than their straight-chain isomers. This results in weaker attractions between instantaneous and induced dipoles on neighbouring molecules with weaker overall London dispersion forces and lower boiling points.

The existence of these different straight-chain and branched-chain isomers of the alkanes is of great significance in the petroleum industry. It has been found that branched-chain isomers generally burn more smoothly in internal combustion engines than straight-chain isomers, and so oil fractions with a higher proportion of these branched-chain isomers are considered to be of 'better grade'. This is often referred to as a higher 'octane number' and means that it fetches a higher price at the pump.

Worked example

Draw full structural formulas for the five different structural isomers of hexane, C_6H_{14}.

Solution

When identifying structural isomers for a given formula it is best to start with the longest possible chain and then progress through the possibilities with successively shorter chain lengths. C_6H_{14} has one straight chain isomer:

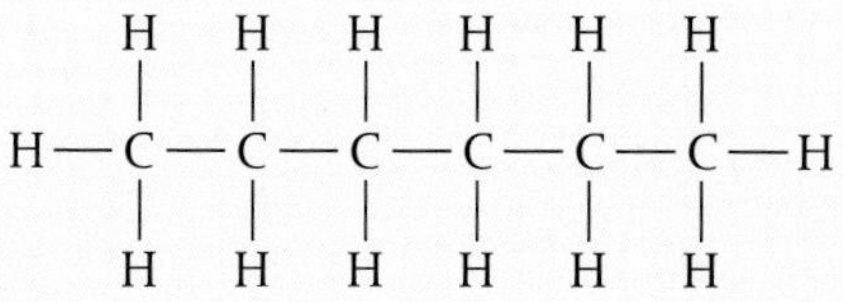

There are two isomers with a five-carbon longest chain and one methyl branch:

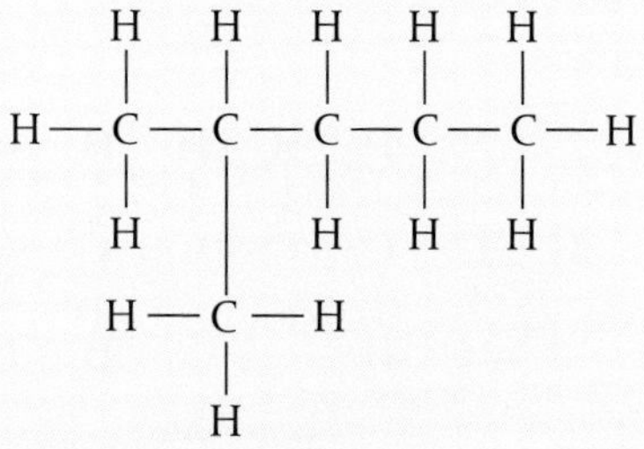

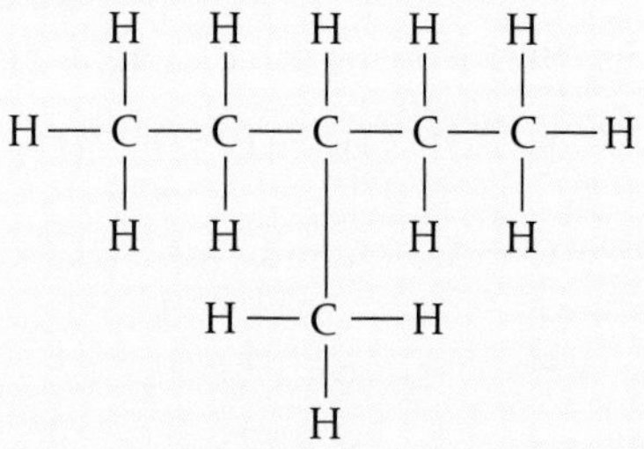

There are two isomers with a four-carbon longest chain and two methyl branches:

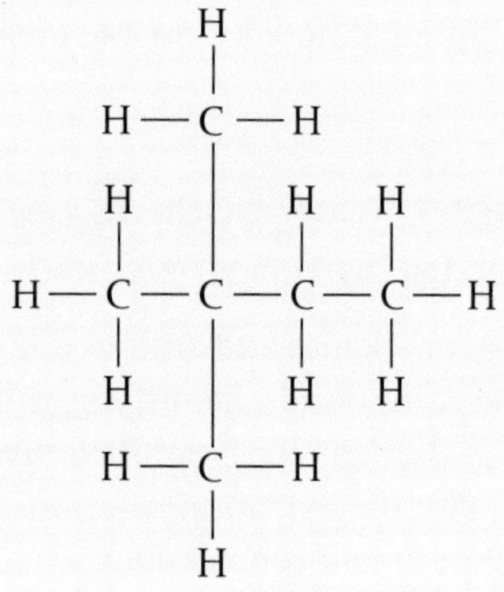

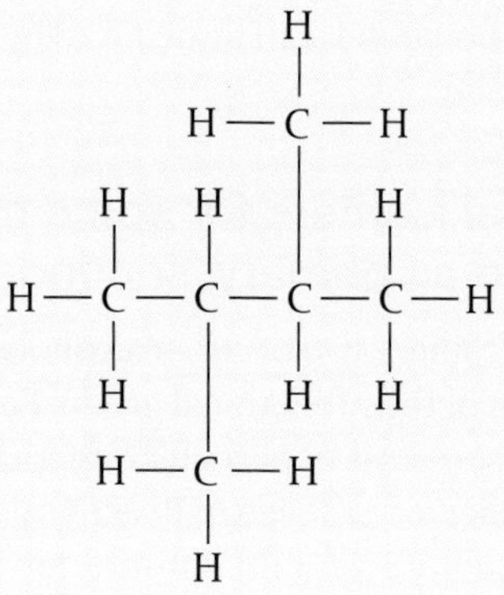

Challenge yourself

2. The Subject Guide only requires you to determine isomers for compounds containing up to six carbons in the longest chain. However, it can be beneficial to apply your knowledge to larger compounds. See if you can identify the nine structural isomers of heptane (C_7H_{16}).

Alkenes can have positional structural isomers

Alkenes are a class of organic compounds that contain at least one carbon–carbon double bond. For alkanes, we saw that structural isomers can occur due to branching of the carbon chains. For alkenes, a different type of structural isomer can also occur, where the double bond is found in different positions.

C_4H_8 has the following straight-chain isomers:

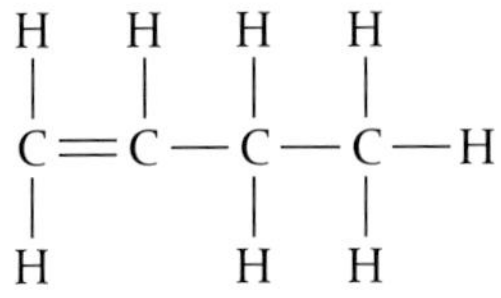

but-1-ene

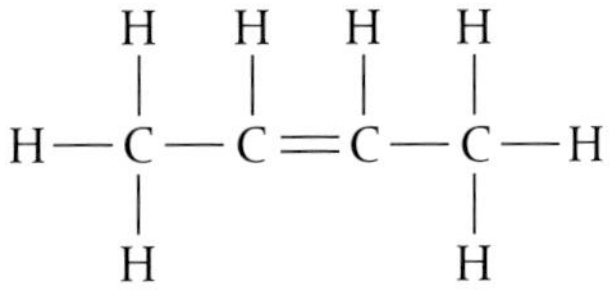

but-2-ene

C_5H_{10} has the following two straight-chain isomers:

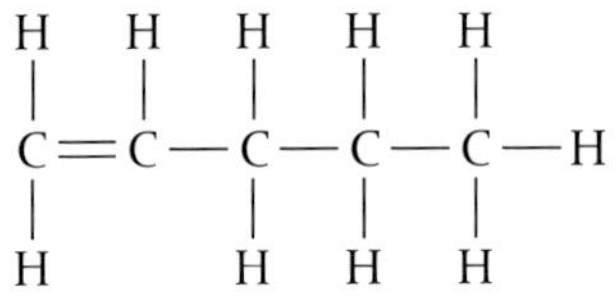

pent-1-ene

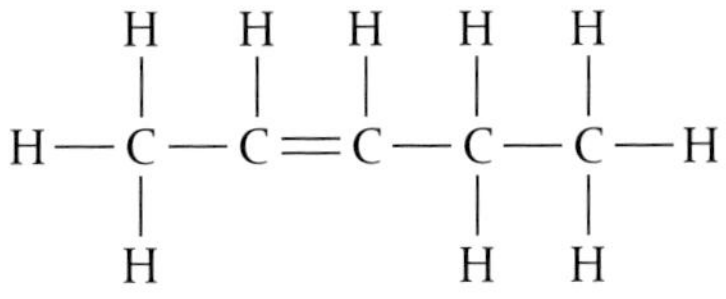

pent-2-ene

C_6H_{12} has the following three straight-chain isomers:

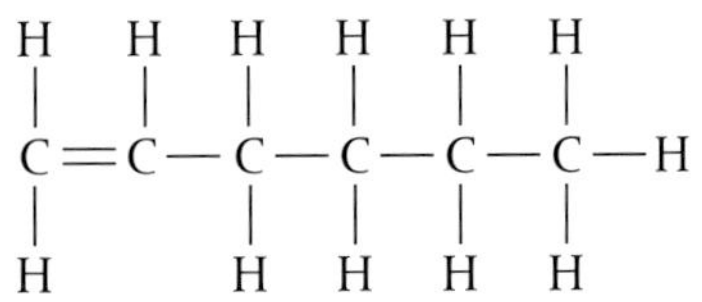

hex-1-ene

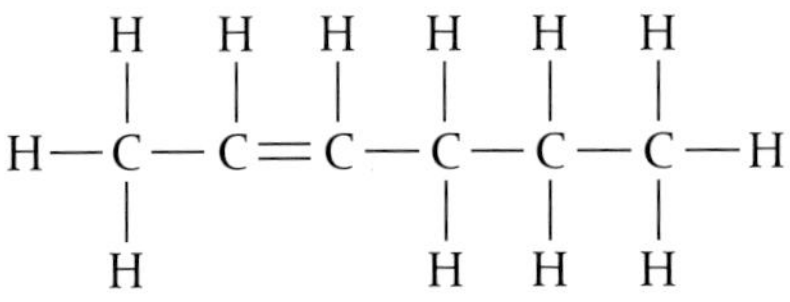

hex-2-ene

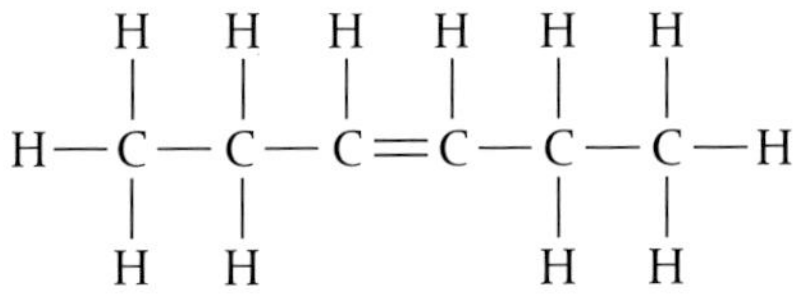

hex-3-ene

Similarly, the structures of the isomers of the alkynes, which contain at least one triple carbon–carbon bond, can be deduced by considering different possible positions for the triple bond.

Positional isomers for some classes of compounds can be classified as primary, secondary or tertiary compounds

Structural isomers also occur in alcohols, halogenoalkanes and amines due to the position of their functional group. For these compounds, identifying different positional isomers is of particular importance as they can have significantly different chemical reactivity, as seen in Reactivity 3.2–3.4 for the oxidation reactions of alcohols and in Reactivity 3.4 for the substitution reactions of halogenoalkanes.

A **primary carbon atom** is attached to the functional group and also to at least two hydrogen atoms. Molecules with this arrangement are known as primary molecules.

For example, ethanol, C_2H_5OH, is a primary alcohol and chloroethane, C_2H_5Cl, is a primary halogenoalkane.

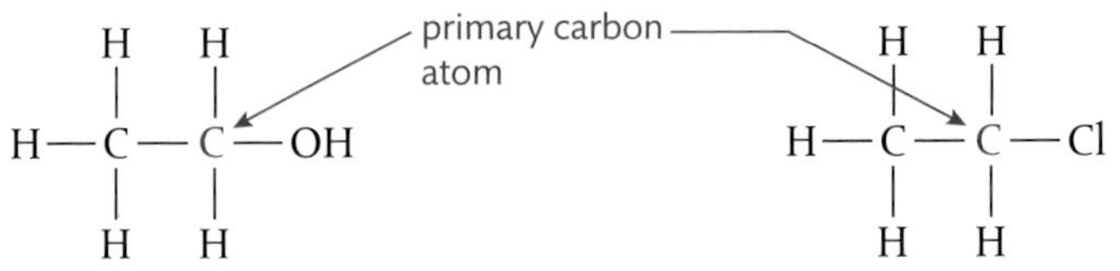

primary carbon atom

A **secondary carbon atom** is attached to the functional group and also to one hydrogen atom and two alkyl groups. These molecules are known as secondary molecules. For example, propan-2-ol, $CH_3CH(OH)CH_3$, is a secondary alcohol, and 2-chloroethane, $CH_3CHClCH_3$, is a secondary halogenoalkane.

H H H   secondary carbon
 | | | atom
H—C—C—C—H
 | | |
 H OH H

H H H
 | | |
H—C—C—C—H
 | | |
 H Cl H

A **tertiary carbon atom** is attached to the functional group and is also bonded to three alkyl groups and so has no hydrogen atoms. These molecules are known as tertiary molecules. For example, 2-methylpropan-2-ol, $C(CH_3)_3OH$, is a tertiary alcohol, and 2-chloro-2-methylpropane, $C(CH_3)_3Cl$, is a tertiary halogenoalkane.

      H
      |
  H—C—H
  H   H
  | |
H—C—C—C—H   tertiary carbon
  | | | atom
  H OH H

      H
      |
  H—C—H
  H   H
  | |
H—C—C—C—H
  | | |
  H Cl H

▲
Butan-2-ol (left) and 2-methylpropan-2-ol (right) are two structural isomers with the formula C_4H_9OH. With molecular models it is easy to visualize their different structures and identify that butan-2-ol is a secondary alcohol and 2-methylpropan-2-ol is a tertiary alcohol. Carbon atoms are shown in black, hydrogen atoms are white and oxygen atoms are red.

In amines, a similar classification can be applied according to the number of alkyl groups and hydrogen atoms bonded to the nitrogen atom.

Type of amine	Primary	Secondary	Tertiary
Number of H atoms attached to central N	2	1	0
Number of C atoms attached to central N	1	2	3

primary
nitrogen atom
         H
         |
H_3C—N
         \
          H

secondary
nitrogen atom
         H
         |
H_3C—N
         \
          CH_3

tertiary
nitrogen atom
          CH_3
         /
H_3C—N
         \
          CH_3

Structural isomers can have different functional groups

There are many different types of isomers, including those in which the molecules have different functional groups and so are members of different classes of compounds with very different reactivities. For example, the molecular formula C_2H_6O describes both an alcohol and an ether, with models of these **functional group isomers** shown in the picture below.

Nature of Science, Tool 2 – How useful are 3D models (real or virtual) to visualize the invisible?

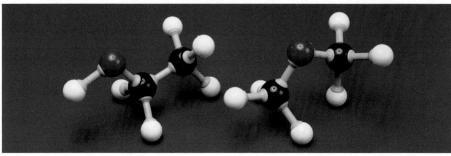

▲
Methoxymethane (left) and ethanol (right) are functional group isomers, both with the molecular formula C_2H_6O. With the use of molecular models, it is easy to visualize the different functional groups in these molecules and explain their different physical and chemical properties. Ethanol is a liquid widely used in alcoholic drinks while methoxymethane is a gas used in aerosol propellants. Carbon atoms are represented here as black, hydrogen atoms are white and oxygen atoms are red.

Exercise

Q15. Which of the following pairs of compounds are structural isomers?

    I. $CH_3CH_2CH_2CH_3$, $CH_3CH(CH_3)CH_3$

    II. $CH_3CH_2CH_2CH_3$, $CH_3CH_2CH_3$

    III. $CH_3CH_2CH_2CH_2OH$, $CH_3CH(CH_3)CH_2OH$,

    A I only   **B** I, II and III   **C** III only   **D** I and III

Q16. Pentyne, C_5H_8, contains a triple bond.

    (a) Draw two positional isomers of pentyne.

    (b) Draw a third structural isomer of pentyne that is not a positional isomer of the two isomers you drew for part **(a)**.

Q17. Which of the following compounds contains a secondary carbon?

    A $CH_3CH_2CH_2CH_2OH$ **B** $CH_3Cl(CH_3)CH_2CH_3$

    C $CH_2BrCH_2CH_2CH_2OH$ **D** $CH_3CH_2CHClCH$

Q18. Draw and name the structural isomers of bromobutane, C_4H_9Br, and identify each isomer as being primary, secondary or tertiary.

Q19. Draw and name two different functional group isomers that can have the following molecular formulas:

    (a) C_3H_8O **(b)** C_3H_6O **(c)** $C_4H_8O_2$

Guiding Question revisited

How does the classification of organic molecules help us to predict their properties?

In this chapter we have learnt that:

- Organic compounds can be grouped into classes based on their functional groups.
- The presence of different functional groups means that different classes of compounds have different physical and chemical properties.
- Homologous series exist for each class of compound and successive members of a series differ by a CH_2 unit.
- Applying IUPAC nomenclature to generate unique names for individual compounds allows for further classification of organic compounds.
- Structural isomers are compounds that have the same molecular formula but different connectivities. They have unique IUPAC names as well as different chemical and physical properties.
  - Chain isomers occur when C-C bonds are connected differently to form structures that are a straight chain or branched.
  - Positional isomers contain the same functional group but it is at different positions on the carbon chain.
  - Functional group isomers contain different functional groups.

Practice questions

1. Which of the following compounds are structural isomers?

   I. $CH_3CH_2COCH_3$ II. $CH_3CH_2CH_2CHO$ III. $CH_3CH(CH_3)CHO$

   A I and III   **B** I, II and III   **C** I and II   **D** II and III

2. Which of the following compounds is a secondary alcohol?

   A $CH_3CH_2CH_2OH$ **B** $(CH_3)_2CHCH_2OH$

   C $(CH_3)_2C(OH)CH_3$ **D** $CH_3CH_2CH(OH)CH_3$

3. Which molecule has a secondary nitrogen?

   A CH_3NHCH_3 **B** $H_2NCH_2CH_2NH_2$

   C $CH_3CH_2NH_2$ **D** $(CH_3)_3N$

4. Which of the following has the molecules ordered by decreasing volatility?

   A butan-1-ol, butan-2-ol, 2-methylpropan-2-ol

   B butanal, butan-1-ol, butanone

   C 2-methylpropan-2-ol, butanone, butan-1-ol

   D butanone, 2-methylpropan-2-ol, butan-1-ol

5. Below are four structural isomers with molecular formula C_4H_9Br.
State the name of each of the isomers A, B, C and D.

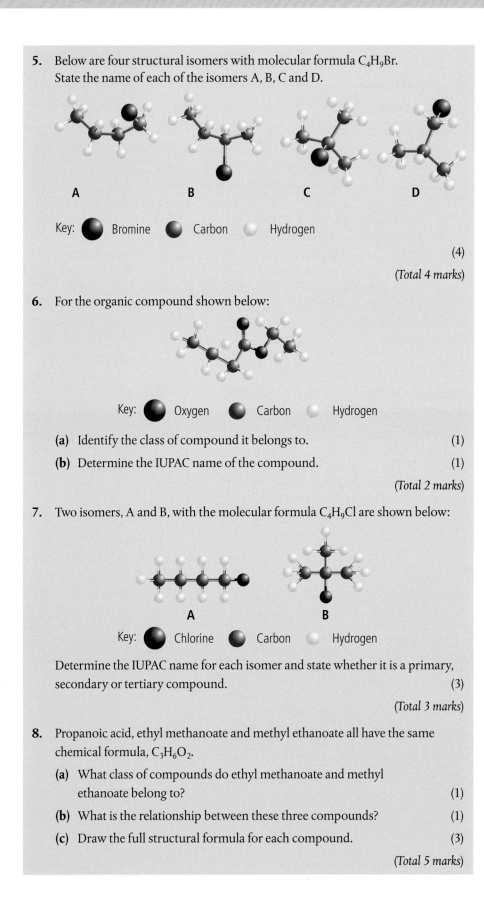

A   B   C   D

Key: ⬤ Bromine ⬤ Carbon ○ Hydrogen

(4)

(Total 4 marks)

6. For the organic compound shown below:

Key: ⬤ Oxygen ⬤ Carbon ○ Hydrogen

   (a) Identify the class of compound it belongs to. (1)

   (b) Determine the IUPAC name of the compound. (1)

(Total 2 marks)

7. Two isomers, A and B, with the molecular formula C_4H_9Cl are shown below:

A   B

Key: ⬤ Chlorine ⬤ Carbon ○ Hydrogen

Determine the IUPAC name for each isomer and state whether it is a primary, secondary or tertiary compound. (3)

(Total 3 marks)

8. Propanoic acid, ethyl methanoate and methyl ethanoate all have the same chemical formula, $C_3H_6O_2$.

   (a) What class of compounds do ethyl methanoate and methyl ethanoate belong to? (1)

   (b) What is the relationship between these three compounds? (1)

   (c) Draw the full structural formula for each compound. (3)

(Total 5 marks)

9. Cyclobutane, C_4H_8, is a colorless gas used in organic synthesis and medicinal chemistry.

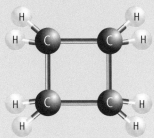

Two structural isomers of cyclobutane are but-1-ene and 2-methylpropene.

(a) What class of compounds do but-1-ene and 2-methylpropene belong to? (1)

(b) Draw the full structural formulas of but-1-ene and 2-methylpropene. (2)

(c) Draw and name one other structural isomer of cyclobutane. (2)

(Total 5 marks)

10. Iodobutane, C_4H_9I, has a number of different structural isomers, each with unique physical and chemical properties.

(a) Draw the structures of three isomers of iodobutane that represent primary, secondary and tertiary isomers. (3)

(b) Provide the IUPAC name for the secondary and tertiary isomers drawn in part (a). (2)

(c) Predict and explain which of the isomers drawn in part (a) will have the lowest boiling point. (2)

(Total 7 marks)

11. Methoxypropane, $CH_3OCH_2CH_2CH_3$, was once used as a general anesthetic.

(a) Name the class of compounds that methoxypropane belongs to. (1)

(b) Draw and name a structural isomer of methoxypropane that belongs to a different class of compounds. (2)

(c) Predict and explain whether the isomer identified in part (b) will have a higher or lower boiling point than methoxypropane. (2)

(Total 5 marks)

12. Amines are organic compounds that contain a nitrogen atom with a lone pair of electrons. Amines can exist as primary, secondary or tertiary isomers.

(a) Draw the structure of a primary, secondary and tertiary amine, each with the chemical formula C_3H_9N. (3)

(b) Predict and explain which of the isomers drawn in part (a) will have the highest boiling point. (2)

(Total 5 marks)

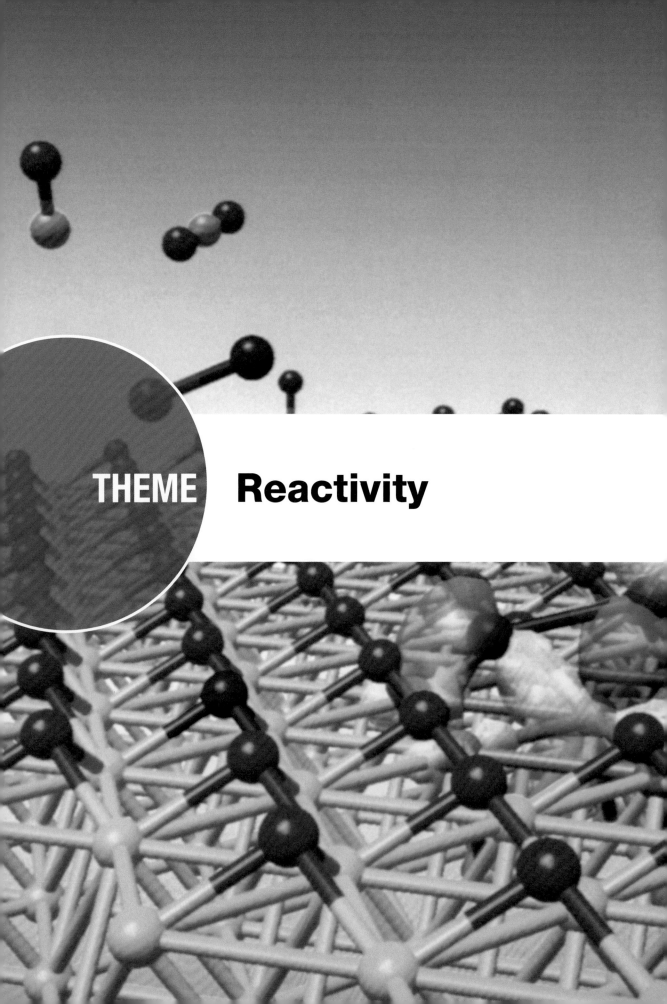

THEME **Reactivity**

Molecular graphic showing a catalytic converter in a vehicle exhaust system. The yellow framework is the catalyst, a lattice of copper atoms. The reactant gases NO and CO are shown adsorbed onto the copper surface. A chemical reaction occurs on the surface of the catalyst, converting these polluting gases into non-polluting products:

$$2NO(g) + 2CO(g) \rightarrow N_2(g) + 2CO_2(g)$$

Catalysts are widely used in research and industry to increase the rate of reactions.

Reactivity

Reactivity refers to the ways in which chemical reactions transform reactants into products. These reactions occur through bond-breaking and bond-making, and so they change structures into new forms. An understanding of what drives these reactions helps us to determine how they can be controlled and how we can influence the product.

In Reactivity 1 we explore the direction of energy transfer in reactions, starting with measurements of temperature change as an indication of heat transfer. The conservation of energy during reactions is an application of the First Law of Thermodynamics, and is illustrated through energy cycles. We consider the release of energy from different fuels to supply our daily energy needs within the context of their relative benefits and environmental impact. Through an investigation of the direction of change, we learn that reactions are spontaneous when they lead to products which have lower free energy than the reactants.

In Reactivity 2 we use experimental data to examine the factors that influence both the extent and the rate of reactions, while stoichiometric principles help us to assess the yield and the atom economy of reactions. The role of catalysts in increasing the efficiency of reactions is explained.

The focus of Reactivity 3 is the mechanisms of reactions and how reactions proceed at the molecular level. Mechanisms are classified on the basis of the transfer or sharing of subatomic particles between reactants. In this way, we can see how our understanding of reactivity is based on the application of our knowledge of structure.

Chemical reactivity is at the heart of chemical industries such as resource extraction, and processes which synthesize novel products such as pharmaceuticals and petrochemicals. These reactions supply many products for our modern world, but can also be responsible for unwanted side-products and harmful effects on the environment. Throughout this study, you are encouraged to consider how the application of Green Chemistry principles can ameliorate some of the negative side-effects of chemical reactivity.

Structure determines reactivity, which in turn transforms structure.

REACTIVITY

1

What drives chemical reactions?

The law of conservation of energy states that energy cannot be created or destroyed. It can, however, be transformed from one form into another, and all chemical reactions are accompanied by energy changes. These energy changes are vital. For example, our bodies' processes are dependent on the energy changes which occur during respiration, when glucose reacts with oxygen; and modern lifestyles are dependent on the transfer of energy that occurs when fuels burn. As we explore the source of these energy changes, we will deepen our understanding of why bonds are broken and formed during a chemical reaction, and why electron transfer can lead to the formation of stable ionic compounds. The questions of why things change will lead to the development of the concept of entropy. We will see that this concept allows us to give the same explanation for a variety of physical and chemical changes: the universe is becoming more disordered as matter and energy are dispersed. This provides us with a signpost for the direction of all change. The distinction between the quantity and quality of energy will lead to the development of the concept of free energy, a useful accounting tool for chemists to predict the feasibility of any hypothetical reaction.

We will see how creative thinking, accurate calculations, and careful observations and measurement can work together to lead to a deeper understanding of the relationship between heat and chemical change.

TOK

The Nobel Prize winning physicist, Richard Feynman, on the scientific idea of energy:
'... the conservation of energy.... says that there is a certain quantity, which we call energy, that does not change in the manifold changes which nature undergoes. That is a most abstract idea, because it is a mathematical principle; it says that there is a numerical quantity, which does not change when something happens.'

'It is important to realize that we have no knowledge of what energy is.'

Why are many of the laws in the natural sciences stated using the language of mathematics? Do you agree that we have no *knowledge* of what energy is?

Measuring enthalpy changes

◄ A polystyrene cup with a lid can act as a calorimeter. The energy changes during aqueous reactions can be determined from the temperature changes of the water solvent. The water increases in temperature when a reaction gives out heat and decreases in temperature when a reaction absorbs heat.

Guiding Question

What can be deduced from the temperature change that accompanies chemical or physical change?

It is important to distinguish between heat and temperature. Heat is a method of energy transfer and temperature is a measure of the average kinetic energy of the constituent particles in an object. Heat transfers can, however, be determined from temperature changes. The heat needed to change the temperature of an object by 1 K is a property of the mass and composition of the object, known as the heat capacity, which can be experimentally determined. Generally, we determine the heat transfers during a chemical reaction by measuring the temperature changes of a known mass of water. For example, the heat from a flame during combustion reactions is transferred to the water in a metal calorimeter. Alternatively, reactions in aqueous solution are investigated in calorimeters made from an insulating material, and the heat is transferred between the reactants and the water solvent. The temperature of the water increases during an exothermic reaction, which releases heat as the reactants change to products with less potential energy. The temperature of the water decreases during the less common endothermic reactions as heat is transferred from the water to the products with more potential energy than the reactants.

Reactivity 1.1.1 – Chemical reactions involve heat transfers

Reactivity 1.1.1 – Chemical reactions involve a transfer of energy between the system and the surroundings, while total energy is conserved.

Understand the difference between heat and temperature.

	Structure 1.1 – What is the relationship between temperature and kinetic energy of particles?

Heat is a process of energy transfer

Energy is a measure of the ability to do **work**, that is, to move an object against an opposing force. Energy can be transferred in different ways including heat, light, sound and electricity, and stored in different forms of potential energy such as chemical potential energy. This chapter focuses on the heat transfers which accompany chemical reactions. Reactions that involve the transfer of electrical energy are explored in Reactivity 3.2.

Heat is a form of energy transfer that occurs as a result of a temperature difference. When heat is transferred to a system, the average kinetic energy of the molecules and temperature are increased and the kinetic energy is more dispersed among the particles. Heat is a process of energy transfer, whereas temperature, a measure of the *average* kinetic energy of the particles, is a property of the object or system.

More heat energy is needed to raise the temperature of water in a swimming pool by 1 K than is needed for the water in a kettle because there are more water molecules in the swimming pool.

The heat needed to boil a kettle will only lead to a very small increase in the temperature of a swimming pool.

Work is another form of energy transfer and is a more ordered process than energy transfer by heat. For example, you do work on a beaker of water when you lift it from a table.

System and surroundings

Chemical and physical changes take place in many different environments such as test tubes, polystyrene cups, industrial plants and living cells. It is useful in these cases to distinguish between the **system** – the area of interest, and the **surroundings** – in theory, everything else in the universe (Figure 1).

surroundings

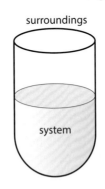

system

R1.1 Figure 1 The system is the sample or reaction vessel of interest. The surroundings are the rest of the universe.

Most chemical reactions take place in an **open system** in which energy and matter can be exchanged with the surroundings. For example, in an open system, gases and heat produced by a reaction in a test tube can be transferred to the surroundings. In a **closed system**, energy can be exchanged with the surroundings but matter cannot. Although energy can be exchanged between a closed system and the surroundings, the total energy cannot change during the process; any energy lost by the system is gained by the surroundings and vice versa. An **isolated system** cannot exchange matter or energy with the surroundings.

Structure 1.1 – What is the relationship between temperature and kinetic energy of particles?

The joule is the unit of energy and work. You do 1 J of work when you exert a force of 1 N over a distance of 1 m. 1 J of energy is used every time the human heart beats.

Energy is conserved in chemical reactions.

An open system can exchange matter and energy with its surroundings. A closed system allows the transfer of energy but not matter with its surroundings. An isolated system cannot exchange matter or energy with its surroundings.

The chemical potential energy of a system is its enthalpy

Although, according to the law of conservation of energy, the total energy of the system and surroundings cannot change during a process, heat can be transferred between a system and its surroundings. The chemical potential energy of a system is called its **enthalpy**, a name which comes from the Greek word for 'heat inside'. A system acts like a reservoir of chemical potential energy or enthalpy. When heat is added to a system from the surroundings, the enthalpy of the system increases and when a system gives out heat its enthalpy decreases. Changes in enthalpy are denoted by ΔH. ΔH is positive when heat is added to the system (Figure 2).

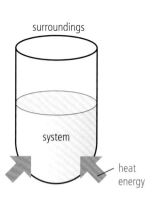
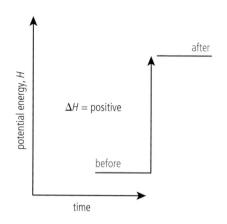

R1.1 Figure 2 When heat is added from the system to the surroundings, the enthalpy of the system increases and ΔH is positive.

> **Enthalpy (H) is a measure of the chemical potential energy stored in a system. The enthalpy changes as chemical bonds and intermolecular forces change during a reaction. This results in a heat transfer which can be observed.**

When heat is released from the system to the surroundings, the chemical potential energy or enthalpy of the system decreases and ΔH is negative (Figure 3).

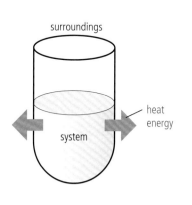
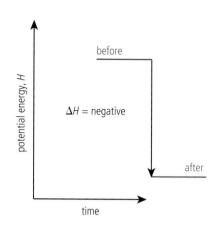

TOK How important are technical terms such as enthalpy in different areas of knowledge? To what extent does the technical language we use help or hinder the acquisition of knowledge?

R1.1 Figure 3 When heat is given out from the system to the surroundings, the enthalpy or potential energy of the system decreases and ΔH is negative.

Exercise

Q1. A large iron cube at high temperature, T_H, is brought into contact with a small iron cube at lower temperature, T_C. Discuss the changes that occur, assuming that both objects are insulated from all other objects.

Q2. The temperature of a volume of water increases from 300 K to 310 K. State the temperature increase in °C.

Reactivity 1.1.2 – Endothermic and exothermic reactions

Reactivity 1.1.2 – Reactions are described as endothermic or exothermic, depending on the direction of energy transfer between the system and the surroundings.

Understand the temperature change (decrease or increase) that accompanies endothermic and exothermic reactions, respectively.

	Tool 1, Inquiry 2 – What observations would you expect to make during an endothermic and an exothermic reaction?

SKILLS

Observe endothermic and exothermic processes in action. Full details of how to carry out this experiment with a worksheet are available in the eBook.

Exothermic and endothermic reactions involve heat transfers between the system and the surroundings

The enthalpy (H) of the system changes as the chemical and intermolecular forces change. Most chemical reactions, including most combustion and all neutralization reactions, are exothermic. They give out heat to the surroundings and $\Delta H_{reaction}$ is negative (Figure 4a). Some reactions are **endothermic** as they result in a heat transfer from the surroundings to the system. In this case, the products have more stored potential energy than the reactants and $\Delta H_{reaction}$ is positive (Figure 4b).

R1.1 Figure 4 (a)
An exothermic reaction. The stored chemical energy, or enthalpy, of the products is less than the enthalpy of the reactants.
R1.1 Figure 4 (b)
An endothermic reaction. The enthalpy of the products is greater than the enthalpy of the reactants. The change from reactants to products is represented on the horizontal axis and labelled as the reaction coordinate.

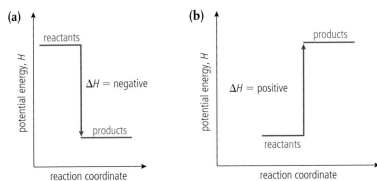

Temperature changes during reactions in solution

When a reaction occurs in aqueous solution, any enthalpy changes in the reaction are best observed by changes in the temperature of the water, which is acting as the solvent. In an exothermic reaction, the heat released by the chemical reaction is first transferred to the solvent. The temperature increases as the water molecules increase their average kinetic energy. Heat is eventually transferred from the water to the wider surroundings as the system returns to its initial temperature. The net result is transfer from the system to the surroundings, and the enthalpy of the system decreases (Figure 5).

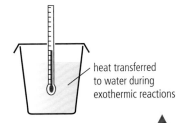
heat transferred to water during exothermic reactions

temperature of the water increases

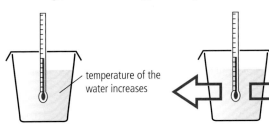

Heat is transferred to the surroundings as the temperature returns to its initial value. The enthalpy of the system has decreased.

R1.1 Figure 5 Temperature changes during an exothermic reaction.

The oxidation of glycerin by potassium permanganate is an exothermic reaction.

Conversely, any heat needed for an endothermic reaction is transferred from the water, leading to a decrease in temperature of the aqueous solvent. The system returns to its initial temperature when heat is transferred to the water from the wider surroundings.

Nature of Science

Scientists share their knowledge using a precise language. In everyday language, heat and work are both nouns and verbs, whereas in science they are nouns which describe energy transfer processes. Heat is often said to 'flow' from high temperature to low temperature. This image originates from the incorrect view that heat was a liquid, *calorique*, which was included in Lavoisier's list of chemical elements. Heat is now more correctly characterized as a process of energy transfer.

An endothermic reaction. The reaction between ammonium thiocyanate and barium hydroxide absorbs so much heat from water placed between the flask and the metal container that the water freezes.

Tool 1, Inquiry 2 – What observations would you expect to make during an endothermic and an exothermic reaction?

Exercise

Q3. When a sample of $NH_4SCN(s)$ is mixed with $Ba(OH)_2.8H_2O(s)$ in a glass beaker, the mixture changes to a liquid and the temperature drops sufficiently to freeze the beaker to the table. Which statement is true about the reaction?

    A The process is endothermic and ΔH is negative.

    B The process is endothermic and ΔH is positive.

    C The process is exothermic and ΔH is negative.

    D The process is exothermic and ΔH is positive.

Q4. Which one of the following statements is true of all exothermic reactions?

    A They produce gases.

    B They give out heat.

    C They occur quickly.

    D They involve combustion.

Q5. Which is the best explanation of why the temperature of the liquid decreases when it evaporates?

    A The number of liquid molecules is decreasing.

    B The mean kinetic energy of the liquid molecules is decreasing.

    C The pressure above the liquid surface is increasing.

    D The rate of evaporation is increasing.

Q6. Two objects made of different masses of aluminium have the same temperature.

Which is the best description of the relative average kinetic energy and total energies of the aluminium atoms in the different objects?

	Average kinetic energy of the atoms	Total energy of the atoms
A	greater in object of larger mass	same
B	less in object of larger mass	same
C	same	greater in object of larger mass
D	same	less in object of larger mass

Reactivity 1.1.3 – Energetic stability and the direction of change

Reactivity 1.1.3 – The relative stability of reactants and products determines whether reactions are endothermic or exothermic.

Sketch and interpret potential energy profiles for endothermic and exothermic reactions.

Axes for energy profiles should be labelled as reaction coordinate x, potential energy y.	Structure 2.2 – Most combustion reactions are exothermic; how does the bonding in N_2 explain the fact that its combustion is endothermic?

Enthalpy changes as the direction of a reaction

There is a natural direction for change. When we slip on a ladder, we go down, not up (Figure 6). The direction of change is in the direction of lower stored, or potential, energy. In a similar way, we expect methane to burn when we strike a match and form carbon dioxide and water. The chemicals are changing in a way which reduces their chemical potential energy or enthalpy.

R1.1 Figure 6 An exothermic reaction can be compared to a person falling off a ladder. Both changes lead to a decrease in potential energy. The state of lower potential energy is more stable. The mixture of carbon dioxide and water is more stable than a mixture of methane and oxygen.

There are many examples of exothermic reactions, and we generally expect a reaction to occur if it leads to a reduction in the potential energy or enthalpy. In the same way that a ball is more stable on the ground than in mid-air, we can say that the products in an exothermic reaction are more stable than the reactants. Most combustion reactions, for example, are exothermic as the energy needed to break the bonds in the reactants is less than the energy produced as the bonds form in the products. The enthalpy changes that occur as bonds are broken and formed are discussed in more detail in Reactivity 1.2.

It is important to realize that stability is a relative term. Hydrogen peroxide, for example, is stable with respect to its elements but unstable relative to its decomposition to water and oxygen (Figure 7).

Structure 2.2 – Most combustion reactions are exothermic; how does the bonding in N_2 explain the fact that its combustion is endothermic?

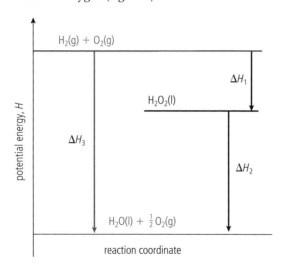

R1.1 Figure 7 Hydrogen peroxide is stable relative to hydrogen and oxygen but unstable relative to water. Note $\Delta H_3 = \Delta H_1 + \Delta H_2$ in agreement with Hess's law discussed in Reactivity 1.2.

The sign of ΔH is a guide for the likely direction of change but it is not a complete guide. We do not expect a person to fall up a ladder, and likewise reactions tend to proceed in the direction of lowering energy/increasing stability. But this is not always the case and some endothermic reactions do occur spontaneously. For example, the reaction below is extremely endothermic:

$$SOCl_2(l) + FeCl_3.6H_2O(s) \rightarrow FeCl_3(s) + 6SO_2(g) + 12HCl(g)$$

Endothermic reactions are less common and can only occur when there is an increase in the dispersal of the system, for example, owing to the production of sulfur dioxide and hydrogen chloride gas in the reaction above. This is explained more fully in Reactivity 1.4.

The other limitation of using ΔH values as a guide to change is that it does not indicate anything about the rate of reaction. Some reactions, which should take place on energetic grounds, do not occur at a noticeable rate in practice as the reactants need to be given some initial energy to react. The minimum kinetic energy to react is known as the **activation energy**, E_a. For example, a lighted match provides the necessary activation energy for the combustion of methane in oxygen. This energy is needed as some bonds in the reactants must be broken before new bonds in the products can form. Typical energy profiles are given in Figure 8.

R1.1 Figure 8 Energy profiles of an endothermic and exothermic reaction.

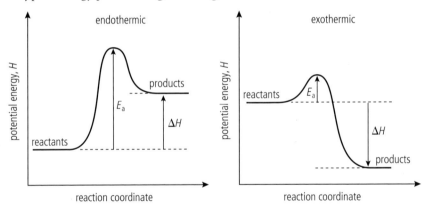

Activation energy is discussed in more detail in Reactivity 2.2.

Diamond is a naturally occurring form of carbon that has crystallized under great pressure. It is energetically unstable relative to graphite, but the reaction does not take place at an observable rate.

Exercise

Q7. Which statement is consistent with the potential energy profile shown?

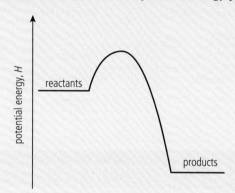

A Heat is produced during the reaction and the products are more stable than the reactants.

B Heat is taken in during the reaction and the products are more stable than the reactants.

C Heat is produced during the reaction and the reactants are more stable than the products.

D Heat is taken in during the reaction and the reactants are more stable than the products.

Q8. Identify the endothermic reaction.

	Enthalpy	Direction of heat transfer
A	$H_{\text{reactants}} < H_{\text{products}}$	system $\rightarrow$ surroundings
B	$H_{\text{reactants}} < H_{\text{products}}$	surroundings $\rightarrow$ system
C	$H_{\text{reactants}} > H_{\text{products}}$	system $\rightarrow$ surroundings
D	$H_{\text{reactants}} > H_{\text{products}}$	surroundings $\rightarrow$ system

Q9. The potential energy profile of a reaction is shown.

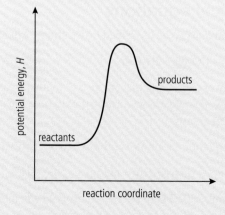

What can be deduced about the nature of the reaction and the relative stability of the reactants and products?

	Nature of reaction	Most stable
A	endothermic	reactants
B	endothermic	products
C	exothermic	reactants
D	exothermic	products

Reactivity 1.1.4 – Measuring enthalpy changes

Reactivity 1.1.4 – The standard enthalpy change for a chemical reaction, $\Delta H^\ominus$, refers to the heat transferred at constant pressure under standard conditions and states. It can be determined from the change in temperature of a pure substance.

Apply the equations $Q = mc\Delta T$ and $\Delta H = -\dfrac{Q}{n}$ in the calculation of the enthalpy change of a reaction.

The units of $\Delta H^\ominus$ are $kJ\,mol^{-1}$. The equation $q = mc\Delta T$ and the value of c, the specific heat capacity of water, are given in the IB data booklet.	Tool 1, Inquiry 1, 2, 3 – How can the enthalpy change for combustion reactions, such as for alcohols or food, be investigated experimentally? Tool 1, Inquiry 3 – Why do calorimetry experiments typically measure a smaller change in temperature than is expected from theoretical values?

The standard conditions for enthalpy changes are as follows: a pressure of 100 kPa, concentrations of 1 mol dm⁻³ for all solutions, all the substances in their standard states.

The standard state of a substance is the pure form of the substance under standard conditions of 298 K (25 °C) and 1.00×10^5 Pa.

Standard enthalpy changes

The **standard enthalpy change**, $\Delta H^\ominus$, given in the literature, is measured under the following standard conditions:

- a pressure of 100 kPa
- concentration of 1 mol dm⁻³ for all solutions
- all substances in their standard states.

Temperature is not part of the definition of standard state, but 298 K is usually given as the specified temperature.

Thermochemical equations

The combustion of methane can be described by the thermochemical equation:

$$CH_4(g) + 2O_2(g) \rightarrow CO_2(g) + 2H_2O(l) \qquad \Delta H^\ominus_{reaction} = -890\,kJ\,mol^{-1}$$

This is a shorthand way of expressing that *one mole* of methane gas reacts with *two moles* of oxygen gas to give *one mole* of gaseous carbon dioxide and *two moles* of liquid water and *releases* 890 kJ of heat energy.

The thermochemical equation for photosynthesis can be represented as:

$$6CO_2(g) + 6H_2O(l) \rightarrow C_6H_{12}O_6(aq) + 6O_2(g) \qquad \Delta H^\ominus_{reaction} = +2802.5\,kJ\,mol^{-1}$$

which means that overall, 2802.5 kJ of energy is absorbed when one mole of glucose is formed under standard conditions from gaseous carbon dioxide and liquid water.

It is important to give the state symbols in thermochemical equations because the enthalpy changes depend on the state of the reactants and the products.

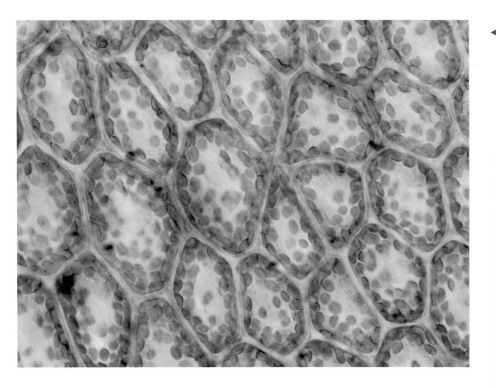

◀ Light micrograph showing
a section through a leaf.
The green chloroplasts
contain chlorophyll, which is
responsible for the absorption
of light during photosynthesis.
Energy is needed for
photosynthesis as the
products have more potential
energy than the reactants.

Enthalpy changes can be calculated from temperature changes

In Structure 1.1, we discuss the relationship between temperature and kinetic energy. The absolute temperature in K is a measure of the average kinetic energy E_k of particles. If the same amount of heat energy is added to two different objects made of the same material but with a different number of particles, the temperature change will not be the same because the average kinetic energy of the particles will not increase by the same amount. The object with the smaller number of particles will experience the larger temperature increase because the same energy is shared among a smaller collection of particles. As equal masses of different materials have different numbers of particles, the temperature change will also depend on the chemical nature of the material. In general, the increase in temperature (ΔT) when an object is heated depends on:

- the mass (m) of the object
- the heat added (q)
- the nature of the substance.

The **specific heat capacity** (c) gives the heat needed to increase the temperature of a unit mass of a substance by 1 K. The specific heat capacity (c) depends on the number of particles present in a unit mass sample, which in turn will depend on the mass of the individual particles.

heat added (q) = mass (m) × specific heat capacity (c) × temperature change (ΔT)

heat added (q) = m × c × ΔT

$q(J) = m(g) \times c(J\,g^{-1}\,K^{-1}) \times \Delta T(K)$

When the heat is absorbed by water, $c = 4.18\,J\,g^{-1}\,K^{-1}$
This value is given in the data booklet.

A temperature rise of
1 K is the same as a
temperature rise of 1 °C
so there is no need
to change the units
of temperature when
using the equation
$q = m \times c \times \Delta T$.

The specific heat capacity (c) is the heat needed to increase the temperature of a unit mass of material by 1 K.

$$\text{specific heat capacity } (c) = \frac{\text{heat added } (q)}{\text{mass } (m) \times \text{temperature change } (\Delta T)}$$

Nature of Science

Enthalpy is a word rarely used in non-scientific contexts; it is an abstract entity with a precise mathematical definition. At this level we need not concern ourselves with absolute enthalpy values but only enthalpy changes which can be determined from temperature changes at constant pressure. The use of appropriate terminology is a key issue with scientific literacy and the public understanding of science, and scientists need to take this into account when communicating with the public.

TOK

Our shared knowledge is passed on from one generation to the next by language. The language we use today is often based on the shared knowledge of the past which can sometimes be based on outdated concepts. What do such phrases as 'keep the heat in and the cold out' tell us about previous concepts of heat and cold? How does the use of language hinder the pursuit of knowledge?

Worked example

How much heat is released when 10.0 g of copper with a specific heat capacity of $0.385\,\text{J}\,\text{g}^{-1}\,°\text{C}^{-1}$ is cooled from 85.0 °C to 25.0 °C?

Solution

$q = m \times c \times \Delta T$

$\quad = 10.0\,\text{g} \times 0.385\,\text{J}\,\text{g}^{-1}\,°\text{C}^{-1} \times -60.0\,°\text{C}$ (the value is negative because the Cu has lost heat)

$\quad = -231\,\text{J}$

Challenge yourself

1. Explain the particularly high specific heat capacity of water.

Measuring enthalpy changes of combustion

SKILLS

Comparing the enthalpy of isopropanol and ethanol. Full details of how to carry out this experiment with a worksheet are available in the eBook.

For liquids such as ethanol, the enthalpy change of combustion, ΔH_c, can be determined using the simple apparatus shown in Figure 9. The heat given out by the flame during combustion reactions is used to heat a known mass of water in a metal calorimeter. Copper is often used as it is a good conductor of heat. The heat absorbed by the water can be calculated from its temperature change and its mass. The heat absorbed by the calorimeter can also be calculated from its heat capacity.

R1.1 Figure 9 The heat produced by the combustion of the fuel is calculated from the temperature change of the water in the metal calorimeter. Copper is a good conductor of heat, so heat from the flame can be transferred to the water.

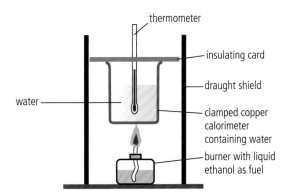

The temperature of the water increases due to the heat released by the combustion reaction (Figure 10). This heat is released as the ethanol and oxygen are converted to carbon dioxide and water. There is a decrease in enthalpy during the reaction.

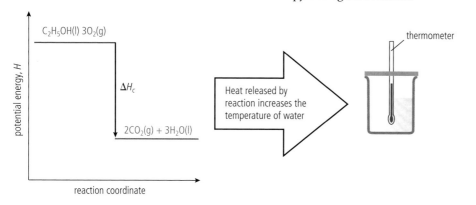

► **R1.1 Figure 10** The enthalpy of the system decreases and the enthalpy of the water increases. The enthalpy change of the water can be calculated if it is assumed that all the heat from the reaction is transferred to the water.

Worked example

Calculate the enthalpy of combustion of ethanol from the following data. Assume all the heat from the reaction is absorbed by the water. Compare your value with the IB data booklet value and suggest reasons for any differences.

Mass of water in copper calorimeter / g	200.00
Temperature increase in water / °C	13.00
Mass of ethanol burnt / g	0.45

Solution

$$\text{number of moles} = \frac{\text{mass } (m)}{\text{molar mass } (M)}$$

$$\text{number of moles of ethanol} = \frac{m(C_2H_5OH)}{M(C_2H_5OH)}$$

$$M(C_2H_5OH) = (12.01 \times 2) + (6 \times 1.01) + 16.00 = 46.08 \text{ g mol}^{-1}$$

$$\Delta H_{\text{reaction}} = -m(H_2O) \times c(H_2O) \times \Delta T(H_2O)$$

Note the negative sign. The enthalpy change of the water is opposite to that in the reaction. ΔT is positive when the reaction is exothermic and ΔH is negative, and ΔT is negative when the reaction is endothermic and ΔH is positive.

$\Delta H_c = \Delta H_{\text{reaction}}$ (for one mole of ethanol)

$$= \frac{-m(H_2O) \times c(H_2O) \times \Delta T(H_2O)}{\left(\dfrac{m(C_2H_5OH)}{M(C_2H_5OH)} \right)}$$

$$= \frac{200 \text{ g} \times 4.18 \text{ J g}^{-1}\,°C^{-1} \times 13.00\,°C}{\left(\dfrac{0.45}{46.08} \right) \text{ g mol}^{-1}}$$

$$= -1\,112\,883 \text{ J mol}^{-1}$$

$$= -1100 \text{ kJ mol}^{-1}$$

The precision of the final answer is limited by the precision of the mass of the ethanol.

It is important to state all assumptions when processing data such as all the heat released is absorbed by the water. The heat absorbed by the copper calorimeter can also be considered in more sophisticated calculations.

SKILLS

You can investigate the energy content of different foods and fuels.

Tool 1, Inquiry 1, 2, 3 – How can the enthalpy change for combustion reactions, such as for alcohols or food, be investigated experimentally?

Some organic compounds burn with a sooty yellow flame. This is due to incomplete combustion.

Combustion reactions are generally exothermic, so ΔH_c values are generally negative.

The data booklet value is $-1367 \, \text{kJ mol}^{-1}$. The difference between the values can be accounted for by some of the following factors:

- Not all the heat produced by the combustion reaction is transferred to the water.
- Some heat is absorbed by the copper calorimeter, and some has passed to the surroundings.
- The combustion of the ethanol is incomplete owing to the limited oxygen available (rather than complete as is assumed by the literature value).
- The experiment was not performed under standard conditions.

Nature of Science

Quantitative measurements are more objective than qualitative observations and are amenable to mathematical analysis. All measurements are limited in precision and accuracy, however, and qualitative observations have a role in chemistry. The lower than expected exothermic value for an enthalpy of combustion of a fuel, for example, is often as a result of incomplete combustion. The evidence for this comes from any black soot and residue observed during the experiment.

The combustion of fossil fuels such as coal produces carbon dioxide which is a greenhouse gas. It is important we are aware of how our lifestyle contributes to global warming. It is a global problem, but we need to act locally to solve it. The greenhouse effect is discussed in more detail in Reactivity 1.3.

Enthalpy changes of a reaction in solution

The enthalpy changes of a reaction in solution can be calculated by carrying out the reaction in an insulated system, for example, a polystyrene cup (Figure 11). The heat released or absorbed by the reaction can be measured from the temperature change in the water which is acting as a solvent. Note in this case the calorimeter is made from an insulator to maximize the amount of heat that is transferred between the reaction and the water in the system.

R1.1 Figure 11 A simple calorimeter. Polystyrene is a very good thermal insulator with a low heat capacity.

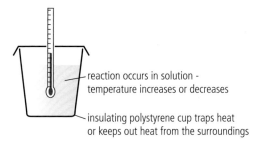

reaction occurs in solution - temperature increases or decreases

insulating polystyrene cup traps heat or keeps out heat from the surroundings

Coffee-cup calorimeter. Full details of how to carry out this experiment with a worksheet are available in the eBook.

SKILLS

In the previous calculation, we assumed that all the heat produced in the reaction is absorbed by the water. One of the largest sources of error in experiments conducted in a polystyrene cup is heat loss to the surroundings. Consider, for example, the exothermic reaction between zinc and aqueous copper sulfate (Figure 12):

$$Cu^{2+}(aq) + Zn(s) \rightarrow Cu(s) + Zn^{2+}(aq)$$

R1.1 Figure 12 A known volume of copper sulfate solution is added to a calorimeter and its temperature measured every 25 s. Excess zinc powder is added after 100 s and the temperature starts to rise until a maximum after which it falls approximately linearly with time.

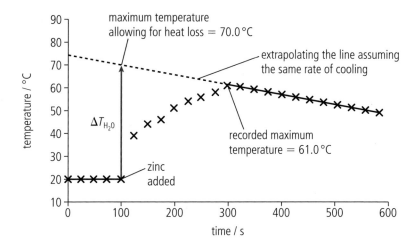

Heat is lost from the system as soon as the temperature rises above the temperature of the surroundings, in this case 20 °C, and so the maximum recorded temperature is lower than the true value obtained in a perfectly insulated system. We can make some allowance for heat loss by extrapolating the cooling section of the graph to the time when the reaction started (100 s). To proceed we can make the following assumptions:

1. no heat loss from the system

2. all the heat goes from the reaction to the water

3. the solution is dilute: $V(CuSO_4) = V(H_2O)$

4. water has a density of $1.00 \, g \, cm^{-3}$.

$\Delta H_{(system)} = 0$ (Assumption 1)

$\Delta H_{(system)} = \Delta H_{(water)} + \Delta H_{reaction}$ (Assumption 2)

$\Delta H_{reaction} = -\Delta H_{(water)}$

For an exothermic reaction, $\Delta H_{reaction}$ is negative because heat has passed from the reaction into the water.

$$\Delta H_{(water)} = m_{(H_2O)} \times c_{(H_2O)} \times \Delta T_{(H_2O)}$$

The limiting reactant (see Reactivity 2.1) must be identified in order to determine the molar enthalpy change of the reaction.

$$\Delta H_{reaction} = \frac{-m(H_2O) \times c(H_2O) \times \Delta T(H_2O)}{\text{moles of limiting reactant}}$$

As the zinc was added in excess, the copper sulfate is the limiting reactant.

From Structure 1.4:

$$\text{number of moles } (n) = \text{concentration (mol dm}^{-3}) \times \text{volume } (V \text{ dm}^3)$$

There are 1000 cm^3 in 1 dm^3.

$$\text{volume } (V \text{ dm}^3) = \frac{\text{volume } (V \text{ cm}^3)}{1000 \text{ (cm}^3 \text{ dm}^{-3})}$$

$$\text{number of moles of CuSO}_4 (n(\text{CuSO}_4)) = \frac{[\text{CuSO}_4] \times V(\text{CuSO}_4)}{1000}$$

$$\Delta H_{reaction} = \frac{-m(H_2O) \times c(H_2O) \times \Delta T(H_2O)}{n(\text{CuSO}_4)} \text{ J mol}^{-1}$$

$$= \frac{-m(H_2O) \times c(H_2O) \times \Delta T(H_2O)}{\left(\dfrac{[\text{CuSO}_4] \times V(\text{CuSO}_4)}{1000 \text{ J mol}^{-1}} \right)}$$

$$= \frac{-m(H_2O) \times c(H_2O) \times \Delta T(H_2O)}{[\text{CuSO}_4] \times V(\text{CuSO}_4)} \text{ kJ mol}^{-1}$$

$$= \frac{-m(H_2O) \times c(H_2O) \times \Delta T(H_2O)}{[\text{CuSO}_4] \times V(H_2O)} \text{ kJ mol}^{-1} \qquad \text{from assumption 3}$$

$$\Delta H_{reaction} = \frac{-c(H_2O) \times \Delta T(H_2O)}{[\text{CuSO}_4]} \text{ kJ mol}^{-1} \qquad \text{from assumption 4}$$

Worked example

Calculate the molar enthalpy change for the reaction between zinc and copper sulfate from the data in Figure 12. The copper sulfate has a concentration of 1.00 mol dm^{-3} and a volume of 100 cm^3.

Solution

From the equation we derived earlier:

$$\Delta H_{reaction} = \frac{-c(H_2O) \times \Delta T(H_2O)}{[\text{CuSO}_4]} \text{ kJ}$$

$$= \frac{-4.18 \times (70.0 - 20.0)}{1.00} \text{ kJ}$$

$$= -209 \text{ kJ mol}^{-1} \text{ (for 1 mole of 1 mol dm}^{-3} \text{ solution)}$$

Tool 1, Inquiry 3 – Why do calorimetry experiments typically measure a smaller change in temperature than is expected from theoretical values?

Worked example

The neutralization reaction between solutions of sodium hydroxide and sulfuric acid was studied by measuring the temperature changes when different volumes of the two solutions were mixed. The total volume was kept constant at $120.0\ cm^3$ and the concentrations of the two solutions were both $1.00\ mol\ dm^{-3}$ (see figure below).

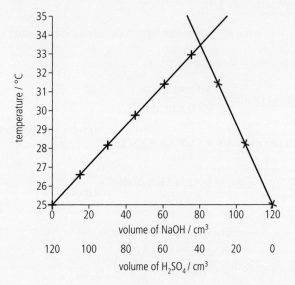

Key

✕ Volume of NaOH added

✕ Volume of H_2SO_4 added

(a) Determine the volumes of the solutions which produce the largest increase in temperature.

(b) Calculate the heat produced by the reaction when the maximum temperature was produced.

(c) Calculate the heat produced for one mole of sodium hydroxide.

(d) The literature value for the enthalpy of neutralization is $= -57.5\ kJ\ mol^{-1}$.

Calculate the percentage error value and suggest a reason for the discrepancy between the experimental and literature values.

Solution

(a) From the graph:

$V(NaOH) = 80.0\ cm^3$

$V(H_2SO_4) = 40.0\ cm^3$

(b) Assuming $120.0\ cm^3$ of the solution contains $120.0\ g$ of water and all the heat produced by the neutralization reaction passes into the water:

$\Delta H_{reaction} = -m(H_2O) \times c(H_2O) \times \Delta T(H_2O)\ J$

$= -120.0\ g \times 4.18\ J\ g^{-1}\ K^{-1} \times (33.5 - 25.0)\ K$

$= -4264\ J$

A common error is to miss out or incorrectly state the units and to miss out the negative sign for ΔH.

(c) $\Delta H_{\text{reaction}} = \dfrac{-4264}{n(\text{NaOH})}\,\text{J mol}^{-1}$

$\qquad = \dfrac{-4264}{\left(1.00 \times \dfrac{80.0}{1000\,\text{J mol}^{-1}}\right)}$

$\qquad = \dfrac{-4264}{80.0}\,\text{kJ mol}^{-1}$

$\qquad = -53.3\,\text{kJ mol}^{-1}$

(d) % error $= \dfrac{-57.5 - (-53.3)}{-57.5} \times 100\% = 7\%$

The calculated value assumes:

- no heat loss from the system
- all heat is transferred to the water
- the solutions contain 120 g of water
- there are also uncertainties in the temperature, volume, and concentration measurements.

The literature value assumes standard conditions.

Nature of Science

The accurate determination of enthalpy changes involves making careful observations and measurements, which inevitably have experimental uncertainties. The calculated enthalpy values are also dependent on the assumptions made. All these elements should be considered when reporting and evaluating experimental enthalpy values. Mathematics is a powerful tool in the sciences, but scientific systems are often too complex to treat rigorously. These underlying assumptions, which make a problem solvable, should not be ignored when evaluating quantitative data generated by a mathematical model.

Challenge yourself

2. A piece of brass is held in the flame of a Bunsen burner for several minutes. The brass is then quickly transferred into an aluminium calorimeter which contains 200.00 g of water.

Determine the temperature of the Bunsen flame from the following data.

m(water) / ±0.01 g	200.00
m(brass) / ±0.01 g	212.10
m(aluminium calorimeter) / ±0.01 g	80.00
c(brass) / J g⁻¹ K⁻¹	0.400
c(Al) / J g⁻¹ K⁻¹	0.900
Initial temperature of water / ±0.1 °C	24.5
Final temperature of water / ±0.1 °C	77.5

TOK

What criteria do we use in judging whether discrepancies between experimental and theoretical values are due to experimental limitations or theoretical assumptions? Being a risk taker is one attribute of the IB Learner Profile. When is a scientist justified in rejecting the generally accepted literature value in favour of their experimentally determined value?

SKILLS

The temperature of the Bunsen flame and the enthalpy of fusion can be investigated in the lab.

Exercise

Q10. If 500 J of heat is added to 100.0 g samples of each of the substances below, which will have the largest temperature increase?

	Substance	Specific heat capacity / $J g^{-1} K^{-1}$
A	gold	0.129
B	silver	0.237
C	copper	0.385
D	water	4.180

Q11. Calculate the amount of heat needed to increase the temperature of a 5.0 g sample of copper from 27 °C to 29 °C.
(Specific heat capacity of Cu is 0.385 $J g^{-1} °C^{-1}$)

    A 0.770 J   **B** 1.50 J   **C** 3.00 J   **D** 3.85 J

Q12. Consider the specific heat capacity of the following metals.

Metal	Specific heat capacity / $J g^{-1} K^{-1}$
Al	0.897
Be	1.820
Cd	0.231
Cr	0.449

1 kg samples of the metals at room temperature are heated by the same electrical heater for 10 minutes.

Identify the metal which has the highest final temperature.

    A Al   **B** Be   **C** Cd   **D** Cr

Q13. In an experiment to determine the enthalpy of combustion of an alcohol, the mass of a burner plus its contents, and the temperature of a known mass of water in a calorimeter, are measured before and after the experiment. What are the expected results?

	Mass of burner and contents	Reading on thermometer
A	decreases	increases
B	decreases	stays the same
C	increases	increases
D	increases	stays the same

Q14. The experimental arrangement in Figure 9 is used to determine the enthalpy of combustion of an alcohol.

Which of the following would lead to an experimental result which is **less** exothermic than the literature value?

I.   Heat loss from the sides of the copper calorimeter.

II.   Evaporation of alcohol during the experiment.

III. The thermometer touches the bottom of the calorimeter.

    A I and II only   **B** I and III only   **C** II and III only   **D** I, II and III

Q15. A copper calorimeter was used to determine the enthalpy of combustion of butan-1-ol. The experimental value obtained was $-2100 \pm 200\,\text{kJ mol}^{-1}$, and the data booklet value is $-2676\,\text{kJ mol}^{-1}$.

Which of the following accounts for the difference between the two values?

I.   random measurement errors

II. incomplete combustion

III. heat loss to the surroundings

A	I and II only	**B**	I and III only
C	II and III only	**D**	I, II and III

Q16. The thermochemical equation for the combustion of propane is:

$$C_3H_8(g) + 5O_2(g) \rightarrow 3CO_2(g) + 4H_2O(l) \qquad \Delta H^{\ominus} = -2219\,\text{kJ mol}^{-1}$$

Identify the true statements.

I.   If 1.00 g of propane is burnt completely, 2219 kJ of energy are produced.

II. The reactants have more potential energy than the products.

III. The reaction is exothermic.

A	I and II only	**B**	I and III only
C	II and III only	**D**	I, II and III

Q17. An experiment is carried out to determine the enthalpy of combustion of methanol. A known mass of tap water is heated by burning a sample of methanol.

The following data were collected.

Mass of water m_{water}	Temperature change of the tap water ΔT
Initial mass of methanol $m_{initial}$	Final mass of methanol m_{final}

(a) Which formula can be used to calculate an experimental value for the enthalpy change of combustion of methanol?

A $m_{water} \times 4.18 \times \Delta T$

B $\dfrac{m_{water} \times 4.18 \times \Delta T \times 32.05}{(m_{initial} - m_{final})}$

C $(m_{initial} - m_{final}) \times 4.18 \times \Delta T$

D $\dfrac{m_{water} \times 4.18 \times \Delta T \times (m_{initial} - m_{final})}{32.05}$

(b) Identify the assumptions made in using the formula in part **(a)**.

I.   The density of tap water is $1.0\,\text{g cm}^{-3}$.

II. All the energy from the combustion heats the beaker of water.

III. The specific heat capacity of the beaker is negligible.

A	I and II only	**B**	I and III only
C	II and III only	**D**	I, II and III

Q18. $20.0\,cm^3$ of $2.0\,mol\,dm^{-3}$ $HNO_3(aq)$ is mixed with $40\,cm^3$ of $1.0\,mol\,dm^{-3}$ KOH(aq) at $25.0\,°C$. The temperature of the resulting solution increases by $9\,°C$.

Predict the temperature change when $5.0\,cm^3$ of $2.0\,mol\,dm^{-3}$ $HNO_3(aq)$ is mixed with $10.0\,cm^3$ of $1.0\,mol\,dm^{-3}$ KOH(aq) at the same temperature.

    A $4.5\,°C$ **B** $9\,°C$ **C** $18\,°C$ **D** $27°C$

Q19. The specific heat capacity of metallic mercury is $0.138\,J\,g^{-1}\,°C^{-1}$. What is the final temperature if $100.0\,J$ of heat is added to a $100.0\,g$ sample of mercury at $25.0\,°C$?

Q20. $1.10\,g$ of glucose was completely burnt and the heat produced increased the temperature of the water in a copper calorimeter from $25.85\,°C$ to $36.50\,°C$.

(a) Calculate the enthalpy of combustion of glucose from the data below.

Mass of water / g	200.00
Specific heat capacity of water / $g^{-1}\,K^{-1}$	4.18
Mass of copper / g	120.00
Specific heat capacity of copper / $g^{-1}\,K^{-1}$	0.385

(b) Draw an enthalpy level diagram to represent this reaction.

Q21. The heat released from the combustion of $0.0500\,g$ of white phosphorus increases the temperature of $150.00\,g$ of water from $25.0\,°C$ to $31.5\,°C$. Calculate a value for the enthalpy change of combustion of phosphorus. Discuss possible sources of error in the experiment.

Q22. A student wanted to find the enthalpy of combustion of propan-1-ol and collected the following data.

Initial mass of burner and propan-1-ol / g	45.65
Final mass of burner and propan-1-ol / g	45.05
Mass of water heated / g	210
Initial temperature of water / °C	23.7
Final temperature of water / °C	41.5
Specific heat capacity of water / $J\,K^{-1}\,g^{-1}$	4.18

(a) Calculate the amount, in mol, of propan-1-ol burnt.

(b) Calculate the heat absorbed, in kJ, by the water.

(c) Determine the enthalpy change, in $kJ\,mol^{-1}$, for the combustion of one mole of propan-1-ol.

(d) Suggest why this value differs from the literature value of $-2021\,kJ\,mol^{-1}$.

Guiding Question revisited

What can be deduced from the temperature change that accompanies chemical or physical change?

In this chapter we have seen that although energy is always conserved, many chemical reactions involve a transfer of energy between the system and the surroundings. We introduced the concept of enthalpy and saw that the enthalpy change of a reaction gives an indication of the relative stability of the reactants and products. These ideas can be summarized:

- A system is the reaction container and its contents. The surroundings include everything outside the system of interest. Exothermic reactions release heat energy to the surroundings, and endothermic reactions absorb heat energy from the surroundings.

- The heat is transferred from the temperature change using a conversion factor known as the specific heat capacity. The specific heat capacity (c) of a pure material is the heat needed to change the temperature of a unit mass of the material by 1 K.

$$Q = mc\Delta T \text{ for a pure substance.}$$

- Generally, heat changes in the laboratory are determined by the temperature changes of a known mass of water. In experiments to determine the enthalpy of combustion, the water is heated in a copper calorimeter. In experiments to determine enthalpy changes in solution, the water is the solvent in an insulated calorimeter.

- The heat change transferred at constant pressure is known as the enthalpy change ΔH. ΔH is negative for exothermic reactions because they release heat, and positive for endothermic reactions which absorb heat.

- A thermochemical equation gives the equation for the reaction and the corresponding enthalpy change.

Practice questions

1. When equal masses of X and Y absorb the same amount of energy, their temperatures rise by 5 °C and 10 °C respectively. Which is correct?

    A The specific heat capacity of X is twice that of Y.

    B The specific heat capacity of X is half that of Y.

    C The specific heat capacity of X is one fifth that of Y.

    D The specific heat capacity of X is the same as Y.

2. A pure aluminium block with a mass of 10 g is heated so that its temperature increases from 20 °C to 50 °C. The specific heat capacity of aluminium is $8.99 \times 10^{-1} \text{J g}^{-1} \text{K}^{-1}$. Which expression gives the heat energy change in kJ?

    A $10 \times 8.99 \times 10^{-1} \times 303$

    B $10 \times 8.99 \times 10^{-1} \times 30$

    C $\dfrac{10 \times 8.99 \times 10^{-1} \times 303}{1000}$

    D $\dfrac{10 \times 8.99 \times 10^{-1} \times 30}{1000}$

3. Two $100\,cm^3$ aqueous solutions, one containing 0.010 mol NaOH and the other containing 0.010 mol HCl, are at the same temperature.

When the two solutions are mixed the temperature rises by $y\,°C$.

Assume the density of the final solution is $1.00\,g\,cm^{-3}$.

Specific heat capacity of water = $4.18\,J\,g^{-1}\,K^{-1}$

What is the enthalpy change of neutralization in $kJ\,mol^{-1}$?

A $\dfrac{200 \times 4.18 \times y}{1000 \times 0.020}$

B $\dfrac{200 \times 4.18 \times y}{1000 \times 0.010}$

C $\dfrac{100 \times 4.18 \times y}{1000 \times 0.010}$

D $\dfrac{200 \times 4.18 \times (y + 273)}{1000 \times 0.010}$

Questions 4 and 5 are about an experiment to measure the enthalpy of combustion, ΔH_c, of ethanol, using the apparatus and setup shown.

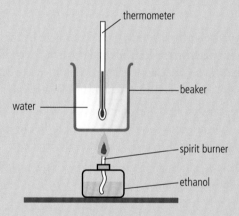

Maximum temperature of water: 30.0 °C
Initial temperature of water: 20.0 °C
Mass of water in beaker: 100.0 g
Loss in mass of ethanol: 0.230 g
M_r (ethanol): 46.08
Specific heat capacity of water: $4.18\,J\,g^{-1}\,K^{-1}$
$q = mc\Delta T$

4. What is the enthalpy of combustion, ΔH_c, of ethanol in $kJ\,mol^{-1}$?

A $-\dfrac{100.0 \times 4.18 \times (10.0 \times 273)}{\frac{0.230}{46.08} \times 1000}$

B $-\dfrac{0.0230 \times 4.18 \times 10.0}{\frac{100.0}{46.08} \times 1000}$

C $-\dfrac{100.0 \times 4.18 \times 10.0}{\frac{0.230}{46.08} \times 1000}$

D $-\dfrac{100.0 \times 4.18 \times 10.0}{\frac{0.230}{46.08}}$

5. Which quantity is likely to be the most inaccurate due to the sources of error in this experiment?

A mass of ethanol burnt

B molecular mass of ethanol

C mass of water

D temperature change

6. Methanol is made in large quantities because it is used in the production of polymers and in fuels. The enthalpy of combustion of methanol can be determined theoretically or experimentally.

$$CH_3OH(l) + 1\tfrac{1}{2}O_2(g) \rightarrow CO_2(g) + 2H_2O(g)$$

(a) The enthalpy of combustion of methanol can be determined experimentally in a school laboratory. A burner containing methanol was weighed and used to heat water in a test tube, as illustrated below.

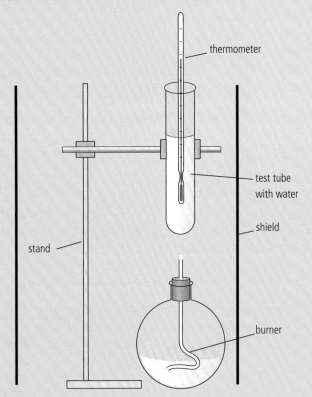

The following data were collected.

Initial mass of burner and methanol / g	80.557
Final mass of burner and methanol / g	80.034
Mass of water in test tube / g	20.000
Initial temperature of water / °C	21.5
Final temperature of water / °C	26.4

(i) Calculate the amount, in mol, of methanol burnt. (2)

(ii) Calculate the heat absorbed, in kJ, by the water. (3)

(iii) Determine the enthalpy change, in kJ mol^{-1}, for the combustion of 1 mole of methanol. (2)

(b) The data booklet value for the enthalpy of combustion of methanol is −726 kJ mol^{-1}. Suggest why this value differs from the value calculated in part (a). (1)

(Total 8 marks)

7. The data below is from an experiment to measure the enthalpy change for the reaction of aqueous copper(II) sulfate, $CuSO_4(aq)$, and zinc, $Zn(s)$.

$$Cu^{2+}(aq) + Zn(s) \rightarrow Cu(s) + Zn^{2+}(aq)$$

50.0 cm^3 of 1.00 mol dm^{-3} copper(II) sulfate solution was placed in a polystyrene cup and zinc powder was added after 100 s. The temperature–time data was taken from a data logging software program. The table shows the initial 19 readings.

	A	B	C	D	E	F	G	H
1	time/s	temperature/°C						
2	0.0	24.8						
3	1.0	24.8						
4	2.0	24.8						
5	3.0	24.8						
6	4.0	24.8						
7	5.0	24.8						
8	6.0	24.8						
9	7.0	24.8						
10	8.0	24.8						
11	9.0	24.8						
12	10.0	24.8						
13	11.0	24.8						
14	12.0	24.8						
15	13.0	24.8						
16	14.0	24.8						
17	15.0	24.8						
18	16.0	24.8						
19	17.0	24.8						
20	18.0	24.8						
21								
22								
23								
24								

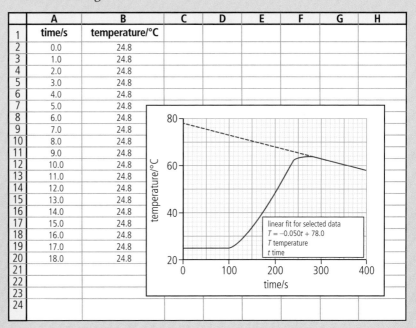

A straight line has been drawn through some of the data points. The equation for this line is given by the data logging software as

$$T = -0.050t + 78.0$$

where T is the temperature at time t.

(a) The heat produced by the reaction can be calculated from the temperature change, ΔT, using the expression below.

heat change = volume of $CuSO_4(aq)$ × specific heat capacity of H_2O × ΔT

Describe **two** assumptions made in using this expression to calculate heat changes. (2)

(b) (i) Use the data presented by the data logging software to deduce the temperature change, ΔT, which would have occurred if the reaction had taken place instantaneously with no heat loss. (2)

(ii) State the assumption made in part **(b)(i)**. (1)

(iii) Calculate the heat, in kJ, produced during the reaction using the expression given in part **(a)**. (1)

(c) The colour of the solution changed from blue to colourless. Deduce the amount, in moles, of zinc which reacted in the polystyrene cup. (1)

(d) Calculate the enthalpy change, in $kJ \text{ mol}^{-1}$, for this reaction. (1)

(Total 8 marks)

8. **(a)** Calculate the enthalpy of neutralization based on the following data.

Initial temperature of solutions / °C	24.5
Concentration of KOH(aq) / mol dm^{-3}	0.950
Concentration of HCl(aq) / mol dm^{-3}	1.050
Volume of HCl(aq) / cm^3	50.00
Volume of KOH(aq) / cm^3	50.00
Final temperature of mixture / °C	30.3

(4)

   (b) State the assumptions you have made in your calculation. (1)

(Total 5 marks)

9. A student added 5.35 g of ammonium chloride to 100.00 cm^3 of water. The initial temperature of the water was 25.55 °C but it decreased to 21.79 °C. Calculate the enthalpy change that would occur when 1 mol of the solute is added to 1.000 dm^3 of water. (3)

(Total 3 marks)

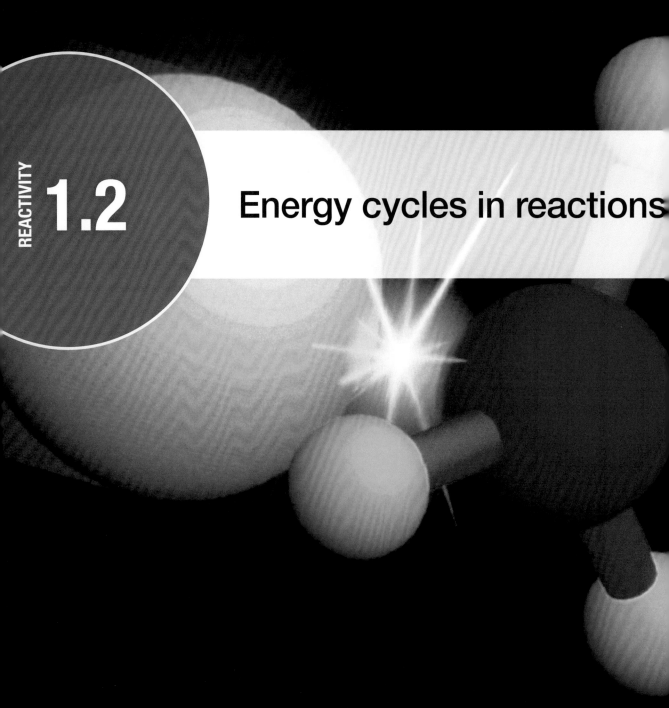

Energy cycles in reactions

◀ A chemical reaction involves the breaking and making of bonds. 300–500 kJ of heat are typically needed to break one mole of covalent bonds. This image shows a change in the bond between the atoms represented by the yellow (on the left) and the dark blue (in the centre).

Guiding Question

How does application of the law of conservation of energy help us to predict energy changes during reactions?

Energy changes occur during chemical reactions as bonds are broken and new bonds are formed. Energy is needed to separate particles held together by chemical bonds, and energy is released when the same particles come together in a different arrangement with new bonds. The net enthalpy change is the difference between the two energy contributions of bond-breaking and bond-making. The reaction is exothermic overall when bond formation releases more energy than is needed for bond-breaking.

The law of conservation of energy tells us that the total energy of the universe is constant and cannot be changed. The energy can be converted from one form to another, but no process can change its amount. This law imposes restrictions on energy changes when a chemical change can occur via different routes.

Consider the energy cycle below (Figure 1) in which A can be converted to C either directly or indirectly.

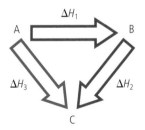

◀ **R1.2 Figure 1** An energy cycle.

Going around the cycle in a clockwise route from A → B → C → A, according to the law of conservation of energy, the total energy change is zero:

$$\Delta H_1 + \Delta H_2 - \Delta H_3 = 0$$

Otherwise, the cycle could form the basis of a perpetual motion machine with energy being generated continually by going around the cycle in one direction.

The construction of the cycle allows us to calculate enthalpy changes to reactions indirectly. ΔH_3 can be calculated from ΔH_1 and ΔH_2:

$$\Delta H_1 + \Delta H_2 = \Delta H_3$$

We can then determine enthalpy changes of hypothetical reactions and make predictions about the relative stability of different compounds. The law of conservation of energy imposes restrictions but also adds order. Energy cycles are a key tool in understanding any chemical change.

Reactivity 1.2.1 – Bond enthalpy

Reactivity 1.2.1 – Bond-breaking absorbs and bond-forming releases energy.

Calculate the enthalpy change of a reaction from the given average bond enthalpy data.

Include an explanation of why bond enthalpy data are average values and may differ from those measured experimentally.	Structure 2.2 – How would you expect bond enthalpy data to relate to bond length and polarity?
Average bond enthalpy values are given in the data booklet.	Reactivity 3.4 – How does the strength of a carbon–halogen bond affect the rate of a nucleophilic substitution reaction?

Breaking bonds is an endothermic process

In Structure 2.2, we see that a covalent bond is due to the electrostatic attraction between a shared pair of electrons and the positive nuclei of the bonded atoms. Energy is needed to separate the atoms in a bond.

The **bond enthalpy** is the energy needed to break one mole of bonds in gaseous molecules under standard conditions.

The energy change, for example, during the formation of two moles of chlorine atoms from one mole of chlorine molecules can be represented in a thermochemical equation.

$$Cl_2(g) \rightarrow 2Cl(g) \qquad\qquad \Delta H^\ominus = +242 \, kJ \, mol^{-1}$$

The enthalpy change is positive as it is an endothermic process.

The situation is complicated in molecules which contain more than two atoms. More energy is needed to break the first O–H bond in a water molecule than the second bond:

$$H_2O(g) \rightarrow H(g) + OH(g) \qquad\qquad \Delta H^\ominus = +502 \, kJ \, mol^{-1}$$

$$OH(g) \rightarrow H(g) + O(g) \qquad\qquad \Delta H^\ominus = +427 \, kJ \, mol^{-1}$$

Similarly, the energy needed to break the O–H bond in other molecules such as ethanol, C_2H_5OH, differs from either of these values. To compare bond enthalpies for bonds which occur in different environments, **average bond enthalpies** are used.

The two previous equations can be combined:

$$H_2O(g) \rightarrow H(g) + H(g) + O(g) \qquad\qquad \Delta H^\ominus = (+502) + (+427) \, kJ \, mol^{-1}$$

As there are two O–H bonds we can determine the average value $(E(O–H))$:

$$E(O–H) = \frac{+502 + 427}{2} \, kJ \, mol^{-1}$$

$$= \frac{929}{2} = 464.5 \, kJ \, mol^{-1}$$

This value should be compared with the bond enthalpies given in the table on the next page, which are calculated for a larger range of molecules.

Bond	E(X–Y) / kJ mol⁻¹	Bond length / 10⁻¹² m
H–H	+436	74
C–C	+346	154
C–H	+414	108
O–H	+463	97
C–O	+358	143
Cl–Cl	+242	199
C–Cl	+324	177
H–Cl	+431	128
F–F	+159	142
C–F	+492	138
H–F	+567	92
Br–Br	+193	228
H–Br	+366	141
I–I	+151	267
H–I	+298	160
O–O	+144	148
C=C	+614	134
O=O	+498	121
C=O	+804	122

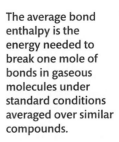

The average bond enthalpy is the energy needed to break one mole of bonds in gaseous molecules under standard conditions averaged over similar compounds.

As expected, multiple bonds which involve more bonding electrons generally have higher bond enthalpies and shorter bond lengths than single bonds.

All bond enthalpies refer to reactions in the gaseous state, so any enthalpy changes resulting from the formation or breaking of **intermolecular** forces are not included.

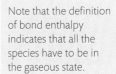

Note that the definition of bond enthalpy indicates that all the species have to be in the gaseous state.

Bond enthalpy and polarity

You can improve your understanding by comparing the average bond enthalpy of the H–H and X–X bonds in the halogens and then comparing them to those in H–X.

X	Average of E(H–H) and E(X–X) / kJ mol⁻¹	E(H–X) / kJ mol⁻¹	$E(H-X) - \left(\dfrac{E(H-H) + E(X-X)}{2} \right)$ / kJ mol⁻¹
Cl	$\dfrac{436 + 242}{2} = +339$	+431	92
Br	$\dfrac{436 + 193}{2} = +314.5$	+366	51.5
I	$\dfrac{436 + 151}{2} = +293.5$	+298	4.5

We can see that the difference in bond enthalpies between the hydrogen halide bonds and the average bonds in the corresponding elements, as shown in the fourth column, is greatest with chlorine and smallest with iodine. These differences can be related to the polarity of the molecule with the largest difference occurring with the more polar H–Cl bond. More generally we can say that the polar X–Y bond between two elements X and Y is stronger than the non-polar X–X and Y–Y bonds.

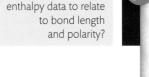

Electronegativities on the Pauling scale are calculated from bond enthalpy differences.

$$\chi_A - \chi_B = \sqrt{(E(A-B)} - \left(\frac{E(A-A) + (B-B)}{2}\right)$$

The units of energy used are eV not $kJ\,mol^{-1}$. You are not required to know the details.

$$1\ eV = 1.6 \times 10^{-19} \times 6.02 \times 10^{23}\ J\,mol^{-1} = 96\ kJ\,mol^{-1}$$

Structure 2.2 – How would you expect bond enthalpy data to relate to bond length and polarity?

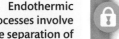

Challenge yourself

1. Fluorine shows some anomalous properties due to its small atomic radius.

   (a) Comment on the bond enthalpy of F−F compared to the X−X bond in the other halogens.

   (b) Compare the bond enthalpy of H−F with the average bond enthalpies of H_2 and F_2.

Making bonds is an exothermic process

The same amount of energy is absorbed when a bond is broken as is given out when a bond is formed (Figure 2). For example:

$$H(g) + H(g) \rightarrow H_2(g) \qquad\qquad \Delta H^\ominus = -436\ kJ\,mol^{-1}$$

The negative enthalpy change shows that bond-making is an exothermic process.

Endothermic processes involve the separation of particles which are held together by a force of attraction. Exothermic processes involve the bringing together of particles which are mutually attracted.

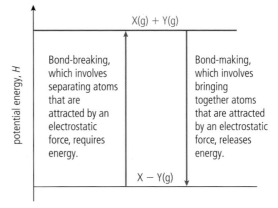

▲ **R1.2 Figure 2** The energy changes that occur when bonds are broken and formed.

Worked example

Identify the endothermic process.

A $2Cl(g) \rightarrow Cl_2(g)$

B $Na(g) \rightarrow Na^+(g) + e^-$

C $Na^+(g) + Cl^-(g) \rightarrow NaCl(s)$

D $Na(g) \rightarrow Na(s)$

Solution

Only one of the processes involves the separation of particles:

$$Na(g) \rightarrow Na^+(g) + e^-$$

A negatively charged electron e^- is separated from a positive ion $Na^+(g)$.

Answer = B

Energy changes in reactions

To understand the energy changes that occur in chemical reactions, consider, for example, the complete combustion of methane:

$$H-\underset{\underset{H}{|}}{\overset{\overset{H}{|}}{C}}-H \ + \ 2O{=}O \ \rightarrow \ O{=}C{=}O \ + \ 2H-O-H$$

Energy is needed to break the C–H and O=O bonds in the reactants and is given out when the C=O and O–H bonds are formed. The reaction is exothermic overall as the bonds formed are stronger than those broken. Conversely, a reaction is endothermic when the bonds broken are stronger than the bonds formed.

Worked example

Use bond enthalpies to calculate the enthalpy of combustion of methane, the principal component of natural gas.

Solution

1. Write down the equation for the reaction showing all the bonds. (This has already been done above.)
2. Construct a table to show the bonds which are broken and formed during the reaction with the corresponding enthalpy changes.

Bonds broken	$\Delta H^{\ominus}$ / kJ mol^{-1} (endothermic)	Bonds formed	$\Delta H^{\ominus}$ / kJ mol^{-1} (exothermic)
4 C–H	4 (+414)	2 C=O	2 (−804)
2 O=O	2 (+498)	4 O–H	4 (−463)
Total	= +2652		= −3460

$$\Delta H^{\ominus} = +2652 - 3460 \text{ kJ mol}^{-1} = -808 \text{ kJ mol}^{-1}$$

The value calculated from the bond enthalpies should be compared with the experimental value of −891 kJ mol^{-1} measured under standard conditions given in Section 14 of the data booklet. The values are significantly different because the standard state of water is liquid and the bond enthalpy calculation assumes that the reaction occurs in the gaseous state. The use of average bond enthalpies is an additional approximation.

Challenge yourself

2. Compare the value of the enthalpy of combustion of methane obtained in the worked example to that in Section 14 of the data booklet and estimate the strength of a hydrogen bond.

The worked example illustrates a general result:

$$\Delta H^\ominus = \Sigma E(\text{bonds broken}) - \Sigma E(\text{bonds formed})$$

$$\Delta H^\ominus = \Sigma E(\text{bonds broken}) - \Sigma E(\text{bonds formed})$$

Worked example

Use bond enthalpy data given earlier to calculate the enthalpy change of reaction between ethene and chorine:

$$C_2H_4(g) + Cl_2(g) \rightarrow CH_2ClCH_2Cl(g)$$

Solution

Follow the same strategy as the previous example.

Start with a balanced chemical equation showing all the bonds:

$$
\begin{array}{ccc}
\text{H} \quad \text{H} & & \text{H} \quad \text{H} \\
| \quad\ \ | & & | \quad\ \ | \\
\text{C}=\text{C} \ + \ \text{Cl}-\text{Cl (g)} & \rightarrow & \text{H}-\text{C}-\text{C}-\text{H} \\
| \quad\ \ | & & | \quad\ \ | \\
\text{H} \quad \text{H} & & \text{Cl} \quad \text{Cl}
\end{array}
$$

Identify the bonds broken and formed and the corresponding enthalpy changes.

Bonds broken	$\Delta H / \text{kJ mol}^{-1}$ (endothermic)	Bonds formed	$\Delta H / \text{kJ mol}^{-1}$ (exothermic)
Cl–Cl	+242	2 C–Cl	−2(324)
C=C	+614	C–C	−(346)
Total	= +242 +614 = +856		= −2(324) − (346) = −994

$$\Delta H^\ominus = \Sigma E(\text{bonds broken}) - \Sigma E(\text{bonds formed})$$
$$= +856 - 994 = -138 \text{ k mol}^{-1}$$

As discussed in Reactivity 1.1, some reactants need to be given an initial energy known as the activation energy before they will react. This energy is needed because some bonds in the reactants must break before new bonds in the products can form. The rate of some reactions can be explained by the relative bond enthalpy of the bonds broken. The relative strength of the carbon–halogen bond, for example, can be used to explain the relative reactivity of the halogenoalkanes discussed in Reactivity 3.4. The iodoalkanes are the most reactive as the C–I bond is relatively easy to break compared with the other carbon–halogen bonds.

Reactivity 3.4 – How does the strength of a carbon–halogen bond affect the rate of a nucleophilic substitution reaction?

The Earth is unique among the planets in having an atmosphere that is chemically active and rich in oxygen. Oxygen, O_2, and ozone, O_3, both play a key role in protecting life. The oxygen-to-oxygen bond in ozone is weaker than the double bond in O_2 and so is broken by long-wavelength ultraviolet radiation. The absorption of harmful ultraviolet (UV) radiation.in the ozone layer protects the Earth's surface.

Challenge yourself

3. The bond enthalpy in ozone is $+363\,kJ\,mol^{-1}$.

    (a) Compare this bond enthalpy with the oxygen–oxygen bond enthalpies tabulated on page 329 and comment on its value.

    (b) Explain why ozone (O_3) can be decomposed by ultraviolet radiation with a longer wavelength than that required to decompose oxygen (O_2).

Exercise

Q1. Which of the following processes are endothermic?

    I. $H_2O(s) \rightarrow H_2O(g)$

    II. $CO_2(g) \rightarrow CO_2(s)$

    III. $O_2(g) \rightarrow 2O(g)$

    A I and II only   **B** I and III only   **C** II and III only   **D** I, II and III

Q2. Identify the equation that represents the bond enthalpy for the H−Cl bond.

    A $HCl(g) \rightarrow H(g) + Cl(g)$

    B $HCl(g) \rightarrow \frac{1}{2}H_2(g) + \frac{1}{2}Cl_2(g)$

    C $HCl(g) \rightarrow H^+(g) + Cl^-(g)$

    D $HCl(aq) \rightarrow H^+(aq) + Cl^-(aq)$

Q3. Identify the bonds that are broken in the following process.

$$C_2H_6(g) \rightarrow 2C(g) + 6H(g)$$

Q4. Which of the following enthalpy changes corresponds to the bond enthalpy of the carbon–oxygen bond in carbon monoxide?

    A $CO(g) \rightarrow C(s) + O(g)$

    B $CO(g) \rightarrow C(g) + O(g)$

    C $CO(g) \rightarrow C(s) + \frac{1}{2}O_2(g)$

    D $CO(g) \rightarrow C(g) + \frac{1}{2}O_2(g)$

Q5. The hydrogenation of the alkene double bond in unsaturated oils is an important reaction in margarine production. Calculate the enthalpy change when one mole of C=C bonds is hydrogenated from the bond energy data given earlier.

    A $-290\,kJ\,mol^{-1}$ **B** $-124\,kJ\,mol^{-1}$

    C $+124\,kJ\,mol^{-1}$ **D** $+290\,kJ\,mol^{-1}$

▲

Weather balloon being inflated by a meteorologist. A measuring device attached below the balloon will measure ozone distribution to an altitude of 35–40 kilometres, allowing study of the ozone layer at 20–30 kilometres. Ozone at this level (formed by sunlight) absorbs the dangerous types of UV radiation in sunlight.

Q6. Use bond enthalpy data to calculate the enthalpy change of the reaction between methane and fluorine.

$$C_2H_4(g) + F_2(g) \rightarrow CH_2FCH_2F(g)$$

A $+557\,kJ\,mol^{-1}$ **B** $+65\,kJ\,mol^{-1}$

C $-65\,kJ\,mol^{-1}$ **D** $-557\,kJ\,mol^{-1}$

Q7. Use bond enthalpy data to calculate $\Delta H^{\ominus}$ for the reaction:

$$2H_2(g) + O_2(g) \rightarrow 2H_2O(g)$$

Q8. Use the bond enthalpies given earlier to estimate the enthalpy of combustion of ethanol and comment on the reliability of your result.

Reactivity 1.2.2 – Hess's law

Reactivity 1.2.2 – Hess's law states that the enthalpy change for a reaction is independent of the pathway between the initial and final states.

Apply Hess's law to calculate enthalpy changes in multistep reactions.

Hess's law is a consequence of the law of conservation of energy

Consider again the energy level diagram below.

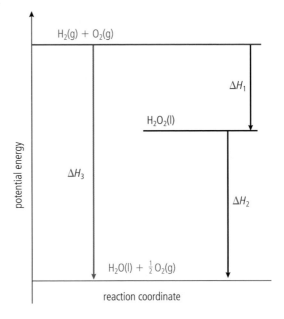

The thermochemical equations for the three reactions are:

$$H_2(g) + O_2(g) \rightarrow H_2O_2(l) \qquad\qquad \Delta H_1 = -188\,kJ\,mol^{-1}$$

$$H_2O_2(l) \rightarrow H_2O(l) + \tfrac{1}{2}O_2(g) \qquad\qquad \Delta H_2 = -98\,kJ\,mol^{-1}$$

$$H_2(g) + \tfrac{1}{2}O_2(g) \rightarrow H_2O(l) \qquad\qquad \Delta H_3 = -286\,kJ\,mol^{-1}$$

We can see from the figure on the previous page that $\Delta H_1 + \Delta H_2 = \Delta H_3$ and can confirm this by adding the first two equations:

$$H_2(g) + O_2(g) + \cancel{H_2O_2(l)} \rightarrow \cancel{H_2O_2(l)} + H_2O(l) + \tfrac{1}{2}O_2(g) \qquad \Delta H_3 = -188 - 98\,\text{kJ mol}^{-1}$$

$$H_2(g) + \tfrac{1}{2}O_2(g) \rightarrow H_2O(l) \qquad\qquad\qquad\qquad \Delta H_3 = -286\,\text{kJ mol}^{-1}$$

This shows that the overall enthalpy change that occurs when hydrogen and oxygen combine to form water is the same whether the chemical change occurs in one step (ΔH_3) or two steps ($\Delta H_1 + \Delta H_2$).

This result can be generalized and is known as **Hess's law**: the enthalpy change for any chemical reaction is independent of the route, provided the starting conditions and final conditions, and reactants and products, are the same.

These reactions form an energy cycle. In the clockwise direction shown in green in Figure 3, the starting and end points are $H_2(g) + O_2(g)$ so there is no net chemical reaction or enthalpy change.

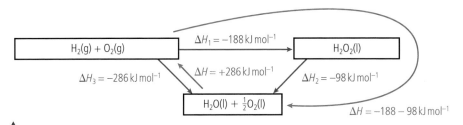

▲
R1.2 Figure 3 There is no net chemical or enthalpy change as the start and end points are the same.

As per the law of conservation of energy, the enthalpy change in a complete cycle = 0.

$$0 = \Delta H_1 + \Delta H_2 - \Delta H_3$$

This leads to:

$$\Delta H_3 = \Delta H_1 + \Delta H_2$$

This shows that Hess's law is a consequence of the law of conservation of energy; otherwise it would be possible to devise cycles in which energy was created or destroyed.

Using Hess's law

Hess's law states that the enthalpy change for any chemical reaction is independent of the route, provided the starting and final conditions, and reactants and products, are the same. The importance of Hess's law is that it allows us to calculate the enthalpy changes of reactions that we cannot measure directly in the laboratory. For example, although the elements carbon and hydrogen do not combine directly to form propane, the enthalpy change for the reaction:

$$3C(\text{graphite}) + 4H_2(g) \rightarrow C_3H_8(g)$$

can be calculated from the enthalpy of combustion data of the elements and the compound (Figure 4).

R1.2 Figure 4 ▶
$\Delta H_1 + \Delta H_2 = \Delta H_3$, therefore
$\Delta H_1 = \Delta H_3 - \Delta H_2$.
Although ΔH_1 cannot be
measured directly, it can
be calculated from the
enthalpies of combustion of
carbon and hydrogen (ΔH_3)
and propane (ΔH_2).

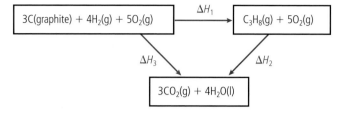

The steps in an enthalpy cycle may be hypothetical. The only requirement is that the individual chemical reactions in the cycle must balance. The relationship between the different reactions is clearly shown in an energy level diagram (Figure 5).

R1.2 Figure 5 Energy level
diagram used to obtain the
enthalpy of formation of
propane indirectly.

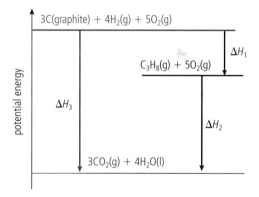

The general use of enthalpies of combustion in determining enthalpies of reaction is discussed in more detail in the next section.

> **Reversing the
> direction of a
> reaction reverses
> the sign of ΔH.**

Worked example

Calculate the standard enthalpy change, $\Delta H^\ominus$, for the formation of sulfur dioxide from the thermochemical equations (1) and (2).

$$S(s) + O_2(g) \rightarrow SO_2(g)$$

$$S(s) + 1\tfrac{1}{2}O_2(g) \rightarrow SO_3(g) \qquad \Delta H^\ominus = -395\,kJ\,mol^{-1} \qquad (1)$$

$$SO_2(g) + \tfrac{1}{2}O_2(g) \rightarrow SO_3(g) \qquad \Delta H^\ominus = -98\,kJ\,mol^{-1} \qquad (2)$$

Solution

We can think of the reaction as a journey from $S(s)$ to $SO_2(g)$. As the standard enthalpy change cannot be measured directly, we must go by an alternative route suggested by the equations given.

Reaction 1 starts from the required reactant:

$$S(s) + 1\tfrac{1}{2}O_2(g) \rightarrow SO_3(g) \qquad \Delta H^\ominus = -395\,kJ\,mol^{-1} \qquad (1)$$

Reaction 2 relates $SO_3(g)$ to $SO_2(g)$. To finish with the required product, we reverse the chemical change and the sign of enthalpy change:

$$SO_3(g) \rightarrow SO_2(g) + \tfrac{1}{2}O_2(g) \qquad \Delta H^\ominus = -(-98)\,kJ\,mol^{-1} = +98\,kJ\,mol^{-1} \qquad (2)$$

The enthalpy change
for the decomposition
of metal carbonates
can be determined in
the school laboratory
by reacting the
metal carbonate and
the metal oxide or
hydrogen carbonate
with dilute acids using
Hess's law.

SKILLS

We can now combine these equations:

$$S(s) + 1\tfrac{1}{2}O_2(g) + SO_3(g) \rightarrow SO_3(g) + SO_2(g) + \tfrac{1}{2}O_2(g) \quad \Delta H^\ominus = -395 + 98\,kJ\,mol^{-1}$$

Simplifying:

$$S(s) + 1\tfrac{1}{2}O_2(g) + \cancel{SO_3(g)} \rightarrow \cancel{SO_3(g)} + SO_2(g) + \cancel{\tfrac{1}{2}O_2(g)} \quad \Delta H^\ominus = -297\,kJ\,mol^{-1}$$

$$S(s) + O_2(g) \rightarrow SO_2(g) \quad \Delta H^\ominus = -297\,kJ\,mol^{-1}$$

The heat produced when water is added to anhydrous copper(II) sulfate (white) is enough to produce steam and disturb the powder.

SKILLS

The heat produced when an anhydrous salt is converted to a hydrated salt can be determined by adding the anhydrous and hydrated salt to separate samples of water. Full details of how to carry out this experiment with a worksheet are available in the eBook.

Challenge yourself

4. Explain why it is difficult to determine the enthalpy change of thermal decomposition reactions directly.

Exercise

Q9. The diagram illustrates the enthalpy changes for a set of reactions.

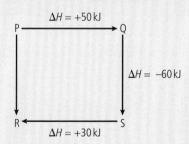

Which of the following statements are correct?

I. $P \rightarrow S \quad \Delta H = -10\,kJ$

II. $R \rightarrow Q \quad \Delta H = +90\,kJ$

III. $P \rightarrow R \quad \Delta H = +20\,kJ$

A I and II only
B I and III only
C II and III only
D I, II and III

Q10. The thermochemical equations for three related reactions are shown.

$$CO(g) + \tfrac{1}{2}O_2(g) \rightarrow CO_2(g) \qquad\qquad \Delta H_1 = -283 \, \text{kJ mol}^{-1}$$

$$2H_2(g) + O_2(g) \rightarrow 2H_2O(l) \qquad\qquad \Delta H_2 = -572 \, \text{kJ mol}^{-1}$$

$$CO_2(g) + H_2(g) \rightarrow CO(g) + H_2O(l) \qquad\qquad \Delta H_3 = ?$$

What is the value of ΔH_3?

A $+289 \, \text{kJ mol}^{-1}$

B $-3 \, \text{kJ mol}^{-1}$

C $-289 \, \text{kJ mol}^{-1}$

D $-855 \, \text{kJ mol}^{-1}$

Q11. Calculate the enthalpy change, $\Delta H^{\ominus}$, for the reaction:

$$C(\text{graphite}) + \tfrac{1}{2}O_2(g) \rightarrow CO(g)$$

from the information below:

$$C(\text{graphite}) + O_2(g) \rightarrow CO_2(g) \qquad\qquad \Delta H^{\ominus} = -394 \, \text{kJ mol}^{-1}$$

$$CO(g) + \tfrac{1}{2}O_2(g) \rightarrow CO_2(g) \qquad\qquad \Delta H^{\ominus} = -283 \, \text{kJ mol}^{-1}$$

Q12. Calculate the enthalpy change, $\Delta H^{\ominus}$, for the reaction:

$$2NO(g) + O_2(g) \rightarrow 2NO_2(g)$$

from the information below:

$$N_2(g) + O_2(g) \rightarrow 2NO(g) \qquad\qquad \Delta H^{\ominus} = +180.5 \, \text{kJ mol}^{-1}$$

$$N_2(g) + 2O_2(g) \rightarrow 2NO_2(g) \qquad\qquad \Delta H^{\ominus} = +66.4 \, \text{kJ mol}^{-1}$$

Q13. Calculate the enthalpy change for the dimerization of nitrogen dioxide:

$$2NO_2(g) \rightarrow N_2O_4(g)$$

from the following data:

$$\tfrac{1}{2}N_2(g) + O_2(g) \rightarrow NO_2(g) \qquad\qquad \Delta H^{\ominus} = +33.2 \, \text{kJ mol}^{-1}$$

$$N_2(g) + 2O_2(g) \rightarrow N_2O_4(g) \qquad\qquad \Delta H^{\ominus} = +9.16 \, \text{kJ mol}^{-1}$$

How does the application of the law of conservation of energy help us to predict energy changes during reactions?

In this chapter we introduced the concept of energy cycles and saw how they can be used to relate the enthalpy changes of related reactions. You can construct energy cycles for covalent compounds with the making and breaking of covalent bonds.

Energy cycles can be used to predict the enthalpy changes of real or hypothetical reactions and so offer insight into underlying principles of chemical change. These ideas can be summarized as follows:

- Hess's law states that the total enthalpy change for a reaction is independent of the route taken. It is a special case of the law of conservation of energy.
- Energy cycles allow for the calculation of values corresponding to certain enthalpy changes which cannot be determined directly.
- The average bond enthalpy is the energy required to break one mole of the same type of bonds in the gaseous state averaged over a variety of similar compounds.
- Bond-breaking absorbs energy and is endothermic. Bond-making releases energy and is exothermic.
- Enthalpy of reactions calculated from bond enthalpies do not include intermolecular interaction and so are only accurate for reactions in the gaseous state.

Practice questions

1. Consider the following reactions.

$$Cu_2O(s) + \tfrac{1}{2}O_2(g) \rightarrow 2CuO(s) \qquad \Delta H^{\ominus} = -144\,kJ$$

$$Cu_2O(s) \rightarrow Cu(s) + CuO(s) \qquad \Delta H^{\ominus} = +11\,kJ$$

What is the value of $\Delta H^{\ominus}$, in kJ, for this reaction?

$$Cu(s) + \tfrac{1}{2}O_2(g) \rightarrow CuO(s)$$

A $-144 + 11$ B $+144 - 11$

C $-144 - 11$ D $+144 + 11$

2. Which equation best represents the bond enthalpy of HCl?

A $HCl(g) \rightarrow H^+(g) + Cl^-(g)$ B $HCl(g) \rightarrow H(g) + Cl(g)$

C $HCl(g) \rightarrow \tfrac{1}{2}H_2(g) + \tfrac{1}{2}Cl_2(g)$ D $2HCl(g) \rightarrow H_2(g) + Cl_2(g)$

3. Consider the equations below.

$$CH_4(g) + O_2(g) \rightarrow HCHO(l) + H_2O(l) \qquad \Delta H^\ominus = x$$

$$HCHO(l) + \tfrac{1}{2}O_2(g) \rightarrow HCOOH(l) \qquad \Delta H^\ominus = y$$

$$2HCOOH(l) + \tfrac{1}{2}O_2(g) \rightarrow (COOH)_2(s) + H_2O(l) \qquad \Delta H^\ominus = z$$

What is the enthalpy change of the reaction below?

$$2CH_4(g) + 3\tfrac{1}{2}O_2(g) \rightarrow (COOH)_2(s) + 3H_2O(l)$$

A $x + y + z$

B $2x + y + z$

C $2x + 2y + z$

D $2x + 2y + 2z$

4. Which process has an enthalpy change that corresponds to the bond enthalpy for the O−H bond?

A $H_2O(l) \rightarrow H_2(g) + \tfrac{1}{2}O_2(g)$

B $\tfrac{1}{2}H_2O(l) \rightarrow \tfrac{1}{2}H_2(g) + O(g)$

C $\tfrac{1}{2}H_2O(g) \rightarrow H(g) + O(g)$

D $\tfrac{1}{2}H_2O(g) \rightarrow H(g) + \tfrac{1}{2}O(g)$

5. What is the energy, in kJ, released when 1.0 mol of carbon monoxide is burned according to the following equation?

$$2CO(g) + O_2(g) \rightarrow 2CO_2(g) \qquad \Delta H^\ominus = -564\,kJ$$

A 141

B 282

C 564

D 1128

6. Which enthalpy changes **cannot** be calculated accurately using only bond enthalpy data?

I. $C_2H_6(g) + \tfrac{7}{2}O_2(g) \rightarrow 3H_2O(l) + 2CO_2(g)$

II. $C_2H_4(g) + Br_2(l) \rightarrow CH_2ClCH_2Cl(l)$

III. $2H_2(g) + O_2(g) \rightarrow H_2O(g)$

A I and II only

B I and III only

C II and III only

D I, II and III

7. Which combination will give you the enthalpy change for the hydrogenation of ethene to ethane, ΔH_3?

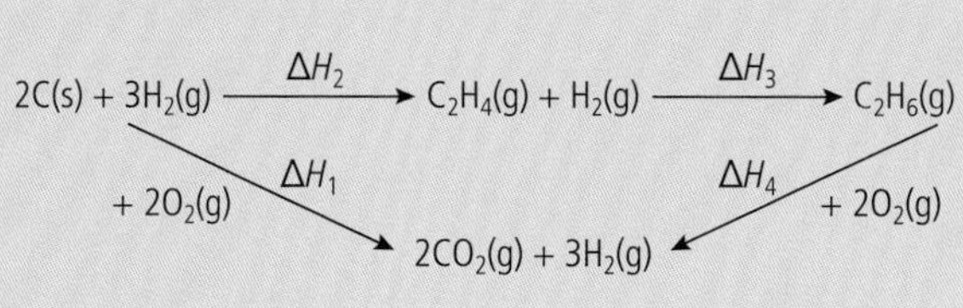

$$2C(s) + 3H_2(g) \xrightarrow{\Delta H_2} C_2H_4(g) + H_2(g) \xrightarrow{\Delta H_3} C_2H_6(g)$$

with $+ 2O_2(g)$, ΔH_1 down to $2CO_2(g) + 3H_2(g)$, and ΔH_4 with $+ 2O_2(g)$

A $-H_2 + H_1 - H_4$

B $H_2 - H_1 + H_4$

C $H_2 + H_1 - H_4$

D $-H_2 - H_1 + H_4$

8. Consider the following thermochemical equations.

$$N_2(g) + 3H_2(g) \rightarrow 2NH_3(g) \qquad \Delta H_1^{\ominus} = x$$

$$H_2(g) \rightarrow 2H(g) \qquad \Delta H_2^{\ominus} = y$$

$$N_2(g) \rightarrow 2N(g) \qquad \Delta H_3^{\ominus} = z$$

Deduce an expression, in terms of x, y and z, for the average bond enthalpy for the $N-H$ bond in ammonia.   (3)

(Total 3 marks)

9. Hydrazine is a valuable rocket fuel. The equation for the reaction between hydrazine and oxygen is given below.

$$N_2H_4(l) + O_2(g) \rightarrow N_2(g) + 2H_2O(l)$$

Use the bond enthalpy values from Section 12 of the data booklet to determine the enthalpy change for this reaction.   (3)

(Total 3 marks)

10. Consider the reaction:

$$CH_4(g) + H_2O(g) \rightleftharpoons CO(g) + 3H_2(g)$$

Determine the enthalpy change, ΔH, in kJ. Use Section 12 of the data booklet.

Bond enthalpy of CO = 1077 kJ mol^{-1}

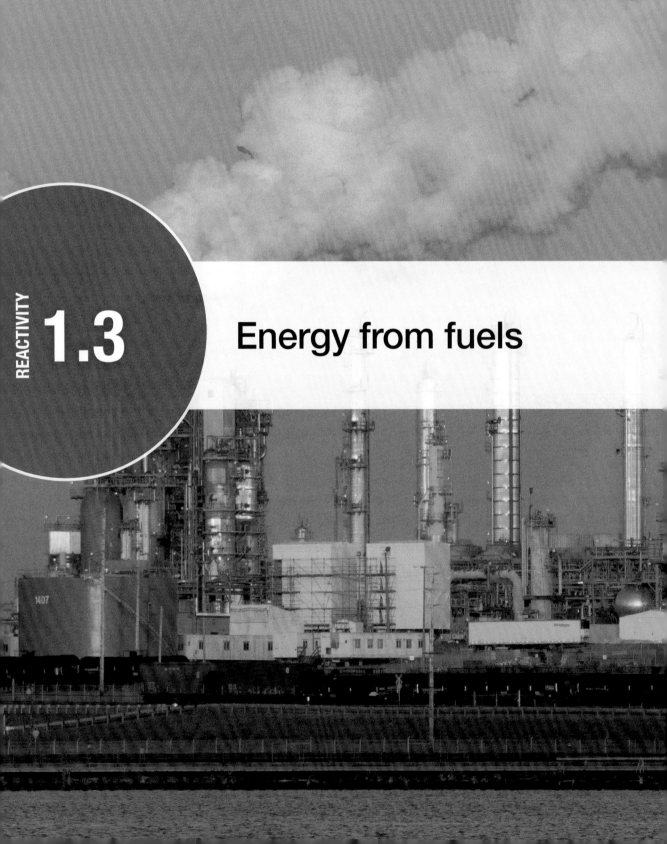

1.3

Energy from fuels

 Oil refinery at dusk. Oil currently provides the world's economy with the most energy. It is, however, a limited resource which has a serious negative impact on our environment. We need to find alternative sources of energy.

Guiding Question

What are the challenges of using chemical energy to address our energy needs?

We are completely dependent on our energy resources. Our ability to harness energy has allowed us to transform our world and give us control of what we do and when and where we do it. The burning of fuels to provide heat is probably one of the oldest human technologies and still accounts for most of our primary energy. All this energy can be traced back to the formation of the Earth and the solar system. We are currently particularly dependent on the use of fossil fuels, which are a chemical store of solar energy from millions of years ago. Carbon dioxide (CO_2), released during the complete combustion of fossil fuels, is a greenhouse gas and a major cause of global warming. Incomplete combustion has resulted in more local pollutants: carbon monoxide and soot particulates. We need to find and use alternative high-quality energy sources, and the Sun again offers many options. Life itself is driven by the chemical changes of photosynthesis, which has always been the ultimate source of our food but can also be used to provide biofuels. Whatever energy source we use in the future it will need to be more sustainable and more environmentally friendly than those we currently use. It is difficult to predict the future, but it is clear our use of energy will have to change. Hydrogen is a possible alternative to fossil fuels, and fuel cells offer another source of electrical energy without the need for combustion.

TOK

The choice of energy source is controversial and complex. How can we navigate issues that involve questions and claims from different areas of knowledge?

View of New York City from the International Space Station (ISS). This image highlights our ability to harness energy supplies to allow us to control our environment.

Reactivity 1.3.1 – Combustion reactions

Reactivity 1.3.1 – Reactive metals, non-metals and organic compounds undergo combustion reactions when heated in oxygen.

Deduce equations for reactions of combustion, including hydrocarbons and alcohols.

| | Reactivity 2.2 – Why is high activation energy often considered to be a useful property of a fuel? |
| | Reactivity 3.2 – Which species are the oxidizing and reducing agents in a combustion reaction? |

Many substances undergo combustion reactions when heated in oxygen

It was originally thought that fire was an element, but we now know it is a process of chemical transformation and change. Many substances such as metals, non-metals and organic compounds undergo combustion reactions when heated in oxygen.

Nature of Science

Early ideas to explain chemical change in combustion and rusting included the 'phlogiston' theory. This proposed the existence of a fire-like element that was released during these processes. The theory seemed to explain some of the observations of its time, although these were purely qualitative. It could not explain later quantitative data showing that substances actually gain rather than lose mass during burning. In 1783, Lavoisier's work on oxygen confirmed that combustion and rusting involve combination with oxygen from the air, so overturning the phlogiston theory. This is a good example of how the evolution of scientific ideas, such as how chemical change occurs, is based on the need for theories that can be tested by experiment. Where results are not compatible with the theory, a new theory must be put forward, which must then be subject to the same rigour of experimental test.

TOK

Chemistry was a late developer as a physical science. Newton was working on the laws of physics more than a century before the work of the French chemist Antoine Lavoisier (1743–94) brought chemistry into the modern age. Chemical reactions involve changes in smell, colour and texture and these are difficult to quantify. Lavoisier appreciated the importance of attaching numbers to properties and recognized the need for precise measurement. His use of the balance allowed changes in mass to be used to analyze chemical reactions. A quantitative approach helped chemistry develop beyond the pseudoscience of alchemy. What differentiates the scientific from the non-scientific or 'pseudo-scientific'?

Combustion of metals and non-metals

Combustion is a reaction in which an element or compound burns in oxygen. The s block metals form ionic oxides which are basic and p block non-metals generally form covalent oxides which are acidic. As discussed in Structure 3.1, some non-metals can show a range of oxidation states and form different oxides when they undergo combustion.

Magnesium, for example, forms magnesium oxide and sulfur forms sulfur dioxide:

$$Mg(s) + \tfrac{1}{2}O_2(g) \rightarrow MgO(s)$$

$$S(s) + O_2(g) \rightarrow SO_2(g)$$

(a) Magnesium burns with an intensely brilliant white flame, leaving a residue of magnesium oxide. The metal is sometimes used in flares and flashbulbs.

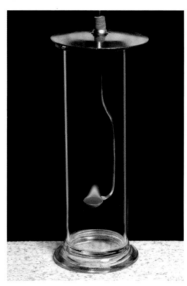

(b) Sulfur burns with a blue flame in oxygen and the reaction produces the toxic gas sulfur dioxide.

Magnesium also reacts with nitrogen when it burns in the air. As discussed in Reactivity 3.2, this is also an oxidation reaction as the magnesium loses two electrons.

Elemental phosphorus exists as two allotropes: white phosphorus and red phosphorus. White phosphorus reacts when exposed to air to produce two oxides: $P_4O_{10}(s)$ and $P_4O_6(s)$.

Worked example

White phosphorus is an allotrope of phosphorus and exists as $P_4(s)$.
Write an equation for the reaction of white phosphorus (P_4) with oxygen gas to form phosphorus(V) oxide.

Solution

Start with the unbalanced equation showing the reactants including $O_2(g)$ and products.

$$P_4(s) + \mathbf{5}O_2(g) \rightarrow P_4O_{10}(s)$$

Balance the Os on the left-hand side as shown in bold.

Most combustion reactions are exothermic, but nitrogen combines with oxygen only at high temperatures in an endothermic reaction to produce nitrogen monoxide.

TOK

A number of sources claim that combustion reactions are exothermic, but the reaction

$N_2(g) + O_2(g) \rightleftharpoons 2NO(g)$

is endothermic. Does this falsify the claim or is it 'an exception to a rule'?

Is the rule still useful or does it need to be reconsidered? What is the difference between such rules and scientific laws?

Reactivity 3.2 – Which species are the oxidizing and reducing agents in a combustion reaction?

As discussed in Reactivity 3.2, these combustion reactions involve electron transfers to oxygen. Oxygen is an oxidizing agent in these cases because it accepts electrons. The metal or non-metal element is the reducing agent because it loses electrons.

▲
The rusting of iron is also an oxidation reaction but not a combustion reaction. There is no flame involved.

Combustion of organic compounds breaks the carbon chain

Many hydrocarbons and alcohols are used as fuels because their combustion reactions release energy at a reasonable rate to be useful. Although the compounds are energetically unstable relative to their combustion products, they are kinetically stable. They do not spontaneously combust because the reaction has a sufficiently high activation energy. The high activation energy also allows the fuels to be safely transported and stored. Activation energy is discussed in more detail in Reactivity 2.2.

Spontaneous combustion. The oxidation of glycerin by potassium manganate(VII) is very exothermic and after a few seconds the released heat causes the glycerin to ignite and burst into flame.

Reactivity 2.2 – Why is high activation energy often considered to be a useful property of a fuel?

The complete combustion of organic compounds breaks the carbon chain and results in the production of carbon dioxide and water and the release of a lot of heat energy.

For example, propane burns in the presence of excess oxygen to produce carbon dioxide and water:

$$C_3H_8(g) + 5O_2(g) \rightarrow 3CO_2(g) + 4H_2O(l)$$

This is known as complete combustion because the products are fully oxidized.

When the oxygen supply is limited, *incomplete* combustion occurs, as discussed in the next section.

Similarly, the alcohols with the general formula $C_nH_{2n+1}OH$ can be burnt to produce similar products:

$$C_2H_5OH(g) + 3O_2(g) \rightarrow 2CO_2(g) + 3H_2O(l)$$

As discussed in Reactivity 3.2, the carbon atom attached to the OH group can be selectively oxidized by oxidizing agents such as potassium dichromate(VI).

Worked example

Write equations for the complete combustion of pentane.

Solution

Start with the unbalanced equation showing the reactants including $O_2(g)$ and products.

$$C_5H_{12}(g) + \_O_2(g) \rightarrow \mathbf{5}CO_2(g) + \mathbf{6}H_2O(l)$$

Balance the Cs and the Hs on the right-hand side as shown in **bold** above.

Balance the Os on the left-hand side as shown in bold below:

$$C_5H_{12}(g) + \mathbf{8}O_2(g) \rightarrow 5CO_2(g) + 6H_2O(l)$$

Worked example

Write the equation for the complete combustion of ethanol.

Solution

Following the same strategy:

$$C_2H_5OH(l) + \_O_2(g) \rightarrow \mathbf{2}CO_2(g) + \mathbf{3}H_2O(l)$$

Balance the Os on the left-hand side as shown but do not forget that the alcohol has an O atom:

$$C_2H_5OH(l) + \mathbf{3}O_2(g) \rightarrow 2CO_2(g) + \mathbf{3}H_2O(l)$$

The combustion of organic molecules involves the formation and reaction of many radicals. The colour of a flame is due to the presence of carbon radicals which eventually all break up to form carbon dioxide.

The combustion of organic compounds that contain elements other than carbon, hydrogen and oxygen produces a range of other products. Compounds containing nitrogen for example can release nitrogen gas or different nitrogen oxides. Combustion analysis can be used to determine the empirical formula of a compound as discussed in Structure 1.4.

(a) Ethanol burns with a relatively clean (not too sooty) flame, having a relatively short chain of carbon atoms for each molecule.

(b) Glucose is a carbohydrate which burns to produce carbon dioxide and water vapour, leaving behind carbon. The process releases a large amount of energy. Metabolism releases an equivalent amount of energy to burning, but in a slower, more controlled, and less direct way.

Challenge yourself

1. A compound containing carbon, hydrogen and nitrogen was analyzed by combustion. 0.20 g of the compound produced 0.456 g of carbon dioxide and 0.248 g of water. Identify the empirical formula of the compound.

Exercise

Q1. Identify the element in the third period which reacts with the least amount of oxygen when 1.0 mol of the element undergoes complete combustion.

    A aluminium   **B** magnesium   **C** sodium   **D** sulfur

Q2. 1 mol of an organic compound X was found to require 2.5 mol of molecular oxygen for complete combustion. Which of the following could be X?

    I. C_2H_2 II. CH_3CHO III. C_2H_5OH

    A I and II only   **B** I and III only   **C** II and III only   **D** I, II and III

Q3. Hydrazine, N_2H_4, is used as a rocket fuel. It reacts according to the thermochemical equation:

$$N_2H_4(l) + O_2(g) \rightarrow N_2(g) + 2H_2O(l) \qquad \Delta H^\ominus = -623 \text{ kJ mol}^{-1}$$

Hydrazine does not, however, burn spontaneously in oxygen. What is the best explanation for this?

    A Hydrazine is a liquid.   **B** The $N\equiv N$ triple bond is too strong.

    C The activation energy is too high.   **D** The reaction is endothermic.

Q4. What mass of carbon dioxide is produced when 2.9 g of butane undergoes complete combustion?

Q5. The complete combustion of 0.60 g of a liquid compound, X, produces 0.88 g of carbon dioxide gas and 0.36 g of water vapour. What is the empirical formula of X?

Q6. Deduce the general equation for complete combustion of a hydrocarbon C_xH_y

Reactivity 1.3.2 – Incomplete combustion of organic compounds

Reactivity 1.3.2 – Incomplete combustion of organic compounds, especially hydrocarbons, leads to the production of carbon monoxide and carbon.	
Deduce equations for the incomplete combustion of hydrocarbons and alcohols.	
	Inquiry 2 – What might be observed when a fuel such as methane is burnt in a limited supply of oxygen?
	Reactivity 2.1 – How does limiting the supply of oxygen in combustion affect the products and increase health risks?

Incomplete combustion of organic compounds produces carbon monoxide and carbon

Carbon dioxide and water are produced in complete combustion when there is a plentiful supply of air, but carbon monoxide or carbon in the form of soot are produced during incomplete combustion. This happens if the supply of air is limited or when the compound has a high percentage of carbon content.

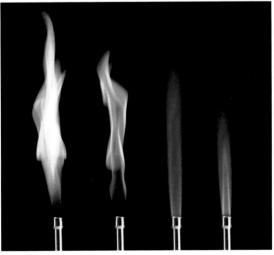

(a) A Bunsen burner burns methane. The flame is controlled with a vent that changes the fuel : air mixture. Complete combustion produces a hot, blue flame. Incomplete combustion produces a yellow flame.

(b) Combustion of hydrocarbons with a benzene ring produces a sooty yellow flame due to their high percentage of carbon content.

Limiting the amount of oxygen limits the extent to which the carbon in organic compounds can be oxidized. Soot is produced when there is a very limited amount of oxygen available.

For example, the incomplete combustion of propane produces soot, C(s), or carbon monoxide.

$$2C_3H_8(g) + 7O_2(g) \rightarrow 6CO(g) + 8H_2O(g)$$

$$C_3H_8(g) + 2O_2(g) \rightarrow 3C(g) + 4H_2O(g)$$

Compounds with higher carbon content generally undergo more incomplete combustion and produce more soot when the oxygen supply is limited. This can be used to distinguish different organic compounds.

Worked example

How could samples of cyclohexane (C_6H_{12}), cyclohexene (C_6H_{10}) and methylbenzene ($CH_3C_6H_5$) be distinguished by observing their combustion?

Solution

SKILLS

The combustion of an organic liquid can be observed by placing a few drops of the liquid on a watch glass.

Calculate the percentage carbon composition by mass in each of the three compounds.

	C_6H_{12}	C_6H_{10}	$CH_3C_6H_5$
M / g mol^{-1}	84	82	92
% of C	86	88	91

Inquiry 2 – What might be observed when a fuel such as methane is burnt in a limited supply of oxygen?

Methylbenzene will produce the sootiest flame and cyclohexane the least sooty flame.

Worked example

Give the balanced equation for the formation of carbon monoxide (CO) in the internal combustion engine caused by the incomplete combustion of octane, $C_8H_{18}(l)$.

Solution

First write the unbalanced equation with $H_2O(g)$, produced at the temperature of the engine, and CO(g) as products.

$$C_8H_{18}(l) + \_O_2(g) \rightarrow \mathbf{8}CO(g) + \mathbf{9}H_2O(g)$$

Balance the carbon atoms and hydrogen atoms from left to right as shown in bold above.

Then balance the oxygen atoms (from right to left) as shown in bold below.

$$C_8H_{18}(l) + \mathbf{8\tfrac{1}{2}}O_2(g) \rightarrow 8CO(g) + 9H_2O(g)$$

Multiply ×2 if an equation with only integers is needed:

$$2C_8H_{18}(l) + 17O_2(g) \rightarrow 16CO(g) + 18H_2O(g)$$

▲ This household device is designed to detect excessive levels of carbon monoxide (CO) which can be produced by malfunctioning gas boilers or fires. After six hours of exposure, 35 ppm of CO produces headaches and dizziness. Within hours, 800 ppm of CO causes unconsciousness, with higher levels causing death.

As mentioned in Structure 3.1, catalytic converters are used in cars to fully oxidize carbon monoxide to carbon dioxide:

$$2CO(g) + O_2(g) \rightarrow 2CO_2(g)$$

and to convert unburnt and partially burnt fuel hydrocarbons to carbon dioxide and water.

Three-way catalytic converters also reduce nitrogen oxides to nitrogen and oxygen.

$$2CO(g) + 2NO(g) \rightarrow 2CO_2(g) + N_2(g)$$

Carbon monoxide (CO) is toxic to humans because it affects oxygen uptake in the blood. It is absorbed by the lungs and binds to haemoglobin (Hb) in red blood cells more effectively than oxygen.

Hb   +   CO   ⇌   COHb
haemoglobin   carboxyhemoglobin

This prevents oxygen from being transported around the body. It can cause dizziness at low concentrations and can be fatal at high levels. As it is a colourless, odourless gas, it can rise to dangerous levels without being detected.

Carbon or soot produced from the incomplete combustion of hydrocarbons and coal in power stations is an example of a particulate. Particulates are solid particles of carbon or dust, or liquid droplets suspended in the air. They generally have a diameter in the range 0.001–10 μm, which is just large enough to be seen. They enter the body during breathing and can affect the respiratory system and cause lung diseases such as emphysema, bronchitis and cancer. Although they are sometimes inert solids, they are dangerous pollutants because they can act as catalysts in the production of secondary pollutants and increase the harmful effects of gaseous pollutants.

Challenge yourself

2. Explain how nitrogen oxides are formed in an internal combustion engine and suggest an effect of these pollutants.

Exercise

Q7. Deduce the general equation for the incomplete combustion of a hydrocarbon C_xH_y which produces carbon monoxide.

Q8. Deduce the equation for the incomplete combustion of hexane which produces equal amounts of carbon dioxide and carbon monoxide.

Q9. (a) State the equation for the complete combustion of octane.

(b) Calculate the mass of oxygen which reacts with 1.00 g of octane during complete combustion.

(c) Assume that 20% by mass of the air is made up of oxygen. Calculate the mass of air which reacts with 1.00 g of octane.

(d) Explain why carbon monoxide is a dangerous pollutant.

(e) State one method of controlling carbon monoxide emissions from automobile engines.

Parts per million (ppm) is a measure of concentration often used for pollutants present at low but dangerous levels.

Reactivity 2.1 – How does limiting the supply of oxygen in combustion affect the products and increase health risks?

Reactivity 1.3.3 – Fossil fuels

Reactivity 1.3.3 – Fossil fuels include coal, crude oil and natural gas, which have different advantages and disadvantages.

Evaluate the amount of carbon dioxide added to the atmosphere when different fuels burn.

Understand the link between carbon dioxide levels and the greenhouse effect.

The tendency for incomplete combustion and energy released per unit mass should be covered.	Structure 3.2 – Why do larger hydrocarbons have a greater tendency to undergo incomplete combustion?
	Nature of Science, Reactivity 3.2 – What are some of the environmental, economic, ethical and social implications of burning fossil fuels?

An ideal fuel releases significant amounts of energy at a reasonable rate and produces minimal pollution

An energy source needs to be cheap, plentiful, and readily accessible, and provide high-quality energy at a reasonable rate – not too fast or too slow. It should do all this in a way that has minimal effect on the environment. We all use a great range of energy sources. Some of these, such as wood, are ancient technologies, while others, such as nuclear power and solar cells, are very recent. Most of our energy derives from the Sun, which drives the climate and photosynthesis. Humans harness this energy indirectly by burning fossil fuels. Oil is the fossil fuel that currently provides the world's economy with the most energy, but this will change. Its use has had a very negative impact on our environment.

Fossil fuels are non-renewable and wood is a renewable source

Fossil fuels are non-renewable and non-sustainable energy sources as it takes millions of years for fossil fuels to form. They are used at a rate faster than they can be replaced and our finite resources will run out. Wood, by contrast, is a renewable source as trees can be grown to replace those chopped down to provide wood as fuel. Biofuels are discussed in more detail in the next section.

The properties of four fuels are summarized in the table below. The energy density of a fuel is the energy produced per unit volume, and the specific energy is the energy produced per unit mass.

The liquid fuels have significantly greater energy densities than the gases.

Formula	Energy released per unit mass / $kJ\,g^{-1}$	Energy released per unit / $kJ\,cm^{-3}$
$H_2(g)$	142	0.0129
$CH_4(g)$	55.5	0.0401
$C_8H_{18}(l)$	47.9	33.7
$CH_3OH(l)$	22.7	18.0

One advantage of using gasoline or petrol, which has octane as a main component, in car engines is its high energy density. Methane is preferred to hydrogen in pipelines for the same reason. When the mass of the fuel is important, for example in rockets, hydrogen is a better choice with its high specific energy. Hydrogen is the only fuel here which does not produce the greenhouse gas carbon dioxide.

Fossil fuels were formed by the reduction of biological compounds

The fossil fuels have been described as 'sunshine in the solid, liquid and gaseous form', as their energy comes from sunlight which was trapped by green plants millions of years ago. Fossil fuels are produced by the slow and partial decomposition of plant and animal matter that is trapped in the absence of air. Oxygen is lost from the biological molecules at a faster rate than other elements (carbon, hydrogen, nitrogen and sulfur). This results in reduced biological compounds which are generally hydrocarbons.

Coal is the most abundant fossil fuel

Coal is a combustible sedimentary rock formed from the remains of plant life which have been subject to geological heat and pressure. This action changed plant material in stages to increase the percentage of the carbon content. Anthracite, for example, is formed under conditions of very high heat and pressure and is almost pure carbon, but coal generally contains between 80% and 90% carbon by mass. Coal occurs in many areas, though the majority of the world's supplies are in the northern hemisphere. It is by far the most plentiful of the Earth's fossil fuels.

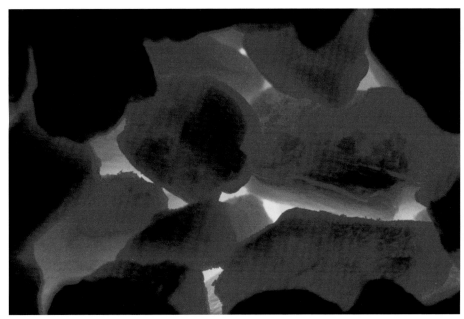

Burning anthracite. Anthracite has a fixed carbon content of 92–98%. It burns with a short blue flame and no noticeable smoke.

Nature of Science

Coal fueled much of the Industrial Revolution, which led to important advances in both science and technology. The development of steam power was crucial to a country's economy, but to understand the link between heat and work scientists had to unravel the nature of energy – its quality and its quantity and the underlying molecular basis for the laws governing the transfer of heat into work. Whatever the field, whether it is pure science or engineering technology, there is boundless scope for creative and imaginative thinking.

Crude oil is a valuable fuel and chemical feedstock

Crude oil is one of the most important raw materials in the world today. It is a complex mixture of straight-chain and branched-chain saturated alkanes, cycloalkanes and aromatic compounds, and, in smaller quantities, compounds of nitrogen, oxygen and sulfur. It supplies us with the fuels we need for transport and electricity generation and is an important chemical feedstock for the production of important organic compounds such as polymers, pharmaceuticals, dyes and solvents.

Crude oil was formed over millions of years from the remains of marine animals and plants, which were trapped under layers of rock. Under these conditions of high temperature and high pressure, organic matter decays in the presence of bacteria and the absence of oxygen. As it is a limited resource chemists will need to consider other sources of carbon, both as a fuel and as a chemical feedstock.

The Organization of the Petroleum Exporting Countries (OPEC) is an intergovernmental organization set up to stabilize oil markets in order to secure an efficient, economic, and regular supply of petroleum. There are political implications in being dependent on imported oil and gas. Price fluctuations have a significant effect on national economies; supplies and energy resources are therefore an important bartering tool in political disputes.

A petrol pump supplies energy to a car at a rate of about 34 MW (34 MJ s^{-1}).

Oil and gas are easier to extract than coal since they are fluids and can be pumped up from underground reserves. It is possible to extract oil from beneath the sea. There are risks and benefits in using oil as a source of energy or as a source of carbon to make new products.

Petrol is a highly concentrated and convenient energy source for use in transport. It could be argued, however, that burning hydrocarbons, with their resulting environmental side-effects such as smog and global warming, is a misuse of this valuable resource. When the great Russian chemist Dmitri Mendeleev (Structure 3.1) visited the oils fields of Azerbaijan at the end of the 19th century, he is said to have likened the burning of oil as a fuel to 'firing up a kitchen stove with banknotes'.

Natural gas is mainly methane

Methane is the primary constituent of natural gas, which also contains nitrogen and sulfur compounds (hydrogen sulfide) as impurities. Natural gas was formed millions of years ago by the action of heat, pressure, and perhaps bacteria on buried organic matter. The gas is trapped in geological formations capped by impermeable rock. It is also formed from the decomposition of crude oil and coal deposits. It can occur almost entirely on its own, dissolved under pressure in oil, or in a layer above the oil in a reservoir. Natural gas can also be found associated with coal when it is a major hazard as it forms an explosive mixture with air.

Natural gas is the cleanest of the fossil fuels to burn due to its low percentage carbon content. Impurities can easily be removed, and the combustion of natural gas produces minimal amounts of carbon monoxide, hydrocarbons, and particulates. Where it is available as mains gas, it flows through pipes to wherever it is needed, so little energy is needed to transport the gas from the ground to the consumer. Setting up such a distribution network, however, involves a massive capital investment, and in some countries liquefied gas (butane or propane) is used instead for domestic heating and cooking when no network is available.

Constructing international pipelines has financial costs and political risks.

Natural gas pipelines.

The past and future of fossil fuels

As civilization has advanced, the carbon content of fuel has decreased. Coal was the first fossil fuel to be used. It is the most abundant and easiest to obtain and most similar in properties to charcoal, which was generally used as a fuel by primitive cultures. Coal has mainly been replaced by gasoline and natural gas as the fuel of choice as they have higher specific energies and energy densities and are easier to transport. The two fluids, gas and oil, are also cleaner burning and produce less carbon dioxide per unit energy.

The choice of fossil fuel used by different countries depends on historical, geological and technological factors. Different societies have different priorities in their energy choices

The advantages and disadvantages of the three fossil fuels are summarized below.

	Coal	Petroleum	Natural gas
Advantages	Cheap and plentiful throughout the world. Longest lifespan compared to other fossil fuels. Can be converted into synthetic liquid fuels and gases. Safer than nuclear power. Ash produced can be used in making roads.	Easily transported in pipelines or by tankers. Convenient fuel for use in cars as volatile and burns easily. High enthalpy density. Sulfur impurities can be easily removed.	Higher specific energy. Clean; produces fewer pollutants per unit energy. Easily transported in pipelines and pressurized containers. Does not contribute to acid rain.
Disadvantages	Contributes to global warming due to CO_2. Contributes to acid rain, SO_2. Produces particulates (electrostatic preceptors can remove most of these). Difficult to transport. Waste can lead to visual and chemical pollution. Mining is dangerous.	Limited lifespan and uneven world distribution. Contributes to acid rain and global warming. Transport can lead to pollution. Carbon monoxide is a local pollutant produced by incomplete combustion of gasoline in internal combustion engines. Photochemical smog is produced as a secondary pollutant due to reactions of the primary pollutants (nitrogen oxides and hydrocarbons) released from internal combustion engines.	Limited supplies. Contributes to global warming. Risk of explosion due to leaks.

All fossil fuels are non-renewable and produce the greenhouse gas carbon dioxide.

One measure of the impact our activities have on the environment is given by our **carbon footprint**. The carbon footprint is a measurement of all the greenhouse gases we individually produce. It has units of mass of carbon dioxide equivalent and depends on the amount of greenhouse gases we produce in our day-to-day activities through the use of fossil fuels – such as heating, transport and electricity. 'Carbon footprint' has become a widely used term in the public debate on our responsibilities in the fight against global climate change.

Nature of Science

There is some limited consensus on how to quantify a carbon footprint. Some questions are:

- Should the carbon footprint include just carbon dioxide (CO_2) emissions or include other greenhouse gas emissions such as methane as well?
- Should it be restricted to carbon-based gases or can it include substances that do not have carbon in their molecules such as dinitrogen monoxide, which is another greenhouse gas?
- Should it include other gases such as carbon monoxide (CO), which is a toxic primary pollutant that can be oxidized to carbon dioxide in the atmosphere?

Scientific terms which have an impact on the public must be clearly communicated.

Challenge yourself

3. On a typical winter's day, 1.33×10^6 kJ of energy is needed in a home.

   The specific energy densities and approximate empirical formulae of wood and coal and the efficiency of the heating systems are tabulated below.

Fuel	Formula of fuel	Specific energy / kJ g^{-1}	Efficiency of heating / %
Coal	CH	31	65
Wood	$C_5H_9O_4$	22	70

   (a) Calculate the percentage mass of carbon in the two fuels.

   (b) Determine the carbon footprint from using the two forms of heating in terms of the mass of carbon dioxide produced by the two fuels.

   (c) Suggest a reason why the carbon footprint calculation for wood does not give a full account of its impact on the environment.

Combustion of the alkanes in detail

Here we look at the combustion of the alkanes in a little more detail.

Carbon content and incomplete combustion in the alkanes

As discussed earlier, the incomplete combustion of a hydrocarbon results in the incomplete combustion of the carbon in the compound with the production of carbon monoxide and carbon. The higher the carbon content of the compound, the greater the tendency for incomplete combustion. Consider the carbon content of the alkanes.

Alkane	CH_4	C_2H_6	C_3H_8	C_4H_{10}	C_5H_{12}	C_6H_{14}	C_7H_{16}	C_8H_{18}
% Carbon content	74.8	79.9	81.7	82.6	83.2	83.6	83.9	84.1
Mass of CO_2/g when 1 g of fuel is burnt	2.74	2.93	2.99	3.03	3.05	3.06	3.07	3.08
Energy released per unit mass / kJ g^{-1}	61.1	51.9	50.3	49.5	48.6	48.3	0.0	47.9

Structure 3.2 – Why do larger hydrocarbons have a greater tendency to undergo incomplete combustion?

The increase in percentage carbon content down the homologous series suggests that incomplete combustion increases with the length of the carbon chain.

This supports the claim that natural gas, which is mostly methane, is the cleanest of the fossil fuels to burn, and coal, which has a very high carbon content, is the dirtiest.

Carbon content and carbon dioxide production during complete combustion

The amount of carbon dioxide gas produced by a fuel is another important factor when considering the use of different fuels. Complete combustion of a hydrocarbon produces carbon dioxide and water, and the mass of carbon dioxide produced per unit mass of fuel increases with the percentage carbon content as shown in the previous table.

This also supports the claim that natural gas is the cleanest of the fossil fuels to burn and coal is the dirtiest.

Worked example

Calculate the mass of carbon dioxide produced when 1.00 g of pentane undergoes complete combustion.

Solution

$$C_5H_{12}(l) + 8O_2(g) \rightarrow 5CO_2(g) + 6H_2O(l)$$
$$1 \text{ mol} \qquad\qquad 5 \text{ mol}$$

$$n(CO_2(g)) = 5n(C_5H_{12}(l))$$

$$= 5 \times \frac{1.00}{(12.01 \times 5) + (1.01 \times 12)} = 5 \times \frac{1.00}{72.17}$$

$$mass(CO_2(g)) = 5 \times \frac{44.01}{72.17} = 3.05 \text{ g}$$

Structure 3.2 – Why is carbon dioxide described as a greenhouse gas?

The mass of carbon dioxide produced for the other alkanes is calculated in the same way.

Carbon content and specific energy in the alkanes

We saw earlier that hydrogen has higher specific energy than methane. Comparing the specific energy of the alkanes we see that the higher the percentage carbon content (and the lower the percentage hydrogen), the lower the specific energy.

Comparing the enthalpy of pentane and hexane. Full details of how to carry out this experiment with a worksheet are available in the eBook.

SKILLS

The energy obtained per gram of CO_2 produced during complete combustion (Figure 1) is another important consideration when choosing fuels.

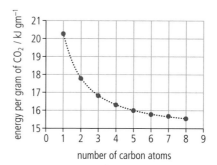

R1.3 Figure 1 A graph showing how the energy obtained per gram of CO_2 produced varies with the length of the carbon chain in the alkanes.

The graph in Figure 1 shows that natural gas (which is mainly CH_4) produces the most energy per gram of CO_2 added to the atmosphere. The higher alkanes found in petroleum and carbon in coal produce less energy for the same level of carbon dioxide pollution.

Fossil fuels and the greenhouse effect

The level of carbon dioxide produced from burning fossil fuels has been increasing steadily over the last 200 years.

Carbon dioxide levels and the greenhouse effect

Levels of carbon dioxide have been measured at Mauna Loa in Hawaii since the International Geophysical Year in 1957 (Figure 2).

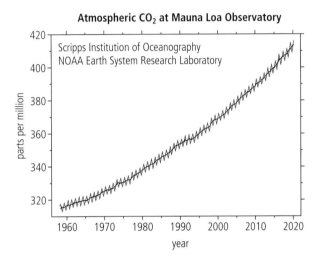

R1.3 Figure 2 Graph showing the rising concentration of atmospheric CO_2 between 1958–2021 measured 4170 m up on Mauna Loa, Hawaii. The graph shows the steady rise of CO_2 levels in the atmosphere each year due to increasing fossil fuel consumption. The regular fluctuations reflect seasonal plant growth in the spring and decay in the autumn in the northern hemisphere each year.

We can compare the increase in levels of carbon dioxide to changes in average global temperatures during the same period (Figure 3). It is estimated that carbon dioxide contributes about 50% to global warming.

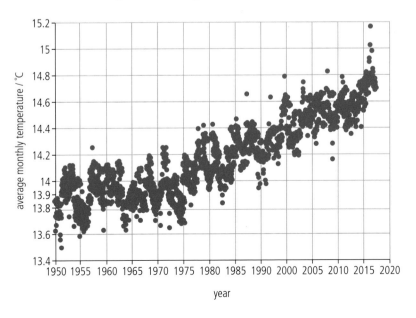

R1.3 Figure 3 Global average temperatures from 1950 to 2016.

The Mauna Loa Observatory monitors all atmospheric constituents that may contribute to climatic change, such as greenhouse gases and aerosols.

359

Nature of Science

Data are not enough to prove the causal link between CO_2 concentration and temperature. Both changes could be due to changes in a third variable. The ideas of correlation and cause are very important in science. A correlation is a statistical link or association between one variable and another. To establish a causal relationship, scientists need to have a plausible mechanism linking the factors. This has been proposed in the case of the greenhouse effect. Ideally, however, the relationship between the variables needs to be experimentally investigated with all other possible key variables being controlled. This is often impossible; levels of carbon dioxide cannot be added to the environment in such a controlled way, and so we need to rely on computer models.

Greenhouse gases allow shortwave radiation from the Sun to pass through the atmosphere but absorb the longer wave infrared radiation re-radiated from the Earth's surface. As discussed in Structure 3.2, carbon dioxide is a greenhouse gas as its molecules increase their vibrational energy by absorbing infrared radiation. Three of the vibrational modes of the carbon dioxide molecule are IR active – the dipole changes as it vibrates (Figure 4). The molecules then re-radiate this energy back to the Earth's surface, contributing to global warming.

R1.3 Figure 4 Three of the vibrational modes of the carbon dioxide molecule are IR active. The symmetric stretch produces no change in dipole and so is IR inactive.

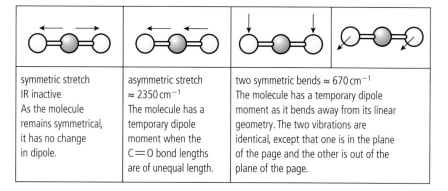

symmetric stretch IR inactive As the molecule remains symmetrical, it has no change in dipole.	asymmetric stretch $\approx 2350\,cm^{-1}$ The molecule has a temporary dipole moment when the $C{=}O$ bond lengths are of unequal length.	two symmetric bends $\approx 670\,cm^{-1}$ The molecule has a temporary dipole moment as it bends away from its linear geometry. The two vibrations are identical, except that one is in the plane of the page and the other is out of the plane of the page.

Challenge yourself

4. (a) Suggest why the first measurements of CO_2 levels were taken at Mauna Loa in Hawaii and at the South Pole.
   (b) Explain the annual fluctuations in CO_2 levels at Mauna Loa.
   (c) Identify two different means by which CO_2 can be removed from the atmosphere naturally and give balanced equations for both of these.
   (d) Suggest how the depletion of tropical forests can lead to an increase in CO_2 levels.

The impact of increasing amounts of greenhouse gases on the climate

There is little doubt that, since the 19th century, the amount of carbon dioxide and other anthropogenic (resulting from the impact of humans on nature) greenhouse gases in the atmosphere have increased dramatically and that the average temperature of the world has also increased, even if rather erratically. It has been suggested that levels of carbon dioxide will double in about 100 years. Allowing for the effect of all the gases, the temperature of the Earth could rise by 2 °C within 50 years. These temperature changes could have a drastic effect on the climate, for example:

- changes in agriculture such as crop yields
- changes in biodistribution due to desertification and loss of cold-water fish habitats
- rising sea levels because of thermal expansion and the melting of polar ice caps and glaciers.

 Retreating glacier in Iceland most probably caused by global warming.

TOK

Some people question the reality of climate change and question the motives of scientists who have 'exaggerated' the problem. How do we assess the evidence collected and the models used to predict the impact of human activities? What effect does a highly sensitive political context have on objectivity? Can politicians exploit the ambiguity of conclusions coming from the scientific community for their own ends?

Nature of science, Reactivity 3.2 – What are some of the environmental, economic, ethical and social implications of burning fossil fuels?

Exercise

Q10. Consider the following specific energies of different fuels.

Fuel	Specific energy / kJ g^{-1}	Carbon content by mass / %
Coal	32	94
Oil	42	83
Natural gas	55	75
Wood	15	70
Hydrogen	142	0

(a) Suggest an explanation for the trend in specific energies between coal, oil, natural gas and hydrogen.

(b) The choice of fuel also depends on other factors besides specific energies. One of these is the amount of carbon dioxide produced. Explain why this is an important factor to consider.

(c) Calculate the carbon dioxide production for 1000 kJ of energy from each source and identify the best and worst fuel on this basis.

Q11. An analytical chemist determines the percentage composition by mass of a sample of coal.

C	H	O	N	S
84.96	5.08	7.55	0.73	1.68

(a) Determine the empirical formula of the coal.

(b) Explain how the combustion of the coal could produce acid rain. Give equations to illustrate how the elements in the coal can produce acidic gases.

Q12. The coal burnt in a 500 MW (500×10^3 kJ s^{-1}) power station has a specific energy of 33.0 kJ g^{-1}.

    (a) Determine the mass of coal burnt each second when the power station is operating at full capacity, assuming an efficiency of 38%.

    (b) The empirical formula of the coal can be approximated as CH. Calculate the mass of carbon dioxide produced each second.

Q13. State three reasons for not relying on carbon-containing fuels in the future.

Q14. Crude oil currently provides a greater percentage of the world's energy than any other energy source.

    (a) State three reasons for its wide use as an energy source.

    (b) Describe briefly how crude oil is formed in nature.

Q15. State three reasons why, worldwide, gas is preferred to oil as a fuel in power stations.

Reactivity 1.3.4 – Biofuels

Reactivity 1.3.4 – Biofuels are produced from the biological fixation of carbon over a short period of time through photosynthesis.

Understand the difference between renewable and non-renewable energy sources.

Consider the advantages and disadvantages of biofuels.

The reactants and products of photosynthesis should be known.	

Photosynthesis converts light energy into chemical energy

Energy from the Sun fuels the sequence of redox reactions that is photosynthesis. **Chlorophyll**, the green pigment in green plants, absorbs solar energy which is then used to convert carbon dioxide and water into glucose. The reactions can be summarized by the equation:

$$6CO_2(l) + 6H_2O(l) \rightarrow C_6H_{12}O_6(s) + 6O_2(g)$$

The products of photosynthesis can be used for food, as a primary fuel such as wood, or converted to other fuels such as ethanol. **Biofuels**, produced from the biological fixation of carbon over a short period of time, are a renewable and sustainable energy source. Photosynthesis is not very efficient, however, so relatively little of the available solar energy is trapped.

Challenge yourself

5. One part of the Earth's surface receives 1.25×10^6 J of solar energy. Green plants such as algae absorb 1.25×10^4 J of this energy.

(a) Determine the percentage of the Sun's energy absorbed by green plants.

(b) Suggest two reasons why the remainder of the Sun's energy is not absorbed by green plants.

Wood is mainly cellulose, a polymer made up of glucose molecules. Comparing coal with wood is very useful as the latter has not been subject to the same geological conditions.

Worked example

Coal and wood are used domestically to provide heating.

	Percentage composition				Specific energy / kJ g^{-1}
	C	**H**	**O**	**N**	
Wood	50	6	43	1	10–18
Bituminous coal	88	5	5–15	1	30
Anthracite	95	2–3	2–3	Trace	31

(a) Describe the relationship between the oxygen content of a fuel and its specific energy.

(b) Describe how the compositions of the fuels change as they are exposed to high pressure.

(c) State two advantages of coal use over wood.

(d) State one advantage of the use of wood.

(e) Discuss some of the disadvantages of the use of coal.

(f) What harmful pollutants are produced when wood is burnt in an enclosed space?

Solution

(a) The lower the oxygen content the greater its specific energy.

(b) Oxygen (and water vapour) are lost as the pressure increases.

(c) Coal is the most abundant fossil fuel and has a high specific energy.

(d) Wood is a renewable and sustainable fuel.

(e) Difficult and expensive to mine and transport; produces sulfur oxides, which lead to acid rain, particulates (soot) and carbon dioxide, a greenhouse gas which leads to global warming; mining has an environmental impact.

(f) Particulates (soot), hydrocarbons and carbon monoxide are produced due to incomplete combustion.

Wood-burning stove. Wood has a lower specific energy than fossil fuels due to its relatively high oxygen content.

Ethanol can be used as a biofuel

Ethanol is a liquid biofuel, which can be used in internal combustion engines. It can be made from biomass by fermenting plants high in starches and sugars.

$$C_6H_{12}O_6 \rightarrow 2C_2H_5OH + 2CO_2$$

This process can be carried out at approximately 37 °C in the absence of oxygen by yeast, which provides an enzyme which catalyzes the reaction.

Worked example

(a) State the equation for the complete combustion of ethanol.

(b) The enthalpy of combustion of ethanol is −1367 kJ mol⁻¹.
Calculate the specific energy of ethanol in kJ g⁻¹.

(c) Compare this value with the corresponding value for octane and explain the difference.

Solution

(a) $C_2H_5OH(l) + 3O_2(g) \rightarrow 3H_2O(l) + 2CO_2(g)$

(b) heat produced by one mole of ethanol (46.08 g) = 1367 kJ

$$\text{heat produced by } 1.00\,g = \frac{-1367}{46.08}\,kJ\,g^{-1} = 29.67\,kJ\,g^{-1}$$

(c) heat produced by one mole of octane (114.26 g) = 5470 kJ

$$\text{heat produced by } 1.00\,g = \frac{-5470}{114.26}\,kJ\,g^{-1} = 47.87\,kJ\,g^{-1}$$

Less heat is produced per unit mass with ethanol because it is already partially oxidized.

Gasohol is a mixture of 10% ethanol and 90% unleaded gasoline. Ethanol can be produced from several types of plant material and gasohol can generally be substituted for gasoline with only minor changes in fuel consumption and performance. The advantages of using ethanol include:

- It is renewable.
- It produces lower emissions of carbon monoxide and nitrogen oxides.
- It decreases a country's dependence on oil.

One disadvantage is that ethanol absorbs water as it can form hydrogen bonds with the water molecules in the atmosphere. This leads to the ethanol separating from the hydrocarbon components in the fuel. It can also cause corrosion.

Methanol can also be used; it is produced by heating wood in the presence of steam and limited amounts of oxygen.

Challenge yourself

6. A possible incomplete equation for the conversion of wood into hydrogen, carbon monoxide, and carbon dioxide is:

$$2C_{16}H_{23}O_{11} + 19H_2O + O_2 \rightarrow xH_2 + yCO + zCO_2$$

(a) Identify the oxidation number of all the elements in the equation.

(b) Identify the elements which are oxidized and reduced in the reaction.

(c) Deduce the value of x by balancing the hydrogen atoms.

(d) Determine the values of y and z by balancing the changes of oxidation numbers and the number of carbon atoms.

(e) Deduce how many methanol molecules can be produced from one wood molecule.

Methane can be made from the bacterial breakdown of plant material in the absence of oxygen. The process can be controlled to produce a clean burning biogas such as natural gas.

Carbohydrates, for example, produce a mixture which is about 50% methane:

$$C_6H_{12}O_6(s) \rightarrow 3CO_2(g) + 3CH_4(g)$$

Fats with a lower oxygen content produce 72% methane:

$$2C_{15}H_{31}COOH + 14H_2O \rightarrow 9CO_2 + 23CH_4$$

The advantages and disadvantages of using biofuels

Advantages	Disadvantages
Cheap and readily available.	Uses land which could be used for other purposes, such as growing food.
If crops/trees are replanted, they can be a renewable and sustainable source.	High cost of harvesting and transportation in large volumes.
Less polluting than fossil fuels.	Takes nutrients from soil/uses large amounts of fertilizers.
	Lower specific energy than fossil fuels.

Q16. **(a)** Scientists have described solar energy as the ultimate energy source. Identify energy sources which derive from photosynthesis.

(b) Discuss the advantages and disadvantages of the use of biomass as a fuel.

Q17. **(a)** State the chemical equation for the photosynthesis of glucose.

(b) Identify the molecule needed by plants to photosynthesize.

(c) Glucose can be converted to ethanol, which can be used as a fuel. State the name and equation of the process and describe the conditions needed.

Q18. **(a)** When biomass decomposes in the absence of oxygen, biogas is formed. Name the main gas present in biogas.

(b) When wood is burnt in an enclosed space, the combustion produces harmful pollutants. Identify the pollutants.

(c) Suggest why biomass is likely to become more important in the future as a fuel.

Reactivity 1.3.5 – Fuel cells

Reactivity 1.3.5 – A fuel cell can be used to convert chemical energy from a fuel directly to electrical energy.

Deduce half-equations for the electrode reactions in a fuel cell.

Hydrogen and methanol should be covered as fuels for fuel cells. The use of proton exchange membranes will not be assessed.	Reactivity 3.2 – What are the main differences between a fuel cell and a primary (voltaic) cell?

The hydrogen fuel cell

As we have seen, hydrogen is a possible alternative fuel to fossil fuels as it does not produce carbon dioxide. One mole of hydrogen can release 286 kJ of heat energy when it combines directly with oxygen:

$$H_2(g) + \tfrac{1}{2}O_2(g) \rightarrow H_2O(l) \qquad \Delta H^\ominus = -286\,\text{kJ mol}^{-1}$$

As this is a redox reaction that involves the transfer of electrons from hydrogen to oxygen, it can be used to produce an electric current directly if the reactants are physically separated. This is the basis of a **fuel cell**, where the reactants are continuously supplied to different electrodes. The hydrogen–oxygen fuel cell operates with either an acidic or alkaline electrolyte.

The hydrogen–oxygen fuel cell with an alkaline electrolyte

The most commonly used hydrogen–oxygen fuel cell has an alkaline electrolyte (Figure 5).

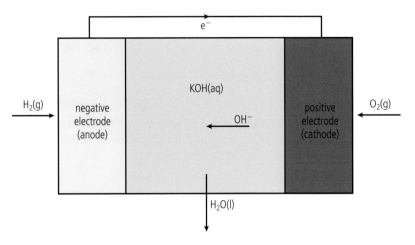

The balanced equation can be deduced from changes in oxidation states as discussed in Reactivity 3.2.

Worked example

Deduce the two half-reactions that occur at the anode and the cathode in the hydrogen–oxygen fuel cell with aqueous sodium hydroxide as the electrolyte.

Solution

$H_2(g)$ is oxidized at the anode. Start with a simplified equation and the oxidation state of the hydrogen in the reactant and product showing the oxidation states in blue as in Structure 3.2 and Reactivity 3.2.

$$H_2(g) + \_ \rightarrow H_2O(l) + \_$$
$$2(0) \qquad 2(+1) \qquad\qquad\qquad \text{oxidation state}$$

2 electrons are produced as both Hs are oxidized: $0 \rightarrow +1$

$$H_2(g) + \rightarrow H_2O(l) + 2e^-$$

As the conditions are alkaline, add $2OH^-$ on the left to balance the negative charges.

Balance the Hs and Os by adding another H_2O on the right-hand side.

$$H_2(g) + 2OH^-(aq) \rightarrow 2H_2O(l) + 2e^-$$

$O_2(g)$ is reduced at the cathode. Start with a simplified equation and the oxidation state of the oxygen in the reactant and product.

$$O_2(g) + \_ \rightarrow 2H_2O(l) + \_$$
$$2(0) \qquad 2(-2) \qquad\qquad\qquad \text{oxidation state}$$

4 electrons are needed as both Os are reduced: $0 \rightarrow -2$

$$O_2(g) + 4e^- \rightarrow 2H_2O(l) + \_$$

As the conditions are alkaline, add $4OH^-$ on the right to balance the negative charges.

Balance the Hs and Os by adding another $4H_2O$ on the left-hand side and simplify.

$$O_2(g) + 4e^- + 4H_2O \rightarrow 2H_2O(l) + 4OH^-$$

$$O_2(g) + 4e^- + 2H_2O \rightarrow 4OH^-(aq)$$

Preparing a simple fuel cell: electrolysis of water. Full details of how to carry out this experiment with a worksheet are available in the eBook.

SKILLS

The overall reaction is the sum of the oxidation and reduction half-reactions:

$$2H_2(g) + \cancel{4OH^-(aq)} + 2H_2O(l) + O_2(g) + \cancel{4e^-} \rightarrow 4H_2O(l) + \cancel{4e^-} + \cancel{4OH^-(aq)}$$

$$2H_2(g) + O_2(g) \rightarrow 2H_2O(l)$$

The fuel cell will function as long as hydrogen and oxygen are supplied. The electrodes are often made of porous carbon with added transition metals such as nickel. The potassium hydroxide provides the hydroxide ions that are transferred across the cell.

The equations for the hydrogen fuel cell under acid conditions are discussed in Reactivity 3.2.

One of the problems with the hydrogen fuel cell is that hydrogen gas is almost never found as the element in nature and needs to be extracted from other sources. Hydrocarbons, including fossil fuels and biomass (waste organic matter), for example, can be processed to break down into hydrogen and carbon dioxide. The alternative is to electrolyze water, but this of course needs electricity and then only becomes a way of storing energy at times of low demand.

For the whole process to be environmentally clean, the hydrogen should be generated using renewable resources such as wind power.

Challenge yourself

7. Explain the increased efficiency of a hydrogen–oxygen fuel cell which produces steam compared with a cell that produces water.

Methanol is oxidized using catalysts in the methanol fuel cell

Methanol is an alternative fuel used in fuel cells. Methanol is a stable liquid at normal environmental conditions, has a high energy density, and is easy to transport. In the direct methanol fuel cell (DMFC), the fuel is oxidized under acidic conditions on a catalyst surface to form carbon dioxide. The H^+ ions formed are transported across a proton exchange membrane from the anode to the cathode where they react with oxygen to produce water, and electrons are transported through an external circuit from the anode to the cathode. Water is consumed at the anode and is produced at the cathode.

Worked example

Deduce the two half-reactions that occur at the anode and the cathode in the methanol–oxygen fuel cell with an acidic electrolyte.

Solution

$CH_3OH(g)$ is oxidized to carbon dioxide at the anode. Start with a simplified equation and the oxidation state of the carbon in the reactant and product.

$$CH_3OH(g) + \_ \rightarrow CO_2(g) + \_$$
$$\quad (-2) \qquad\qquad (+4) \qquad\qquad\qquad \text{oxidation state}$$

6 electrons are produced as C is oxidized: $-2 \rightarrow +4$

$$CH_3OH(g) + \_ \rightarrow CO_2(g) + 6e^-$$

As the conditions are acidic, add $6H^+$ on the right to balance the charges.

Balance the Hs and Os by adding another H_2O on the left-hand side.

$$CH_3OH(g) + H_2O(l) \rightarrow CO_2(g) + 6e^- + 6H^+$$

As in the hydrogen fuel cell, $O_2(g)$ is reduced at the cathode.

$$O_2(g) + 4e^- \rightarrow 2H_2O(l)$$

As the conditions are acidic, add $4H^+$ on the left-hand side to balance the charges.

$$O_2(g) + 4e^- + 4H^+ \rightarrow 2H_2O(l)$$

The overall equation for the methanol fuel cell can be obtained by balancing the number of electrons in the two half-reactions:

$$CH_3OH(g) + H_2O(l) \rightarrow CO_2(g) + 6e^- + 6H^+(aq)$$

$$\tfrac{3}{2}O_2(g) + 6e^- + 6H^+(aq) \rightarrow 3H_2O(l)$$

$$CH_3OH(g) + \tfrac{3}{2}O_2(g) \rightarrow CO_2(g) + 2H_2O(l)$$

Microbial fuel cells (MFC) may one day offer a cheap and renewable source of energy. They can be used to harness the energy produced from the microbial oxidation of a huge range of organic substances. Bacteria are very small organisms which oxidize many organic compounds into CO_2 and H_2O with the production of energy. Some of this energy is needed by the organisms to grow and maintain their metabolism, but the rest can be harvested and used to generate electricity. Microbial fuel cells offer many potential advantages over hydrogen fuel cells. For example, hydrogen fuel cells require a very pure source of a highly explosive gas that is difficult to store and distribute. Furthermore, hydrogen is generally derived mainly from fossil fuels rather than renewable sources. In contrast, the energy sources for microbial fuel cells are renewable organics, including some that are, literally, 'dirt cheap'. The details of the microbial fuel cell will not be assessed in this course.

Researcher holding a two-chambered microbial fuel cell filled with a solution of organic acids. Microbial fuel cells may one day offer a cheap and renewable source of energy.

The main difference between a fuel cell and a primary (voltaic) cell is that fuel cells do not run out. The fuel is supplied continuously to the cell as it is oxidized.
The chemicals are contained within a primary cell and so it stops working when the redox reaction which generates the current is complete.

The advantages and disadvantages of different cells are summarized in Reactivity 3.2.

Reactivity 3.2 – What are the main differences between a fuel cell and a primary (voltaic) cell?

Exercise

Q19. One type of fuel cell uses hydrogen as the fuel under acid conditions with a proton exchange membrane. The overall equation for the process is:

$$2H_2(g) + O_2(g) \rightarrow 2H_2O(l)$$

(a) State the two half-equations for the reactions involving each reactant.

(b) Hydrogen is a more expensive source of chemical energy than gasoline. Explain why fuel cells are considered to be more economical than gasoline engines.

Guiding Question revisited

What are the challenges of using chemical energy to address our energy needs?

In this chapter we have explored how potential energy released during combustion as heat can address our energy needs. We have also explored how it can generate electrical energy directly:

- Combustion is a reaction in which an element or compound burns in oxygen.

- Fossil fuels originate from living organisms that died millions of years ago. They include coal, crude oil or petroleum, and natural gas.

- Coal is a black rock that is mainly carbon.

- Natural gas is mainly methane, and is the cleanest of the fossil fuels.

- Crude oil (petroleum) is a black liquid mixture of mainly hydrocarbons.

- The different hydrocarbons in crude oil can be separated by fractional distillation. Longer chain molecules have stronger intermolecular dispersion forces and higher boiling points.

- Petrol (gasoline) is separated from crude oil by fractional distillation and is a mixture of hydrocarbons with between five and twelve carbon atoms.

- Complete combustion of hydrocarbons breaks the carbon chain and produces carbon dioxide and water. Carbon monoxide and carbon in the form of soot are formed during incomplete combustion.

- Carbon dioxide is a greenhouse gas as it absorbs infrared radiation. Its large-scale production due to the combustion of fossil fuels has led to global warming.

- Carbon monoxide is a poisonous gas that binds to haemoglobin. This limits the ability of the blood to transport oxygen around the body.

- Most of our energy comes from the Sun. Photosynthesis is an endothermic reaction that stores solar energy as chemical energy.

- Plants can be grown specifically for use as fuels. Burning biomass can also involve burning paper or other waste organic matter.

- Fuel cells generate electricity directly by reacting a fuel such as hydrogen or methanol with oxygen in an electrochemical cell.

- Fuels are evaluated according to several criteria including energy density, specific energy or the environmental and economic impact of their use.

Practice questions

1. It is now widely accepted that the increased production of carbon dioxide is leading to global warming.

    (a) Describe how carbon dioxide acts as a greenhouse gas. (2)

    (b) Outline why oxygen and nitrogen, which are more abundant in the air, are not greenhouse gases. (1)

    (c) Identify two pollutants, other than carbon dioxide, that are produced in the combustion engines of cars and outline their polluting effect. (2)

    (Total 5 marks)

2. (a) Most of the world's energy comes from fossil fuels. State two reasons for this. (2)

    (b) The 'specific energy' is the energy produced per unit mass. Identify a fuel which has a higher specific energy than fossil fuels. (1)

    (Total 3 marks)

3. (a) Determine the percentage of carbon by mass in the fuels, coal (CH), petrol (C_8H_{18}) and natural gas (CH_4). (3)

    (b) Suggest two reasons why coal has been replaced by petrol and natural gas in many countries. (2)

    (c) One fuel that has been proposed as a replacement for carbon-containing substances is hydrogen. State two advantages of using hydrogen as a fuel. (2)

    (Total 7 marks)

4. The enthalpies of combustion of propane and butane are given in Section 14 of the data booklet.

    (a) Calculate the heat produced by the combustion of 1 g of C_3H_8 and C_4H_{10} in kJ g^{-1}. (2)

    (b) Use the ideal gas equation to calculate the density of the two gases at STP and hence calculate the heat produced when 1 cm³ of the two gases are burnt. (3)

    (Total 5 marks)

5. (a) State the name of the process by which green plants capture the Sun's energy and convert carbon dioxide and water into glucose. (1)

    (b) State the overall equation for the full redox reaction. (1)

    (c) Glucose can be converted into ethanol, which can be used in fuel cells. Identify this process by name and write the equation for the chemical reaction. (2)

    (Total 4 marks)

6. (a) Use Section 14 of the data booklet to calculate the specific energy (the energy produced when 1 g of fuel is burnt in excess oxygen) of methane and methanol. (2)

    (b) Deduce the average oxidation number of carbon in the two compounds. (1)

    (c) Suggest a possible relationship between the oxidation number and specific energy. (1)

    (d) Calculate the specific energy of glucose and discuss the validity of your hypothesis in part (c). (3)

    (Total 7 marks)

7. Fuel cells may be twice as efficient as the internal combustion engine. Although fuel cells are not yet in widespread use, NASA has used a basic hydrogen–oxygen fuel cell as the energy source for space vehicles.

    Deduce the half-equations occurring at each electrode in the hydrogen–oxygen fuel cell in an alkaline medium. (2)

    (Total 2 marks)

8. (a) One method of comparing fuels is by considering their specific energies.

    Calculate the specific energy of octane, C_8H_{18}, in $kJ\,kg^{-1}$ using Sections 7 and 14 of the data booklet. (2)

    (b) A typical wood has a specific energy of $17 \times 10^3\,kJ\,kg^{-1}$. Comment on the usefulness of octane and wood for powering a moving vehicle, using your answer to part (a).

    If you did not work out an answer for part (a), use $45 \times 10^3\,kJ\,kg^{-1}$, but this is not the correct answer. (1)

    (Total 3 marks)

9. (a) Determine the specific energy (heat produced when 1 g is burnt) and energy density (heat produced from $1\,cm^3$) of petrol (gasoline), using data from Sections 7 and 14 of the data booklet. Assume petrol is pure octane, C_8H_{18}. Octane: molar mass = $114.26\,g\,mol^{-1}$; density = $0.703\,g\,cm^{-3}$. (2)

    (b) Outline why the energy available from an engine will be less than these theoretical values. (1)

    (Total 3 marks)

10. (a) Identify the major constituent of natural gas and write a balanced chemical equation for its combustion. (2)

(b) Compare natural gas with other fossil fuels as a source of air pollution for a given amount of energy and account for any differences. (3)

(c) Compare the relative supplies of natural gas and other fossil fuels. (1)

(Total 6 marks)

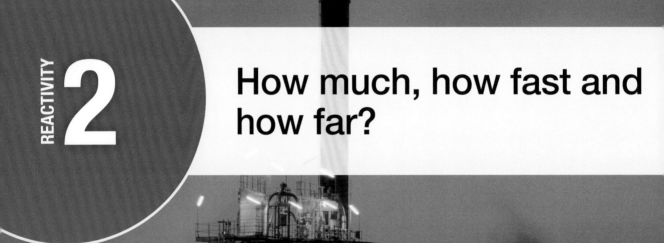

How much, how fast and how far?

A Taiwanese petrochemical plant. This plant is used for the production of ethene (C_2H_4) via a process called cracking. Larger organic molecules found in natural gas and the naptha component of crude oil are split into smaller molecules for use in a number of chemical processes such as polymerisation to make plastics. Ethene is one of the most widely manufactured organic compounds with about 200 million tonnes produced anually.

Between 1800 and the mid 2020s, the global population grew from 1 billion to over 8 billion. To support this growth, large-scale production of chemicals has been required to meet the demand for essential feedstock in industries such as agriculture, construction, oil and gas, and pharmaceuticals. For example, each year, 250 million tonnes of sulfuric acid is produced world-wide for use in the production of fertilizers, steel processing, petroleum refining, and as a solvent in many industrial processes. In order to effectively support this demand, chemists must understand the nature of chemical change so that they are able to quantify the amounts of chemicals involved and maximize the rate of and yield in their production.

Reactivity 2.1 explores the reacting ratios in chemical reactions, a study known as stoichiometry, and how these ratios can be used to determine the quantities of reactants and products. In Reactivity 2.2, we consider a range of techniques for measuring the rate of a chemical process and the theory behind the factors that affect it, including the role of catalysts, The application of this knowledge greatly enhances the efficiency with which we can carry out chemical reactions.

In Reactivity 2.3, we investigate the reversible nature of chemical processes and determine how we are able to increase the yield of desired products by manipulating reaction conditions. Again, the use of mathematical approaches will play a key role in our ability to quantify the composition of reaction mixtures and analyse chemical change.

Central to this chapter are the principles of Green Chemistry. Maximizing atom economy, rate and yield whilst minimizing waste and energy consumption are key tenets of this framework, and allow us to make more sustainable decisions in relation to the design of chemical processes.

SKILLS

Much of the content in Reactivity 2 uses the mathematical tools given in the Skills in the study of chemistry chapter. There are also links to related lab work using these tools in each chapter.

2.1

How much? The amount o chemical change

The Eni oil refinery located in Venice, Italy. In 2014, the global energy company Eni were the first to convert a traditional oil refinery into a biorefinery that produces hydrogenated vegetable oils (HVO) from raw materials such as cooking oils, animal fats and waste vegetable oil. HVO can be added to diesel, also produced at the refinery, to create biodiesel in accordance with European regulations that require an increasing amount of renewable sources in the production of fuels.

Nature of Science

The negative impacts of large-scale chemical industries such as fuel production place a responsibility on scientists to continually strive for greater efficiency and sustainability in the processes and products we use. This approach is described by the principles of Green Chemistry that identify ways to reduce or eliminate the use and generation of hazardous substances.

Guiding Question

How are chemical equations used to calculate reacting ratios?

We can trace the birth of chemistry as a true science back to the first successful attempts to quantify (or measure) chemical change. At the heart of this endeavour is the relationship between the amount of reactants and products in a chemical equation. This is known as **stoichiometry**.

The term stoichiometry is derived from two Greek words – *stoicheion* for element and *metron* for measure. Stoichiometry describes the **reacting ratios** during chemical reactions. As we know that matter is conserved during chemical change, stoichiometry is a form of book-keeping at the atomic level. It enables chemists to determine what amounts of substances they should react together and enables them to predict how much product they will obtain. The application of stoichiometry closes the gap between what is happening on the atomic scale and what can be measured.

In many ways, you can consider this chapter as a toolkit for the mathematical content in much of the course. It covers the universal language of chemistry, **chemical equations**, and their application to predict the masses, volumes and concentrations of reactants and products involved, as well as tools for quantifying the efficiency of these processes.

Reactivity 2.1.1 – Chemical equations

Reactivity 2.1.1 – Chemical equations show the ratio of reactants and products in a reaction.

Deduce chemical equations when reactants and products are specified.

Include the use of state symbols in chemical equations.	Reactivity 3.2 – When is it useful to use half-equations?

Antoine-Laurent Lavoisier, French chemist (1743–1794).

A chemical equation shows:

direction of change
reactants ⟶ products

Chemical equations summarize chemical change

Antoine-Laurent Lavoisier (1743–1794) is often called the 'father of chemistry'. Among his many accomplishments, he established an understanding of combustion as a process involving combination with oxygen from the air, and recognized that matter retains its mass through chemical change, leading to the law of conservation of mass. He also compiled the first extensive list of elements in his book *Elements of Chemistry* (1789). In short, he changed chemistry from a qualitative to a quantitative science. But, as an unpopular tax collector in France during the French Revolution and Terror, he was tried for treason and guillotined in 1794. One and a half years after his death he was exonerated, and his early demise was recognized as a major loss to France.

The formation of compounds from elements is an example of **chemical change** and can be represented by a chemical equation. A chemical equation is a representation, using chemical symbols, of the simplest ratio of atoms, as elements or in compounds, undergoing chemical change. The left-hand side of the equation shows the **reactants** and the right-hand side shows the **products**.

For example:

$$\text{copper} + \text{chlorine} \rightarrow \text{copper(II) chloride}$$
$$\text{Cu} + \text{Cl}_2 \rightarrow \text{CuCl}_2$$

As atoms are neither created nor destroyed during a chemical reaction, *the total number of atoms of each element must be the same on both sides of the equation.* This is known as **balancing the equation**, and uses numbers called **stoichiometric coefficients** to denote the number of units of each term in the equation. A correctly balanced equation indicates the ratio of reactants and products in the reaction.

For example:

$$\text{hydrogen} + \text{oxygen} \rightarrow \text{water}$$
$$2\text{H}_2 + \text{O}_2 \rightarrow 2\text{H}_2\text{O}$$
$$\underset{\text{coefficients}}{\uparrow \qquad\qquad\qquad \uparrow}$$

We now have an identical number of each atom on both sides of the chemical equation.

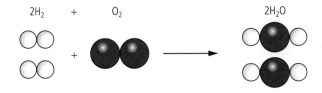

$$2\text{H}_2 \quad + \quad \text{O}_2 \qquad\qquad\qquad 2\text{H}_2\text{O}$$

When hydrogen and oxygen react to form water, the atoms are rearranged, but the number of atoms of each element remains the same.

	total on left side	total on right side
hydrogen	4	4
oxygen	2	2

Note that when the coefficient is 1, this does not need to be explicitly stated.

Chemical equations are used to show all types of reactions in chemistry, including reactions of decomposition, combustion, neutralization, and so on. Examples of these are given below and you will come across very many more during this course. Learning to write equations is an important skill in chemistry, which develops quickly with practice.

Worked example

Write an equation for the decomposition reaction of sodium hydrogencarbonate ($NaHCO_3$) into sodium carbonate (Na_2CO_3), water (H_2O), and carbon dioxide (CO_2) on heating.

Solution

First write the information from the question in the form of an equation, and then check the number of atoms of each element on both sides of the equation.

$$NaHCO_3 \rightarrow Na_2CO_3 + H_2O + CO_2$$

	total on left side	total on right side
sodium	1	2
hydrogen	1	2
carbon	1	2
oxygen	3	6

In order to balance this, we introduce a coefficient, 2, on the left.

$$2NaHCO_3 \rightarrow Na_2CO_3 + H_2O + CO_2$$

Finally check that the equation is now balanced for all elements.

	total on left side	total on right side
sodium	2	2
hydrogen	2	2
carbon	2	2
oxygen	6	6

Note that **state symbols** are commonly included in chemical equations as an additional layer of detail to indicate the physical state of a substance: solid (s), liquid (l), gas (g) or an aqueous solution (aq).

An equation, by definition, has to be *balanced*, so do not expect this to be specified in a question. After you have written an equation, *always* check the numbers of atoms of each element on both sides to ensure the equation is correctly balanced.

When a question refers to 'heating' a reactant or to 'thermal decomposition', this does not mean the addition of oxygen, only that heat is the source of energy for the reaction. If the question refers to 'burning' or 'combustion', this indicates that oxygen is a reactant.

Reactivity 3.2 – When is it useful to use half-equations?

We can also present some chemical equations as two half-equations. In this case, each half-equation identifies elements that have lost and gained electrons during a chemical reaction. These processes are called oxidation and reduction respectively. Species that have not been involved in electron transfer are ignored. This is discussed in more detail in Reactivity 3.2.

The four state symbols are as follows:

Physical state	State symbol
solid	(s)
liquid	(l)
gas	(g)
aqueous	(aq)

Nature of Science

Early ideas to explain chemical change in combustion and rusting included the 'phlogiston' theory. This proposed the existence of a fire-like element that was released during these processes. The theory seemed to explain some of the observations of its time, although these were purely qualitative. It could not explain later quantitative data showing that substances actually gain rather than lose mass during burning. In 1783, Lavoisier's work on oxygen confirmed that combustion and rusting involve combination with oxygen from the air, so overturning the phlogiston theory. This is a good example of how the evolution of scientific ideas, such as how chemical change occurs, is based on the need for theories that can be tested by experiment. Where results are not compatible with the theory, a new theory must be put forward, which must then be subject to the same rigour of experimental test.

When balancing an equation, change the stoichiometric coefficient but never change the subscript in a chemical formula.

Exercise

Q1. Write balanced chemical equations, including state symbols, for the following reactions:

  (a) The decomposition of copper(II) carbonate powder ($CuCO_3$) into copper(II) oxide powder (CuO) and carbon dioxide gas (CO_2).

  (b) The combustion of a magnesium strip (Mg) in oxygen gas (O_2) to form magnesium oxide powder (MgO).

  (c) The neutralization of sulfuric acid solution (H_2SO_4) with sodium hydroxide solution ($NaOH$) to form aqueous sodium sulfate (Na_2SO_4) and water (H_2O).

  (d) The synthesis of ammonia gas (NH_3) from nitrogen gas (N_2) and hydrogen gas (H_2).

  (e) The combustion of methane gas (CH_4) to produce gaseous carbon dioxide (CO_2) and water vapour (H_2O).

Q2. Write balanced chemical equations for the following reactions:

  (a) $K + H_2O \rightarrow KOH + H_2$

  (b) $C_2H_5OH + O_2 \rightarrow CO_2 + H_2O$

  (c) $Cl_2 + KI \rightarrow KCl + I_2$

  (d) $CrO_3 \rightarrow Cr_2O_3 + O_2$

  (e) $Fe_2O_3 + C \rightarrow CO + Fe$

Q3. Use the same processes to balance the following more difficult examples:

(a) $C_4H_{10} + O_2 \rightarrow CO_2 + H_2O$

(b) $NH_3 + O_2 \rightarrow NO + H_2O$

(c) $Cu + HNO_3 \rightarrow Cu(NO_3)_2 + NO + H_2O$

(d) $H_2O_2 + N_2H_4 \rightarrow N_2 + H_2O + O_2$

(e) $C_2H_7N + O_2 \rightarrow CO_2 + H_2O + N_2$

When balancing equations, start with the most complex species, and leave terms that involve a single element to last.

Reactivity 2.1.2 – Using mole ratios in equations

Reactivity 2.1.2 – The mole ratio of an equation can be used to determine:
- **the masses and/or volumes of reactants and products**
- **the concentrations of reactants and products for reactions occurring in solution.**

Calculate reacting masses and/or volumes and concentrations of reactants and products.

Avogadro's law and definitions of molar concentration are covered in Structure 1.4. The values for A_r given in the data booklet to two decimal places should be used in calculations.	Structure 1.5 – How does the molar volume of a gas vary with changes in temperature and pressure? Nature of Science, Structure 1.4 – In what ways does Avogadro's law help us to describe, but not explain, the behaviour of gases?

The mole ratio of an equation can be used to determine the masses of reactants and products

As we have seen in the previous section, a chemical equation is simply an expression of reactants combining in fixed ratios to form products. The most convenient way to express this ratio is as moles, as that gives us a means of relating the number of particles that react to the mass that we can measure. So, for example, when methane, CH_4, burns in air, we can conclude the following, all from the balanced chemical equation:

$$CH_4(g) + 2O_2(g) \rightarrow CO_2(g) + 2H_2O(g)$$

Reacting ratio by mole:	1 mole	2 moles	1 mole	2 moles
Reacting ratio by mass:	16.05 g	64.00 g	44.01 g	36.04 g

80.05 g reactant   80.05 g product

(The figures for total mass of reactant and product are just a check, as we know something would be wrong if they did not equate.)

This simple interpretation of equations, going directly from coefficients to molar ratios, opens the door to a wide range of calculations involving reacting masses.

The term **carbon footprint** refers to the mass of carbon dioxide and other greenhouse gases, such as methane, that an individual emits in a one-year period. It is often expressed as the carbon dioxide equivalent, or CO_2e, to represent the total climate change impact of all the greenhouse gases caused by an item or activity. It includes emissions from fuels used in transport, services such as heating, production and consumption of food, and the direct and indirect emissions from manufactured goods and construction. It is extremely difficult to measure all sources accurately, but the fundamental concept uses the type of calculation shown here. The carbon footprint is a measure of an individual's consumption of resources, and suggests the link between this and the enhanced greenhouse effect.

Remember to convert between mass and moles using

$$n = \frac{m}{M} \quad \text{or} \quad m = nM$$

It is common for students to waste time solving questions like this by doing the mole to gram conversions for all the species represented in the equation. You can save yourself a lot of trouble by focusing only on the terms indicated in the question, as shown in this example.

When using A_r values given in the data booklet to calculate M, make sure to use the two decimal places provided.

In exam questions, it is appropriate to state the final calculated value to a similar number of significant figures as those stated in the data provided.

Worked example

Calculate the mass of carbon dioxide produced from the complete combustion of 1.00 g of methane.

Solution

Write the balanced equation and deduce the mole ratio as above. Then pick out from the question the terms that we need to analyze, here they are marked in red; these are the species where we need to convert moles to grams.

$$CH_4(g) + 2O_2(g) \rightarrow CO_2(g) + 2H_2O(g)$$

Reacting ratio by mole:	1 mole	1 mole
Reacting ratio by mass / g:	16.05	44.01
For 1.00 g methane:	1.00	x

Now solve the ratio, shown here using cross-multiplication, to determine the value of x.

$$\frac{gCH_4}{gCO_2} = \frac{16.05}{44.01} = \frac{1.00}{x}$$

$$x = \frac{1.00 \times 44.01}{16.05} = 2.74\,g\ CO_2$$

All questions on reacting ratios involve a variation of this approach:
- write the balanced equation
- work out the mole ratio for the species identified in the question
- work out the reacting ratio by mass for these species, using $m = nM$
- insert the data from the question and solve the ratio.

Worked example

Iodine chloride, ICl, can be made by the following reaction:

$$2I_2 + KIO_3 + 6HCl \rightarrow 5ICl + KCl + 3H_2O$$

Calculate the mass of iodine, I_2, needed to prepare 28.60 g of ICl by this reaction.

Solution

The relevant terms from the question are I_2 and ICl, so these are our focus.

$$2I_2 + KIO_3 + 6HCl \rightarrow 5ICl + KCl + 3H_2O$$

reacting ratio by mole:	2 moles	5 moles
reacting ratio by mass:	$2 \times (126.90 \times 2)$	$5 \times (126.90 + 35.45)$
	$= 507.60\,g$	$= 811.75\,g$
For 28.60 g ICl:	x	28.60 g

$$\frac{gI_2}{gICl} = \frac{507.60}{811.75} = \frac{x}{28.60}$$

$$x = \frac{507.60 \times 28.60}{811.75}\,g\ I_2 = 17.88\,g\ I_2$$

The mole ratio of an equation can be used to determine the volumes of gaseous reactants and products

All the examples above use mass as a way to measure the amount, the number of moles. But in the laboratory, we often work with gases, where *volume* is a more convenient measure. As explained in Structure 1.4, gas volume is determined only by the number of particles and by the temperature and pressure. This understanding is known as Avogadro's law, which states that:

Equal volumes of all gases measured under the same conditions of temperature and pressure contain equal numbers of molecules.

Alternatively, it can be stated that equal numbers of molecules of all gases, when measured at the same temperature and pressure, occupy equal volumes.

Using V for volume and n for number of moles:

$$V \propto n$$

This relationship enables us to relate gas volume (of any gas) to the number of moles, and so to reacting ratios in equations.

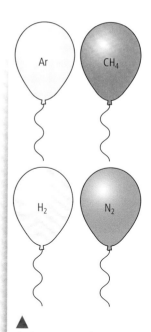

▲ At the same conditions of temperature and pressure, each balloon contains the same number of molecules of gas.

Nature of Science

Avogadro's hypothesis was not widely accepted initially. Experiments led him to suggest that equal volumes of all gases at the same temperature and pressure contain the same number of molecules, but data to confirm this was somewhat lacking. In addition, his ideas conflicted with Dalton's atomic theory, which suggested that particles in gases could be only single atoms, not molecules as Avogadro proposed. It took the logical argument of Cannizzaro, nearly 50 years later, to show that Avogadro's hypothesis could be explained, and moreover used as a means to determine molecular weight. Following this, the relationship between gas volume and number of molecules became widely accepted and known as Avogadro's law. History has shown that the acceptance of scientific ideas by the scientific community is sometimes influenced by the time and manner of their presentation, as well as by their power to explain existing ideas.

Avogadro's law states that equal volumes of all gases at the same conditions of temperature and pressure contain equal numbers of molecules: $V \propto n$

The descriptive nature of Avogadro's law allows us to accurately predict the behaviour of gases in mathematical terms but does not offer an explanation as to why they behave in this manner. To do so, we must consider the underlying model of ideal gases, which identifies characteristics of particles in the gaseous state.

Nature of Science, Structure 1.4 – In what ways does Avogadro's law help us to describe, but not explain, the behaviour of gases?

Worked example

$10 \, cm^3$ of nitrogen is reacted with excess hydrogen in the reaction:

$$N_2(g) + 3H_2(g) \rightarrow 2NH_3(g)$$

What volume of ammonia would you expect to be produced? (Assume all volumes are measured at the same temperature and pressure.)

Solution

'Excess' hydrogen indicates that there is enough to react with *all* of the nitrogen.

First identify the mole ratios in the equation:

$$N_2(g) + 3H_2(g) \rightarrow 2NH_3(g)$$

1 mole   excess   2 moles

As the ratio of moles is equivalent to the ratio of reacting volumes, we can deduce:

$$N_2(g) + 3H_2(g) \rightarrow 2NH_3(g)$$

$10 \, cm^3$ $20 \, cm^3$

Therefore, we expect $20 \, cm^3$ of ammonia to be produced. (Although it is not asked for in this question, we could also deduce that $30 \, cm^3$ of H_2 would be expected to react.)

Note that in reality, this is a reversible reaction so we would be unlikely to collect $20 \, cm^3$ ammonia experimentally. More detail can be found on reversible reactions in Reactivity 2.3.

Worked example

When $10 \, cm^3$ of a gaseous hydrocarbon (a compound containing only carbon and hydrogen) is burnt in excess oxygen, the products consist of $30 \, cm^3$ of carbon dioxide and $30 \, cm^3$ of water vapour, measured under the same conditions of temperature and pressure. Determine the molecular formula of the hydrocarbon.

Solution

'Excess' oxygen indicates that the combustion reaction is complete. Therefore *all* of the hydrocarbon would react to form *only* CO_2 and H_2O.

	C_xH_y	$+$ O_2	$\rightarrow$	CO_2	$+$ H_2O
volumes:	$10 \, cm^3$	excess		$30 \, cm^3$	$30 \, cm^3$
ratio of volumes / moles:	1			3	3

$\therefore$ 1 molecule hydrocarbon $\rightarrow$ 3 molecules CO_2 + 3 molecules H_2O

3 C atoms   6 H atoms

The molecular formula is C_3H_6.

All gases under the same conditions have the same molar volume

On the basis of Avogadro's law, the volume occupied by one mole of any gas, known as the **molar volume** (V_m), must be the same for *all* gases when measured under the same conditions of temperature and pressure.

At standard temperature and pressure (STP), one mole of a gas has a volume of $2.27 \times 10^{-2}\,m^3\,mol^{-1}$ (= $22.7\,dm^3\,mol^{-1}$). The conditions at STP are a temperature of $0\,°C$ (273 K) and a pressure of 100 kPa. The value of the molar volume and the conditions for STP are given in Section 2 of the data booklet.

It is important to note that when temperature and pressure deviate from STP, the molar volume will also change. An increase in temperature will increase the molar volume whereas an increase in pressure will lead to a decrease in molar volume. These effects are described by the combined gas law seen in Structure 1.5.

The molar volume can be used to calculate the amount of gas in a similar way to the use of molar mass earlier in this chapter. Here though, the calculations are easier, as all gases have the same molar volume.

$$\text{number of moles of gas } (n) = \frac{\text{volume } (V)}{\text{molar volume } (V_m)}$$

Structure 1.5 – How does the molar volume of a gas vary with changes in temperature and pressure?

Note that standard temperature and pressure (STP) is not the same as the 'standard state', which is used in thermodynamic data and is explained in Reactivity 1.1.

STP refers to a temperature of 273 K and a pressure of 100 kPa.

The SI unit of pressure is Pa ($N\,m^{-2}$), but the bar ($10^5\,Pa$) is now widely used as it is conveniently close to 1 atmosphere. Different countries continue to use a variety of units for pressure, including millimetres of mercury (mm Hg), torr (Torr), pounds per square inch (psi), and atmosphere (atm). The SI unit of volume is m^3, although litre (l or L) is widely used to represent $1\,dm^3$.

Worked example

Calculate the volume occupied by 0.0200 g He at standard temperature and pressure.

Solution

First convert the mass of He to moles.

$$n = \frac{m}{M} = \frac{0.0200}{4.00} = 0.00500\,mol = 0.00500\,mol = 5.00 \times 10^{-3}\,mol$$

volume = number of moles of gas × molar volume
$$= (5.00 \times 10^{-3}\,mol) \times 22.7\,dm^3\,mol^{-1} = 0.114\,dm^3$$

Worked example

What volume of oxygen at standard temperature and pressure would be needed to completely burn 1.00 mole of butane, C_4H_{10}?

Solution

As always, start with the balanced equation and pick out the terms from the question.

$$2C_4H_{10}(g) + 13O_2(g) \rightarrow 8CO_2(g) + 10H_2O(g)$$

mole / volume ratio: 1   6.5

So 1.00 mol butane will react completely with 6.50 mol O_2.

6.50 moles of gas at STP have volume = $6.50\,mol \times 22.7\,dm^3\,mol^{-1} = 148\,dm^3$

Nature of Science

Committee on Data for Science and Technology (CODATA) is an interdisciplinary scientific committee of the International Council of Science. It was established in 1966 to promote the worldwide compilation and sharing of reliable numerical data, such as the molar volume of a gas. What might be some of the responsibilities held by bodies such as CODATA?

Challenge yourself

1. The combustion of both ammonia, NH_3, and hydrazine, N_2H_4, in oxygen gives nitrogen and water only. When a mixture of ammonia and hydrazine is burnt in pure oxygen, the volumetric N_2:H_2O ratio in the product gas is 0.40. Calculate the % by mass of ammonia in the original mixture. What assumptions are being made here?

The litre (l or L) is widely used in place of dm^3, and millilitre (ml or mL) in place of cm^3. You will not be penalized for the use of these terms in an examination, but they will not be used in examination questions, so it is essential that you know the correct use of m^3, dm^3, and cm^3.

A note about units of volume

The metric unit m^3 is widely used in industrial and engineering calculations, but is too large to be convenient for many volume measurements in the laboratory. Instead, dm^3 and cm^3 are commonly used, so it is important to be able to interconvert these.

$$\text{cm}^3 \underset{\text{multiply by 1000}}{\overset{\text{divide by 1000}}{\rightleftarrows}} \text{dm}^3 \underset{\text{multiply by 1000}}{\overset{\text{divide by 1000}}{\rightleftarrows}} \text{m}^3$$

Nature of Science

Early ideas on gas behaviour were suggested from the postulates of the kinetic theory, but could not advance without scientific evidence. This was provided by experimental work, mainly that of Boyle and Mariotte, Charles, and Gay-Lussac who contributed quantitative data based on testable predictions of how gases would respond to changes in temperature, volume and pressure. In a fairly classic example of scientific process, the data supported the theory, and the theory explained the data. As a result, there was wide acceptance of what became known as 'the gas laws' by the 18th century.

The mole ratio of an equation can be used to determine the concentrations of reactants and products in solutions

A standard solution is a solution of accurately known concentration.

Suppose we have an unlabelled bottle of hydrochloric acid, HCl, and want to know its concentration. We can find out the concentration by reacting the acid with a standard solution of an alkali such as sodium hydroxide, NaOH, and determining the exact volumes that react together. From the stoichiometry of the reaction, when we know the volumes of both solutions and the concentration of one of them, we can use the mole ratio to calculate the unknown concentration as follows.

$$HCl(aq) \quad + \quad NaOH(aq) \; \rightarrow \; NaCl(aq) + H_2O(l)$$

mole ratio:	1	:	1
volume:	known		known (by titration)
conc.:	unknown = x		known (standard solution)

- moles NaOH can be calculated as follows:
  $n(NaOH) = c(NaOH) \times V(NaOH)$

- use the mole ratio in the equation:
  $n(NaOH) = n(HCl)$

- $\therefore$ concentration of HCl, x, can be calculated from:
  $n(HCl) = x \times V(HCl)$

This is an example of a process called **volumetric analysis**. Most commonly, a technique called **titration** is used to determine the reacting volumes precisely.

A **pipette** is used to measure a known volume of one of the solutions into a **conical flask**. The other solution is put into a **burette**, a calibrated glass tube that can deliver precise volumes into the conical flask by opening the tap at the bottom. The point at which the two solutions have reacted completely, the **equivalence point**, is usually determined by an indicator that is added to the solution in the conical flask. The indicator changes colour at its **end point**. Specific indicators are chosen for different titrations to change colour at the equivalence point. This is explained more fully in Reactivity 3.1.

Titration usually involves multiple trials to obtain a more accurate result of the volume required to reach the equivalence point; this volume is known as the **titre**. A good titration result gives readings which are consistent with each other relative to the precision of the burette used. Most commonly, this means within 0.05 cm³ of each other. Remember, the uncertainty in the measurement of analogue instruments is half the smallest division, usually ± 0.05 cm³ in a burette. Titration is widely used in acid–base chemistry and also in redox chemistry (discussed in Reactivity 3.2). Its applications include industrial chemical processing, chemical research, quality control checks in the food and pharmaceutical industries, and aspects of environmental monitoring.

When recording volume measurements from a burette, it is important to remember that the total volume of solution added, the titre, is calculated using an initial and final measurement. Therefore, the absolute uncertainty is 0.05 cm³ + 0.05 cm³ = ± 0.10 cm³.

▲ Acid–base titration. The burette tap controls the flow of a standard solution of sodium hydroxide into the conical flask which contains hydrochloric acid solution of unknown concentration and a few drops of phenolphthalein indicator. The equivalence point occurs when the exact volumes of the two solutions have reacted completely. This is indicated by a color change from colourless to pink.

Worked example

25.00 cm³ of 0.100 mol dm⁻³ sodium hydrogencarbonate, NaHCO₃, solution were titrated with dilute sulfuric acid, H₂SO₄.

$$2NaHCO_3(aq) + H_2SO_4(aq) \; \rightarrow \; Na_2SO_4(aq) + 2H_2O(l) + 2CO_2(g)$$

15.20 cm³ of the acid were needed to neutralize the solution. Calculate the concentration of the acid.

Solution

We can calculate the amount of $NaHCO_3$ as we are given both the volume and the concentration.

$$n = cV$$

$$n(NaHCO_3) = 0.100 \, \text{mol dm}^{-3} \times \frac{25.00}{1000} \, \text{dm}^3 = 2.50 \times 10^{-3} \, \text{mol}$$

Look at the mole ratio in the equation:

$$2n(NaHCO_3) = n(H_2SO_4)$$

$$\therefore n(H_2SO_4) = 0.5 \times (2.50 \times 10^{-3} \, \text{mol}) = 1.25 \times 10^{-3} \, \text{mol}$$

$$\therefore c = \frac{n}{V} = \frac{1.25 \times 10^{-3} \, \text{mol}}{15.20/1000 \, \text{dm}^3} = 0.0822 \, \text{mol dm}^{-3}$$

$$\therefore [H_2SO_4] = 0.0822 \, \text{mol dm}^{-3}$$

Here is a summary of the steps in volumetric analysis calculations:

- If it is not already given, write a balanced equation for the reaction.
- Look for the reactant whose volume and concentration are given and calculate its number of moles from $n = cV$.
- Use this answer and the mole ratio in the equation to determine the number of moles of the other reactant.
- Use the number of moles and volume of the second reactant to calculate its concentration from $c = \frac{n}{V}$.

Back titration

As the name implies, a **back titration** is done in reverse by returning to the end point after it is passed. Back titration is used when the end point is hard to identify or when one of the reactants is impure. A known excess of one of the reactants is added to the reaction mixture, and the unreacted excess is then determined by titration against a standard solution. By subtracting the amount of unreacted reactant from the original amount used, the reacting amount can be determined.

Worked example

An antacid tablet with a mass of 0.300 g and containing $NaHCO_3$ was added to 25.00 cm³ of 0.125 mol dm⁻³ hydrochloric acid. After the reaction was complete, the excess hydrochloric acid required 3.50 cm³ of 0.200 mol dm⁻³ NaOH to reach the equivalence point in a titration. Calculate the percentage of $NaHCO_3$ in the tablet.

Solution

original reaction: $NaHCO_3(s) + HCl(aq) \rightarrow NaCl(aq) + H_2O(l) + CO_2(g)$

mole ratio: 1   :   1

First calculate the total amount of HCl added, which is a known excess.

$$n(HCl \, total) = \frac{25.00}{1000} \, \text{dm}^3 \times 0.125 \, \text{mol dm}^{-3} = 0.00313 \, \text{mol HCl total}$$

titration reaction: $HCl(aq) + NaOH(aq) \rightarrow NaCl(aq) + H_2O(l)$

mole ratio: 1   :   1

$$n(NaOH) = 0.00350\,dm^3 \times 0.200\,mol\,dm^{-3} = 0.000700\,mol$$

$n(NaOH) = n(HCl\ unreacted) = 0.000700\,mol\ HCl\ unreacted$

$\therefore n(HCl\ reacted) = 0.00313 - 0.000700 = 0.00243\,mol$

$\therefore$ from the mole ratio in the first equation $n(NaHCO_3) = n(HCl) = 0.00243\,mol$

$M(NaHCO_3) = 22.99 + 1.01 + 12.01 + (16.00 \times 3) = 84.01\,g\,mol^{-1}$

$m = nM = 0.00243\,mol \times 84.01\,g\,mol^{-1} = 0.204\,g$

percentage by mass in tablet $= \dfrac{0.204}{0.300} \times 100 = 68.0\%$

Note that there are several assumptions made in this calculation. These include the fact that all the $NaHCO_3$ did react with the acid, and that the only component of the tablet that reacted with HCl was $NaHCO_3$. You may like to think about how you could test the validity of these assumptions in the laboratory.

Exercise

Q4. Iron ore can be reduced to iron by the following reaction:

$$Fe_2O_3(s) + 3H_2(g) \rightarrow 2Fe + 3H_2O(l)$$

  (a) How many moles of Fe can be made from 1.25 moles of Fe_2O_3?

  (b) How many moles of H_2 are needed to make 3.75 moles of Fe?

  (c) If the reaction yields 12.50 moles of H_2O, what mass of Fe_2O_3 was used up?

Q5. Lighters commonly use butane, C_4H_{10}, as the fuel.

  (a) Formulate the equation for the combustion of butane.

  (b) Determine the mass of butane that burnt when 2.46 g of water were produced.

Q6. The fuel used in booster rockets for the Space Shuttle is a mixture of aluminum and ammonium perchlorate. The reaction equation is as follows:

$$3Al(s) + 3NH_4ClO_4(s) \rightarrow Al_2O_3(s) + AlCl_3(s) + 3NO(g) + 6H_2O(g)$$

Calculate the mass of NH_4ClO_4 that should be added to this fuel mixture to react completely with every kilogram of Al.

Q7. Limestone is mostly calcium carbonate, $CaCO_3$, but it also contains other minerals. When heated, the $CaCO_3$ decomposes into CaO and CO_2. A sample of 1.605 g of limestone was heated and gave off 0.657 g of CO_2.

  (a) Formulate the equation for the thermal decomposition of calcium carbonate.

  (b) Determine the percentage mass of $CaCO_3$ in the limestone.

  (c) State the assumptions you are making in this calculation.

Q8. How many moles are present in each of the following at STP?

(a) $54.5 \, dm^3 \, CH_4$

(b) $250.0 \, cm^3 \, CO$

(c) $1.0 \, m^3 \, O_2$

Q9. What is the volume of each of the following gases at STP?

(a) $44.00 \, g \, N_2$

(b) $0.25 \, mol \, NH_3$

Q10. Pure oxygen gas was first prepared by heating mercury(II) oxide, HgO.

$$2HgO(s) \rightarrow 2Hg(l) + O_2(g)$$

What volume of oxygen, at STP, is released by heating $12.45 \, g$ of HgO?

Q11. At STP, which sample contains more molecules, $3.14 \, dm^3$ of bromine, Br_2; or $11.07 \, g$ of chlorine, Cl_2?

Q12. Calcium reacts with water to produce hydrogen.

$$Ca(s) + 2H_2O(l) \rightarrow Ca(OH)_2(aq) + H_2(g)$$

Calculate the volume of hydrogen gas produced, at STP, when $0.200 \, g$ of calcium reacts completely with water.

Q13. Dinitrogen oxide, N_2O, is a greenhouse gas produced from the decomposition of artificial nitrate fertilizers. Calculate the volume of N_2O produced from $1.0 \, g$ of ammonium nitrate when it reacts, at STP, according to the equation:

$$NH_4NO_3(s) \rightarrow N_2O(g) + 2H_2O(l)$$

Q14. In a titration, a $15.00 \, cm^3$ sample of H_2SO_4 required $36.42 \, cm^3$ of $0.147 \, mol \, dm^{-3}$ NaOH solution for complete neutralization. What was the concentration of the H_2SO_4?

Q15. Gastric juice contains hydrochloric acid, HCl. A $5.00 \, cm^3$ sample of gastric juice required $11.00 \, cm^3$ of $0.0100 \, mol \, dm^{-3}$ KOH for neutralization in a titration. What was the concentration of HCl in this fluid? If we assume a density of $1.00 \, g \, cm^{-3}$ for the fluid, what was the percentage by mass of HCl?

Q16. Sodium sulfate, Na_2SO_4, reacts in aqueous solution with lead nitrate, $Pb(NO_3)_2$, as follows:

$$Na_2SO_4(aq) + Pb(NO_3)_2(aq) \rightarrow PbSO_4(s) + 2NaNO_3(aq)$$

In an experiment, $35.30 \, cm^3$ of a solution of sodium sulfate reacted exactly with $32.50 \, cm^3$ of a solution of lead nitrate. The precipitated lead sulfate was dried and found to have a mass of $1.13 \, g$. Determine the concentrations of the original solutions of lead nitrate and sodium sulfate. State what assumptions are made.

Reactivity 2.1.3 – The limiting reactant and theoretical yield

Reactivity 2.1.3 – The limiting reactant determines the theoretical yield.	
Identify the limiting and excess reactants from given data.	
Distinguish between the theoretical yield and the experimental yield.	Tool 1, Inquiry 1, 2, 3 – What errors may cause the experimental yield to be (i) higher and (ii) lower than the theoretical yield?

The theoretical yield is determined by the limiting reactant

Imagine that you are following a recipe to make 12 cookies. It calls for you to mix two eggs with four cups of flour. The problem is that you only have one egg. You will quickly realize that this means you can use only two cups of flour and end up with only six cookies. We could say that the number of eggs *limited* the amount of product.

In many chemical reactions, the relative amounts of reactants available to react together will similarly affect the amount of product. The reactant that determines the quantity of product is known as the **limiting reactant**. Other reactants will not be fully used, and are said to be in **excess**. Identifying the limiting reactant is therefore a crucial step before we can calculate the expected quantity of product. The **theoretical yield**, which is usually expressed in grams or moles, refers to the maximum amount of product obtainable, assuming 100% of the limiting reactant is converted to product.

The concept of limiting reactant is often useful in the design of experiments and synthetic processes. By deliberately making one reactant available in a greater amount than that determined by its mole ratio in the balanced equation, it ensures that the other reactant is limiting and so will be fully used up. For example, in order to remove lead ions in lead (II) nitrate, $Pb(NO_3)_2$, from a contaminated water supply, sodium carbonate, Na_2CO_3, is added to precipitate lead carbonate, as shown in the equation:

$$Pb(NO_3)_2(aq) + Na_2CO_3(aq) \rightarrow PbCO_3(s) + 2NaNO_3(aq)$$
$$\text{limiting} \qquad \text{excess}$$

By using excess Na_2CO_3, this ensures that *all* the lead ions react and so are removed from the water supply.

Note that identification of the limiting reactant depends on the *mole ratios* in the balanced chemical equation for the reaction. This means that if reactant quantities are given in grams, they must first be converted to moles.

A medical worker taking a blood sample. Blood samples can be used to test for lead exposure. During the Flint water crisis in the US state of Michigan in 2014, after Flint switched its water supply to save money during a financial state of emergency, tens of thousands of people were exposed to elevated levels of lead ions in their drinking water. In addition to other methods, salts such as Na_2CO_3 can be added in excess to contaminated water to precipitate and then remove heavy metal ions such as Pb^{2+} and Pb^{4+}.

The limiting reactant determines the amount of product that can form. The theoretical yield is the quantity of product that can form from the complete conversion of the limiting reactant.

Worked example

Nitrogen gas (N_2) can be prepared according to this reaction:

$$2NH_3(g) + 3CuO(s) \rightarrow N_2(g) + 3Cu(s) + 3H_2O(g)$$

If 18.1 g NH_3 are reacted with 90.4 g CuO, determine the mass of N_2 that can be formed.

Solution

First we must determine the limiting reactant. We convert the mass of reactants to moles, and then compare the mole ratio in the balanced equation with the mole ratio of reactants given.

$$n(NH_3) = \frac{18.1}{14.01 + (3 \times 1.01)} = 1.06 \, \text{mol} \, NH_3$$

$$n(CuO) = \frac{90.4}{63.55 + 16.00} = 1.14 \, \text{mol} \, CuO$$

mole ratio from equation: $\quad \dfrac{NH_3}{CuO} = \dfrac{2}{3} = 0.67$

mole ratio from given masses: $\quad \dfrac{NH_3}{CuO} = \dfrac{1.06}{1.14} = 1.14$

As the ratio NH_3 : CuO of the given masses is *larger* than the required ratio in the equation, it means NH_3 is in excess and CuO is the limiting reactant.

This means that the amount of N_2 that can form will be determined by the amount of CuO. This is now similar to the earlier questions, where we write out the equation and focus on the terms identified in the question.

$$2NH_3(g) + 3CuO(s) \rightarrow N_2(g) + 3Cu(s) + 3H_2O(g)$$

reacting ratio by mole: 3 moles   1 mole

for 1.14 moles CuO: 1.14 x

mole ratio: $\quad \dfrac{CuO}{N_2} = \dfrac{3}{1} = \dfrac{1.14}{x}$

$$x = \frac{1.14 \times 1}{3} = 0.380 \, \text{mol} \, N_2$$

$$\therefore \text{mass } N_2 = 0.380 \times M(N_2) = 0.380 \times 28.02 = 10.7 \, \text{g} \, N_2$$

There are alternative approaches to determining the limiting reactant, such as calculating which given amount of reactant would yield the smallest amount of product. But in essence, all questions on limiting reactant and theoretical yield involve comparing the mole ratio of given quantities of reactants with the coefficients in the equation. This is a summary of the steps:

- Write the balanced equation and focus on the mole ratio of reactants.
- Convert the given mass of reactants to moles.
- Compare the given mole ratios with the ratio of coefficients in the equation.
- Identify the limiting reactant from the above ratios.
- Calculate the moles of product from the given moles of *limiting* reactant.

Tool 1, Inquiry 1, 2, 3 – What errors may cause the experimental yield to be (i) higher and (ii) lower than the theoretical yield?

Sometimes it is useful to measure how much excess reactant will remain when all the limiting reactant has been used up and the reaction stops. One example of this is a technique called back titration, seen in the previous section, which analyses excess acid or alkali after a reaction is complete, and so indirectly measures the amount of a limiting reactant. A simple example of how to calculate the excess is shown below, using an example of burning CH_4 when 1 mole of CH_4 and 1 mole of O_2 are supplied.

$$CH_4(g) + 2O_2(g) \rightarrow CO_2(g) + 2H_2O(g)$$

reacting mole ratio in equation: 1   2

reactant mole ratio given: 1   1

$\Rightarrow O_2$ is the limiting reactant

mole ratio at the end of reaction: 0.5   0   0.5   1

As the 1 mole of O_2 has been identified at the limiting reactant, only 0.5 moles of CH_4 will react, leaving 0.5 moles of CH_4 unreacted at the end of the reaction.

Exercise

Q17. 0.40 moles of magnesium are mixed with 0.20 moles of hydrochloric acid.

$$Mg(s) + 2HCl(aq) \rightarrow MgCl_2(aq) + H_2(g)$$

Which is correct?

	Limiting reactant	Theoretical yield of H_2 (mol)
A	HCl	0.20
B	Mg	0.40
C	HCl	0.10
D	Mg	0.20

Q18. 0.30 moles of hydrochloric acid are mixed with 0.30 moles of calcium carbonate.

$$2HCl(aq) + CaCO_3(s) \rightarrow CaCl_2(aq) + H_2O(g) + CO_2(g)$$

Which is correct?

	Limiting reactant	Theoretical yield of CO_2 (mol)
A	HCl	0.30
B	HCl	0.15
C	$CaCO_3$	0.15
D	$CaCO_3$	0.30

Q19. 5 cm³ of nitrogen are reacted with 20 cm³ of hydrogen in the reaction:

$$N_2(g) + 3H_2(g) \rightarrow 2NH_3(g)$$

Assuming that all volumes are measured at the same temperature and pressure:

(a) What volume of ammonia would you expect to be produced?

(b) What volume of H_2 would remain unreacted?

Q20. Methanol, CH_3OH, is a useful fuel that can be made as follows:

$$CO(g) + 2H_2(g) \rightarrow CH_3OH(l)$$

A reaction mixture used 12.0 g of H_2 and 74.5 g of CO.

(a) Determine the theoretical yield of CH_3OH.

(b) Calculate the amount of the excess reactant that remains unchanged at the end of the reaction.

Q21. The dry-cleaning solvent 1,2-dichloroethane, $C_2H_4Cl_2$, is prepared by the following reaction:

$$C_2H_4(g) + Cl_2(g) \rightarrow C_2H_4Cl_2(l)$$

Determine the mass of product that can be formed from 15.4 g of C_2H_4 and 3.74 g of Cl_2.

Reactivity 2.1.4 – Percentage yield

Reactivity 2.1.4 – The percentage yield is calculated from the ratio of experimental yield to theoretical yield.

Solve problems involving reacting quantities, limiting and excess reactants, theoretical, experimental and percentage yields.

The percentage yield can be calculated from the experimental and theoretical yields

So far in this chapter, worked examples have focused on chemical reactions occurring with no loss, impurities present, wastage, or incomplete reaction. In reality, all of these happen to different extents in most chemical reactions, and so the theoretical yield is usually different from the actual or **experimental yield**. Below are some sources of error that may cause the experimental yield to be lower or higher than the theoretical yield.

A separating funnel being used to extract iodine from an aqueous solution containing other impurities (the upper layer). As iodine is more soluble in dichloromethane (the lower layer), I_2 molecules will migrate into it upon gentle shaking of the separating funnel, causing a pink solution. However, some I_2 may remain in the aqueous layer so the experimental yield will be less than the theoretical value.

Factors that may cause the experimental yield to be *lower* than the theoretical yield	Factors that may cause the experimental yield to be *higher* than the theoretical yield
Side reactions occurring.	Impurities in a product.
The decomposition of reactants and/or products.	When a product has not been fully dried.
Loss of product during purification.	
Reversible chemical reactions preventing process completion. (See Reactivity 2.3 for more detail.)	
An incomplete reaction (this could impact the experimental yield in both directions depending on the type of reaction used).	

When we compare the experimental yield with the theoretical yield, we get a measure of the efficiency of the conversion of reactants to products. This is usually expressed as the **percentage yield**, defined as follows:

$$\text{percentage yield} = \frac{\text{experimental yield}}{\text{theoretical yield}} \times 100$$

In your own experiments, you may often be required to calculate the percentage yield of product in evaluating the results. In industry, this is a very important calculation used to determine the efficiency of a process, such as the synthesis of a drug in the pharmaceutical industry. Many aspects of Green Chemistry focus on ways to increase the yield of product by reducing wastage.

Worked example

The synthesis of N_2 from NH_3 and CuO has a theoretical yield of 10.65 g N_2 from the starting amounts of reactants. Under the same conditions, an experiment produced 8.35 g N_2. Determine the percentage yield of the experiment.

Solution

$$\text{percentage yield} = \frac{\text{experimental yield}}{\text{theoretical yield}} \times 100$$

$$\therefore \text{percentage yield} = \frac{8.35}{10.65} \times 100 = 78.4 \%$$

Industrial plants such as this oil refinery in Bangkok, Thailand, need to be able to track the efficiency of the chemical reactions taking place. Measuring the yield of product is an essential part of this.

Exercise

Q22. Calcium carbonate, $CaCO_3$, is able to remove sulfur dioxide, SO_2, from waste gases by a reaction in which the two substances react in a 1:1 stoichiometric ratio to form equimolar amounts of $CaSO_3$ and CO_2. When 255 g of $CaCO_3$ reacted with 135 g of SO_2, 198 g of $CaSO_3$ were formed. Determine the percentage yield of $CaSO_3$.

Q23. Pentyl ethanoate, $CH_3COOC_5H_{11}$, which smells like bananas, is produced from the esterification reaction:

$$CH_3COOH(aq) + C_5H_{11}OH(aq) \rightarrow CH_3COOC_5H_{11}(aq) + H_2O(l)$$

A reaction uses 3.58 g of CH_3COOH and 4.75 g of $C_5H_{11}OH$ and has a yield of 45%. Determine the mass of ester that forms.

Q24. A student has to make a 100 g sample of chlorobenzene, C_6H_5Cl, from the following reaction:

$$C_6H_6 + Cl_2 \rightarrow C_6H_5Cl + HCl$$

Determine the minimum quantity of benzene, C_6H_6, that can be used to achieve this with a yield of 65%.

SKILLS

Synthesis and purification of Aspirin. Full details of how to carry out this experiment with a worksheet are available in the eBook.

Reactivity 2.1.5 – Atom economy

Reactivity 2.1.5 – The atom economy is a measure of efficiency in Green Chemistry.

Calculate the atom economy from the stoichiometry of a reaction.

Include discussion of the inverse relationship between atom economy and wastage in industrial processes. The equation for calculation of the atom economy is given in the data booklet.	Structure 2.4, Reactivity 2.2 – The atom economy and the percentage yield both give important information about the 'efficiency' of a chemical process. What other factors should be considered in this assessment?

The atom economy is a measure of efficiency in Green Chemistry

Green Chemistry is the sustainable design of chemical products and chemical processes. It aims to minimize the use and generation of chemical substances that are hazardous to human health and the environment. Traditionally, the efficiency of a reaction has been measured by calculating the percentage yield (the proportion of the desired product obtained compared to the theoretical maximum), but this is only a limited measure as it gives no indication of the quantity of waste produced. A synthetic route should maximize the **atom economy** by incorporating as many of the atoms of the reactants as possible into the desired product.

The atom economy of a reaction can be calculated from the equation provided in Section 1 of the data booklet:

$$\% \text{ atom economy} = \frac{\text{molar mass of desired product}}{\text{molar mass of all reactants}} \times 100$$

Efficient processes have high atom economies and are important for sustainable development, as they use fewer natural resources and create less waste. Addition polymerization reactions, for example, have an atom economy of 100% as all the reacting atoms end up in the polymer and there are no side products. Many real-world processes, for example, in drug synthesis, use a deliberate excess of reactants to increase the yield but have low atom economies.

The second principle of Green Chemistry states that synthesis design should aim to maximize atom economy in a chemical process

Worked example

Nickel can be produced by heating nickel oxide and carbon in the following reaction:

$$NiO(s) + C(s) \rightarrow Ni(s) + CO(g)$$

Calculate the atom economy for the production of nickel. Give your answer to 3 significant figures.

Solution

Molar mass of desired product = 58.69

Molar mass of all reactants = (58.69 + 16.00) + 12.01 = 86.70

$$\% \text{ atom economy} = \frac{58.69}{86.70} \times 100 = 67.7\%$$

Challenge yourself

2. Percentage yield and atom economy are different concepts, but both can be used to assess aspects of the overall efficiency of a chemical process. See if you can find a reaction that has a high percentage yield under certain conditions, but a low atom economy.

In addition to atom economy and percentage yield, there are a number of other factors that can be used to gauge the efficiency of a chemical process. We might, for example, consider the rate at which a chemical reaction takes place (explored in Reactivity 2.2) or the quantities of reagents, such as solvents and catalysts, which are not consumed in the reaction. In industry, it is also important to evaluate energy usage and the economic efficiency of a process.

i

Structure 2.4, Reactivity 2.2 – The atom economy and the percentage yield both give important information about the 'efficiency' of a chemical process. What other factors should be considered in this assessment?

Exercise

Q25. Copper(II) sulfate, $CuSO_4$, can be produced by reacting copper(II) carbonate and sulfuric acid according to the following equation:

$$CuCO_3 + H_2SO_4 \rightarrow CuSO_4 + H_2O + CO_2$$

Calculate the atom economy for the reaction. Give your answer to 3 significant figures.

Q26. Chloroethene, C_2H_3Cl, is the monomer used in the manufacture of polyvinyl chloride. It can be produced from 1,2-dichloroethane, $C_2H_4Cl_2$, by the following reaction.

$$C_2H_4Cl_2 \rightarrow C_2H_3Cl + HCl$$

Calculate the atom economy for the reaction. Give your answer to 3 significant figures.

Q27. Identify which of these methods of producing hydrochloric acid, HCl, has the highest atom economy.

Method 1: $2NaCl(s) + H_2SO_4(l) \rightarrow 2HCl(g) + Na_2SO_4(s)$

Method 2: $H_2(g) + Cl_2(g) \rightarrow 2HCl(g)$

Challenge yourself

3. The fertilizer triammonium phosphate is made from 'phosphate rock' by:

   1. reacting the phosphate rock with sulfuric acid, H_2SO_4, to produce phosphoric acid, H_3PO_4

   2. reacting the phosphoric acid with ammonia, NH_3, to give tri-ammonium phosphate, $(NH_3)_3PO_4$

   If the phosphate rock contains 90% by mass $Ca_3(PO_4)_2$ from which the overall yield of triammonium phosphate is 95%, calculate the mass of phosphate rock required to make 1000 tonnes of triammonium phosphate.

4. Sulfuric acid, H_2SO_4, is produced from sulfur in a three-step process:

1. $$S(s) + O_2(g) \rightarrow SO_2(g)$$

2. $$2SO_2(g) + O_2(g) \rightarrow 2SO_3(g)$$

3. $$SO_3(g) + H_2O(l) \rightarrow H_2SO_4(l)$$

Assuming 100% conversion and yield for each step, what is the minimum mass of sulfur needed to produce 980 tonnes of H_2SO_4?

5. The concentration of hydrogen peroxide, H_2O_2, in excess aqueous sulfuric acid, H_2SO_4, can be determined by redox titration using potassium permanganate, $KMnO_4$ as follows:

$$2KMnO_4(aq) + 5H_2O_2(l) + 3H_2SO_4 \rightarrow 2MnSO_4(aq) + K_2SO_4(aq) + 8H_2O(l) + 5O_2(g)$$

A $10.00\,cm^3$ sample of H_2O_2 solution requires $18.00\,cm^3$ of a $0.0500\,mol\,dm^{-3}$ solution of $KMnO_4$ to reach the equivalence point in a titration. Calculate the concentration of H_2O_2 in the solution.

6. Mixtures of sodium carbonate, Na_2CO_3, and sodium hydrogencarbonate, $NaHCO_3$, in aqueous solution are determined by titration with hydrochloric acid, HCl, in a two-step procedure.

1. Titrate to the phenolphthalein end point:

$$Na_2CO_3(s) + HCl(aq) \rightarrow NaHCO_3(aq) + NaCl(aq)$$

2. Continue titration to the methyl orange end point:

$$NaHCO_3(aq) + HCl(aq) \rightarrow NaCl(aq) + H_2O(l) + CO_2(g)$$

For an $X\,cm^3$ sample of a sodium carbonate/sodium hydrogencarbonate mixture titrated with $Y\,mol\,dm^{-3}$ HCl, the respective end points are Step 1 = $P\,cm^3$ HCl and Step 2 = $Q\,cm^3$ HCl. Derive relationships between X, Y, P, and Q to obtain the concentrations of sodium carbonate and sodium hydrogencarbonate in the original mixture.

7. A sealed vessel with fixed total internal volume of $2\,m^3$ contains $0.740\,kg$ pentane, C_5H_{12}, and oxygen only. The pentane is ignited and undergoes 100% conversion to carbon dioxide and water. Subsequently the temperature and pressure in the vessel are respectively $740\,K$ and $400\,kPa$. Calculate the initial amount and mass of oxygen in the vessel.

Guiding Question revisited

How are chemical equations used to calculate reacting ratios?

In this chapter we have used chemical equations to:

- Deduce the mole ratio of reactants and products from their coefficients in the balanced equation.

- Determine masses, volumes and concentrations of reactants and products involved in chemical reactions.
  - When dealing with masses, mole ratios can be converted to a reacting ratio by mass using $m = nM$.
  - For gases, Avogadro's law can be applied to mole ratios to determine a reacting ratio by volume. At STP, we can also convert between moles and volume using $n = \dfrac{V}{V_m}$ where V_m is the molar volume of a gas (provided in the data booklet).
  - The volumetric analysis of solutions, often carried out using a titration, requires the use of $n = cV$ to calculate the number of moles of one reactant before using the mole ratio to determine the number of moles of the other reactant and then its concentration.

- Identify the reactant that will be used up first in a reaction, the limiting reactant, by comparing the mole ratio from the balanced equation to the mole ratio in the quantities being reacted. The reactant not fully used is said to be in excess. As the limiting reactant determines the amount of product being formed, we can use it to calculate the expected amount of product, known as the theoretical yield.

- Quantify the efficiency of a chemical process using percentage yield and atom economy.
  - Percentage yield is calculated from the ratio of experimental yield to theoretical yield using the formula:

$$\text{percentage yield} = \frac{\text{experimental yield}}{\text{theoretical yield}} \times 100$$

  - Atom economy takes into account the amount of waste produced in a reaction by comparing the molar mass of the target product to the molar mass of the reactants used. This formula can be found in the data booklet:

$$\% \text{ atom economy} = \frac{\text{molar mass of desired product}}{\text{molar mass or all reactants}} \times 100$$

Practice questions

1. How many oxygen atoms are in the formula of hydrated copper sulfate, $CuSO_4.5H_2O$?

   A 2   B 4   C 5   D 9

2. What is the sum of the coefficients when the following equation is balanced using whole numbers?

$$\_\_\_ Fe_2O_3(s) + \_\_\_ CO(g) \rightarrow \_\_\_ Fe(s) + \_\_\_ CO_2(g)$$

   A 5   B 6   C 8   D 9

3. In which mixture is NaOH the limiting reactant?

   A $0.10\,\text{mol NaOH} + 0.05\,\text{mol H}_2\text{SO}_4$

   B $0.05\,\text{mol NaOH} + 0.05\,\text{mol H}_2\text{SO}_4$

   C $0.20\,\text{mol NaOH} + 0.05\,\text{mol HNO}_3$

   D $0.20\,\text{mol NaOH} + 0.10\,\text{mol HNO}_3$

4. Which volume, in cm^3, of $0.40\,\text{mol dm}^{-3}$ NaOH(aq) is required to neutralize $0.050\,\text{mol}$ of $H_2S(g)$?

$$H_2S(g) + 2NaOH(aq) \rightarrow Na_2S(aq) + 2H_2O(l)$$

   A 0.25   B 0.50   C 250   D 500

5. What is the maximum volume, in dm^3, of $CO_2(g)$ produced when 1.00 g of $CaCO_3(s)$ reacts with $20.0\,cm^3$ of $2.00\,mol\,dm^{-3}$ $HCl(aq)$?

$$CaCO_3(s) + 2HCl(aq) \rightarrow CaCl_2(aq) + H_2O(l) + CO_2(g)$$

Molar volume of gas = $22.7\,dm^3\,mol^{-1}$; $M(CaCO_3) = 100$

A $\dfrac{1}{2} \times \dfrac{20.0 \times 2.0}{1000} \times 22.7$ B $\dfrac{20.0 \times 2.0}{1000} \times 22.7$

C $\dfrac{1.0}{100} \times 22.7$ D $\dfrac{1.0}{100} \times 2 \times 22.7$

6. Which of the following is correct when $40\,cm^3$ hydrogen is reacted with $10\,cm^3$ oxygen?

$$2H_2(g) + O_2(g) \rightarrow 2H_2O(g)$$

	Limiting reactant	Volume of $H_2O(g)$ produced (cm^3)
A	H_2	40
B	H_2	80
C	O_2	10
D	O_2	20

7. What mass, in g, of hydrogen is formed when 3 moles of aluminum react with excess hydrochloric acid according to the following equation?

$$2Al(s) + 6HCl(aq) \rightarrow 2AlCl_3(aq) + 3H_2(g)$$

A   3.0   B   4.5   C   6.0   D   9.0

8. What is the percentage yield when 7 g ethene produces 6 g of ethanol?

$$C_2H_4(g) + H_2O(g) \rightarrow C_2H_5OH(g)$$

M_r(ethene) = 28 and M_r(ethanol) = 46

A $\dfrac{6 \times 7 \times 100}{28 \times 46}$ B $\dfrac{6 \times 46 \times 100}{7 \times 28}$ C $\dfrac{6 \times 28}{7 \times 46 \times 100}$ D $\dfrac{6 \times 28 \times 100}{7 \times 46}$

9. What is the sum of all coefficients when the following equation is balanced using the smallest possible whole numbers?

$$\_\_\_ C_2H_2 + \_\_\_ O_2 \rightarrow \_\_\_ CO_2 + \_\_\_ H_2O$$

A   5   B   7   C   11   D   13

10. Chloroethene, C_2H_3Cl, reacts with oxygen according to the equation below:

$$2C_2H_3Cl(g) + 5O_2(g) \rightarrow 4CO_2(g) + 2H_2O(g) + 2HCl(g)$$

What is the amount, in mol, of H_2O produced when 10.0 mol of C_2H_3Cl and 10.0 mol of O_2 are mixed together, and the above reaction goes to completion?

A   4.00   B   8.00   C   10.0   D   20.0

11. Airbags are an important safety feature in vehicles. Sodium azide, a toxic compound, undergoes the following decomposition reaction under certain conditions to produce nitrogen gas which fills the airbag.

$$2NaN_3(s) \rightarrow 2Na(s) + 3N_2(g)$$

(a) Calculate the mass of sodium azide, NaN_3, required to produce $80.0\,dm^3$ of nitrogen gas at STP.   (3)

(b) Calculate the mass of sodium produced in this reaction.   (2)

(Total 5 marks)

12. Nitrogen monoxide may be removed from industrial emissions via a reaction with ammonia as shown by the equation below:

$$4NH_3(g) + 6NO(g) \rightarrow 5N_2(g) + 6H_2O(l)$$

$30.0\,dm^3$ of ammonia reacts with $30.0\,dm^3$ of nitrogen monoxide at $100\,°C$. Identify which gas is in excess and by how much, and calculate the volume of nitrogen produced. (2)

(Total 2 marks)

13. The percentage by mass of calcium carbonate in eggshell was determined by adding excess hydrochloric acid to ensure that all the calcium carbonate had reacted. The excess acid left was then titrated with aqueous sodium hydroxide.

(a) A student added $27.20\,cm^3$ of $0.200\,mol\,dm^{-3}$ HCl to $0.188\,g$ of eggshell. Calculate the amount, in mol, of HCl added. (1)

(b) The excess acid requires $23.80\,cm^3$ of $0.100\,mol\,dm^{-3}$ NaOH for neutralization. Calculate the amount, in mol, of acid that is in excess. (1)

(c) Determine the amount, in mol, of HCl that reacted with the calcium carbonate in the eggshell. (1)

(d) State the equation for the reaction of HCl with the calcium carbonate in the eggshell. (2)

(e) Determine the amount, in mol, of calcium carbonate in the sample of the eggshell. (2)

(f) Calculate the mass **and** the percentage by mass of calcium carbonate in the eggshell sample. (3)

(g) Deduce **one** assumption made in arriving at the percentage of calcium carbonate in the eggshell sample. (1)

(Total 11 marks)

14. Melamine has the following structure:

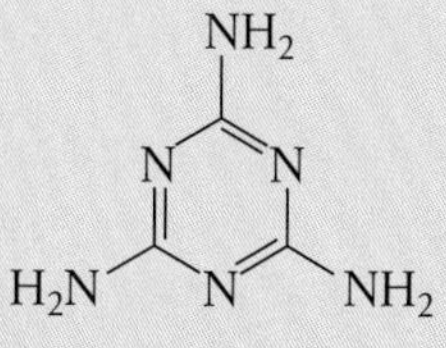

(a) It can be produced from urea, $(NH_2)_2CO$, in the following reaction:

$$6(NH_2)_2CO \rightarrow C_3H_6N_6 + 6NH_3 + 3CO_2$$

Calculate the atom economy for the reaction. Give your answer to 3 significant figures. (2)

(b) Calculate the theoretical yield of melanine in grams produced when $14.0\,g$ of urea is broken down. (2)

(c) Use your answer from part **(b)** to calculate the % yield of the reaction if $4.12\,g$ melanine was produced experimentally. (1)

If you did not get an answer to part **(b)**, *use a theoretical yield of $5.00\,g$ melanine.*

(Total 5 marks)

How fast? The rate of chemical change

◄ The launch of the first Bangladeshi geostationary communications and broadcast satellite, *Bangabandhu-1*, on a Falcon 9 rocket from Cape Canaveral, Florida, in 2018. To produce enough thrust to launch a rocket, an extremely high rate of reaction between a fuel (in this case kerosene) and an oxidizing agent (liquid oxygen) is required to produce a rapid flow of exhaust gases out of the combustion chamber. The ability to control and influence the rate of a chemical reaction is an essential feature of all industrial chemical processes.

Guiding Question

How can the rate of a reaction be controlled?

The **rate**, or kinetics, of a chemical reaction refers to the speed at which reactants are converted into products during a chemical process. We will all have some familiarity with this concept in everyday life. Imagine, for example, you are cooking in the kitchen. As you drop an egg into hot oil in the pan, it immediately changes to a white solid; meanwhile, a container of milk that was left out of the refrigerator is slowly turning sour. We observe a similarly wide variation in the rate of reactions that we study in the laboratory, and these data can be very useful.

Studies of rates of reaction are of prime importance in industry because they give information on how quickly products form and on the conditions that give the most efficient and economic yield. They can also be useful in situations where we want to slow reactions down – for example, the reactions that cause the destruction of ozone in the stratosphere, or the reactions where pollutants in the air combine to produce smog. At other times, it is important to know for how long a certain reaction will continue – for example, the radioactive effect from radioactive waste.

This chapter begins with a study of reaction rates and a consideration of how these are measured in different cases. Through the explanation of **collision theory**, we will come to understand why different factors affect the rate of reactions. These factors will include pressure, concentration, surface area, temperature and presence of a catalyst..

Reactivity 2.2.1 – Rate of reaction

Reactivity 2.2.1 – The rate of reaction is expressed as the change in concentration of a particular reactant or product per unit time.

Determine rates of reaction.

Calculation of reaction rates from tangents of graphs of concentration, volume or mass against time should be covered.	Tool 1, 3, Inquiry 2 – Concentration changes in reactions are not usually measured directly. What methods are used to provide data to determine the rate of reactions?
	Tool 1 – What experiments measuring reaction rates might use time as (i) a dependent variable or (ii) an independent variable?

Rate of reaction is defined as the rate of change in concentration

When we are interested in how quickly something happens, the factor that we usually measure is **time**. For example, in a sports race, the competitors are judged by the time it takes them to reach the finishing line. However, if we want to compare their performance in different races over different distances, we would need to express this as a rate – in other words, how they performed ***per unit time***.

Rate takes the reciprocal value of time, so is expressed *per unit time* or, in SI units, *per second* (symbol = s^{-1}).

$$\text{rate} = \frac{1}{\text{time}} = \frac{1}{s} = s^{-1}$$

Note that because time and rate are reciprocal values, as one decreases, the other increases. So in the example above, the racer with the *shortest time* wins the race, because they had the *fastest rate.*

In the study of chemical reactions, we use the concept of **rate of reaction** to describe how quickly a reaction happens. As the reaction proceeds, reactants are converted into products, and so the concentration of reactants decreases as the concentration of products increases. The graphs in Figures 1 and 2 show sketches of typical data from reactions.

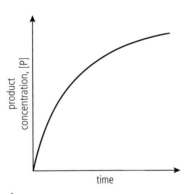

▲ **R2.2 Figure 1** Concentration of product against time.

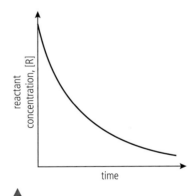

▲ **R2.2 Figure 2** Concentration of reactant against time.

The rate of a reaction depends on how quickly the concentration of either reactant or product changes with respect to time. It can be defined as follows:

$$\text{rate of reaction} = \frac{\text{increase in product concentration}}{\text{time taken}} \quad \text{or} \quad \frac{\text{decrease in reactant concentration}}{\text{time taken}}$$

Using Δ to represent 'change in', [R] for concentration of reactant, and [P] for concentration of product, we can express this as:

$$\text{rate of reaction} = \frac{\Delta[P]}{\Delta t} \text{ or } -\frac{\Delta[R]}{\Delta t}$$

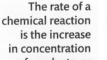

The rate of a chemical reaction is the increase in concentration of products, or the decrease in concentration of reactants, per unit time.

It is worth noting that, by convention, rate should be expressed as a positive value. For this reason, a negative sign has been placed in front of the reactant expression which would otherwise produce a negative value due to [R] decreasing over time.

As rate = change in concentration per time, its units are **$mol\,dm^{-3}\,s^{-1}$**.

Figure 3 presents graphs of two different reactions showing the change in concentration of reactant against time.

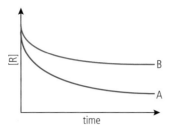

▲
R2.2 Figure 3 Reactant concentration against time for two different reactions, A and B.

We can see that the concentration of reactants is decreasing more quickly in reaction A than in reaction B – the curve is steeper. The steepness, or gradient, of the curve is a measure of the change in concentration per unit time, in other words, the rate of the reaction.

Because the graphs are curves and not straight lines, the gradient is not constant and so can only be given for a particular value of time. We can measure this by drawing a tangent to the curve at the specified time point and measuring its gradient. This is shown in Figure 4 for the time at 120 s. Note that the gradient of this graph is negative as the reactant concentration is decreasing, but rate is expressed as a positive value.

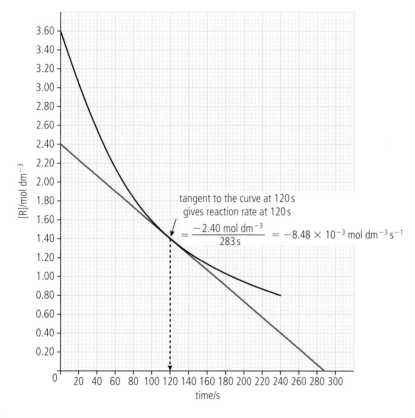

▲
R2.2 Figure 4 Measuring the gradient of the tangent to the curve at time t = 120 s. The measured rate of the reaction is 8.48×10^{-3} mol dm^{-3} s^{-1} at this time. Note that, by convention, rate is always expressed as a positive value.

The shape of the curve indicates that the rate of the reaction is not constant during the reaction, but is greatest at the start and slows down as the reaction proceeds. This is illustrated in Figure 5, which shows the calculation of reaction rate at two different times during a reaction.

$$t = 0\,\text{s; rate} = 2.9 \times 10^{-2}\,\text{mol dm}^{-3}\,\text{s}^{-1}$$

$$t = 90\,\text{s; rate} = 7.2 \times 10^{-3}\,\text{mol dm}^{-3}\,\text{s}^{-1}$$

You will get a more accurate value for the slope of the tangent if you draw it as long as you reasonably can, so that the 'y' value (concentration) and the 'x' value (time) are as large as possible.

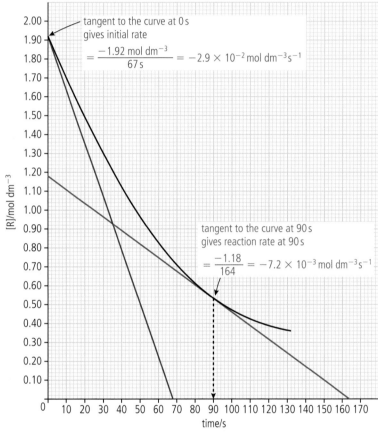

▲ **R2.2 Figure 5** Measuring the gradient of the tangent to the curve at two different times during a reaction. The rate decreases as the reaction proceeds.

The rate is greatest at the start because this is when the reactant concentration is highest. The effect of concentration on reaction rate is discussed later in this chapter. Because of this variation in rate as a reaction proceeds, in order to compare the rates of reactions under different conditions, it is common to compare the **initial rate** of each reaction by taking the tangent to the curve at $t = 0$. As we will see later in this chapter, initial rate data are very useful in analyzing the effect of concentration on rate.

The rate of a reaction is not constant during a reaction, but is greatest at the start and decreases as the reaction proceeds.

Note the difference between the two instructions 'draw a graph' and 'sketch a graph'. *Drawing* is based on actual data so scales must be chosen appropriately and data points and units clearly marked; *sketching* is a way of showing a trend in the variables, without reference to specific data. Note that in both cases, the axes must be clearly labelled with the dependent variable on the y-axis and the independent variable on the x-axis.

Measuring rates of reaction uses different techniques depending on the reaction

Choosing whether to measure the change in concentration of reactants or products, and the technique with which to measure that change, really depends on what is the most convenient for a particular reaction. This is different for different reactions. In most cases the concentration is not measured directly, but is measured by means of a signal that is related to the changing concentration. If, for example, a reaction forms a coloured precipitate as a product, the change in colour could be measured. If a reaction gives off a gas, then the change in volume or change in mass could be measured. The raw data collected using these 'signals' will be in a variety of units, rather than as concentrations measured in mol dm^{-3} directly. This is generally not a problem, as it still enables us to determine the rate of the reaction. Many of the experiments used here can be carried out using data-logging devices. Some examples of common techniques in relation to specific reactions are discussed below.

Tool 1, 3, Inquiry 2 – Concentration changes in reactions are not usually measured directly. What methods are used to provide data to determine the rate of reactions?

Nature of Science

Although the chemical reactions between individual molecules cannot be observed directly, there are many indirect methods for investigating these processes in order to develop explanations for what is happening. As with all of chemistry, the validity of these theories is dependent on the extent to which they are supported by experimental data and their ability to make accurate predictions.

 TOK

Note that the balanced equation of a reaction gives us no information about its rate. We can obtain this information only from experimental (empirical) data. Is there a fundamental difference between knowledge claims based on conclusions from theoretical data and those based on experimental data?

Change in volume of gas produced

This is a convenient method if one of the products is a gas. Collecting the gas and measuring the change in volume at regular time intervals enables a graph to be plotted of volume against time. A **gas syringe** can be used for this purpose. It consists of a ground glass barrel and plunger. The plunger moves outwards as the gas collects and the barrel is calibrated to record the volume directly. Alternatively, the gas can be collected by displacement of water from an inverted burette. Note though that the displacement method can only be used if the gas collected has low solubility in water. Data loggers are also available for continuous monitoring of volume change against time. The rate of reaction of a metal with dilute acid to release hydrogen gas can be followed in this way.

$$Mg(s) + 2HCl(aq) \rightarrow MgCl_2(aq) + H_2(g)$$

Challenge yourself

1. Most gases are less soluble in warm water than in cold water, so using warm water may help to reduce a source of error when collecting a gas by displacement of water. But what new error might this create?

◄ Experiments to measure rate of reaction by following change in volume against time.

When you take a glass of cold water from the refrigerator and leave it by your bed overnight, you may notice bubbles of gas have formed by morning. This is because the gas (mostly dissolved oxygen) has become less soluble as the temperature of the water has increased.

◄ Volume of gas against time. The reaction has ended when the curve reaches a plateau.

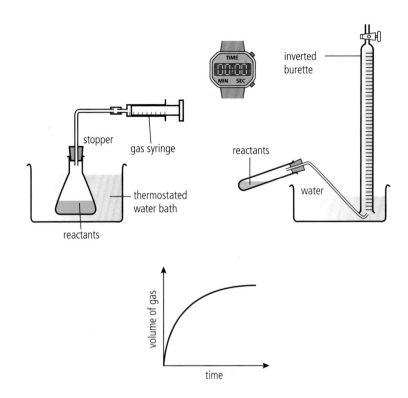

Change in mass

Many reactions involve a change in mass, and it may be convenient to measure this directly. If the reaction is giving off a gas, the corresponding decrease in mass can be measured by standing the reaction mixture directly on a balance. This method is unlikely to work well where the evolved gas is hydrogen, as it is too light to give significant changes in mass. The method allows for continuous readings, so a graph can be plotted directly of mass against time. For example, the release of carbon dioxide from the reaction between a carbonate and dilute acid can be followed in this way.

$$CaCO_3(s) + 2HCl(aq) \rightarrow CaCl_2(aq) + CO_2(g) + H_2O(l)$$

◄ Experiment to measure rate of reaction by following change in mass against time.

reactants
e.g. $CaCO_3 + HCl$

cotton wool to prevent escape of liquid and solid

digital balance

0.895_g

◄ Mass against time. The reaction has ended when the curve reaches a plateau.

Change in transmission of light: colorimetry/spectrophotometry

This technique can be used if one of the reactants or products is coloured and so gives characteristic absorption in the visible region (wavelengths about 320–800 nm). Sometimes an indicator can be added to generate a coloured compound that can then be followed in the reaction. A colorimeter or spectrophotometer works by passing light of a selected wavelength through the solution being studied and measuring the intensity of the light transmitted by the reaction components. As the concentration of the coloured compound increases, it absorbs proportionally more light, so less is transmitted. A photocell generates an electric current according to the amount of light transmitted and this is recorded on a meter or connected to a computer.

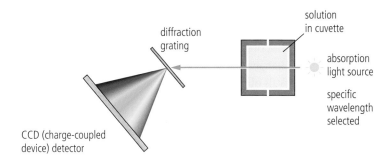

A spectrophotometer measures absorbance, which directly relates to concentration. Light of a specific wavelength is selected from an internal source and passed through the solution placed in the cuvette. The transmitted light passes through a diffraction grating, then the diffracted light is collected and changed into a digital signal by the charge-coupled device (CCD) detector.

The method allows for continuous readings to be made, so a graph of absorbance against time can be plotted directly. It is possible to convert the absorbance values into concentrations using a standard curve based on readings of known concentration. Often, however, it is sufficient to record absorbance itself (or transmittance, which is inversely proportional to absorbance) as a function against time.

For example, the reaction between the dye crystal violet and sodium hydroxide solution can be written in an abbreviated form as follows:

$$CV^+ + OH^- \rightarrow CVOH$$

The initial crystal violet solution (CV^+) is coloured but this slowly changes to a colourless solution as the product (CVOH) is formed. So we can determine the rate of product formation by measuring the *decrease in absorbance* that occurs as the coloured reactant is depleted. The figure shows data from this reaction, with absorbance at 591 nm measured against time.

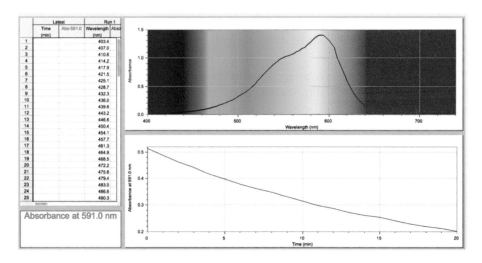

Results of an experiment to determine the rate of reaction by following change in absorbance against time using a digital spectrophotometer.

An example of the use of a calibration curve to determine concentration from absorbance data is outlined in Structure 3.1.

Change in concentration measured using titration

In some reactions it may be possible to measure the concentration of one of the reactants or products by titrating it against a known 'standard' (see Structure 1.4). However, because this technique involves chemically changing the reaction mixture, it cannot be done continuously as the reaction proceeds. Instead, samples must be withdrawn from the reaction mixture at regular time intervals and then analyzed by titration. A problem here is that the process of titration takes time, during which the reaction mixture in the sample will continue to react. To overcome this, a technique known as **quenching** can be used, where a substance is introduced which effectively stops the reaction in the sample at the moment it is withdrawn. It is rather like obtaining a 'freeze frame' shot of the reaction at a particular interval of time. In order to see how concentration changes as the reaction proceeds, it is necessary to repeat this process at several intervals of time.

For example, the reaction between H_2O_2 and acidified KI yields I_2, which can be titrated against sodium thiosulfate, $Na_2S_2O_3$, to determine its concentration. Sodium carbonate, Na_2CO_3, is used to quench the reaction by neutralizing the added acid.

$$H_2O_2(aq) + 2H^+(aq) + 2I^-(aq) \rightarrow I_2(aq) + 2H_2O(l)$$

Change in concentration measured using conductivity

The total electrical conductivity of a solution depends on the total concentration of its ions and on their charges. If this changes when reactants are converted to products, it can provide a convenient method to follow the progress of the reaction. Conductivity can be measured directly using a conductivity meter which involves immersing inert electrodes in the solution. As with colorimetry, the apparatus can be calibrated using solutions of known concentration so that the readings can be converted into the concentrations of the ions present.

For example, in the reaction:

$$BrO_3^-(aq) + 5Br^-(aq) + 6H^+(aq) \rightarrow 3Br_2(aq) + 3H_2O(l)$$

the sharp decrease in the concentration of ions (12 on the reactants side and 0 on the products side) will give a corresponding decrease in the electrical conductivity of the solution as the reaction proceeds.

Non-continuous methods of detecting change during a reaction: 'clock reactions'

Sometimes it is difficult to record the continuous change in the rate of a reaction. In these cases, it may be more convenient to measure the time it takes for a reaction to reach a certain chosen fixed point – that is, something observable which can be used as an arbitrary 'end point' by which to stop the clock. The time taken to reach this point for the same reaction under different conditions can then be compared and used as a means of judging the different rates of the reaction. So in contrast to the first five techniques of measuring rate of reaction where time is the independent variable, detecting a change in a clock reaction uses time as a dependent variable.

Tool 1 – What experiments measuring reaction rates might use time as (i) a dependent variable or (ii) an independent variable?

Note that the limitation of this method is that the data obtained give only an average rate over the time interval.

For example, the following can be measured:

- the time taken for a certain size piece of magnesium ribbon to react completely with dilute acid, until it is no longer visible

$$Mg(s) + 2HCl\,(aq) \rightarrow MgCl_2(aq) + H_2(g)$$

- the time taken for a solution of sodium thiosulfate titrated with dilute acid to become opaque by the precipitation of sulfur, so that a cross viewed through paper is no longer visible

$$Na_2S_2O_3(aq) + 2HCl(aq) \rightarrow 2NaCl(aq) + SO_2(aq) + H_2O(l) + S(s)$$

SKILLS Investigating rates of reaction. Full details of how to carry out this experiment with a worksheet are available in the eBook.

Recording the increase in opaqueness during a reaction. Sodium thiosulfate (left) is a clear liquid, which reacts with hydrochloric acid (upper centre) to form an opaque solution (right). A cross (left) drawn onto paper is placed under the reaction beaker to allow the end point to be confirmed. The experiment is timed from when the reactants are mixed until the cross disappears from sight.

In all these reactions, the goal is to measure change in concentration against time. This allows a comparison of reaction rates under different conditions. But because the rate is dependent on temperature, it is essential to control the temperature throughout these experiments. It is sometimes suitable to carry out the reactions in a thermostatically controlled water bath.

Exercise

Q1. Consider the following reaction.

$$2MnO_4^-(aq) + 5C_2O_4^{2-}(aq) + 16H^+(aq) \rightarrow 2Mn^{2+}(aq) + 10CO_2(g) + 8H_2O(l)$$

Describe three different ways in which you could measure the rate of this reaction.

Q2. Which units are used to express the rate of a reaction?

A $mol\,dm^{-3}$ time   **B** $mol^{-1}\,dm^3\,time^{-1}$

C $mol\,dm^{-3}\,time^{-1}$ **D** $mol\,time^{-1}$

Q3. The reaction between calcium carbonate and hydrochloric acid is carried out in an open flask. Measurements are made to determine the rate of the reaction.

$$CaCO_3(s) + 2HCl(aq) \rightarrow CaCl_2(aq) + H_2O(l) + CO_2(g)$$

(a) Suggest three different types of data that could be collected to measure the rate of this reaction.

(b) Explain how you would expect the rate of the reaction to change with time and why.

Q4. The following data were collected for the reaction
$2H_2O_2(aq) \rightarrow 2H_2O(l) + O_2(g)$ at 390 °C.

$[H_2O_2]$ / mol dm^{-3}	Time / s	$[H_2O_2]$ / mol dm^{-3}	Time / s
0.200	0	0.070	120
0.153	20	0.063	140
0.124	40	0.058	160
0.104	60	0.053	180
0.090	80	0.049	200
0.079	100		

Draw a graph of concentration against time and determine the reaction rate after 60 s and after 120 s.

 SKILLS

Modelling rate of reaction: collisions in the court. Full details of how to carry out this activity with a worksheet are available in the eBook.

TOK

The Kelvin scale of temperature gives a natural measure of the average kinetic energy of molecules in a gas, whereas the Celsius scale is based on the properties of water. Are physical properties such as temperature invented or discovered?

▲ A puddle of water evaporates, even on a cold day, because a small proportion of molecules on the surface have enough kinetic energy to vaporize. This demonstrates that the particles in a substance must have a range of values of kinetic energy as described by kinetic molecular theory.

Reactivity 2.2.2 – Collision theory

Reactivity 2.2.2 – Species react as a result of collisions of sufficient energy and proper orientation.

Explain the relationship between the kinetic energy of the particles and the temperature in kelvin, and the role of collision geometry.

	Structure 1.1 – What is the relationship between the kinetic molecular theory and collision theory?

Species react as a result of collisions

Kinetic energy and temperature

Since the early 18th century, scientists have developed theories to explain the fact that gases exert a pressure. These theories developed alongside a growing understanding of the atomic and molecular nature of matter, and were extended to include the behaviour of particles in all states of matter. Today they are summarized as the **kinetic molecular theory of matter** (see Structure 1.1).

The essence of the kinetic molecular theory is that particles in a substance move randomly as a result of the **kinetic energy** that they possess. Because of the random nature of these movements and collisions, not all particles in a substance at any one time have the same values of kinetic energy, but will instead have a range of values. A convenient way to describe the kinetic energy of a substance is therefore to take the *average* of these values, and this is directly related to its **absolute temperature** – that is, its temperature measured in kelvin.

Increasing temperature therefore means an increase in the average kinetic energy of the particles of a substance. As we supply a substance with extra energy by heating it, we raise the average kinetic energy of the particles and so also raise its temperature. When we compare the behaviour of the particles in the three states of matter from solid, through liquid, to gas, the differences are a result of this increase in the average kinetic energy of the particles.

For a given substance this can be summarized as follows:

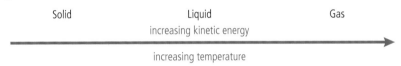

Solid   Liquid   Gas

increasing kinetic energy

increasing temperature

Temperature in kelvin (K) is proportional to the average kinetic energy of the particles in a substance.

The Maxwell–Boltzmann energy distribution curve

The fact that particles in a gas at a specific temperature show a range in their values of kinetic energy is expressed by the **Maxwell–Boltzmann energy distribution curve**.

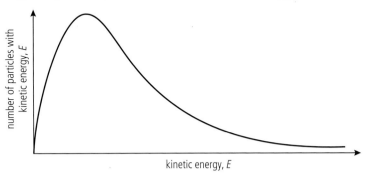

The Maxwell–Boltzmann energy distribution curve.

Although their names are linked in the famous energy distribution curve discussed here, James Clerk Maxwell and Ludwig Boltzmann had very different outlooks on life. Maxwell was a Scottish physicist, known for his curiosity and humour. His wife worked alongside him in many of his experiments. Boltzmann, an Austrian, was prone to depression and took his own life, seemingly believing that his work was not valued. Nonetheless, as peers during the 19th century, they refined and developed each other's ideas, culminating in the distribution curve that bears their two names.

Like other distribution curves, this shows the number of particles having a specific value of kinetic energy (or the probability of that value occurring) plotted against the values for kinetic energy. The area under the curve represents the total number of particles in the sample.

The nature of collisions between particles

When reactants are placed together, the kinetic energy that their particles possess causes them to collide with each other. The energy of these collisions may result in some bonds within the reactants being broken, and some new bonds forming. As a result, products form and the reaction 'happens'. This explanation of chemical reactions is known as collision theory.

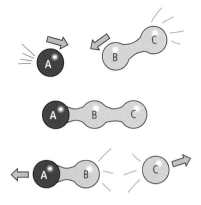

Particles react by colliding with sufficient kinetic energy and in the correct orientation.

It follows that the rate of the reaction will depend on the number of collisions between particles that are 'successful', that is, which lead to the formation of products. But a very important point is that *not all collisions will be successful*. There are two main factors that influence this:

- energy of collision and
- geometry of collision.

Energy of collision

In order for a collision to lead to a reaction, the particles must have a certain minimum value for their kinetic energy, known as the **activation energy**, E_a. This energy is necessary for overcoming repulsion between molecules, and often for breaking some bonds in the reactants before they can react. When this energy is supplied, the reactants achieve the **transition state** from which products can form. The activation energy therefore represents an energy barrier for the reaction, and it has a different value in different reactions.

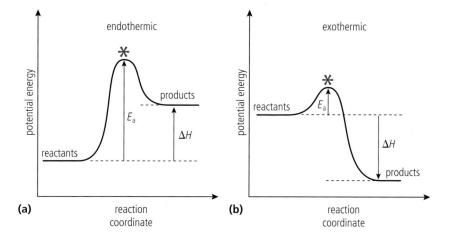

Energy path of a reaction with the activation energy for **(a)** endothermic and **(b)** exothermic reactions. * represents the transition state

Activation energy (E_a) is defined as the minimum value of kinetic energy that particles must have before they are able to react.

We can think of the activation energy as a 'threshold value' – a bit like a pass mark in an examination: values greater than this mark achieve a pass, lower values do not achieve a pass. The activation energy threshold similarly determines which particles react and which do not. So only particles that have a kinetic energy value greater than the activation energy will have successful collisions. Note that particles with lower values of kinetic energy may still collide, but these collisions will not be 'successful' in the sense of causing a reaction.

It therefore follows that the rate of the reaction depends on the proportion of particles that have values for kinetic energy greater than the activation energy.

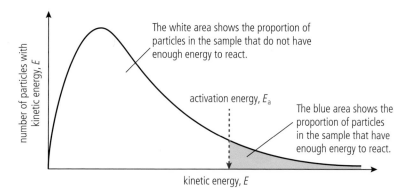

Maxwell–Boltzmann energy distribution curve showing how the activation energy, E_a, distinguishes between particles that have greater or lesser values of the kinetic energy required to react.

The magnitude of the activation energy varies greatly from one reaction to another, and it is an important factor in determining the overall rate of the reaction. In general, reactions with high activation energy proceed more slowly than those with low activation energy as fewer particles will have the required energy for a successful collision.

Geometry of collision

Because collisions between particles are random, they are likely to occur with the particles in many different orientations. In some reactions, this can be crucial in determining whether or not the collisions will be successful, and therefore what proportion of collisions will lead to a reaction.

(a)

reactant molecules approach each other

ineffective collision – no reaction occurs

reactant molecules separate – no product formed

(b)

reactant molecules approach each other

effective collision – particles have correct collision geometry so reaction occurs

product molecules formed

The effect of collision geometry. In **(a)** the particles do not have the correct collision geometry and no reaction occurs. In **(b)** the particles collide with the correct geometry, enabling products to form.

We can now summarize the collision theory as follows. The rate of a reaction depends on the frequency of collisions that occur between particles possessing both:

- values of kinetic energy greater than the activation energy and
- appropriate collision geometry.

Understanding this theory will help us to investigate and explain the factors that increase the rate of reaction.

In order to react, particles must collide with kinetic energy greater than the activation energy and have the correct collision geometry.

Nature of Science

The collision between a small number of reacting species resulting in chemical change has not been observed directly. Instead, collision theory is based on an application of kinetic molecular theory and theory of how chemical reactions occur through bond-breaking and bond-making. In science, theories generally accommodate the assumptions and premises of other theories in this way. The fact that collision theory helps to explain the observed effects of factors influencing the rate of reactions adds to its validity. The theory enables chemists to make predictions about the impact of different factors on the rates of specific reactions, which has important applications in many branches of chemistry such as industry, biochemistry and environmental chemistry.

Structure 1.1 – What is the relationship between the kinetic molecular theory and collision theory?

Exercise

Q5. Which statement is correct for a collision between reactant particles that leads to reaction?

    A Colliding particles must have different energy.

    B Colliding particles must have the same energy.

    C Colliding particles must have kinetic energy greater than the average kinetic energy.

    D Colliding particles must have kinetic energy greater than the activation energy.

Q6. Which of the following determine whether a reaction will occur?

    I.   the orientation of the molecules

    II. the energy of the molecules

    III. the volume of the container

    A I and II   **B** I and III   **C** II only   **D** I, II and III

Q7. If we compare two reactions, one which requires the simultaneous collision of three molecules and the other which requires a collision between two molecules, which reaction would you expect to be faster? Explain why.

Reactivity 2.2.3, 2.2.4 and 2.2.5 – Factors that influence the rate of reaction

Reactivity 2.2.3 – Factors that influence the rate of a reaction include pressure, concentration, surface area, temperature and the presence of a catalyst.

Predict and explain the effects of changing conditions on the rate of a reaction.

	Tool 1 – What variables must be controlled in studying the effect of a factor on the rate of a reaction?
	Nature of Science, Tool 3, Inquiry 3 – How can graphs provide evidence of systematic and random error?

Reactivity 2.2.4 – Activation energy, E_a, is the minimum energy that colliding particles need for a successful collision leading to a reaction.

Construct Maxwell–Boltzmann energy distribution curves to explain the effect of temperature on the probability of successful collisions.

Reactivity 2.2.5 – Catalysts increase the rate of reaction by providing an alternative reaction pathway with lower E_a.

Sketch and explain energy profiles with and without catalysts for endothermic and exothermic reactions.

Construct Maxwell–Boltzmann energy distribution curves to explain the effect of different values for E_a on the probability of successful collisions.

Biological catalysts are called enzymes.	Reactivity 2.3 – What is the relative effect of a catalyst on the rate of the forward and backward reactions?
The different mechanisms of homogeneous and heterogeneous catalysts will not be assessed.	

Be careful not to confuse the question of 'how fast?' a reaction goes with the question of 'how far?' it goes. We are discussing only the first question here. The question of how far a reaction goes, which influences the *yield* of the reaction, will be discussed in Reactivity 2.3.

Temperature, concentration, pressure, surface area and the use of a catalyst can influence the rate of reaction

From the collision theory, we know that any factor which increases the number of successful collisions will increase the rate of the reaction. We will investigate five such factors here.

Temperature

Increasing the **temperature** causes an increase in the average kinetic energy of the particles. We can see this by comparing Maxwell–Boltzmann energy distribution curves of the same sample of particles at two different temperatures.

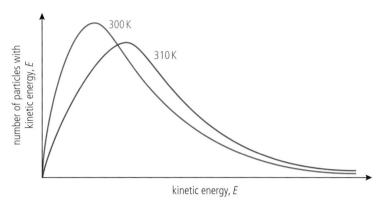

◄ Maxwell–Boltzmann distribution curves for a sample of gas at 300 K and 310 K. At the higher temperature, the curve has shifted to the right and become broader, showing a larger number of particles with higher values for kinetic energy.

The area under the two curves is equal as this represents the total number of particles in the sample. But at the higher temperature, more of the particles have higher values for kinetic energy and the peak of the curve shifts to the right. In Figure 6, we can see how this shift increases the proportion of particles that have values for kinetic energy greater than that of the activation energy.

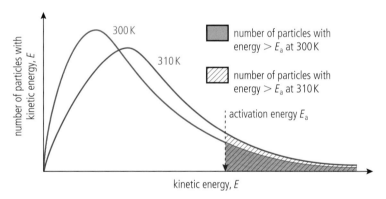

number of particles with energy $> E_a$ at 300 K

number of particles with energy $> E_a$ at 310 K

activation energy E_a

R2.2 Figure 6 Maxwell–Boltzmann energy distribution curves for a sample of gas at two different temperatures, showing the higher proportion of particles with kinetic energy greater than the activation energy at the higher temperature.

So, as temperature increases, there is an increase in collision frequency due to the higher kinetic energy, but more importantly, there is an increase in the number of collisions involving particles with the necessary activation energy to overcome the activation energy barrier. Consequently, there is an increase in the frequency of *successful* collisions and so an increase in the rate of reaction. Many reactions approximately double their reaction rate for every 10 °C increase in temperature.

Concentration

Increasing the **concentration** of reactants increases the rate of reaction. This is because as concentration increases, the frequency of collisions between reactant particles increases, as shown in Figure 7. The frequency of successful collisions therefore increases too.

Specimens of the extinct mammal the mammoth, dated as 10 000 years old, have been found perfectly preserved in the Arctic ice, whereas for individual specimens of a similar age found in California, only bones remain. This is an illustration of the effect of the cold temperature in the Arctic decreasing the rate of the reactions of decay. The same concept is used in the process of refrigerating or freezing food to preserve it.

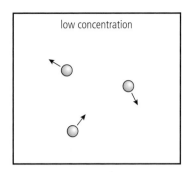
low concentration

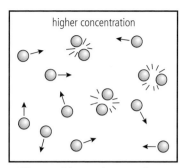
higher concentration

When particles are closer together
they have a greater chance of colliding.

We can see the effect of concentration by following the rate of a reaction as it
progresses. As reactants are used up, their concentration falls and the rate of the
reaction decreases, giving the typical rate curves we saw in Figures 1 and 2.

The effect of concentration
on the rate of reaction
between zinc and sulfuric
acid. The tube on the left
has a more concentrated
solution of the acid, the
tube on the right has a more
dilute solution. The product,
hydrogen gas, is seen
collecting much more quickly
in the presence of the more
concentrated acid.

Pressure

For reactions involving gases, increasing **pressure** increases the rate of reaction.
This is because the higher pressure compresses the gas, effectively increasing its
concentration. This will increase the frequency of collisions.

At the same temperature,
the box on the right has a
higher pressure and a greater
collision frequency.

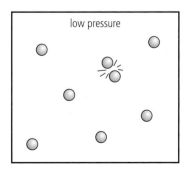
low pressure

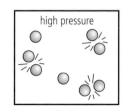
high pressure

In this example, the pressure has
been increased by decreasing
the size of the container

Surface area

Increasing the **surface area** of reactants increases the rate of reaction. This is important in heterogeneous reactions where the reactants are in different phases, such as a solid reacting with a solution. Subdividing a large particle into smaller parts increases the total surface area and therefore allows more contact and a higher probability of collisions between the reactants. You may know, for example, how much easier it is to start a fire using small pieces of wood, rather than a large log – it is because with the small pieces there is more contact between the wood and the oxygen with which it is reacting. In reactions involving solids in solutions, stirring may help to increase the surface area of the solid by ensuring individual particles are spread throughout the solution and do not aggregate at the bottom of the container. This increases the rate of reaction.

The effect of particle size can be demonstrated in the reaction between marble ($CaCO_3$) and hydrochloric acid. When marble chips are replaced with powder, the effervescence caused by carbon dioxide release is much more vigorous.

This effect of particle size on reaction rate can be quite dramatic. It has been responsible for many industrial accidents involving explosions of flammable dust powders – for example, coal dust in mines and flour in mills.

Coal dust explosion experiment at Altofts Colliery, Yorkshire, UK. Health and safety research in the UK started with investigations in the early 20th century into safety in mines. Experiments were carried out at locations such as Altofts Colliery, helping to determine that coal dust was explosive under certain circumstances.

Catalyst

A **catalyst** is a substance that increases the rate of a chemical reaction without itself undergoing chemical change. Most catalysts work by providing an alternative route for the reaction that has a lower activation energy.

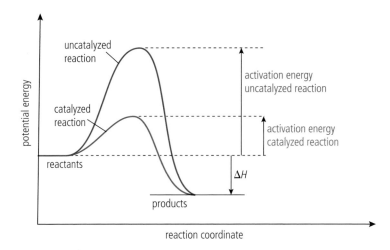

This means that without increasing the temperature, a larger number of particles will now have values of kinetic energy greater than the activation energy, and so will be able to undergo successful collisions. Think again of activation energy as being like the pass mark in an examination: the effect of the catalyst is like lowering the pass mark. This means that with the same work in the tests, a higher number of people would be able to achieve a pass!

Figure 8 uses the Maxwell–Boltzmann distribution to show how a catalyst increases the proportion of particles having values of kinetic energy greater than the activation energy.

R2.2 Figure 8 Effect of catalyst on increasing the proportion of particles able to react. ▶

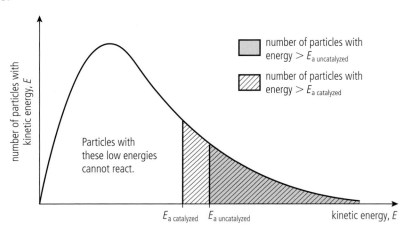

A catalyst is a substance that increases the rate of a chemical reaction without itself undergoing permanent chemical change.

Catalysts bring about an equal reduction in the activation energy of both the forward and the reverse reactions, so they do not change the position of equilibrium or the yield (seen in Reactivity 2.3). However, because of their ability to increase the *rate* of reactions, they influence the rate of formation of products and so play an essential role in the efficiency of many industrial processes. Without catalysts, many reactions would proceed too slowly, or would have to be conducted at such high temperatures that they would simply not be worthwhile. This is why the discovery of the 'best' catalyst for a particular reaction is a very active area of research, and often the exact specification of a catalyst used in an industrial process is a matter of secrecy.

Reactivity 2.3 – What is the relative effect of a catalyst on the rate of the forward and backward reactions?

▲
A catalytic converter containing transition metal catalysts such as platinum, palladium and rhodium. This device catalyzes reactions that convert the toxic emissions from an internal combustion engine into less harmful ones. It is estimated that catalytic converters can reduce pollution emissions by 90% without loss of engine performance or fuel economy.

Every *biological* reaction is controlled by a catalyst, known as an **enzyme**. Thousands of different enzymes exist as they are each specific for a particular biochemical reaction. Enzymes are finding increasingly widespread uses in many domestic and industrial processes, from biological detergents to acting as 'biosensors' in medical studies. Some of these applications, such as cheese making, are centuries old, but others are developing rapidly, constituting the field known as **biotechnology**.

Biological catalysts are called enzymes.

◀ A microscopic cross-section of glands found in the stomach. The pepsinogen digestive enzyme is shown in brown. It is converted by gastric acid into the active enzyme pepsin which catalyzes the breakdown of proteins into small units of amino acids called peptides.

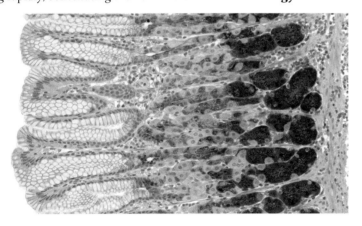

Catalysis is also an important aspect of Green Chemistry, which seeks to reduce the negative impact of chemical processes on the environment. Catalysts can replace stoichiometric reagents, and can greatly enhance the selectivity of processes. As they are effective in very small quantities and can frequently be reused, catalysts do not contribute to the chemical waste, and so they increase the atom economy.

The effect of a catalyst. Full details of how to carry out this experiment with a worksheet are available in the eBook.

Challenge yourself

2. Catalysts can be categorized as 'homogeneous' or 'heterogeneous'. A homogeneous catalyst is in the same phase as the reactants. For example, gaseous chlorine radicals catalyze the breakdown of ozone in the stratosphere. A heterogeneous catalyst, such as solid vanadium(V) oxide in the reaction between gaseous sulfur dioxide and oxygen, is in a different phase to the reactants. Can you identify which type of catalyst might be more difficult to recover after a reaction and suggest a possible approach for doing so?

Nature of Science

Depletion of the protective ozone layer in the stratosphere has been caused largely by the catalytic effects of CFCs and nitrogen oxides. These chemicals are produced from a variety of sources and often have their most destructive effect far from the country of their origin. International action and cooperation have played an important part in reducing the effects of ozone depletion, and continuing adherence by all countries to international treaties is essential.

Inquiry-based skills for investigating rate of reaction
Identification of controlled variables

Given the range of factors that can affect the rate of reaction, we must be careful to identify the relevant controlled variables when planning an investigation. For example, when investigating the effect of temperature on the reaction between hydrochloric acid, HCl(aq), and magnesium, Mg(s), we would need to ensure that the same concentration of acid is used in each trial as well as using the same sized magnesium strip to control the surface area.

When you are identifying variables to be controlled, it can be helpful to first consider the factors that affect rate of reaction, such as concentration, surface area and temperature, and then the specific details of your measurement technique. For example, when using a light sensor to measure the rate of formation of a precipitate in a reaction, you would need to ensure that the position of the sensor and light source are the same during all trials. This approach is crucial in obtaining valid data from an investigation.

Tool 1 – What variables must be controlled in studying the effect of a factor on the rate of a reaction?

Evidence for systematic and random errors in graphs

As well as identifying the effect of a given variable on the rate of reaction, graphs can also be used to provide evidence of systematic and random errors.

We can evaluate the fit of a trend line or curve by applying the coefficient of determination, R^2 (pronounced 'R squared'). An R^2 value will fall between 0 and 1, where 1 indicates that the trend line or curve passes perfectly through all data points. This value can be calculated automatically on most graph-plotting software.

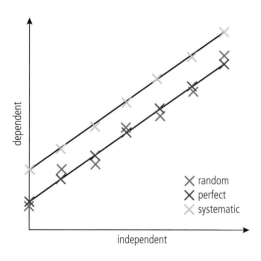

In a perfect data set, all data points would be positioned on the trend line. A systematic error produces a displaced straight line above or below the expected position. Random errors lead to points on both sides of the trend line.

Systematic errors occur when all measurements are higher, or all measurements are lower, than the expected value. This leads to inaccuracy in the final results. Random errors occur in individual measurements due to limitations in measurement equipment (uncertainty), changes in the surroundings, misinterpreted measurements or an insufficient number of repeat trials. These can lead to imprecision in collected data.

Nature of Science, Tool 3, Inquiry 3 – How can graphs provide evidence of systematic and random error?

Exercise

Q8. Which of the following statements is correct?

   A   A catalyst increases the rate of the forward reaction only.

   B   A catalyst increases the rate of the forward and backward reactions.

   C   A catalyst increases the yield of product formed.

   D   A catalyst increases the activation energy of a reaction.

Q9. Which statements are correct for the effects of catalyst and temperature on the rate of reaction?

	Adding a catalyst	**Increasing the temperature**
A	collision frequency increases	collision frequency increases
B	activation energy decreases	collision frequency increases
C	collision frequency increases	activation energy increases
D	activation energy increases	activation energy decreases

Q10. In the reaction between marble (calcium carbonate) and hydrochloric acid, which set of conditions would give the highest rate of reaction?

$$CaCO_3(s) + 2HCl \rightarrow CaCl_2(aq) + CO_2(g) + H_2O(l)$$

   A   marble chips and $1.0\,mol\,dm^{-3}$ HCl

   B   marble powder and $1.0\,mol\,dm^{-3}$ HCl

   C   marble chips and $0.1\,mol\,dm^{-3}$ HCl

   D   marble powder and $0.1\,mol\,dm^{-3}$ HCl

Q11. A sugar cube cannot be ignited with a match, but a sugar cube coated in ashes will ignite. Suggest a reason for this observation.

Q12. Catalytic converters are now used in most cars to convert some components of exhaust gases into less environmentally damaging molecules. One of these reactions converts carbon monoxide and nitrogen monoxide into carbon dioxide and nitrogen. The catalyst usually consists of metals such as platinum or rhodium.

   (a) Write an equation for this reaction.

   (b) Explain why it is important to reduce the concentrations of carbon monoxide and nitrogen monoxide released into the atmosphere.

   (c) Why do you think the converter sometimes consists of small ceramic beads coated with the catalyst?

   (d) Suggest why the converter usually does not work effectively until the car engine has warmed up.

   (e) Discuss whether the use of catalytic converters in cars solves the problem of car pollution.

In this chapter we have explored collision theory and investigative techniques to show:

- The rate of reaction is expressed as the change in concentration of a particular reactant or product per unit time. It has units of $mol\,dm^{-3}\,s^{-1}$.
- Rate of reaction can be measured indirectly using different experimental techniques. The choice of technique will depend on the changes that occur during a specific reaction such as the production of a gas, the formation of a precipitate, or changes in the colour or electrical conductivity of a solution.
- The rate at a specific point in a reaction can be calculated from the gradient of a tangent on a graph of concentration, volume, absorbance or mass against time.
- For a reaction to occur, species must collide with sufficient kinetic energy and proper orientation. These collisions are known as 'successful' collisions.
- The minimum energy required for a successful collision is called the activation energy, E_a.
- The use of a catalyst and increases in pressure, concentration, surface area and temperature lead to an increase in the frequency of successful collisions per unit time and so increase the rate of a reaction.
- Catalysts increase the rate of reaction by providing an alternative reaction pathway with a lower E_a. This effect can be represented in a reaction energy profile.
- Maxwell–Boltzmann distribution curves can be used to explain the effect of changing temperature and the use of a catalyst on the probability of successful collisions occurring.

Practice questions

1. Curve X on the graph shows the volume of oxygen formed during the catalytic decomposition of a $1.0\,mol\,dm^{-3}$ solution of hydrogen peroxide:

$$2H_2O_2(aq) \rightarrow O_2(g) + 2H_2O(l)$$

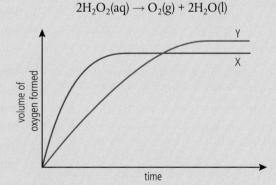

Which change would produce curve Y?

A   adding water

B   adding some $0.1\,mol\,dm^{-3}$ hydrogen peroxide solution

C   using a different catalyst

D   lowering the temperature

2. Which changes increase the rate of the reaction below?

$$C_4H_{10}(g) + Cl_2(g) \rightarrow C_4H_9Cl(l) + HCl(g)$$

I. increase of pressure

II. increase of temperature

III. removal of $HCl(g)$

A	I and II only	**B**	I and III only
C	II and III only	**D**	I, II and III

3. Which experimental procedure could be used to determine the rate for the following reaction?

$$BrO_3^-(aq) + 5Br^-(aq) + 6H^+(aq) \rightarrow 3Br_2(aq) + 3H_2O(l)$$

A measure the change in electrical conductivity in a given time

B measure the change in mass in a given time

C record the time taken for the formation of a precipitate

D measure the change in volume of gas produced in a given time

4. The catalyst manganese(IV) oxide, $MnO_2(s)$, increases the rate of the decomposition reaction of hydrogen peroxide, $H_2O_2(aq)$. Which statements about MnO_2 are correct?

I. The rate is independent of the particle size of MnO_2.

II. MnO_2 provides an alternative reaction pathway for the decomposition with a lower activation energy.

III. All the MnO_2 is present after the decomposition of the hydrogen peroxide is complete.

A	I and II only	**B**	I and III only
C	II and III only	**D**	I, II and III

5. Which of the following statements is incorrect?

A The rate of a chemical reaction depends on the temperature.

B Rate and time are directly proportional.

C The rate of most chemical reactions decreases with time.

D A catalyst for a reaction increases the rate of both its forward and backward reactions.

6. A student measures the rate of a reaction by timing the appearance of a precipitate that forms in aqueous solution. Which of the following factors would increase the time required for the precipitate to form?

A raising the temperature

B adding a catalyst

C diluting the solution

D stirring the reaction mixture

Questions 7–9 refer to the reaction between magnesium carbonate and hydrochloric acid, which is as follows:

$$MgCO_3(s) + 2HCl(aq) \rightarrow MgCl_2(aq) + H_2O(l) + CO_2(g)$$

7. Which of the conditions described below will produce the fastest rate of reaction?

    A 2.0 mol dm^{-3} HCl and MgCO$_3$ lumps

    B 1.0 mol dm^{-3} HCl and MgCO$_3$ powder

    C 2.0 mol dm^{-3} HCl and MgCO$_3$ powder

    D 1.0 mol dm^{-3} HCl and MgCO$_3$ lumps

8. Which of the following measurements would not be a suitable means to follow the rate of this reaction?

    A increase in mass of reaction mixture

    B increase in volume of gas produced at constant pressure

    C increase in pH of reaction mixture

    D increase in pressure of gas produced at constant volume

9. The sketch graph below represents the result of an experiment to measure the rate of this reaction.

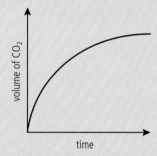

Which of the following is the best explanation for the shape of the graph?

    A The rate of reaction increases with time because the acid becomes more dilute.

    B The rate of reaction increases with time because the calcium carbonate pieces become smaller.

    C The rate of reaction decreases with time because the acid becomes more dilute.

    D The rate of reaction decreases with time because the calcium carbonate pieces become smaller.

10. Collision theory states that collisions between reactant molecules do not always lead to the formation of product. Which of the following is the best explanation for this statement?

    A The reactant molecules are at too low a concentration.

    B The reaction is at equilibrium.

    C The reaction needs a catalyst in order to occur.

    D The reactant molecules do not have sufficient energy.

11. It is found that in the reaction:

$$NO_2(g) + CO(g) \rightarrow NO(g) + CO_2(g)$$

an increase in temperature increases the rate of reaction.
Which of the statements below is the main reason for this?

A The molecules collide more frequently.

B The proportion of molecules with energy greater than the activation energy increases.

C The pressure exerted by the molecules increases.

D The proportion of molecules with the correct collision geometry increases.

12. Zinc metal reacts with copper(II) sulfate solution as follows:

$$Zn(s) + CuSO_4(aq) \rightarrow Cu(s) + ZnSO_4(aq)$$

Which of the following factors will increase the rate of reaction?

I.   increasing the concentration of $CuSO_4(aq)$

II. decreasing the size of the zinc pieces

III. carrying out the reaction at a higher temperature

A I and II

B I and III

C II only

D I, II and III

13. Which of the following units could not be used to express the rate of a reaction?

A $mol\,dm^{-3}\,s$

B $g\,cm^{-3}\,s^{-1}$

C $mol\,dm^{-3}\,s^{-1}$

D $g\,dm^{-3}\,s^{-1}$

14. The atmosphere consists mostly of the gases nitrogen, N_2, and oxygen, O_2. Under normal conditions these two gases react together to form nitrogen monoxide, NO, extremely slowly.

$$N_2(g) + O_2(g) \rightarrow NO(g)$$

Which statement below is the best explanation for the low rate of this reaction?

A The atmosphere does not contain a catalyst for this reaction.

B The reaction between nitrogen and oxygen has a very high activation energy.

C Oxygen molecules are more likely to collide with themselves than with nitrogen molecules.

D The simultaneous collision of three molecules is unlikely.

15. The graph below was obtained when zinc carbonate reacted with dilute hydrochloric acid under two different conditions, denoted as experiments A and B.

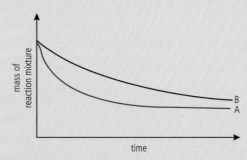

(a) Write an equation including state symbols for the reaction occurring. (2)

(b) Explain why the mass of the reaction mixture decreases in both cases. (1)

(c) Make reference to collision theory to explain the shape of curve A. (3)

(d) Describe the measurements that could be made from the graph to compare the initial rates of the reactions in A and B. Comment on the results expected. (3)

(e) The concentration of hydrochloric acid used in experiments A and B was the same. Suggest three possible differences in the conditions of experiments A and B. (3)

(f) For each of the conditions given in (e), explain why it would affect the rate of the reaction. (6)

(g) The experiment was repeated using zinc metal in place of zinc carbonate.

$$Zn(s) + 2HCl(aq) \rightarrow ZnCl_2(aq) + H_2(g)$$

Describe the differences you would expect in the results, and evaluate whether this is likely to be a satisfactory method for following the rate of reaction. (2)

(Total 20 marks)

16. The figure below shows a Maxwell–Boltzmann distribution curve for a sample of a gas at two different temperatures, T1 and T2.

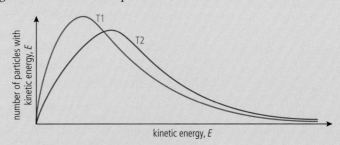

(a) Deduce the relative values of T1 and T2. (1)

(b) Make reference to your answer to (a) to explain the differences in the shape of the two graphs. (2)

(c) 'A catalyst provides a reaction route with a lower activation energy, and so increases the rate of reaction.'

Justify this statement by means of a suitably labelled Maxwell-Boltzmann distribution curve, showing the proportion of reacting particles with and without a catalyst. (4)

(d) Explain why catalysts increase the rate of reaction but have no effect on:

(i) the enthalpy change

(ii) the stoichiometric yield of a product (4)

(Total 11 marks)

17. The decomposition of hydrogen peroxide, H_2O_2, into water and oxygen gas is an exothermic reaction.

$$2H_2O_2(aq) \rightarrow 2H_2O(l) + O_2(g)$$

(a) On the axes below, sketch an energy profile for an exothermic reaction and annotate the key features. (3)

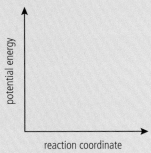

(b) The enzyme 'catalase' can be used to increase the rate of decomposition of hydrogen peroxide.

(i) Outline how the energy profile would change if catalase was added to the reaction mixture. (1)

(ii) Explain why the addition of catalase would have no effect on the enthalpy change of the reaction. (1)

(Total 5 marks)

How far? The extent of
chemical change

◀ An agricultural landscape in the Netherlands. Our ability to feed a growing global population has been dependent on our ability to mass produce plant fertilizer. Industrial scale production of ammonia, an essential starting material in the production of fertilizer, was made possible through an understanding of reversible reactions and the factors that can be used to influence them. Specifically, the Haber process uses high pressures and a temperature of about 450 °C to maximize the conversion of nitrogen and hydrogen gas into ammonia.

Guiding Question

How can the extent of a reversible reaction be influenced?

Imagine that you are part of the way along an escalator (a moving staircase) that is moving up, and you decide to run down it. If you can run down at exactly the same speed as the escalator is moving up, you will have no *net* (overall) movement. So, if someone were to take a picture of you at regular time intervals, it would seem as if you were not moving at all. Of course, in reality both you and the escalator *are* moving, but because there is no net change neither movement is observable. In chemical reactions a similar phenomenon occurs when a reaction takes place at the same rate as its reverse reaction, so no net change is observed. This is known as a **state of dynamic equilibrium**.

In this chapter, we explore some of the features of the equilibrium state and learn how this enables us to predict how far reactions will proceed under different conditions. Industrial processes rely significantly on this type of study to determine the conditions that will maximize the yield of product.

We will also see how **equilibrium law** can be used to determine the **equilibrium constant** from the stoichiometry of a reaction. This temperature-dependent constant will allow us to quantify the **extent** of a reaction at equilibrium.

Equilibrium studies are also important in many biochemical and environmental processes, such as predicting the solubility of gases in the blood and knowing how certain chemicals in the atmosphere may react together to form pollutants that contribute to climate change.

Reactivity 2.3.1 – Dynamic equilibrium

Reactivity 2.3.1 – A state of dynamic equilibrium is reached in a closed system when the rates of forward and backward reactions are equal.

Describe the characteristics of a physical and chemical system at equilibrium.

A state of dynamic equilibrium is reached in a closed system

In this section we will think about how **dynamic equilibrium** is reached in both a **physical system,** in which we consider changes of state, and a **chemical system**, in which we consider **reversible** chemical reactions.

▲ Snow and ice-covered peak in the Cordillera Huayhuash, Peru. Equilibrium considerations help explain the relationship between water vapour in the clouds and precipitations in the form of liquid (rain) and solid (snow) at different temperatures and pressures.

Physical systems

Consider what happens when some bromine, Br_2, is placed in a sealed container at room temperature.

Bromine is the only non-metallic element that is liquid at room temperature. It is extremely toxic and takes its name from a Greek word meaning stench.

As bromine is a **volatile** liquid, with a boiling point close to room temperature, a significant number of particles (molecules of Br_2) will have enough energy to escape from the liquid state and form vapour in the process known as **evaporation**. As the container is sealed, the bromine vapour cannot escape and so its concentration will increase. Some of these vapour molecules will collide with the surface of the liquid, lose energy, and become liquid in the process known as **condensation**.

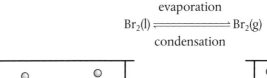

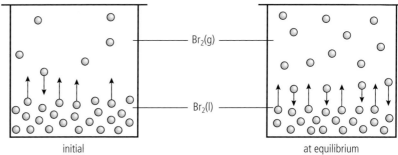

Establishing dynamic equilibrium in the evaporation of bromine. Equilibrium is established when the rate of evaporation equals the rate of condensation.

Bromine stored in a sealed jar. The system is in dynamic equilibrium, so the concentrations of liquid and vapour do not change at constant temperature.

The rate of condensation increases with the increase in concentration of vapour, as more vapour particles collide with the surface of the liquid. Eventually, the rate of condensation is equal to the rate of evaporation, and at this point there will be no net change in the amounts of liquid and gas present (Figure 1). We say that the system has reached **equilibrium**. This will only occur in a **closed system**, where the $Br_2(g)$ cannot escape as vapour but may condense back into the liquid.

R2.3 Figure 1 The rate of evaporation and condensation as liquid–vapour equilibrium is established in a closed system. The rate of evaporation is constant as the surface area remains the same, while the rate of condensation increases as the concentration of vapour increases. Equilibrium is established when the two rates are equal.

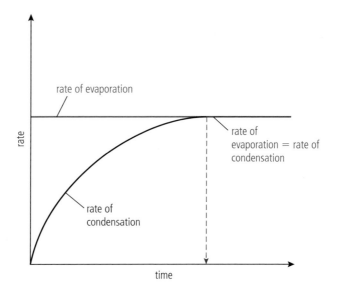

Chemical systems

Consider the reaction of dissociation between hydrogen iodide (HI) and its elements hydrogen (H_2) and iodine (I_2). Hydrogen and hydrogen iodide are both colourless, whereas iodine is released as a purple gas.

$$2HI(g) \rightleftharpoons H_2(g) + I_2(g)$$
colourless gas   colourless gas   purple gas

If we carry out this reaction starting with hydrogen iodide in a sealed container, there will at first be an increase in the purple colour owing to the production of iodine gas. But after a while this increase in colour will stop, and it may appear that the reaction too has stopped. In fact, what has happened is that the rate of the dissociation of HI is fastest at the start when the concentration of HI is greatest and falls as the reaction proceeds. Meanwhile, the reverse reaction, which initially has a zero rate because there is no H_2 or I_2 present, starts slowly and increases in rate as the concentrations of H_2 and I_2 increase. Eventually, the rate of dissociation of HI becomes equal to the rate of the reverse reaction of association between H_2 and I_2, so the concentrations remain constant. This is why the colour in the flask remains the same. At this point, equilibrium has been reached. The equilibrium is described as **dynamic** because both forward and backward reactions are still occurring.

If we were to analyze the contents of the flask at this point, we would find that HI, H_2 and I_2 would all be present and that if there were no change in conditions, their concentrations would remain constant over time. We refer to this as the **equilibrium mixture**.

If we reversed the experiment and started with H_2 and I_2 instead of HI, we would find that eventually an equilibrium mixture would again be achieved in which the concentrations of H_2, I_2 and HI would remain constant. These relationships are shown in Figure 2.

Heated iodine crystals in a stoppered flask. Iodine is a crystalline solid at room temperature, but sublimes on heating to form a purple gas.

TOK

The study of chemical change often involves both the macroscopic and submicroscopic scales. What might this suggest about the role of imagination and intuition in the creation of hypotheses in the natural sciences?

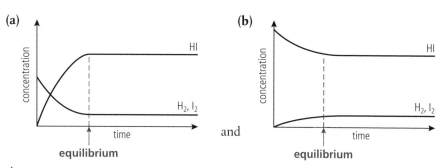

R2.3 Figure 2 Equilibrium is reached when the concentrations of reactants and products become constant. Note that the same equilibrium mixture is reached whether you start (a) from a mixture of H_2 and I_2 or (b) from pure HI.

The equilibrium state has specific characteristics

In studies of equilibria, we are dealing with reversible reactions – that is, reactions that occur in both directions. The convention is to describe the reaction from left to right (reactants to products) as the **forward reaction**, and the reaction from right to left (products to reactants) as the **backward** or **reverse reaction**. The symbol $\rightleftharpoons$ is used to show that the reaction is an equilibrium reaction.

Strictly speaking, all reactions can be considered as equilibrium reactions. However, in many cases, the equilibrium mixture consists almost entirely of products – that is, it is considered to have gone virtually to completion. By convention we use the symbol → rather than the equilibrium symbol in these cases. Combustion reactions, for example, are considered irreversible given the relative stability of the products compared to the reactants. In other reactions there may be so little product formed that it is undetectable and we consider the reaction to have effectively not happened.

Make sure that you use the equilibrium symbol ⇌ when writing equations for reactions where the reverse reactions are significant. For example, it must be used when explaining the behaviour of weak acids and bases (Reactivity 3.1).

At equilibrium the rate of the forward reaction is equal to the rate of the backward reaction.

$$\text{reactants} \underset{\text{backward reaction}}{\overset{\text{forward reaction}}{\rightleftharpoons}} \text{products}$$

▲ The Dragon spacecraft, first launched by the American company SpaceX in 2010, uses the combustion reaction between monomethyl hydrazine and the oxidizing agent dinitrogen tetroxide to produce thrust and orientate the spacecraft during a mission.

$$4CH_3NHNH_2 + 5N_2O_4 \rightarrow 9N_2 + 4CO_2 + 12H_2O$$

The strong N≡N bonds and C=O bonds found in the products make them significantly more stable than the reactants, and so the reaction can be considered non-reversible.

The examples of physical and chemical systems discussed above have shown that at equilibrium the rate of the forward reaction is equal to the rate of the backward reaction.

These reactions have also shown some of the main features of the equilibrium state, and these can now be summarized as they apply to *all* reactions at equilibrium.

	Characteristics of a physical and a chemical system at equilibrium	Explanation
1	Equilibrium is dynamic	The reaction has not stopped, but both forward and backward reactions are still occurring at the same rate.
2	Equilibrium is achieved in a closed system	A closed system has no exchange of matter with the surroundings, so equilibrium is achieved where both reactants and products can react and recombine with each other.
3	The concentrations of reactants and products remain constant at equilibrium	They are being produced and destroyed at an equal rate.
4	At equilibrium there is no change in macroscopic properties	Macroscopic properties are observable properties such as colour and density. These do not change as they depend on the concentrations of the components of the mixture.
5	Equilibrium can be reached from either direction	The same equilibrium mixture will result under the same conditions, no matter whether the reaction is started with all reactants, all products, or a mixture of both.

It is important to understand that even though the concentrations of reactant and product are *constant* at equilibrium, this in no way implies that they are *equal*. In fact, most commonly there will be a much higher concentration of either reactant or product in the equilibrium mixture, depending both on the reaction and on the conditions. For example, we can see in Figure 2 that when the dissociation of HI reaches equilibrium, there is a higher concentration of HI than of H_2 and I_2.

Challenge yourself

1. A closed system is defined differently in different disciplines. In thermodynamics, it means that no matter can be exchanged with the surroundings, but energy can flow. To what extent can the Earth can be considered a closed system?

Think back to the moving staircase analogy in the introduction to this chapter, where the top and bottom of the escalator represent reactants and products respectively. It would be possible for you to be 'at equilibrium' near the top of the staircase, near the bottom, or anywhere in between. As long as you were moving at the same speed as the staircase, you would still have no net change in position.

TOK What are the differences between theories and analogies as forms of explanation?

SKILLS Conceptualize equilibrium with 'water races'. Full details of how to carry out this experiment with a worksheet are available in the eBook.

The proportion of reactant and product in the equilibrium mixture is referred to as its **equilibrium position**. Reactions where the mixture contains predominantly products are said to 'lie to the right' and reactions with predominantly reactants are said to 'lie to the left'.

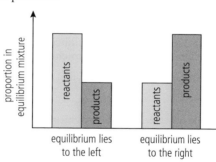

equilibrium lies to the left   equilibrium lies to the right

Describing the position of equilibrium

It is, however, often useful to be able to capture this information mathematically to compare the equilibrium mixtures of different reactions and the effect of different conditions. In the next section we will look at how this is done.

 Sometimes you may see the equilibrium sign written with unequal arrows such as $\rightleftharpoons$. This is used to represent the reaction that lies in favour of products. Likewise, $\rightleftharpoons$ is used to represent a reaction that lies in favour of reactants.

Nature of Science

At equilibrium, no change is observed on the macroscopic level, although particles are reacting at the submicroscopic level. These changes can be deduced using techniques such as isotopic labelling, which allow the progress of a specific reactant to be followed. Our power of understanding is enhanced by contributions from instrumentation and sensors that may gather information beyond human sense perception.

Exercise

Q1. Which statements are correct for a reaction at equilibrium?

I.   The forward and reverse reactions both continue.

II. The rates of the forward and reverse reactions are equal.

III. The concentrations of reactants and products are equal.

A I and II only   B I and III only   C II and III only   D I, II and III

Q2. Which statement is always true for a chemical reaction that has reached equilibrium at constant temperature?

   A   The yield of product(s) is greater than 50%.

   B   The rate of the reverse reaction is lower than that of the forward reaction.

   C   The amounts of reactants and products do not change.

   D   Both forward and reverse reactions have stopped.

Q3. Which statement is *not* true for a mixture of ice and water at equilibrium at constant temperature?

   A   The rates of melting and freezing are equal.

   B   The amounts of ice and water are equal.

   C   The same position of equilibrium can be reached by cooling water and by heating ice.

   D   There is no observable change in the system.

Reactivity 2.3.2 and 2.3.3 – Equilibrium law

Reactivity 2.3.2 – The equilibrium law describes how the equilibrium constant, K, can be determined from the stoichiometry of a reaction.

Deduce the equilibrium constant expression from an equation for a homogeneous reaction.

Reactivity 2.3.3 – The magnitude of the equilibrium constant indicates the extent of a reaction at equilibrium and is temperature dependent.

Determine the relationships between K values for reactions that are the reverse of each other at the same temperature.

Include the extent of reaction for:	Reactivity 3.1 – How does the value of K for the dissociation of an acid convey information about the strength of the acid?
$K \ll 1, K < 1, K = 1, K > 1, K \gg 1$.	

The equilibrium constant, *K*, can be predicted from the reaction stoichiometry

Consider the reaction

$$H_2(g) + I_2(g) \rightleftharpoons 2HI(g)$$

If we were to carry out a series of experiments on this reaction with different starting concentrations of H_2, I_2, and HI, we could wait until each reaction reached equilibrium and then measure the composition of each equilibrium mixture. Here are some typical results obtained at 440 °C.

Experiment I

	Initial concentration / $mol\,dm^{-3}$	Equilibrium concentration / $mol\,dm^{-3}$
H_2	0.100	0.0222
I_2	0.100	0.0222
HI	0.000	0.1560

Experiment II

	Initial concentration / mol dm^{-3}	Equilibrium concentration / mol dm^{-3}
H_2	0.0000	0.0350
I_2	0.0100	0.0450
HI	0.3500	0.2800

Experiment III

	Initial concentration / mol dm^{-3}	Equilibrium concentration / mol dm^{-3}
H_2	0.0150	0.0150
I_2	0.0000	0.0135
HI	0.1270	0.1000

At a glance, the data may not appear to show any pattern. However, there is a predictable relationship among the different compositions of these equilibrium mixtures, and the key to discovering it is in the stoichiometry of the reaction equation.

$$H_2(g) + I_2(g) \rightleftharpoons 2HI(g)$$

If we take the equilibrium concentrations and process them in the following way:

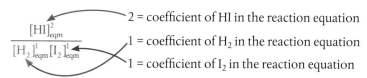

$$\frac{[HI]^2_{eqm}}{[H_2]^1_{eqm}[I_2]^1_{eqm}}$$

2 = coefficient of HI in the reaction equation

1 = coefficient of H_2 in the reaction equation

1 = coefficient of I_2 in the reaction equation

we find the following results:

Experiment I $\dfrac{(0.156)^2}{0.0222 \times 0.0222} = 49.4$

Experiment II $\dfrac{(0.280)^2}{0.035 \times 0.0450} = 49.8$

Experiment III $\dfrac{(0.100)^2}{0.0150 \times 0.0135} = 49.4$

Clearly, this way of processing the equilibrium data produces a constant value within the limits of experimental accuracy. The constant is known as the equilibrium constant, K. It has a fixed value for this reaction at a specified temperature.

In fact, every reaction has its own particular value of K which can be derived in a similar way. This finding is described by the equilibrium law.

- First we use the balanced reaction equation to write the **equilibrium constant expression**.

  For the reaction:

  $$aA + bB \rightleftharpoons cC + dD$$

  the equilibrium constant expression is

  $$K = \frac{[C]^c_{eqm}[D]^d_{eqm}}{[A]^a_{eqm}[B]^b_{eqm}}$$

- The value for K can then be determined by substituting the equilibrium concentrations of all reactants and products into this equation.

Remember square brackets [ ] are commonly used to show concentration in mol dm^{-3}.

The equilibrium law states that the equilibrium constant, K, has a fixed value for a particular reaction at a specified temperature. The only thing that changes the value of K for a reaction is the temperature.

As K is calculated using the concentrations of reactants and products, it is commonly stated as K_c in other sources. The 'c' stands for 'concentration'.

Many sources give units for K which are a multiple of $mol\,dm^{-3}$, depending on the stoichiometry of the reaction. In fact, this is not fully correct, as the terms in the equilibrium expression are really a thermodynamic quality known as 'activity' that has no units. For this reason, we are omitting units in the values of K here, and you will not be required to include them in IB examination answers.

The equilibrium constant expression will only give the value of K when the concentrations used in the equation are the *equilibrium* concentrations for all reactants and products. Strictly speaking, the subscript 'eqm' should always be used in the equation, but by convention this is generally left out. However, make completely sure that the only values you substitute into an equation to calculate K are the equilibrium concentrations.

The equilibrium constant expressions described here apply to homogeneous reactions, that is, reactions where reactants and products are in the same phase, as gases, liquids, or in solution. It is good practice always to include state symbols in your equations to ensure this is being applied correctly.

Note the following:

- The equilibrium constant expression has the concentrations of products in the numerator and the concentrations of reactants in the denominator.
- Each concentration is raised to the power of its coefficient in the balanced equation. (Where the power is equal to one, it is not shown.)
- Where there is more than one reactant or product, the terms are multiplied together.

Worked example

Write the equilibrixm constant expression for the following reactions.

(i) $2H_2(g) + O_2(g) \rightleftharpoons 2H_2O(g)$

(ii) $Cu^{2+}(aq) + 4NH_3(aq) \rightleftharpoons [Cu(NH_3)_4]^{2+}(aq)$

Solution

(i) $K = \dfrac{[H_2O]^2}{[H_2]^2[O_2]}$

(ii) $K = \dfrac{[[Cu(NH_3)_4]^{2+}]}{[Cu^{2+}][NH_3]^4}$

The magnitude of K gives information on the extent of reaction

Different reactions have different values of K. What does this value tell us about a particular reaction?

As the equilibrium constant expression puts products in the numerator and reactants in the denominator, a high value of K will mean that at equilibrium there are proportionately more products than reactants. In other words, such an equilibrium mixture lies to the right and the reaction goes almost to completion. By contrast, a low value of K must mean that there are proportionately fewer products with respect to reactants, so the equilibrium mixture lies to the left and the reaction has barely taken place.

Consider the following three reactions and their K values measured at $550\,K$.

$$H_2(g) + I_2(g) \rightleftharpoons 2HI(g) \qquad K = 2$$

$$H_2(g) + Br_2(g) \rightleftharpoons 2HBr(g) \qquad K = 10^{10}$$

$$H_2(g) + Cl_2(g) \rightleftharpoons 2HCl(g) \qquad K = 10^{18}$$

The large range in their K values tells us about the differing **extents** of these reactions. We can deduce that the reaction between H_2 and Cl_2 has taken place the most fully at this temperature, while H_2 and I_2 have reacted the least.

Challenge yourself

2. Why do you think the reactions of the three halogens, Cl_2, Br_2 and I_2, with H_2 have such different values for their equilibrium constant at the same temperature? What can you conclude about the strength of bonding in the three hydrogen halides?

A good rule of thumb to apply to these situations is that if $K \gg 1$, the reaction is considered to go almost to completion (very high conversion of reactants into products), and if $K \ll 1$, the reaction hardly proceeds.

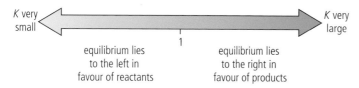

K very small ⟵⟶ *K* very large

1

equilibrium lies to the left in favour of reactants

equilibrium lies to the right in favour of products

▲
The smaller the value of *K*, the further the equilibrium mixture lies to the left. The larger the value of *K*, the further the equilibrium mixture lies to the right.

This can be summarized as follows.

Value of K	Extent of reaction
$K \ll 1$	Reaction hardly proceeds
$K < 1$	Equilibrium lies towards the reactants
$K = 1$	Significant amounts of both reactants and products at equilibrium
$K > 1$	Equilibrium lies towards the products
$K \gg 1$	Reaction goes almost to completion

Note that the magnitude of K does not give us any information on the *rate* of the reaction. It informs us of the nature of the equilibrium mixture, but not how quickly the equilibrium state will be achieved.

The value of K can be particularly useful to consider the dissociation of an acid in solution. Weak acids undergo very little dissociation so will have very low values of K whereas a stronger acid will dissociate to a greater extent, producing a larger value of K.

Consider the values of K_a for the dissociation of methanoic acid, HCOOH, and ethanoic acid, CH_3COOH, in aqueous solution at 298 K:

$$HCOOH(aq) \rightleftharpoons HCOO^-(aq) + H+(aq) \qquad K_a = 1.78 \times 10^{-4}$$
$$CH3COOH(aq) \rightleftharpoons CH_3COO^-(aq) + H+(aq) \qquad K_a = 1.74 \times 10^{-5}$$

Here we can see that although the equilibrium lies to the left for both acids, HCOOH undergoes a greater degree of dissociation as it has a larger value of K_a. Note that the 'a' refers specifically to the equilibrium mixture formed by the dissociation of an acid.

Relationships between K values for reactions that are the reverse of each other at the same temperature

As K is defined with products in the numerator and reactants in the denominator, each raised to the power of their stoichiometric coefficients in the balanced equation, we can manipulate the value of K according to changes made to these terms. For this discussion we will consider the generic reaction:

$$aA + bB \rightleftharpoons cC + dD$$

for which the equilibrium constant is:

$$K = \frac{[C]^c[D]^d}{[A]^a[B]^b}$$

Reactivity 3.1 – How does the value of *K* for the dissociation of an acid convey information about its strength?

The magnitude of the equilibrium constant *K* gives information about the extent of reaction at a particular temperature, but not about how fast it will achieve the equilibrium state.

The reverse reaction

$$cC + dD \rightleftharpoons aA + bB$$

defines the products as reactants and vice versa. We will denote its equilibrium constant as K'.

$$K' = \frac{[A]^a[B]^b}{[C]^c[D]^d}$$

We can see that

$$K' = \frac{1}{K} \quad \text{or} \quad K' = K^{-1}.$$

In other words, the equilibrium constant for a reaction is the reciprocal of the equilibrium constant for its reverse reaction.

Worked example

The equilibrium constant for the reaction

$$2HI(g) \rightleftharpoons H_2(g) + I_2(g)$$

is 0.040 at a certain temperature. What would be the value of the equilibrium constant, K', for the reverse reaction

$$H_2(g) + I_2(g) \rightleftharpoons 2HI(g)$$

at the same temperature?

Solution

The value of K' is the reciprocal of K.

So $K' = \dfrac{1}{K} \quad$ or $\quad K^{-1}$

$\therefore K' = \dfrac{1}{0.040} = 25$

This manipulation of the value of K is summarized below.

	Effect on equilibrium expression	Effect on K
reversing the reaction	inverts the expression	$\dfrac{1}{K}$ or K^{-1}

Challenge yourself

3. We can also express the relationship between K and K' for multiples of a reaction. Doubling the reaction coefficients squares the equilibrium expression, $K' = K^2$, and halving the reaction coefficients square roots the equilibrium expression, $K'' = \sqrt{K}$.

   At a given temperature, the reaction

   $$2SO_2(g) + O_2(g) \rightleftharpoons 2SO_3(g)$$

   has a value of $K = 278$. Determine values of K for the following reactions at this temperature.

   (a) $\qquad\qquad 4SO_2(g) + 2O_2(g) \rightleftharpoons 4SO_3(g)$

   (b) $\qquad\qquad SO_3(g) \rightleftharpoons SO_2(g) + \frac{1}{2}O_2(g)$

   (c) $\qquad\qquad 6SO_3(g) \rightleftharpoons 6SO_2(g) + 3O_2(g)$

Exercise

Q4. Write the equilibrium constant expression for the following reactions:

(a) $2NO(g) + O_2(g) \rightleftharpoons 2NO_2(g)$

(b) $4NH_3(g) + 7O_2(g) \rightleftharpoons 4NO_2(g) + 6H_2O(g)$

(c) $CH_3Cl(aq) + OH^-(aq) \rightleftharpoons CH_3OH(aq) + Cl^-(aq)$

Q5. Write the equations for the reactions represented by the following equilibrium constant expressions:

(a) $K = \dfrac{[NO_2]^2}{[N_2O_4]}$

(b) $K = \dfrac{[CO][H_2]^3}{[CH_4][H_2O]}$

Q6. Write the equilibrium constant expressions for the following chemical reactions:

(a) fluorine gas and chlorine gas combine to form $ClF_3(g)$

(b) NO dissociates into its elements

(c) methane, CH_4, and steam react to form carbon monoxide and hydrogen.

Q7. When the following reactions reach equilibrium at a given temperature, does the equilibrium mixture contain mostly reactants or mostly products? Assume that the values for K are applicable at the given temperature.

(a) $N_2(g) + 2H_2(g) \rightleftharpoons N_2H_4(g)$ $\qquad K = 7.4 \times 10^{-26}$

(b) $N_2(g) + O_2\ (g) \rightleftharpoons 2NO(g)$ $\qquad K = 2.7 \times 10^{-18}$

(c) $2NO(g) + O_2(g) \rightleftharpoons 2NO_2(g)$ $\quad K = 6.0 \times 10^{13}$

Q8. At 298 K, the equilibrium concentrations in the reaction mixture:

$$N_2O_4(g) \rightleftharpoons 2NO_2(g)$$

are $[N_2O_4] = 1.62\,mol\,dm^{-3}$ and $[NO_2] = 1.80\,mol\,dm^{-3}$.

Determine the value of the equilibrium constant, K.

Q9. At a given temperature, the reaction

$$2SO_2(g) + O_2(g) \rightleftharpoons 2SO_3(g)$$

has a value of $K = 278$.

Determine the value of K for the *reverse* reaction at this temperature.

Reactivity 2.3.4 – Le Châtelier's principle

SKILLS

Reactivity 2.3.4 – Le Châtelier's principle enables the prediction of the qualitative effects of changes in concentration, temperature and pressure to a system at equilibrium.

Apply Le Châtelier's principle to predict and explain responses to changes of systems at equilibrium.

Include the effects on the value of K and on the equilibrium composition. Le Châtelier's principle can be applied to heterogeneous equilibria such as: $X(g) \rightleftharpoons X(aq)$	Reactivity 2.2 – Why do catalysts have no effect on the value of K or on the equilibrium composition?

Observations of shifts in the position of chemical equilibria. Full details of how to carry out this experiment with a worksheet are available in the eBook.

When a system at equilibrium is subjected to a change, it will respond in such a way as to minimize the effect of the change.

Experiment to show the effect of changing the concentration of chloride ions on the cobalt chloride equilibrium reaction:

$[Co(H_2O)_6]^{2+}(aq) + 4Cl^-(aq) \rightleftharpoons$ $CoCl_4^{2-}(aq) + 6H_2O(l)$

The flask on the left has a low concentration of chloride ions, giving the pink colour of the complex ion with water. As the concentration of chloride ions is increased, the equilibrium shifts to the right, changing the colour from pink to blue. Adding water would shift the equilibrium in the opposite direction. Cobalt chloride is often used to test for the presence of water because of this colour change.

A system at equilibrium will respond when subjected to a change

A system remains at equilibrium so long as the rate of the forward reaction equals the rate of the backward reaction. But as soon as this balance is disrupted by any change in conditions that unequally affects the rates of these reactions, the equilibrium condition will no longer be met. It has been shown, however, that equilibria respond in a predictable way to such a situation based on a principle known as **Le Châtelier's principle**. This states that *a system at equilibrium when subjected to a change will respond in such a way as to minimize the effect of the change*. Simply put, this means that whatever we do to a system at equilibrium, the system will respond in the opposite way. Add something, and the system will react to remove it; remove something, and the system will react to replace it. After a while, a new equilibrium will be established and this will have a different composition from the earlier equilibrium mixture. Applying the principle, therefore, enables us to predict the qualitative effect of typical changes that occur to systems at equilibrium.

Changes in concentration

Suppose an equilibrium is disrupted by an increase in the concentration of one of the reactants. This will cause the rate of the forward reaction to increase while the backward reaction will not be affected, so the reaction rates will no longer be equal. When equilibrium re-establishes itself, the mixture will have new concentrations of all reactants and products, and the equilibrium will have shifted in favour of products. The value of K will be unchanged. This is in keeping with the prediction from Le Châtelier's principle: the addition of reactant causes the system to respond by removing reactant – this favours the forward reaction and so shifts the equilibrium to the right.

Similarly, the equilibrium could be disrupted by a decrease in the concentration of product by removing product from the equilibrium mixture. As the rate of the backward reaction is now decreased, there will be a shift in the equilibrium in favour of the products. A different equilibrium position will be achieved, but the value of K will be unchanged. Again, this confirms the prediction from Le Châtelier's principle: removal of product causes the system to respond by making more product – this favours the forward reaction and so shifts the equilibrium to the right.

The graph in Figure 3 illustrates these disruptions to equilibrium for the reaction

$$N_2(g) + 3H_2(g) \rightleftharpoons 2NH_3(g)$$

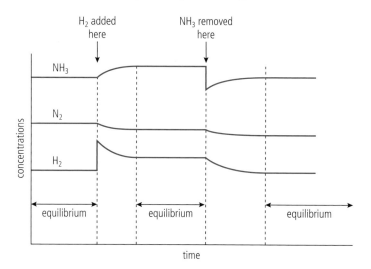

R2.3 Figure 3 Effects of the addition of reactant and removal of product on the equilibrium $N_2(g) + 3H_2(g) \rightleftharpoons 2NH_3(g)$. When H_2 is added, the rate of the forward reaction with N_2 increases and more NH_3 is formed. The equilibrium shifts to the right. When NH_3 is removed, the rate of the backward reaction decreases, and again, the equilibrium shifts to the right producing more NH_3. After each change, a new equilibrium mixture is achieved but K remains constant.

After H_2 is added, the concentrations of N_2 and H_2 decrease (in a 1:3 ratio in keeping with their reaction stoichiometry), while the concentration of NH_3 rises (in a 2:1 ratio relative to N_2) as the rate of the forward reaction increases and the equilibrium shifts to the right. The new equilibrium mixture has a higher concentration of products. In the second part of the graph, removal of NH_3 from the equilibrium mixture causes a decrease in the concentrations of N_2 and H_2 as they react to form more NH_3.

Often in an industrial process, the product will be removed as it forms. This ensures that the equilibrium is continuously pulled to the right, increasing the yield of product.

Applying Le Châtelier's principle, can you think what concentration changes would cause an equilibrium to shift to the left? The answer is either an increase in concentration of product or a decrease in concentration of reactant.

Changes in pressure

Equilibria involving gases will be affected by a change in pressure if the reaction involves a change in the number of molecules. This is because there is a direct relationship between the number of gas molecules and the pressure exerted by a gas in a fixed volume. So if such a reaction at equilibrium is subject to an increase in pressure, the system responds to decrease this pressure by favouring the side with the smaller number of molecules. Conversely, a decrease in pressure will cause a shift in the equilibrium position to the side with the larger number of molecules of gas. A different equilibrium position will be achieved but the value of K will be unchanged, so long as the temperature remains the same.

For example, consider the reaction used in the production of methanol:

$$CO(g) + 2H_2(g) \rightleftharpoons CH_3OH(g)$$

In total, there are three molecules of gas on the left side and one molecule of gas on the right side. So for this reaction, high pressure will shift the equilibrium to the right, in favour of the smaller number of molecules, which increases the yield of CH_3OH.

Changes in pressure or volume will affect the position of equilibrium of a reaction if it involves a change in the number of gas molecules.

443

Note that many common equilibrium reactions do not involve a change in the number of gas molecules and so are not affected by changes in pressure. For example, the reaction

$$H_2(g) + I_2(g) \rightleftharpoons 2HI(g)$$

has two molecules of gas on both sides of the equation. Changing the pressure for this reaction will affect the rate of the reaction but not the position of equilibrium or the value of K.

Changes in temperature

We have noted that K is temperature dependent, so changing the temperature will change K. However, in order to predict *how* K will change we must examine the enthalpy changes (see Reactivity 1.1) of the forward and backward reactions. Remember that an exothermic reaction releases energy (ΔH is negative), whereas an endothermic reaction absorbs energy (ΔH is positive). The enthalpy changes of the forward and backward reactions are equal and opposite to each other.

So if we apply Le Châtelier's principle, including the energy change in the chemical reaction, we can predict how the reaction will respond to a change in temperature.

Consider the reaction

$$2NO_2(g) \rightleftharpoons N_2O_4(g) \quad \Delta H = -57 \text{ kJ mol}^{-1}$$
$$\text{brown} \qquad \text{colourless}$$

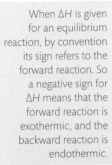

The negative sign of ΔH tells us that the forward reaction is exothermic and so releases heat. If this reaction at equilibrium is subjected to a decrease in temperature, the system will respond by producing heat and it does this by favouring the forward exothermic reaction. This means that the equilibrium will shift to the right, in favour of the product, N_2O_4. A new equilibrium mixture will be achieved and the value of K will increase. So here we can see that the reaction will give a higher yield of products at a lower temperature.

Conversely, increasing the temperature favours the backward, endothermic, reaction, and so shifts the equilibrium to the left, decreasing the value of K. The reaction mixture becomes a darker colour as the concentration of NO_2 increases.

▲ Experiment to show the effect of temperature on the reaction that converts NO_2 (brown) to N_2O_4 (colourless). As the temperature is increased, more NO_2 is produced and the gas becomes darker, as seen in the tube on the left. With a decrease in temperature (the tube on the right), the equilibrium shifts to the right, producing more N_2O_2, and the gas becomes paler.

The table below illustrates the effect of temperature on the value of K for this reaction.

Temperature / K	K for $2NO_2(g) \rightleftharpoons N_2O_4(g)$
273	1300
298	170

Even though, in this case, a lower temperature will produce an equilibrium mixture with a higher proportion of products, remember from Reactivity 2.2 that low temperature also causes a lower rate of reaction. And so, although a higher yield will be produced eventually, it may simply take too long to achieve this for practical and economic considerations if this was an industrial-scale reaction. We will come back to this point later in this chapter.

Now consider the following reaction:

$$N_2(g) + O_2(g) \rightleftharpoons 2NO(g) \quad \Delta H = +181 \text{ kJ mol}^{-1}$$

In this case, we can see that the forward reaction is endothermic and so absorbs heat. So here the effect of a decreased temperature will be to favour the backward exothermic reaction. Therefore, the equilibrium will shift to the left in favour of reactants, and K will decrease. At higher temperatures, the forward, endothermic, reaction is favoured, so the equilibrium shifts to the right and K will increase.

The table below illustrates the effect of temperature on the value of K for this reaction.

Temperature / K	K for $N_2(g) + O_2(g) \rightleftharpoons 2NO(g)$
298	4.5×10^{-31}
900	6.7×10^{-10}
2300	1.7×10^{-3}

Challenge yourself

4. Use information from this section to explain why there is very little NO in the atmosphere under ordinary conditions, and why severe air pollution is often characterized by a brownish haze.

The reaction $N_2(g) + O_2(g) \rightleftharpoons 2NO(g)$ takes place in motor vehicles where the heat released by combustion of the fuel is sufficient to cause the nitrogen and oxygen gases from the air to combine. Unfortunately, the product, NO, is toxic, and, worse still, quickly becomes converted into other toxins that form the components of acid rain and smog. It is therefore of great interest to vehicle manufacturers to find ways of lowering the temperature during combustion in order to reduce the production of NO in the reaction above.

These examples illustrate that, unlike changes in concentration and pressure, changes in temperature *do* cause the value of K to change. An increase in temperature increases the value of K for an endothermic reaction and decreases the value of K for an exothermic reaction. This is because changes in temperature have a different effect on the rates of the forward and backward reactions due to their different activation energies, as discussed in Reactivity 2.2. In the next chapter, we will use this fact to explain why the pH of pure water is temperature dependent.

Addition of a catalyst

As we learn in Reactivity 2.2, a catalyst speeds up the rate of a reaction by providing an alternative reaction pathway that has a transition state with a lower activation energy, E_a. This increases the number of particles that have sufficient energy to react without raising the temperature.

Because the forward and backward reactions pass through the same transition state, a catalyst lowers the activation energy by the same amount for the forward and backward reactions. So the rate of both these reactions will be increased by the same factor, as shown in Figure 4. The catalyst will therefore have no effect on the position of equilibrium, or on the value of K. In other words, the catalyst will not increase the equilibrium yield of product in a reaction. It will, however, speed up the attainment of the equilibrium state and so cause products to form more quickly. Catalysts are generally not shown in the reaction equation or in the equilibrium constant expression as they are not chemically changed at the end of the reaction and have no effect on the equilibrium concentrations.

TOK

Combustion reactions are generally described as exothermic, but the reaction

$N_2(g) + O_2(g) \rightleftharpoons 2NO(g)$

is endothermic. When is it appropriate to refer to an 'exception to a rule', and when does the rule need to be reconsidered?

Increasing the temperature causes an increase in the value of K for an endothermic reaction and a decrease in the value of K for an exothermic reaction.

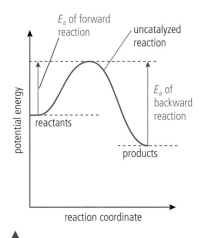
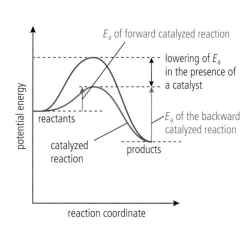

Reactivity 2.2 – Why do catalysts have no effect on the value of K or on the equilibrium composition?

R2.3 Figure 4 The effect of a catalyst in lowering the activation energy of both forward and backward reactions.

Catalysts are widely used in industrial processes to increase the rate of product formation and are involved in every single biochemical reaction, where they are known as enzymes. One of the key principles of Green Chemistry is that catalytic reagents, chosen to be as selective as possible, are superior to stoichiometric reagents as they are not consumed in the reaction.

Catalysts do not change the position of equilibrium or the equilibrium yield of a single reaction, but they enable the equilibrium mixture to be achieved more quickly.

Summary

We can now summarize the effects of concentration, pressure, temperature and catalyst on the position of equilibrium and on the value of K.

Effect of …	Change in position of equilibrium	Change in value of K
1 concentration	changes	no change
2 pressure	changes if reaction involves a change in the number of gas molecules	no change
3 temperature	changes	changes
4 catalyst	no change	no change

A particular reaction at a specified temperature can have many different possible equilibrium positions, but only one value for the equilibrium constant K.

It is worth noting that although we have only seen examples of homogeneous equilibria (in the gaseous phase) in this section, we can also apply Le Châtelier's principle to heterogeneous equilibria involving reactants and products in gaseous and aqueous states.

Consider, for example, a fizzy drink b ottle which contains an equilibrium between dissolved carbon dioxide and gaseous carbon dioxide.

$$CO_2(g) \rightleftharpoons CO_2(aq)$$

When the bottle is opened, releasing some of the gaseous CO_2, the equilibrium shifts to the left. We can see this change occur in the formation of CO_2 bubbles. Once the cap is placed back onto the bottle, a new equilibrium composition will be formed but the equilibrium constant will remain the same.

◄ The fizzing we see when
we open carbonated drinks
is caused by a shift in
equilibrium from aqueous
carbon dioxide to gaseous
carbon dioxide:

$CO_2(g) \rightleftharpoons CO_2(aq)$

If the system remains open, the equilibrium continues to shift to the left until no dissolved carbon dioxide remains and the drink becomes flat.

Applying Le Châtelier's principle

In reactions involving the manufacture of a chemical, it is obviously a goal to obtain as high a conversion of reactant to product as possible. Application of Le Châtelier's principle enables chemists to choose conditions that will cause the equilibrium to lie to the right and so help to achieve this. But the *yield* of a reaction is only part of the consideration. The *rate* is also clearly of great significance as it would be of limited value if a process were able to claim a high equilibrium yield of product, but took several years to achieve this. The economics of the process will depend on considerations of both the equilibrium and the kinetics of the reaction – in other words, on how far and how fast the reaction will proceed. Sometimes these two criteria work against each other, and so the best compromise must be reached. Two case studies of industrial processes are discussed here.

Note that you are
not expected to learn
specific conditions for
any reaction, so there
is no need to focus on
the names of catalysts,
specific temperature
used, etc. But you
should be able to apply
an understanding of
equilibria to any given
example, and predict
the conditions likely to
be effective.

It is estimated that over 150 million tonnes of ammonia, NH_3, are produced worldwide every year. Approximately 80% of this ammonia is used to make fertilizers, notably ammonium nitrate, NH_4NO_3. Other uses include synthesis of textiles such as nylon and powerful explosives. In August 2020, a major explosion occurred in the Port of Beirut after 2750 tonnes of confiscated ammonium nitrate was ignited by a fire in an adjacent warehouse. Accidents such as this in industrial plants raise many questions regarding health and safety, storage, and handling of chemicals, and the balance of responsibility between companies, governments, and individuals.

▲
The Port of Beirut explosion killed over 200 people and caused up to $15 billion in property damage.

Worked example

The Haber process is based on the reaction:

$$N_2(g) + 3H_2(g) \rightleftharpoons 2NH_3(g) \quad \Delta H = -93\,kJ\,mol^{-1}$$

Deduce the conditions that would maximize the conversion of reactants to products in this reaction.

Solution

The following information can be derived from this equation:

- all reactants and products are gases
- there is a change in the number of gas molecules as the reaction proceeds: four gas molecules on the left and two on the right
- the forward reaction is exothermic so releases heat; the backward reaction is endothermic so absorbs heat.

Application of Le Châtelier's principle to this reaction enables us to consider the optimum conditions for converting reactants to products.

- Concentration: we would need to *increase* the concentration of the reactants nitrogen and hydrogen by adding more to the reaction vessel (in the molar ratio 1:3 in accordance with their stoichiometry in the equation) and *decrease* the concentration of ammonia by removing it as it forms. These changes would both shift the equilibrium to the right.
- Pressure: as the forward reaction involves a decrease in the number of gas molecules, it will be favoured by a *high pressure*.
- Temperature: as the forward reaction is exothermic, it will be favoured by a lower temperature. However, too low a temperature would cause the rate of reaction to be uneconomically slow, and so a *moderate temperature* is used as a compromise.
- Catalyst: a catalyst does not affect the position of equilibrium but will speed up the rate of production and so help to compensate for the moderate temperature used. An iron-based catalyst is used in the Haber process.

In fact, the Haber process achieves a conversion of H_2 and N_2 into NH_3 of only about 10–20% per pass through the reactor. After separation of the NH_3 product, the unconverted reactants are recycled to the reactor to obtain an overall yield of about 95%. This recycling of unconverted reactants is commonly used in industrial processes and allows processes with low equilibrium yield to be made commercially viable. Designing processes that prevent or minimize waste is the first of the 12 principles of Green Chemistry.

Tractor applying a chemical solution of fertilizer to the soil. Ammonium salts such as ammonium nitrate and sulfate are particularly effective fertilizers as they supply the nitrogen needed by plants and are readily soluble. It is estimated that without the use of ammonium fertilizers, 2 billion people would starve.

Nature of Science

Scientific discoveries often have significant economic, ethical and political implications. Some of these may be unintended consequences of the discovery, such as the environmental degradation caused by the excess use of nitrate fertilizers as an outcome of the Haber process. This raises the question of who must take moral responsibility for the applications of scientific discoveries. The process of science includes risk-benefit analyses, risk assessment, and ethical considerations, but sometimes the full consequences cannot be predicted and do not become known until much later.

Challenge yourself

5. Consider the atom economy of the Haber process described here using the formula:

$$\% \text{ atom economy} = \frac{\text{molar mass of desired product}}{\text{molar mass of all reactants}} \times 100$$

Explain how the atom economy is different from the reaction yield.

In a similar manner to the Haber process, the production of methanol involves an exothermic homogeneous reaction in the gaseous state with more molecules of gas on the left side of the equation:

$$CO(g) + 2H_2(g) \rightleftharpoons CH_3OH(g) \quad \Delta H = -90 \text{ kJ mol}^{-1}$$

Again, Le Châtelier's principle can be used to consider the pressure and temperature conditions that will optimize the conversion of reactants to products.

Condition used	Reasoning
High pressure	The forward reaction involves a reduction in the number of molecules of gas from three molecules of reactant to one molecule of product. High pressure will shift the equilibrium to the right.
Moderate temperature	The forward reaction is exothermic so a low temperature will increase the equilibrium yield but decrease the rate. A moderate temperature is used as a compromise.

A catalyst will also be used to increase the rate of reaction although it has no impact on the position of equilibrium.

Methanol, CH_3OH, is used as a **chemical feedstock** to make other chemicals. A high proportion is converted into methanal, $HCHO$, which is then converted into plastics, paints, explosives, and plywood. Methanol is used as a laboratory solvent, as an antifreeze agent, and in the process of producing biodiesel fuel from fats. Interest has also focused on the potential of methanol as an energy storage molecule in the so-called 'methanol economy'. In on-going efforts to reduce dependence on imported fossil fuels, China has greatly increased production capacity and consumption of methanol for its transportation sector. In 2021, China had over 20 000 methanol taxis in operation.

Drivers stand beside the first batch of M100 (pure methanol) taxis in China's Shaanxi province in 2018.

▼

The Contact process is an industrially significant reaction used to produce sulfuric acid, H_2SO_4. One step in this process is shown below.

$$2SO_2(g) + O_2(g) \rightleftharpoons 2SO_3(g) \quad \Delta H = -198 \, kJ \, mol^{-1}$$

As we have seen throughout this section, Le Châtelier's principle can be used to predict the conditions that will maximize the production of sulfur trioxide in this step. More sulfuric acid by mass is produced worldwide than any other chemical. It has been found that the production of sulfuric acid closely mirrors historical events such as major wars that affect a country's economy. For this reason, some economists use sulfuric acid production as a measure of a country's industrial strength. Sulfuric acid is used directly or indirectly in nearly all industrial processes, including the production of fertilizers, detergents and paints, and in ore processing, steel production, and water treatment. Approximately 250 million tonnes of sulfuric acid are produced annually across all continents.

Exercise

Q10. The manufacture of sulfur trioxide can be represented by the equation below:

$$2SO_2(g) + O_2(g) \rightleftharpoons 2SO_3(g) \quad \Delta H = -198 \, kJ \, mol^{-1}$$

What happens when a catalyst is added to an equilibrium mixture from this reaction?

A The rate of the forward reaction increases and that of the reverse reaction decreases.

B The rates of both forward and reverse reactions increase.

C The value of ΔH increases.

D The yield of sulfur trioxide increases.

Q11. What will happen to the position of equilibrium and the value of the equilibrium constant when the temperature is increased in the following reaction?

$$Br_2(g) + Cl_2(g) \rightleftharpoons 2BrCl(g) \quad \Delta H = +14 \, kJ \, mol^{-1}$$

	Position of equilibrium	Value of equilibrium constant
A	shifts towards the reactants	decreases
B	shifts towards the reactants	increases
C	shifts towards the products	decreases
D	shifts towards the products	increases

Q12. Which changes will shift the position of equilibrium to the right in the following reaction?

$$2CO_2(g) \rightleftharpoons 2CO(g) + O_2(g)$$

I.   adding a catalyst

II. decreasing the oxygen concentration

III. increasing the volume of the container

A I and II only   B I and III only   C II and III only   D I, II and III

Q13. For each of the following reactions, predict in which direction the equilibrium will shift in response to an increase in pressure:

(a) $2CO_2(g) \rightleftharpoons 2CO(g) + O_2(g)$

(b) $CO(g) + 2H_2 \rightleftharpoons CH_3OH(g)$

(c) $H_2(g) + Cl_2(g) \rightleftharpoons 2HCl(g)$

Q14. How will the equilibrium

$$CH_4(g) + 2H_2S(g) \rightleftharpoons CS_2(g) + 4H_2(g) \quad \Delta H = +ve$$

respond to the following changes?

(a) addition of $H_2(g)$

(b) addition of $CH_4(g)$

(c) a decrease in the volume of the container

(d) removal of $CS_2(g)$

(e) increase in temperature

Q15. The reaction

$$2CO(g) + O_2(g) \rightleftharpoons 2CO_2(g) \quad \Delta H = -566\,kJ\,mol^{-1}$$

takes place in catalytic converters in cars. If this reaction is at equilibrium, will the amount of CO increase, decrease or stay the same when:

(a) the pressure is increased by decreasing the volume?

(b) the pressure is increased by adding $O_2(g)$?

(c) the temperature is increased?

(d) a platinum catalyst is added?

Q16. In the Haber process for the synthesis of ammonia, what effects does the catalyst have?

	Rate of formation of $NH_3(g)$	Amount of $NH_3(g)$ formed
A	increases	increases
B	increases	decreases
C	increases	no change
D	no change	increases

Q17. $$2SO_2(g) + O_2(g) \rightleftharpoons 2SO_3(g) \quad \Delta H = -200\,kJ\,mol^{-1}$$

According to the above information, what temperature and pressure conditions produce the greatest amount of SO_3?

	Temperature	Pressure
A	low	low
B	low	high
C	high	high
D	high	low

Q18. Predict how you would expect the value of K for the exothermic Haber process to change as the temperature is increased. Explain the significance of this in terms of the reaction yield.

Guiding Question revisited

How can the extent of a reversible reaction be influenced?

In this chapter we have considered the following features of reversible reactions:

- A dynamic equilibrium is reached:
  - in a closed physical or chemical system
  - when the rates of forward and backward reactions are equal and the concentrations of reactants and products remain constant.
- The equilibrium constant, K, can be determined from the stoichiometry of a reaction and the equilibrium concentrations of reactants and products.
- The magnitude of K indicates the extent of reaction at a given temperature.
- The equilibrium law:
  - states that every reaction will have a particular value of K at a given temperature
  - can be used to quantify the composition of an equilibrium mixture.
- Le Châtelier's principle can be used to predict changes in equilibrium position when a system is subject to changes in concentration, pressure and temperature. It states that *a system at equilibrium when subjected to a change will respond in such a way as to minimize the effect of the change.* Therefore:
  - an increase in concentration of reactants will shift the equilibrium towards the products and vice versa; K remains unchanged
  - an increase in pressure will shift the equilibrium towards the side of the reaction with the least gaseous molecules and vice versa; K remains unchanged
  - an increase in temperature will shift the equilibrium in the endothermic direction and vice versa; K changes at different temperatures
  - the use of a catalyst has no effect on the position of equilibrium as it increases the forward and backward rate by equal amounts; K remains unchanged.

Practice questions

1. Which statement about chemical equilibria implies they are dynamic?

   A   The position of equilibrium constantly changes.

   B   The rates of forward and backward reactions change.

   C   The reactants and products continue to react.

   D   The concentrations of the reactants and products continue to change.

2. The reaction below represents the equilibrium established between gaseous and aqueous carbon dioxide in a carbonated soft drinks bottle.

   $$CO_2(g) \rightleftharpoons CO_2(aq)$$

   The bottle is opened for a short period of time allowing some $CO_2(g)$ to escape before the lid is placed back on. What is the effect of this change on the position of equilibrium and the concentration of $CO_2(aq)$?

	Position of equilibrium	[CO$_2$(aq)]
A	shifts to the left	decreases
B	shifts to the right	increases
C	shifts to the left	increases
D	shifts to the right	decreases

3. What is the effect of an increase of temperature on the yield and the equilibrium constant for the following reaction?

$$2H_2(g) + CO(g) \rightleftharpoons CH_3OH(l) \quad \Delta H = -128 \text{ kJ mol}^{-1}$$

	Yield	Equilibrium constant
A	increases	increases
B	increases	decreases
C	decreases	increases
D	decreases	decreases

4. Consider the equilibrium between methanol, $CH_3OH(l)$, and methanol vapour, $CH_3OH(g)$.

$$CH_3OH(l) \rightleftharpoons CH_3OH(g)$$

What happens to the position of equilibrium and the value of K as the temperature decreases?

	Position of equilibrium	Value of K
A	shifts to the left	decreases
B	shifts to the left	increases
C	shifts to the right	decreases
D	shifts to the right	increases

5. An increase in temperature increases the amount of chlorine present in the following equilibrium.

$$PCl_5(s) \rightleftharpoons PCl_3(l) + Cl_2(g)$$

What is the best explanation for this?

A The higher temperature increases the rate of the forward reaction only.

B The higher temperature increases the rate of the reverse reaction only.

C The higher temperature increases the rate of both reactions but the forward reaction is affected more than the reverse.

D The higher temperature increases the rate of both reactions but the reverse reaction is affected more than the forward.

6. Consider the following reversible reaction.

$$Cr_2O_7^{2-}(aq) + H_2O(l) \rightleftharpoons 2CrO_4^{2-}(aq) + 2H^+(aq)$$

What will happen to the position of equilibrium and the value of K when more H^+ ions are added at constant temperature?

	Position of equilibrium	Value of K
A	shifts to the left	decreases
B	shifts to the right	increases
C	shifts to the right	does not change
D	shifts to the left	does not change

453

7. Consider this equilibrium reaction in a sealed container:

$$H_2O(g) \rightleftharpoons H_2O(l)$$

What will be the effect on the equilibrium of increasing the temperature from 20°C to 30°C?

   A   More of the water will be in the gaseous state at equilibrium.

   B   More of the water will be in the liquid state at equilibrium.

   C   At equilibrium the rate of condensation will be greater than the rate of evaporation.

   D   At equilibrium the rate of evaporation will be greater than the rate of condensation.

8. Which statement is correct for the equilibrium:

$$H_2O(l) \rightleftharpoons H_2O(g)$$

in a closed system at 100°C?

   A   All the $H_2O(l)$ molecules have been converted to $H_2O(g)$.

   B   The rate of the forward reaction is greater than the rate of the reverse reaction.

   C   The rate of the forward reaction is less than the rate of the reverse reaction.

   D   The pressure remains constant.

9. Consider the following equilibrium:

$$2SO_2(g) + O_2(g) \rightleftharpoons 2SO_3(g) \quad \Delta H = -198\,kJ\,mol^{-1}$$

   (a) Deduce the equilibrium constant expression, K, for the reaction. (1)

   (b) State and explain the effect of increasing the pressure on the yield of sulfur trioxide. (2)

   (c) State and explain the effect of increasing the temperature on the yield of sulfur trioxide. (2)

   (d) State the effects of a catalyst on the forward and reverse reactions, on the position of equilibrium, and on the value of K. (3)

   (Total 8 marks)

10. The Haber process enables the large-scale production of ammonia needed to make fertilizers.

    The equation for the Haber process is given below.

    $$N_2(g) + 3H_2(g) \rightleftharpoons 2NH_3(g)$$

    The percentage of ammonia in the equilibrium mixture varies with temperature.

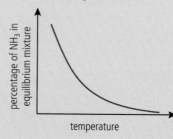

(a) Use the graph to deduce whether the forward reaction is exothermic or endothermic and explain your choice. (2)

(b) State and explain the effect of increasing the pressure on the yield of ammonia. (2)

(c) Deduce the equilibrium constant expression, K, for the reaction. (1)

(Total 5 marks)

11. An example of a homogeneous reversible reaction is the reaction between hydrogen and iodine.

$$H_2(g) + I_2(g) \rightleftharpoons 2HI(g)$$

(a) Outline the characteristics of a homogeneous chemical system that is in a state of equilibrium. (2)

(b) Deduce the expression for the equilibrium constant, K. (1)

(c) Predict what would happen to the position of equilibrium if the pressure was increased from 1 atm to 2 atm. (1)

(d) The value of K at 500 K is 160 and the value of K at 700 K is 54. Deduce what this information tells us about the enthalpy change of the forward reaction. (1)

(e) Determine the value of the equilibrium constant, K', for the reverse reaction at 500 K. (1)

(f) The reaction can be catalyzed by adding platinum metal. State and explain what effect the addition of platinum would have on the value of the equilibrium constant. (2)

(Total 8 marks)

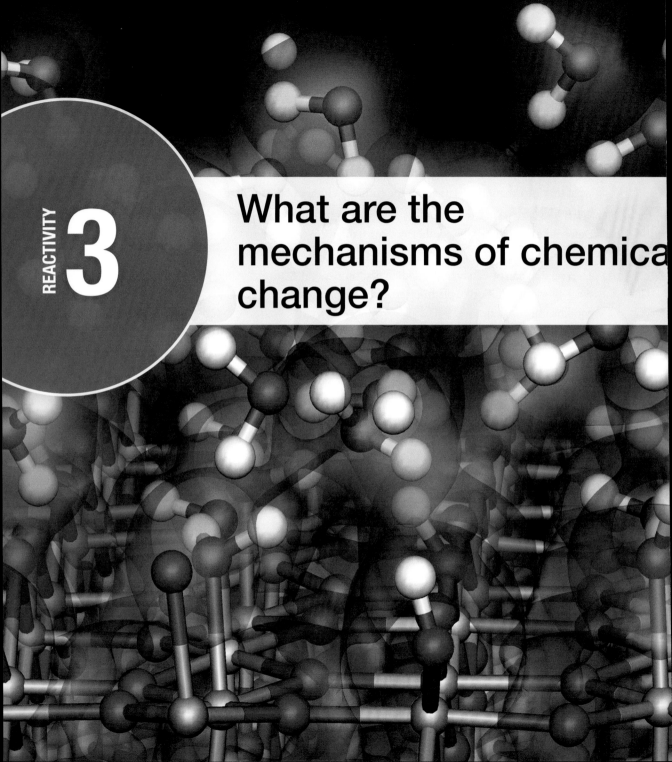

What are the mechanisms of chemical change?

A water-titanium dioxide model produced from molecular dynamics calculations which simulated proton transfer for water. The structure and charge density of H_2O molecules are shown across the centre, and the TiO_2 surface is shown across the bottom. The atoms are colour-coded: hydrogen (white), oxygen (red) and titanium (grey). The charge density (electron cloud) is purple. TiO_2 is a widely used catalyst in solar cells and fuel cells. Understanding the mechanism of its reaction with H_2O is key to further development of its role in photocatalytic mechanisms.

Reaction mechanisms refer to how chemical reactions proceed at the molecular level. Many reactions involve more than one step, and in these cases the mechanism is shown as a sequence where the product of one step is the reactant for the next.

Although there are many ways in which reaction mechanisms can be classified, there are only four fundamental types of reaction based on the behaviour of subatomic particles.

- Proton transfer reactions include the range of acid–base reactions and neutralization.
- Electron transfer reactions include all oxidation–reduction processes.
- Electron-sharing reactions include reactions between radicals.
- Electron-pair-sharing reactions include the reactions of electrophiles and nucleophiles.

We will consider each of these types of reaction in turn in this topic. As you study these examples, try to keep in mind the fundamental processes occurring at the molecular level. It is the particulate nature of the structure of the reactants that determines the reaction mechanisms.

Given the infinite number of different chemical reactions, it is helpful to realize that they can be categorized in this relatively simple way based on the transfer or sharing of subatomic particles between reactants.

Proton transfer reactions

◀ The confluence of two streams in the Andes mountains in Argentina. The stream in the centre is brown due to minerals and precipitates from an upstream mining site, and is more acidic with pH 3.0. The stream on the left is close to neutral with pH 7.5. pH values are a convenient way to express the concentration of H^+ ions in a solution, which is essential to understanding and predicting many acid–base properties.

Guiding Question

What happens when protons are transferred?

Hydrogen atoms are the simplest atoms, containing just one proton and one electron. When they ionize by losing the electron and forming H^+, all that is left is the proton. H^+ is therefore equivalent to a proton, and we can use these two terms interchangeably.

When a proton is transferred in a reaction:

- the reactant loses H^+ so loses a positive charge
- the product gains H^+ and so gains a positive charge.

This type of reaction will therefore only be possible between certain species – reactants that can release H^+, and products that have a lone pair that can accommodate an additional H^+. This behaviour is known as Brønsted–Lowry acid–base behaviour.

In this chapter, we will identify species that can undergo proton transfer reactions, and learn how these reactions can be predicted and followed with reference to the pH scale. Different acids and bases share some common properties, but differ in their strength depending on their equilibria in aqueous solution. It is strongly recommended that you are familiar with the study of equilibria from Reactivity 2.3 before starting this chapter.

Acid–base chemistry is central to topics such as air and water pollution, how climate change affects the chemistry of the oceans, the action of drugs in the body, food science and many aspects of cutting-edge research.

Reactivity 3.1.1 and 3.1.2 – Brønsted–Lowry acids and bases

Reactivity 3.1.1 – A Brønsted–Lowry acid is a proton donor and a Brønsted–Lowry base is a proton acceptor.	
Deduce the Brønsted–Lowry acid and base in a reaction.	
A proton in aqueous solution can be represented as both $H^+(aq)$ and $H_3O^+(aq)$. The distinction between the terms 'base' and 'alkali' should be understood.	Nature of Science, Reactivity 3.4 – Why has the definition of acid evolved over time?

Reactivity 3.1.2 – A pair of species differing by a single proton is called a conjugate acid–base pair.	
Deduce the formula of the conjugate acid or base of any Brønsted–Lowry base or acid.	
	Structure 2.1 – What are the conjugate acids of the polyatomic anions listed in Structure 2.1?

Brønsted–Lowry theory describes acids and bases in terms of proton transfer

Early theories to define acids and bases

The French chemist Lavoisier proposed in 1777 that oxygen was the 'universal acidifying principle'. He believed that an acid could be defined as a compound of oxygen and a non-metal. In fact, the name he gave to the newly discovered gas *oxygen* means 'acid-former'. This theory, however, had to be dismissed when the acid HCl was proven to be made of hydrogen and chlorine only – no oxygen. To hold true, of course, any definition of an acid has to be valid for *all* acids.

A big step forward came in 1887 when the Swedish chemist Arrhenius suggested that an acid could be defined as a substance that dissociates in water to form hydrogen ions (H^+) and anions, while a base dissociates into hydroxide (OH^-) ions and cations. He also recognized that the hydrogen and hydroxide ions could form water, and the cations and anions could form a salt. In a sense, Arrhenius was very close to the theory that is widely used to explain acid and base properties today, but his focus was only on aqueous systems. A broader theory was needed to account for reactions occurring without water, and especially for the fact that some insoluble substances show base properties.

Svante August Arrhenius (1859–1927) wrote up his ideas on acids dissociating into ions in water as part of his doctoral thesis while he was a student at Stockholm University. But his theory was not well received and he was awarded the lowest possible class of degree. Later, his work gradually gained recognition and he received one of the earliest Nobel Prizes in Chemistry in 1903.

Arrhenius may be less well known as the first person documented to predict the possibility of global warming as a result of human activity. In 1896, aware of the rising levels of CO_2 caused by increased industrialization, he calculated the likely effect of this on the temperature of the Earth. Today, over 100 years later, the significance of this relationship between increasing CO_2 and global temperatures is recognized as a major threat to ecosystems and life on Earth.

Nature of Science

The evolution of theories to explain acid–base chemistry and develop general principles is a fascinating tale of the scientific process in action. Some theories have arisen and been disproved, such as Lavoisier's early definition of an acid as a compound containing oxygen. Falsification of an idea is an essential aspect of the scientific process. Other theories have proved to be too limited in application, such as Arrhenius' theory which could not be generalized beyond aqueous solutions.

On the other hand, the Brønsted–Lowry theory has stood the test of time and experimentation. This indicates that it has led to testable predictions which have supported the theory, and enabled wide ranging applications to be made.

Nature of Science, Reactivity 3.4 – Why has the definition of acid evolved over time?

Brønsted–Lowry theory developed from independent work by two chemists

In 1923 two chemists, Martin Lowry of Cambridge, England, and Johannes Brønsted of Copenhagen, Denmark, working independently, published similar conclusions regarding the definitions of acids and bases. Their findings overcame the limitations of Arrhenius' work and have become established as the **Brønsted–Lowry theory**.

This theory focuses on the transfer of H^+ ions during an acid–base reaction: acids donate H^+, while bases accept H^+. For example, in the reaction between hydrogen chloride, HCl, and ammonia, NH_3, to form ammonium chloride, NH_4Cl, we can see that HCl transfers H^+ to NH_3.

$$HCl + NH_3 \rightleftharpoons NH_4^+ + Cl^-$$

NH_4Cl is an ionic compound containing NH_4^+ and Cl^- ions.

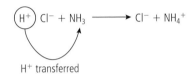

H^+ transferred

HCl acts as an acid and donates H^+, NH_3 acts as a base and accepts H^+.

Reaction between vapors of HCl and NH_3 forming the white smoke of ammonium chloride NH_4Cl.

$HCl(g) + NH_3(g) \rightarrow NH_4Cl(s)$

Challenge yourself

1. From physical properties that you observe in the photo, what might be another way of classifying this reaction?

As we learned on page 459, H^+ is a proton and the two terms can be used interchangeably.

Therefore the Brønsted–Lowry theory can be stated as:

- a Brønsted–Lowry acid is a proton (H^+) donor
- a Brønsted–Lowry base is a proton (H^+) acceptor.

Nature of Science

Brønsted and Lowry's work on acid–base theory is a good example of what is sometimes referred to as 'multiple independent discovery'. This refers to cases where similar discoveries are made by scientists at almost the same time, even though they have worked independently from each other. Examples include the independent discovery of calculus by Newton and Leibniz, the formulation of the mechanism for biological evolution by Darwin and Wallace, and derivation of the pressure–volume relationships of gases by Boyle and Mariotte. Nobel Prizes in Chemistry are commonly awarded to more than one person working in the same field, who may have made the same discovery independently. Multiple independent discoveries are likely increasing as a result of communication technology, which enables scientists who may be widely separated geographically to have access to a common body of knowledge as the basis for their research.

When a discovery is made by more than one scientist at about the same time, ethical questions can arise, such as whether the credit belongs to one individual or should be shared, and whether there are issues regarding intellectual property and patent rights. Historically, these issues have been settled in very different ways in different cases. Nonetheless, the peer review process in science aims to ensure that credit is correctly awarded to the scientist or scientists responsible for a discovery, and that published work represents a new contribution to work in that field.

TOK

Does competition between scientists help or hinder the production of knowledge?

Worked example

Deduce the Brønsted–Lowry acid and base in the following reactions. Consider only the forward reactions here.

1. $NH_3(aq) + H_2O(l) \rightleftharpoons NH_4^+(aq) + OH^-(aq)$

2. $H_2PO_4^-(aq) + HCO_3^-(aq) + \rightleftharpoons HPO_4^{2-}(aq) + H_2CO_3(aq)$

Solution

A Brønsted-Lowry acid is a proton (H⁺) donor.
A Brønsted-Lowry base is a proton (H⁺) acceptor.

1. $NH_3(aq)$ accepts H^+ to become $NH_4^+(aq)$ ∴ $NH_3(aq)$ is the Brønsted–Lowry base.

    $H_2O(l)$ donates H^+ to become $OH^-(aq)$ ∴ $H_2O(l)$ is the Brønsted–Lowry acid.

2. $H_2PO_4^-(aq)$ donates H^+ to become $HPO_4^{2-}(aq)$ ∴ $H_2PO_4^-(aq)$ is the Brønsted–Lowry acid.

    $HCO_3^-(aq)$ accepts H^+ to become $H_2CO_3(aq)$ ∴ $HCO_3^-(aq)$ is the Brønsted–Lowry base.

Reacting species in Brønsted–Lowry acid–base reactions form conjugate pairs

Conjugate acid–base pairs differ by one proton

The act of donating cannot happen in isolation – there must always be something present to play the role of acceptor. In Brønsted–Lowry theory, an acid can therefore only behave as a proton donor if there is also a base present to accept the proton.

Consider the acid–base reaction between a generic acid HA and base B:

$$HA + B \rightleftharpoons A^- + BH^+$$

We can see that HA acts as an acid, donating a proton to B while B acts as a base, accepting the proton from HA. But if we look also at the reverse reaction, we can pick out another acid–base reaction: here, BH^+ is acting as an acid, donating its proton to A^- while A^- acts as a base accepting the proton from BH^+. In other words, acid HA has reacted to form the base A^-, while base B has reacted to form acid BH^+.

$$\text{conjugate acid–base pair}$$
$$HA + B \rightleftharpoons A^- + BH^+$$
$$\text{conjugate acid–base pair}$$

Acids react to form bases and vice versa. The acid–base pairs related to each other in this way are called **conjugate acid–base pairs**, and you can see that they *differ by just one proton*. It is important to be able to recognize these pairs in a Brønsted–Lowry acid–base reaction.

One example of a conjugate pair is H_2O and H_3O^+, which is found in all acid–base reactions in aqueous solution. The reaction:

$$H_2O(l) + H^+(aq) \rightleftharpoons H_3O^+(aq)$$

occurs when a proton released from an acid readily associates with H_2O molecules, forming H_3O^+. In other words, protons become hydrated. H_3O^+ is variously called the hydroxonium ion, the oxonium ion, or the hydronium ion and is always the form of hydrogen ions in aqueous solution. However, for most reactions it is convenient simply to write it as H^+(aq). Note that in this pair H_3O^+ is the conjugate acid and H_2O is its conjugate base.

 Note that H_3O^+(aq) and H^+(aq) are both used to represent a proton in aqueous solution.

 Lowry described the ready hydration of the proton as 'the extreme reluctance of the hydrogen nucleus to lead an isolated existence'.

Worked example

Label the conjugate acid–base pairs in the following reaction:

$$CH_3COOH(aq) + H_2O(l) \rightleftharpoons CH_3COO^-(aq) + H_3O^+(aq)$$

Solution

$$\underbrace{CH_3COOH \,/\, CH_3COO^-}_{\text{conjugate pair}}$$
$$\text{acid} \qquad \text{base}$$

$$\underbrace{H_2O \,/\, H_3O^+}_{\text{conjugate pair}}$$
$$\text{base} \quad \text{acid}$$

 When writing the conjugate acid of a base, add one H^+. When writing the conjugate base of an acid, remove one H^+. Remember to adjust the charge by the 1^+ removed or added.

Deducing the formulas of conjugate acids and bases

In a conjugate pair, the acid always has one proton more than its conjugate base, so this makes it easy to predict the formula of the corresponding conjugate for any given acid or base.

Worked example

1. Write the conjugate base for each of the following.

   (a) H_3O^+ (b) NH_3 (c) H_2CO_3

2. Write the conjugate acid for each of the following.

   (a) NO_2^- (b) OH^- (c) CO_3^{2-}

Solution

The acid and base in a conjugate acid–base pair differ by just one proton.

1. To form the base from these species, remove one H^+

   (a) H_2O (b) NH_2^- (c) HCO_3^-

2. To form the acid from these species, add one H^+

   (a) HNO_2 (b) H_2O (c) HCO_3^-

If we look at the polyatomic ions listed in Structure 2.1, we see that most of the ions can act as Brønsted–Lowry bases by accepting H^+ and forming Brønsted–Lowry acids.

Structure 2.1 – What are the conjugate acids of the polyatomic ions listed in Structure 2.1?

Polyatomic ion	Conjugate acid
hydroxide, OH^-	H_2O
nitrate, NO_3^-	HNO_3
hydrogencarbonate, HCO_3^-	H_2CO_3
carbonate, CO_3^{2-}	HCO_3^-
sulfate, SO_4^{2-}	HSO_4^-
phosphate, PO_4^{3-}	HPO_4^{2-}

The ammonium ion, NH_4^+, which is also a polyatomic ion listed in Structure 2.1, cannot accept H^+ to form a conjugate acid. It instead loses H^+, and so acts as a Brønsted–Lowry acid and forms its conjugate base, NH_3. Which of the polyatomic ions above can also function as a Brønsted–Lowry acid?

Brønsted–Lowry bases are defined as any species which can accept a proton. A smaller group, the **alkalis**, are soluble bases which dissolve in water to release the hydroxide ion OH^-.

For example:

$$K_2O(s) + H_2O(l) \rightarrow 2K^+(aq) + 2OH^-(aq)$$

$$CO_3^{2-}(aq) + H_2O(l) \rightleftharpoons HCO_3^-(aq) + OH^-(aq)$$

$$HCO_3^-(aq) \rightleftharpoons CO_2(g) + OH^-(aq)$$

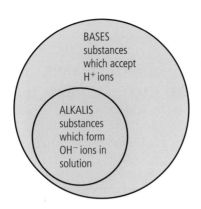

Alkalis are bases that dissolve in water to form the hydroxide ion, OH⁻.

Exercise

Q1. Deduce the formula of the conjugate acid of each of the following:

(a) SO_3^{2-} (b) CH_3NH_2 (c) $C_2H_5COO^-$

(d) NO_3^- (e) F^- (f) HSO_4^-

Q2. Deduce the formula of the conjugate base of each of the following:

(a) H_3PO_4 (b) CH_3COOH (c) H_2SO_3

(d) HSO_4^- (e) OH^- (f) HBr

Q3. For each of the following reactions, identify the Brønsted–Lowry acids and bases and the conjugate acid–base pairs:

(a) $CH_3COOH + NH_3 \rightleftharpoons NH_4^+ + CH_3COO^-$

(b) $CO_3^{2-} + H_3O^+ \rightleftharpoons H_2O + HCO_3^-$

(c) $NH_4^+ + NO_2^- \rightleftharpoons HNO_2 + NH_3$

Reactivity 3.1.3 – Amphiprotic species

Reactivity 3.1.3 – Some species can act as both Brønsted–Lowry acids and bases.

Interpret and formulate equations to show acid–base reactions of these species.

	Structure 3.1 – What is the periodic trend in the acid–base properties of metal and non-metal oxides?
	Structure 3.1 – Why does the release of oxides of nitrogen and sulfur into the atmosphere cause acid rain?

Some species can act as both Brønsted–Lowry acids and bases

You may be surprised to see water described in the answers to the Worked examples on page 464 as a base (Q1, part a), and as an acid (Q2, part b), as you are probably not used to thinking of water as an acid, or as a base, but rather as a neutral substance. The point is that Brønsted–Lowry theory describes acids and bases in terms of how they react together, so it all depends on what water is reacting with.

Consider the following:

$$CH_3COOH + H_2O \rightleftharpoons CH_3COO^- + H_3O^+$$
$$\text{acid} \qquad \text{base} \qquad \text{base} \qquad \text{acid}$$

$$NH_3 + H_2O \rightleftharpoons NH_4^+ + OH^-$$
$$\text{base} \quad \text{acid} \qquad \text{acid} \quad \text{base}$$

So with CH_3COOH, water acts as a Brønsted–Lowry base, and with NH_3 it acts as a Brønsted–Lowry acid.

Notice that water is not the only species that can act as both an acid and as a base. For example, in the Worked example below, we can deduce similar behaviour with HCO_3^-.

Worked example

Write equations to show HCO_3^- acting **(a)** as a Brønsted–Lowry acid and **(b)** as a Brønsted–Lowry base.

Solution

(a) to act as an acid, it donates H^+

$HCO_3^-(aq) + H_2O(l) \rightleftharpoons CO_3^{2-}(aq) + H_3O^+(l)$

(b) to act as a base, it accepts H^+

$HCO_3^-(aq) + H_2O(l) \rightleftharpoons H_2CO_3(aq) + OH^-(aq)$

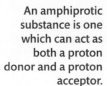
Amphoteros is a Greek word meaning 'both'. For example, *amphi*bians are adapted both to water and to land.

An amphiprotic substance is one which can act as both a proton donor and a proton acceptor.

Substances that can act as both Brønsted–Lowry acids and bases in this way are said to be **amphiprotic**. What are the features that enable these species to have this 'double identity'?

- To act as a Brønsted–Lowry acid, they must be able to dissociate and release H^+.

- To act as a Brønsted–Lowry base, they must be able to accept H^+, which means they must have a lone pair of electrons.

So substances that are amphiprotic according to Brønsted–Lowry theory must possess both a lone pair of electrons and hydrogen that can be released as H^+.

Challenge yourself

2. At ambient temperature, the polyatomic ion $OH^-(aq)$ may not be considered truly amphiprotic as its basic properties dominate – losing H^+ in solution is unlikely due to the instability of O^{2-} in water. However, when a metal hydroxide such as magnesium hydroxide is heated, it decomposes as follows:

$$Mg(OH)_2(s) \rightarrow MgO(s) + H_2O(g)$$

Derive the ionic form of this reaction to analyze how the hydroxide ion is showing amphiprotic behaviour here.

Nature of Science

Terminology in science has to be used appropriately according to the context. *Amphiprotic* specifically relates to Brønsted–Lowry acid–base theory, where the emphasis is on the transfer of a proton. The term *amphoteric,* on the other hand, has a broader meaning as it is used to describe a substance which can act as an acid and as a base, including reactions that do not involve the transfer of a proton. For example, as described in Structure 3.1, aluminium oxide is an amphoteric oxide because it reacts with both dilute acids and alkalis. We will see in Reactivity 3.4 that this acid–base behaviour is best described by a different theory, the Lewis theory. Note that all amphiprotic substances are also amphoteric, but the converse is not true. The fact that the two terms exist reflects the fact that different theories are used to describe acid–base reactions.

Structure 3.1 – What is the periodic trend in the acid–base properties of metal and non-metal oxides?

Structure 3.1 – Why does the release of oxides of nitrogen and sulfur into the atmosphere cause acid rain?

We can recognize trends in acidic and basic behaviour related to the periodic table. Moving left to right across a period, the oxides transition from basic metal oxides through **amphoteric** oxides which are able to react with both acids and bases, to acidic oxides. For example, in period 3:

$$Na_2O \quad MgO \qquad\qquad Al_2O_3 \qquad\qquad SiO_2 \quad P_4O_{10} \quad SO_2 \quad Cl_2O$$

basic metal oxides   amphoteric oxide   acidic non-metal oxides

- The basic metal oxides react with an acid to form a salt and water.

$$Na_2O(s) + 2HCl(aq) \rightarrow 2NaOH(aq) + 2H_2O(l)$$

- Al_2O_3 is amphoteric because it reacts both with an acid and with a base to form a salt and water.

$$Al_2O_3(s) + 3H_2SO_4(aq) \rightarrow Al_2(SO_4)_3(aq) + 3H_2O(l)$$
$$Al_2O_3(s) + 3H_2O(l) + 2OH^-(aq) \rightarrow 2Al(OH)_4^-$$

- The non-metal oxides form weak acids in solution.

$$SO_2(g) + H_2O(l) \rightleftharpoons H_2SO_3(aq)$$
$$H_2SO_3(aq) \rightleftharpoons 2H^+(aq) + SO_3^{2-}(aq)$$

Nitrogen and sulfur both form several different oxides, known as NO_x and SO_x, which are released from the burning of fossil fuels, especially coal and heavy oil. As we see above for SO_2, they react with moisture in the atmosphere to form weak acids such as sulfurous and nitrous acids. These weak acids can react further to form stronger acids, resulting in a complex mixture of pollutants in the atmosphere. For example:

$$2NO_2(g) + H_2O(l) \rightarrow HNO_2(aq) + HNO_3(aq)$$

This lowers the pH causing **acid rain,** defined as when the pH of precipitation has been lowered to less than 5.6. Acid rain is a major environmental problem, causing a negative impact on structural materials, lakes and rivers, and plant life, especially coniferous forests.

Test tubes containing normal rainwater (left) and acid rain (right) with universal indicator solution added to show acidity. The normal rainwater has a pH of above 5.6, while the acid rain has a lower pH: in this example it is pH 4.0.

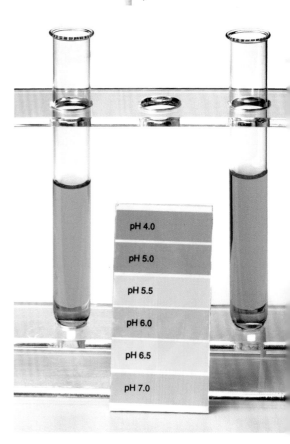

TOK

How might developments in scientific knowledge trigger political controversies or controversies in other areas of knowledge?

Woodland devastated by the effects of acid rain in the Czech Republic, near the border with Germany. Emissions of oxides of sulfur and nitrogen from industry have caused this widespread death of trees. This is one of the most polluted parts of central Europe.

The Danish chemist Søren Peder Lauritz Sorensen (1868–1939) developed the pH concept in 1909, originally proposing that it be formulated as *p*H. He did not account for his choice of the letter 'p' though it has been suggested to originate from the German word *potenz* for power. It could equally well derive from the Latin, Danish, or French terms for the same word.

$$pH = -\log_{10}[H^+]$$
$$[H^+] = 10^{-pH}$$

Exercise

Q4. Show by means of equations how the anion in K_2HPO_4 is amphiprotic.

Q5. (a) Explain with an equation why normal rainwater is acidic because it dissolves CO_2 from the atmosphere.

(b) Formulate equations to show how sulfur dioxide released into the atmosphere can be oxidized to sulfur trioxide, which then dissolves in water to form sulfuric acid.

(c) Suggest why adding lime, calcium oxide, to stream water may help to reduce the effects of acid rain.

Q6. Suggest why the elimination of burning coal as a fuel is an important goal of many countries and international organizations.

Reactivity 3.1.4 – The pH scale

Reactivity 3.1.4 – The pH scale can be used to describe the [H⁺] of a solution:

$$pH = -\log_{10}[H^+] \qquad [H^+] = 10^{-pH}$$

Perform calculations involving the logarithmic relationship between pH and [H⁺].

Include the estimation of pH using universal indicator, and the precise measurement of pH using a pH meter/probe. The equations for pH are given in the data booklet.	Tools 1, 2, 3 – What is the shape of a sketch graph of pH against [H⁺]? Nature of Science, Tool 2 – When are digital sensors (e.g. pH probes) more suitable than analogue methods (e.g. pH paper/solution)?

pH is a logarithmic expression of [H⁺]

Chemists realized a long time ago that it would be useful to have a quantitative scale of acid strength based on the concentration of hydrogen ions. As the majority of acids encountered are weak, the hydrogen ion concentration expressed directly as mol dm⁻³ produces numbers with large negative exponents. For example, the H⁺ concentration in our blood is 4.6×10^{-8} mol dm⁻³. Such numbers are not very user-friendly when it comes to describing and comparing acids. The introduction of the pH scale in 1909 by Sorensen led to wide acceptance owing to its ease of use. It is defined as follows:

$$pH = -\log_{10}[H^+]$$

In other words, pH is the negative number to which the base 10 is raised to give the [H⁺]. This can also be expressed as:

$$[H^+] = 10^{-pH}$$

- A solution that has [H⁺] = 0.1 mol dm⁻³ ⇒ [H⁺] = 10⁻¹ mol dm⁻³ ⇒ pH = 1.
- A solution that has [H⁺] = 0.01 mol dm⁻³ ⇒ [H⁺] = 10⁻² mol dm⁻³ ⇒ pH = 2.

The pH scale has distinct features

pH numbers are usually positive and have no units

Although the pH scale is theoretically an infinite scale and can extend into negative numbers, most common acids and bases will have positive pH values and fall within the range pH 0–14. This corresponds to $[H^+]$ in the range $1.0\ mol\ dm^{-3}$ to $10^{-14}\ mol\ dm^{-3}$.

The pH number is inversely related to [H⁺]

Solutions with a higher $[H^+]$ have a lower pH and vice versa. So stronger and more concentrated acids have a lower pH, while weaker and more dilute acids have a higher pH.

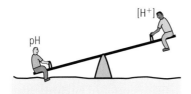

The inverse relationship between pH and H⁺.

A change of one pH unit represents a 10-fold change in [H⁺]

This means that increasing the pH by one unit represents a decrease in $[H^+]$ by 10 times; and decreasing by one pH unit represents an increase in $[H^+]$ by 10 times.

Worked example

If the pH of a solution is changed from 3 to 5, deduce how the hydrogen ion concentration changes.

Solution

$$pH = 3 \Rightarrow [H^+] = 10^{-3}\ mol\ dm^{-3} \qquad pH = 5 \Rightarrow [H^+] = 10^{-5}\ mol\ dm^{-3}$$

So $[H^+]$ has changed by 10^{-2} or decreased by a factor of 100.

The inverse logarithmic relationship between pH and $[H^+]$ can be illustrated by following the change in pH as the $[H^+]$ of a solution is increased. We can observe the following:

- as $[H^+]$ in a solution increases, the pH value decreases
- for each increase of 10 times in $[H^+]$, the pH will decrease by 1 unit.

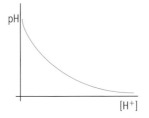

The inverse logarithmic relationship between pH and [H⁺].

Logarithmic scales are used in other disciplines where they can help in the presentation of data. In medicine, the logarithmic decay of levels of drugs in the blood against time is used. In seismic studies, the Richter scale is a logarithmic expression of the relative energy released during earthquakes. In sense perception, the decibel scale of sound intensity and the scale for measuring the visible brightness of stars (their magnitude) are also logarithmic scales.

Tools 1, 2, 3 – What is the shape of a sketch graph of pH against [H⁺]?

Note that a sketch graph shows labelled axes but without units. The graph shows the overall relationship between two variables but without plotted data points.

Nature of Science

The use of a logarithmic scale to represent data enables a wide range of values to be presented as a smaller range of simpler numbers. We see here that the pH scale takes a range of hydrogen ion concentrations from 10^0 to 10^{-14} and effectively compresses it to a much smaller scale of numbers, 0–14. On page 468, we see that the use of a logarithmic scale to represent ionization energies similarly makes a wide range of data easier to interpret. In cases of exponential change, a log base scale shows the data as a straight line, so outliers where the change is greater or less than exponential can be easily identified.

But use of a logarithmic scale can also be misleading, as it can obscure the scale of change if not interpreted correctly. For example, a small change in pH represents a dramatic difference in the hydrogen ion concentration of a solution. Keep this in mind when you read reports of changes in pH, for example, of rainfall as a result of pollution. A reported change from pH 5.5 to pH 4.5 may not sound much, but in fact it represents a *tenfold increase* in the hydrogen ion concentration – hugely significant to the acidic properties. The pH of our blood is carefully controlled by chemicals called buffers to remain at 7.4, and a change of only half a pH unit on either side of this can be fatal. Communication of data in science takes different forms and uses different scales and must always be interpreted in the context used.

The pH scale at 298 K and pH values of some common substances.

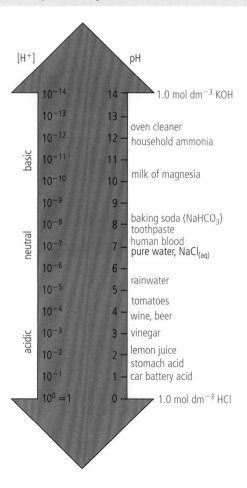

Measuring pH

An easy way to estimate pH is with universal indicator paper or solution. The substance tested will give a characteristic colour, which can then be compared with a colour chart supplied with the indicator. Narrower-range indicators give a more accurate reading than broad-range indicators. The use of indicators is an analogue method which always depends on the user's interpretation of the colour, and so is somewhat subjective. Indicators are useful for estimations and comparisons of pH, rather than for precise measurements.

▲ Digital pH meter being used to measure the pH of a chemical solution. The tip of the probe is immersed in the liquid, which is weakly acidic (pH 5.8). The probe's tip contains electrodes that measure the electrical potential of the hydrogen ions, which is directly related to pH.

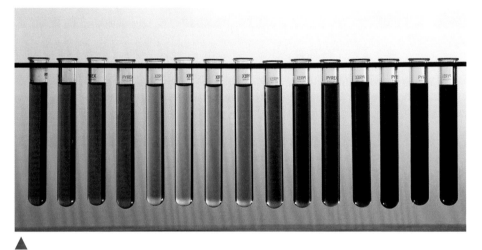

▲ The pH scale and universal indicator solution. The tubes contain solutions of pH 0–14 from left to right, with added universal indicator.

A more objective and usually more accurate means is by using a **pH meter** that directly reads the H^+ concentration through a special electrode. pH meters can record to an accuracy of several decimal places. They must, however, be calibrated before each use with a buffer solution and standardized for the temperature, as pH is a temperature-dependent measurement.

Nature of Science, Tool 2 – When are digital sensors (e.g. pH probes) more suitable than analogue methods (e.g. pH paper/solution)?

SKILLS Use of a pH meter and universal indicator to measure pH. Full details of how to carry out this experiment with a worksheet are available in the eBook.

pH calculations

From the equations of the relationship between pH and $[H^+]$ we can:

- calculate the value of pH from a known concentration of H^+
- calculate the concentration of H^+ from a given pH.

The pH scale is a measure of $[H^+]$ and so may appear to be more suited for the measurement of acids than bases. But we can use the same scale to describe the alkalinity of a solution. As you can see below, this is because the relationship between $[H^+]$ and $[OH^-]$ is inverse in aqueous solutions, and so lower $[H^+]$ means higher $[OH^-]$ and vice versa. Therefore, the scale of pH numbers represents a range of values from strongly acidic through to strongly alkaline.

In pH calculations you will need to be able to work out logarithms to base 10, and anti-logarithms. Make sure that you are familiar with these operations on your calculator.

pH 0	pH 7	pH 14
increasing $[H^+]$ increasing acidity	neutral at 298 K	increasing $[OH^-]$ increasing alkalinity

▲

The colours of the flowers of hydrangea plants (*Hydrangea macrophyll*) depend on the availability of aluminium ions, Al^{3+}, which react with the plant's pigments. A lower pH makes aluminium ions more easily absorbed, and the flowers are blue. A higher pH keeps the aluminium ions locked in insoluble form unavailable to plants, and the flowers are pink. The pink flower on the left was grown at pH 6.5, the blue flower on the right at pH 5.0, and the purple flower in the middle at pH 5.8, so both colours are present.

Challenge yourself

3. Derive equations to show why aluminium sulfate, when added to water, gives a cloudy appearance and has a low pH.

4. Suggest why some gardeners water hydrangea flowers with a weak solution of vinegar.

Note that the logarithms used in all the work on acids and bases are logarithms to base 10 (log). Do not confuse these with natural logarithms, to base e, (ln).

Worked example

A sample of lake water was analyzed at 298 K and found to have $[H^+] = 3.2 \times 10^{-5}\,mol\,dm^{-3}$. Calculate the pH of this water and comment on its value.

Solution

$$pH = -\log_{10}[H^+]$$
$$= -\log_{10}(3.2 \times 10^{-5}) = -(-4.49485)$$
$$pH = 4.49$$

At 298 K this pH < 7, and the lake water is therefore acidic.

The rule to follow for significant figures in logarithms is that *the number of decimal places in the logarithm should equal the number of significant figures in the number*. In other words, the number to the left of the decimal place in the logarithm is not counted as a significant figure (this is because it is derived from the exponential part of the number). In this example, the number given has two significant figures (3.2) and so the answer is given to two decimal places (.49).

Worked example

Human blood has a pH of 7.40. Calculate the concentration of hydrogen ions present.

Solution

$$[H^+] = 10^{-pH}$$
$$= 10^{-7.40} = 4.0 \times 10^{-8}$$
$$[H^+] = 4.0 \times 10^{-8}\,mol\,dm^{-3}$$

Exercise

Q7. What happens to the pH of an acid when $10\,cm^3$ of the acid is added to $90\,cm^3$ of water?

Q8. Beer has a hydrogen ion concentration of $1.9 \times 10^{-5}\,mol\,dm^{-3}$. What is its pH?

Q9. What is the pH of a $0.01\,mol\,dm^{-3}$ solution of HCl which dissociates fully?

$$HCl(aq) \rightarrow H^+(aq) + Cl^-(aq)$$

Q10. Lake water was found to have a concentration of H^+ ions of $3.2 \times 10^{-5}\,mol\,dm^{-3}$. Calculate the pH and determine if the lake water is acidic, basic, or neutral at $298\,K$.

Q11. Soil water found in a limestone area has a pH of 8.3. What is its concentration of H^+ ions?

Q12. Suggest why it is more convenient to express acidity in pH values rather than by concentration of hydrogen ions.

Reactivity 3.1.5 – The ion product constant of water

Reactivity 3.1.5 – The ion product constant of water, K_w, shows an inverse relationship between $[H^+]$ and $[OH^-]$. $K_w = [H^+][OH^-]$

Recognize solutions as acidic, neutral and basic from the relative values of $[H^+]$ and $[OH^-]$.

The equation for K_w and its value at $298\,K$ are given in the data booklet.	Reactivity 2.3 – Why does the extent of ionization of water increase as temperature increases?

The ionization of water produces H⁺ and OH⁻ ions

Victoria Falls on the Zimbabwe–Zambia border. It is estimated that $6.25 \times 10^8\,dm^3$ of water flows over the falls every second. According to the ionization constant discussed here, for every billion of these molecules only two are ionized, and that proportion gets even smaller at lower temperatures.

As the majority of acid–base reactions involve ionization in aqueous solution, it is useful to consider the role of water in more detail. Water itself does ionize, albeit only very slightly at normal temperatures and pressures, so we can write an equilibrium expression for this reaction.

$$H_2O(l) \rightleftharpoons H^+(aq) + OH^-(aq)$$

Therefore

$$K = \frac{[H^+][OH^-]}{[H_2O]}$$

The concentration of water can be considered to be constant due to the fact that so little of it ionizes, and it can therefore be combined with K to produce a modified equilibrium constant known as K_w.

$$K[H_2O] = [H^+][OH^-]$$

Therefore $K_w = [H^+][OH^-]$

K_w is known as the **ionic product constant of water** and has a fixed value at a specified temperature. At 298 K, $K_w = 1.00 \times 10^{-14}$.

The value $K_w = 1.00 \times 10^{-14}$ at 298 K is given in the data booklet.

> We are using $H^+(aq)$ throughout this chapter as a simplified form of $H_3O^+(aq)$. This is acceptable in most situations, but do not forget that H^+ in aqueous solution always exists as $H_3O^+(aq)$.

> Water is a pure liquid so its concentration is just a function of its density, which is $1.00\,g\,cm^{-3}$.
> Using $n = \frac{m}{M}$ gives $n(H_2O) = \frac{1000\,g\,dm^{-3}}{18\,g\,mol^{-1}} = 55\,mol\,dm^{-3}$

In pure water, because $[H^+] = [OH^-]$, it follows that $[H^+] = \sqrt{K_w}$

$K_w = [H^+][OH^-]$

So at 298 K, $[H^+] = 1.00 \times 10^{-7}$, which gives pH = 7.00.

This is consistent with the widely known value for the pH of water at room temperature.

The relationship between H⁺ and OH⁻ is inverse

Because the product $[H^+] \times [OH^-]$ gives a constant value, it follows that the concentrations of these ions must have an inverse relationship. In other words, in aqueous solutions the higher the concentration of H^+, the lower the concentration of OH^- and vice versa. Solutions are defined as acidic, neutral, or basic according to their relative concentrations of these ions as shown below.

		at 298 K
Acidic solutions are defined as those in which	$[H^+] > [OH^-]$	pH < 7
Neutral solutions are defined as those in which	$[H^+] = [OH^-]$	pH = 7
Alkaline solutions are defined as those in which	$[H^+] < [OH^-]$	pH > 7

So if we know the concentration of either H^+ or OH^-, we can calculate the other from the value of K_w.

$$[H]^+ = \frac{K_w}{[OH]^-} \text{ and } [OH]^- = \frac{K_w}{[H]^+}$$

Worked example

A sample of blood at 298 K has $[H^+] = 4.60 \times 10^{-8}$ mol dm^{-3}.

Calculate the concentration of OH$^-$ and state whether the blood is acidic, neutral, or basic.

Solution

At 298 K, $K_w = 1.00 \times 10^{-14} = [H^+][OH^-]$

So $[OH]^- = \dfrac{1.00 \times 10^{-14}}{4.60 \times 10^{-8}} = 2.17 \times 10^{-7}$ mol dm^{-3}

As $[OH^-] > [H^+]$ the solution is basic.

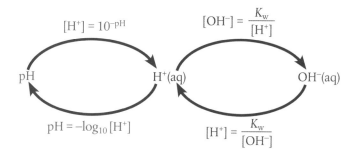

K_w is temperature dependent

The ionic product constant of water, K_w, was derived on page 474.

$$K_w = [H^+][OH^-] = 1.00 \times 10^{-14} \text{ at } 298 \text{ K}$$

As K_w is an equilibrium constant, its value must be temperature dependent. The reaction for the dissociation of water is endothermic as it involves bond breaking. This means that:

- an increase in temperature will shift the equilibrium to the right, which will increase the value of K_w. The concentrations of H$^+$(aq) and OH$^-$(aq) increase, and so pH decreases

- a decrease in temperature will shift the equilibrium to the left, which will decrease the value of K_w. The concentrations of H$^+$(aq) and OH$^-$(aq) decrease, and so pH increases.

Some data to illustrate these trends are given in the table.

Temperature / °C	K_w	[H$^+$] in pure water ($\sqrt{K_w}$)	pH of pure water ($-\log_{10}[H^+]$)
0	1.5×10^{-15}	0.39×10^{-7}	7.47
10	3.0×10^{-15}	0.55×10^{-7}	7.27
20	6.8×10^{-15}	0.82×10^{-7}	7.08
25	1.0×10^{-14}	1.00×10^{-7}	7.00
30	1.5×10^{-14}	1.22×10^{-7}	6.92
40	3.0×10^{-14}	1.73×10^{-7}	6.77
50	5.5×10^{-14}	2.35×10^{-7}	6.63

The concentrations of H$^+$ and OH$^-$ are inversely proportional in an aqueous solution.

 SKILLS

The pH of water at different temperatures. Full details of how to carry out this experiment with a worksheet are available in the eBook.

 Reactivity 2.3 – Why does the extent of ionization of water increase as temperature increases?

 TOK Because many people are so familiar with the value 7.00 as the pH of water, it is often difficult to convince them that water with a pH of greater or less than this is still neutral. Can you think of other examples where entrenched prior knowledge might hinder a fuller understanding of new knowledge? Are we likely to misinterpret experimental data when it does not fit with our expectations based on prior knowledge?

In other words, the pH of pure water is 7.00 only when the temperature is 298 K. Note that at temperatures above and below this, despite changes in the pH value, water is still a neutral substance as its $H^+(aq)$ concentration is equal to its $OH^-(aq)$ concentration. It does not become acidic or basic as we heat it or cool it respectively!

The temperature dependence of K_w means that the temperature should always be stated alongside pH measurements.

Exercise

Q13. A sample of pure water is found to have a pH of 7.3. What can you deduce about the temperature at which the measurement was made?

Q14. An aqueous solution of baking soda, $NaHCO_3$, has a concentration of OH^- ions of $7.8 \times 10^{-6}\,mol\,dm^{-3}$ at 298 K. Determine its concentration of hydrogen ions.

Q15. At body temperature, 37 °C, the value of K_w is 2.4×10^{-14}. Determine the concentration of H^+ ions and the pH of pure water at this temperature.

Q16. An aqueous solution has a pH of 9 at 25 °C. What are its concentrations of H^+ and OH^- ions?

Q17. For each of the following aqueous solutions at 298 K, calculate $[OH^-]$ from $[H^+]$ or $[H^+]$ from $[OH^-]$. Classify each solution as acidic, basic, or neutral.

(a) $[H^+] = 3.4 \times 10^{-9}\,mol\,dm^{-3}$ (b) $[OH^-] = 0.010\,mol\,dm^{-3}$

(c) $[OH^-] = 1.0 \times 10^{-10}\,mol\,dm^{-3}$ (d) $[H^+] = 8.6 \times 10^{-5}\,mol\,dm^{-3}$

Q18. A sample of blood has $[H^+] = 4.6 \times 10^{-8}\,mol\,dm^{-3}$ at 298 K. Determine its concentration of OH^- ions and state whether the sample is acidic, basic or neutral.

Q19. For each of the following biological fluids, calculate the pH from the given concentration of H^+ or OH^- ions at 298 K.

(a) bile: $[OH^-] = 8 \times 10^{-8}\,mol\,dm^{-3}$

(b) gastric juice: $[H^+] = 10^{-2}\,mol\,dm^{-3}$

(c) urine: $[OH^-] = 6 \times 10^{-10}\,mol\,dm^{-3}$

Reactivity 3.1.6 – Strong and weak acids and bases

Reactivity 3.1.6 – Strong and weak acids and bases differ in the extent of ionization.	
Recognize that acid–base equilibria lie in the direction of the weaker conjugate.	
HCl, HBr, HI, HNO_3 and H_2SO_4 are strong acids, and group 1 hydroxides are strong bases.	Reactivity 2.3 – How would you expect the equilibrium constants of strong and weak acids to compare?
The distinction between strong and weak acids or bases and concentrated and dilute reagents should be covered.	Reactivity 1.1 – Why does the acid strength of the hydrogen halides increase down group 17?
	Tool 1, Inquiry 2 – What physical and chemical properties can be observed to distinguish between weak and strong acids or bases of the same concentration?

The strength of an acid or base depends on its extent of ionization

Brønsted–Lowry acids and bases dissociate in solution. Acids produce H^+ ions and bases produce OH^- ions. As a result, their aqueous solutions exist as equilibrium mixtures containing both the undissociated form and the ions. As we will see, the position of this equilibrium is what defines the strength of an acid or a base.

Consider the acid dissociation reaction:

$$HA(aq) + H_2O(l) \rightleftharpoons A^-(aq) + H_3O^+(aq)$$

If this acid dissociates fully, its equilibrium lies to the right and it will exist virtually entirely as ions in solution. It is said to be a **strong acid**. For example:

$$HCl(aq) + H_2O(l) \rightleftharpoons H_3O^+(aq) + Cl^-(aq)$$

Hydrochloric acid, HCl, is a strong acid so it dissociates fully. This means that the reverse reaction can be considered to be negligible and so the reaction is usually written without the equilibrium sign.

$$HCl(aq) + H_2O(l) \rightarrow H_3O^+(aq) + Cl^-(aq)$$

If, on the other hand, the acid dissociates only partially, its equilibrium lies to the left and the undissociated form dominates. It is said to be a **weak acid**. For example:

$$CH_3COOH(aq) + H_2O(l) \rightleftharpoons H_3O^+(aq) + CH_3COO^-(aq)$$

Ethanoic acid, CH_3COOH, is a weak acid so it dissociates only partially. Here it is essential to use the equilibrium sign in its dissociation reaction.

Strong acids are good proton donors. As their dissociation reactions go virtually to completion, their conjugate bases are not readily able to accept a proton. For example, the strong acid HCl reacts to form the conjugate base Cl^- which shows virtually no basic properties.

$HCl(aq)$	+	$H_2O(l)$	$\rightarrow$	$H_3O^+(aq)$	+	$Cl^-(aq)$
strong acid		base		conjugate acid		conjugate base

Weak acids are poor proton donors. As their dissociation reactions are equilibria which lie to the left, in favour of reactants, their conjugate bases are readily able to accept a proton. For example, the weak acid CH_3COOH reacts to form the conjugate base CH_3COO^-, which is a stronger base than Cl^-.

$CH_3COOH(aq)$	+	$H_2O(l)$	$\rightleftharpoons$	$H_3O^+(aq)$	+	$CH_3COO^-(aq)$
weak acid		base		conjugate acid		conjugate base

So acid dissociation reactions favour the production of the weaker conjugate.

In a similar way, bases can be described as strong or weak on the basis of the extent of their ionization. For example, NaOH is a **strong base** because it ionizes fully:

$$NaOH(aq) \rightarrow Na^+(aq) + OH^-(aq)$$

▲ Common household acids and bases. From left to right: Drano contains sodium hydroxide, vitamin C is ascorbic acid, muriatic acid is hydrochloric acid, white vinegar contains ethanoic acid and Windex contains ammonia. If these were all present in the same concentration, what would be the increasing order of their pH values?

Strong acids and strong bases ionize almost completely in solution. Weak acids and weak bases ionize only partially in solution.

Its dissociation is written without equilibrium signs. Note that it is the OH⁻ ions that show Brønsted–Lowry base behaviour by accepting protons.

On the other hand, NH_3 is a **weak base** as it ionizes only partially, so its equilibrium lies to the left and the concentration of ions is low.

$$NH_3(aq) + H_2O(l) \rightleftharpoons NH_4^+(aq) + OH^-(aq)$$

Strong bases are good proton acceptors; they react to form conjugates that do not show acidic properties. Weak bases are poor proton acceptors; they react to form conjugates with stronger acidic properties than the conjugates of strong bases. Base ionization reactions favour the production of the weaker conjugate.

The strength of an acid or base is therefore a measure of how readily it dissociates in aqueous solution. This is an inherent property of a particular acid or base, dependent on its bonding. Do not confuse acid or basic *strength* with its *concentration*, which is a variable depending on the number of moles per unit volume, according to how much solute has been added to the water. Note, for example, that it is possible for an acid or base to be strong but present in a dilute solution, or weak and present in a concentrated solution.

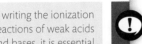

In writing the ionization reactions of weak acids and bases, it is essential to use the equilibrium sign.

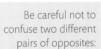

Be careful not to confuse two different pairs of opposites:
• strong and weak acids or bases refer to their extent of dissociation
• concentrated and dilute refer to the ratio of solute and water in the solution.

Weak acids and bases are much more common than strong acids and bases

It is useful to know which of the common acids and bases are strong and which are weak. Fortunately this is quite easy as there are very few common examples of strong acids and bases, so this short list can be committed to memory. You will then know that any other acids and bases you come across are likely to be weak.

	Acid		Base	
common examples of strong forms	HCl	hydrochloric acid	LiOH	lithium hydroxide
	HBr	hydrobromic acid	NaOH	sodium hydroxide
	HI	hydroiodic acid	KOH	potassium hydroxide
	HNO_3	nitric acid		
	H_2SO_4	sulfuric acid		
some examples of weak forms	CH_3COOH and other organic acids	ethanoic acid	NH_3	ammonia
	H_2CO_3	carbonic acid	$C_2H_5NH_2$ and other amines	ethylamine
	H_3PO_4	phosphoric acid		

Note that amines such as ethylamine, $C_2H_5NH_2$, can be considered as organic derivatives of NH_3 in which one of the hydrogen atoms has been replaced by an alkyl (hydrocarbon) group. There are literally hundreds of acids and bases in organic chemistry, nearly all of which are weak in comparison with the strong inorganic acids listed here. Amino acids, the building blocks of proteins, as their name implies, contain both the basic $-NH_2$ amino group and the $-COOH$ carboxylic acid group. The 'A' in DNA, deoxyribonucleic acid, the store of genetic material, stands for 'acid'; in this case, the acidic component is phosphate groups.

Strong and weak acids differ in the extent of their dissociation, and therefore in their equilibrium constants. At equilibrium, strong acids have a high concentration of product, the ionized form, and therefore have large equilibrium constants. Weak acids, with low concentration of product at equilibrium, will have much smaller equilibrium constants.

The strength of a Brønsted–Lowry acid depends on the ease with which it dissociates to release H^+ ions. In turn this depends on the strength of the bond that has to be broken to release H^+. If we consider the hydrogen halides, as the halogen atom increases in size going down group 17, this increases the length of the H–halogen bond. Longer bonds are weaker bonds that need less energy to break. So the acid strength of the hydrogen halides increases down the group: HF < HCl < HBr < HI.

Reactivity 2.3 – How would you expect the equilibrium constants of strong and weak acids to compare?

	Bond length /nm	Bond strength /kJ mol⁻¹	
HF	0.02	568.0	
HCl	0.127	432.0	
HBr	0.141	366.3	
HI	0.161	298.3	

increasing acid strength

Reactivity 1.1 – Why does the acid strength of the hydrogen halides increase down group 17?

Distinguishing between strong and weak acids and bases

Due to their greater ionization in solution, strong acids and strong bases will contain a *higher concentration of ions* than weak acids and weak bases. This then can be used as a means of distinguishing between them. Note though that such comparisons will only be valid when solutions of the same concentration (mol dm⁻³) are compared at the same temperature. We will consider here three properties that depend on the concentration of ions and so can be used for this purpose.

Tool 1, Inquiry 2 – What physical and chemical properties can be observed to help you distinguish between weak and strong acids or bases of the same concentration?

Electrical conductivity

Electrical conductivity of a solution depends on the concentration of mobile ions. Strong acids and strong bases will therefore show higher conductivity than weak acids and bases – as long as solutions of the same concentration are compared. This can be measured using a conductivity meter or probe, or by using the conductivity setting on a pH meter.

SKILLS

Conductivity of strong and weak acid/bases. Full details of how to carry out this experiment with a worksheet are available in the eBook.

The different rates of reaction between calcium carbonate and three acids of different strength at the same concentration. The strongest acid is on the left and the weakest on the right. The carbonate reacts with each acid to produce carbon dioxide gas, which can be seen as bubbles in the solution. The stronger the acid, the more vigorous the reaction – the left test tube fizzes whilst the right one bubbles very gently. What variables must be kept constant in order for this to be a valid comparison of acid strength?

Rate of reaction

The reactions of acids with bases described on page 482 depend on the concentration of H^+ ions. They will therefore happen at an increased rate with stronger acids.

These different rates of reactions may be an important consideration, for example, regarding safety in the laboratory, but they usually do not provide an easy means of quantifying data to distinguish between weak and strong acids.

pH

Because it is a measure of the H^+ concentration, the pH scale can be used directly to compare the strengths of acids, proved they are of equal molar concentration. The stronger the acid of a fixed concentration, the lower the pH value. Universal indicator or a pH meter can be used to measure pH.

Exercise

Q20. Ammonia, NH_3, acts as a weak Brønsted–Lowry base when dissolved in water.

Write an equation for this reaction and outline what is meant by the term 'weak' in the context of the equilibrium position of this reaction.

Q21. Which of the following 1 mol dm^{-3} solutions will be the poorest conductor of electricity?

A   HCl

B CH_3COOH

C   NaOH

D   NaCl

Q22. Which methods will distinguish between equimolar solutions of a strong base and a strong acid?

   I.   Add magnesium to each solution and look for the formation of gas bubbles.

   II.   Add aqueous sodium hydroxide to each solution and measure the temperature change.

   III.   Use each solution in a circuit with a battery and lamp and see how brightly the lamp glows.

A   I and II only

B   I and III only

C   II and II only

D   I, II and III

Q23. Which acid in each of the following pairs has the stronger conjugate base?

   (a) H_2CO_3 or H_2SO_4

   (b)   HCl or HCOOH

Reactivity 3.1.7 – Neutralization reactions

Reactivity 3.1.7 – Acids react with bases in neutralization reactions.
Formulate equations for the reactions between acids and metal oxides, metal hydroxides, hydrogencarbonates and carbonates.

Identify the parent acid and base of different salts. Bases should include ammonia, amines, soluble carbonates and hydrogencarbonates; acids should include organic acids.	Tool 1, Structure 1.1 – How can the salts formed in neutralization reactions be separated? Reactivity 1.1 – Neutralization reactions are exothermic. How can this be explained in terms of bond enthalpies? Reactivity 3.2 – How could we classify the reaction that occurs when hydrogen gas is released from the reaction between an acid and a metal?

Neutralization reactions occur when acids and bases react together

Nature of Science

Classification leading to generalizations is an important aspect of studies in science. Historically, groups of substances such as acids were classified together because they were shown to have similar chemical properties. Over time it was recognized that many substances share these acidic properties, and so the classification broadened. Similarly, grouping of compounds as bases and alkalis occurred on the basis of experimental evidence for their properties.

TOK

To what extent do the classification systems we use in the pursuit of knowledge affect the conclusions that we reach?

Brønsted–Lowry acids release H^+ ions, and soluble bases, alkalis, release OH^- ions. The reaction between H^+ ions and OH^- ions to form H_2O is known as a **neutralization** reaction.

$$H^+(aq) + OH^-(aq) \rightarrow H_2O(l)$$

During the reaction, an ionic compound called a **salt** also forms, as the hydrogen of the acid is replaced by a metal ion or another positive ion.

For example, sodium chloride, NaCl, known as common salt, is derived from the acid HCl by reaction with a base that contains Na, such as NaOH. The terms **parent acid** and **parent base** are sometimes used to describe this relationship between an acid, a base, and their salt.

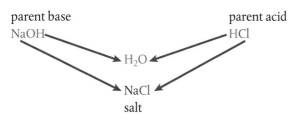

In acid–base theory, the words ionization and dissociation are often used interchangeably as acid dissociation always leads to ion formation.

When acids react with reactive metals, they also form salts as the hydrogen in the acid is replaced by the metal. But in these reactions, there is no transfer of a proton, H^+, as the hydrogen is released as the gas, H_2. Here the hydrogen ions are becoming electrically neutral by accepting electrons while the metal is being ionized by electron loss. For example, in the reaction between magnesium metal and hydrochloric acid:

$$Mg(s) + 2HCl(aq) \rightarrow MgCl_2(aq) + H_2(g)$$

$$Mg(s) \rightarrow Mg^{2+}(aq) + 2e^- \quad \text{electron loss = oxidation}$$

$$2H^+(aq) + 2e^- \rightarrow H_2(g) \quad \text{electron gain = reduction}$$

These reactions are therefore redox reactions, and cannot be described by Brønsted–Lowry acid–base theory.

Reactivity 3.2 – How could we classify the reaction that occurs when hydrogen gas is released from the reaction between an acid and a metal?

acid + base → salt + water

Metal oxides and metal hydroxides are bases which react with acids to produce a salt and water.

For example:

$$HCl(aq) + NaOH(aq) \rightarrow NaCl(aq) + H_2O(l)$$

$$2CH_3COOH(aq) + CuO(s) \rightarrow Cu(CH_3COO)_2(aq) + H_2O(l)$$

$$2HBr(aq) + Mg(OH)_2(aq) \rightarrow 2H_2O(l) + MgBr_2(aq)$$

Ammonia solution, NH_4OH, is also a soluble base which reacts with acids to form a salt and water.

$$HNO_3(aq) + NH_4OH(aq) \rightarrow NH_4NO_3(aq) + H_2O(l)$$

Coloured scanning electron micrograph (SEM) of the end of the abdomen of a bee, showing the needle-like stinger. The stinger is used to inject venom. Bee stings are slightly acidic, and so have traditionally been treated by using a mild alkali such as baking soda ($NaHCO_3$). Wasp stings, on the other hand, are claimed to be alkaline, and so are often treated with the weak acid, ethanoic acid (CH_3COOH), found in vinegar. There is, however, dispute over these claims as the pH of wasp stings is actually very close to neutral.

When you need to choose a base in solution to make a specific salt, it is useful to know some simple solubility rules.
- The only soluble carbonates and hydrogencarbonates are NH_4CO_3, Na_2CO_3, $NaHCO_3$, K_2CO_3, $KHCO_3$, and $Ca(HCO_3)_2$.
- The only soluble hydroxides are NH_4OH, $LiOH$, $NaOH$, and KOH.

Neutralization reactions occur when an acid and base react together to form a salt and water. They are exothermic reactions.

acid + carbonate → salt + water + carbon dioxide

Metal carbonates and hydrogencarbonates react with acids to produce a salt, water and carbon dioxide.

For example:

$$2HCl(aq) + CaCO_3(s) \rightarrow CaCl_2(aq) + H_2O(l) + CO_2(g)$$

$$H_2SO_4(aq) + Na_2CO_3(aq) \rightarrow Na_2SO_4(aq) + H_2O(l) + CO_2(g)$$

$$CH_3COOH(aq) + KHCO_3(aq) \rightarrow KCH_3COO(aq) + H_2O(l) + CO_2(g)$$

These reactions can also be represented as ionic equations:

$$2H^+(aq) + CO_3^{2-}(aq) \rightarrow H_2O(l) + CO_2(g)$$

$$H^+(aq) + HCO_3^-(aq) \rightarrow H_2O(l) + CO_2(g)$$

The reactions involve a gas being given off, so they visibly produce bubbles, known as **effervescence**.

Kitchen chemistry. Baking soda ($NaHCO_3$) and vinegar (CH_3COOH) react together in a powerful acid–base reaction, releasing carbon dioxide gas.
$NaHCO_3(s) + CH_3COOH(aq) \rightarrow NaCH_3COO(aq) + CO_2(g) + H_2O(l)$

Rainwater dissolves some CO_2 from the air to form a weak solution of carbonic acid, H_2CO_3.
$$H_2O(l) + CO_2(g) \rightleftharpoons H_2CO_3(aq)$$
Greater pressure, found underground in beds of limestone rock, $CaCO_3$, causes CO_2 to dissolve more readily, increasing the acidity of the solution. This then reacts with the $CaCO_3$:
$$H_2CO_3(aq) + CaCO_3(s) \rightleftharpoons Ca(HCO_3)_2(aq)$$
$CaCO_3$ is virtually insoluble but calcium hydrogencarbonate, $Ca(HCO_3)_2$, is soluble and washes away, causing caves to form in limestone regions. Inside the cave where the pressure is lower, the reaction above is reversed as less CO_2 dissolves, so $CaCO_3$ comes out of solution and precipitates. This causes the formations known as stalactites and stalagmites.

Stalactites hanging from the roof of a limestone cave in Kangaroo Island, Australia. Stalactites are formed from $CaCO_3$ precipitating out of solution, as the water dripping from the cave dissolves less CO_2 and loses acidity.

Reactions of rainwater on CaCO₃ rock give rise to 'hard water', that is water containing elevated levels of Ca²⁺ ions, in limestone areas.

When rainwater dissolves acidic gases from industrial emissions in the atmosphere, its pH is lowered further and it is known as 'acid rain'. This causes widespread damage including the erosion of carbonates such as marble structures.

The effect of acid rain eroding marble structures on a building in New York, USA, is shown on the right. The image on the left shows the structure after restoration to its original state.

Tool 1, Structure 1.1 – How can the salts formed in neutralization reactions be separated?

Reactivity 1.1 – Neutralization reactions are exothermic. How can this be explained in terms of bond enthalpies?

Neutralization reactions generally form aqueous solutions of salts. The pure salt can be separated by filtration followed by evaporation, as described in Structure 1.1.

In a neutralization reaction between an acid and a base, the net reaction is the formation of H_2O from its ions. (The other ions are called **spectator ions** as they do not change during the reaction and so can be cancelled out.)

$$H^+(aq) + OH^-(aq) \rightarrow H_2O(l)$$

For every mole of H_2O that forms, 2 moles of O–H bonds are made. As bond-making is an exothermic process, the overall reaction is also exothermic, releasing energy. For reactions between all strong acids and strong bases, the enthalpy of neutralization, expressed per mole of H_2O formed, is very similar – approximately $\Delta H = -57\,\text{kJ mol}^{-1}$, reflecting the fact that the overall reaction is the same in all cases, the formation of water.

Challenge yourself

5. Why do neutralization reactions involving weak acids or weak bases give different values for their enthalpy of neutralization compared with strong acids and bases?

Indigestion tablets effervescing as they dissolve in a glass of water to release CO₂. These antacids neutralize acids in the digestive system, which helps to reduce pain and inflammation.

Neutralization reactions can be useful to help reduce the effect of an acid or a base. For example, treatment for acid indigestion often involves using 'antacids' which contain a mixture of weak bases such as magnesium hydroxide and aluminium hydroxide. In places where the soil has become too acidic, the growth of many plants will be restricted. Adding a weak alkali such as lime, CaO, can help to reduce the acidity and hence increase the fertility of the soil.

Exercise

Q24. Write equations for the following reactions:

(a) sulfuric acid and copper oxide

(b) nitric acid and sodium hydrogencarbonate

(c) phosphoric acid and potassium hydroxide

(d) hydrochloric acid and methylamine

Q25. Which of the following is/are formed when a metal oxide reacts with a dilute acid?

I. a metal salt

II. water

III. carbon dioxide gas

A I only   B I and II only   C II and III only   D I, II and III

Q26. Suggest by name a parent acid and parent base that could be used to make the following salts. Write equations for each reaction.

(a) sodium nitrate

(b) ammonium chloride

(c) copper sulfate

(d) potassium methanoate

Reactivity 3.1.8 – pH curves

Reactivity 3.1.8 – pH curves for neutralization reactions involving strong acids and bases have characteristic shapes and features.

Sketch and interpret the general shape of the pH curve.

Interpretation should include the intercept with the pH axis and equivalence point. Only monoprotic neutralization reactions will be assessed.	Structure 1.4 – Why is the equivalence point sometimes referred to as the stoichiometric point? Tools 1 and 3, Structure 1.3 – How can titration be used to calculate the concentration of an acid or base in solution?

In the following section, we will study the pH curves of the reactions between strong acids and strong bases. The reactions between different combinations of strong and weak acids and bases are described on pages 481–484. To make comparisons of the curves easier, these examples all use:

- $0.10 \, mol \, dm^{-3}$ solutions of all acids and bases
- an initial volume of $50.0 \, cm^3$ of acid in the conical flask with base added from the burette
- monoprotic acids and bases which all react in a 1:1 ratio, so that equivalence is achieved at equal volumes for these equimolar solutions (i.e. when $50.0 \, cm^3$ of base has been added to the $50.0 \, cm^3$ of acid).

pH curves can be generated from acid–base titrations

The neutralization reactions between acids and bases described previously can be investigated quantitatively using a technique called **acid–base titration**. Titration is one of the most widely used procedures in chemistry (Figure 1). Quality control of food and drink production, health and safety checks in the cosmetic industry, and clinical analysis in medical services are just some examples of its use. Controlled volumes of one reactant are added gradually from a **burette** to a fixed volume of the other reactant that has been carefully measured using a **pipette** and placed in a conical flask. The reaction between acid and base takes place in the flask until the **equivalence point** or **stoichiometric point** is reached, where they exactly neutralize each other. A pH meter or an indicator is used to detect the exact point at which equivalence is reached.

R3.1 Figure 1 Simple titration apparatus.

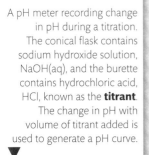

A pH meter recording change in pH during a titration. The conical flask contains sodium hydroxide solution, NaOH(aq), and the burette contains hydrochloric acid, HCl, known as the **titrant**. The change in pH with volume of titrant added is used to generate a pH curve.

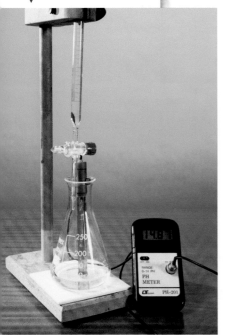

When a base is added to an acid in the neutralization reaction, there is a change in pH as we would expect. But this change does not show a linear relationship with the volume of base added, partly due to the logarithmic nature of the pH scale. The easiest way to follow the reaction is to record pH using a pH meter or data-logging device as a function of volume of base added and plot these values as **pH curves**.

You will find that in most titrations, a big jump in pH occurs at equivalence, and this is known as the **point of inflection**. The equivalence point is determined as being half way up this jump.

The equivalence point in a titration can also be determined using an acid–base indicator, which is chosen to change colour at the pH of the equivalence point.

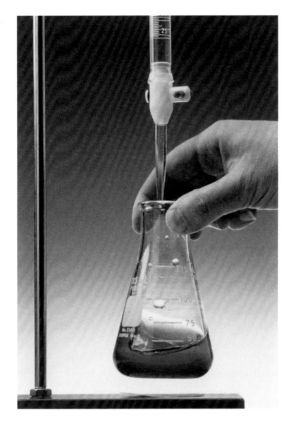

◀ Addition of base from the burette to an acid in the conical flask containing indicator. As the neutralization reaction happens, the change in the pH of the reaction mixture causes a change in the colour of the indicator. Dropwise addition of the base ensures that an accurate measure of the equivalence point can be made.

At the equivalence point, the acid and base have exactly neutralized each other, so the solution contains salt and water only. The reacting ratio of acid to base to reach this point is determined by the *stoichiometry* of the neutralization reaction. For example:

$$HCl(aq) + NaOH(aq) \rightarrow NaCl\ (aq) + H_2O(l)$$

Equivalence point when 1 mole acid : 1 mole base

$$2HCl(aq) + Ba(OH)_2\ (aq) \rightarrow BaCl_2(aq) + H_2O(l)$$

Equivalence point when 2 moles acid : 1 mole base

pH curve of strong acid and strong base

For example:

$$HCl(aq) + NaOH(aq) \rightarrow NaCl(aq) + H_2O(l)$$

pH at equivalence = 7

The calculations in the table on the next page are a sample to show the reasons for the shape of the pH curve. In this example, we assume full dissociation because these are strong acids and bases. Note also that as the base is added to the acid, neutralization of some of the acid occurs, while excess acid remains – until equivalence where the amounts of acid and base have fully reacted. After equivalence, the mixture contains excess base. As the volume changes during the addition, this must be taken into account in determining the concentrations.

As the volume of base added increases, we can see how the resulting pH changes.

Structure 1.4 – Why is the equivalence point sometimes referred to as the stoichiometric point?

Titration can be used to determine the concentration of one solution (acid or base), when the concentration of the other solution in the neutralization reaction is accurately known.

Tools 1 and 3, Structure 1.3 – How can titration be used to calculate the concentration of an acid or a base in solution?

Acid–base titrations using a pH sensor – identifying the equivalence point. Full details of how to carry out this experiment with a worksheet are available in the eBook.

Volume of base added				
0.00 cm³ (flask contains acid only)	25.00 cm³	49.00 cm³	50.00 cm³ (contents of flask at equivalence)	51.00 cm³ (flask contains excess base)
	(flask contains excess acid)			
$[acid]_{initial}$ = 0.10 mol dm⁻³ $[H^+]$ = 1 × 10⁻¹ mol dm⁻³ **pH = 1.0**	n(acid initial) = cV = 0.10 × 0.050 = 0.0050 mol n(base added) = cV = 0.10 × 0.0250 = 0.00250 mol n(acid remaining) = 0.0050 − 0.00250 = 0.00250 mol $n(H^+)$ = 0.00250 mol new volume = 0.0750 dm³ so $[H^+]$ = 0.0333 mol dm⁻³ **pH = 1.5**	n(acid initial) = 0.0050 mol n(base added) = cV = 0.10 × 0.0490 = 0.00490 mol n(acid remaining) = 0.0050 − 0.00490 = 0.00010 mol $n(H^+)$ = 0.00010 mol new volume = 0.0990 dm³ so $[H^+]$ = 0.00101 mol dm⁻³ **pH = 3.0**	**All of the acid has been neutralized by the base; the solution contains NaCl + H₂O only** **pH = 7.0**	n(base added) = cV = 0.10 × 0.0510 = 0.00510 mol n(base remaining) = 0.00510 − 0.0050 = 0.00010 mol $n(OH^-)$ = 0.00010 mol new volume = 0.101 dm³ so $[OH^-]$ = 0.00099 mol dm⁻³ so $[H^+]$ = 10⁻¹¹ **pH = 11.0**

The curve obtained by continuous measurement with a pH probe shows the **pH curve** of these pH changes over the full range of addition of base (Figure 2).

R3.1 Figure 2 pH curve for strong acid–strong base.

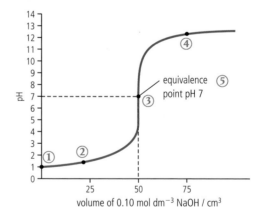

The following points can be deduced from the graph as marked:

① initial pH = 1, pH of strong acid

② pH changes only gradually until equivalence

③ very sharp jump in pH at equivalence: from pH 3 to pH 11

④ after equivalence, the curve flattens out at a high value, pH of strong base

⑤ **pH at equivalence = 7.**

One drop of solution delivered from a burette has a volume of about 0.05 cm³.

SKILLS

Most burettes have 0.1 cm³ as the smallest division, so on this analogue scale you should record all readings to one half of this, that is ±0.05 cm³.

Exercise

Q27. What is the pH of the solution formed when a strong acid is exactly neutralized by a strong base? What is the name of this point in a titration?

Q28. Give the equation for the neutralization reaction that occurs during the titration of aqueous dilute nitric acid with potassium hydroxide solution. If the acid is placed in the conical flask and has a concentration of $0.01 \, mol \, dm^{-3}$, what will be the intercept with the pH axis at the start of the titration?

Q29. In a titration, the equivalence point is reached when $10.00 \, cm^3$ of $0.01 \, mol \, dm^{-3}$ NaOH has been added to $12.00 \, cm^3$ of HCl. What is the concentration of the acid?

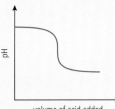

pH curves can equally well be described in terms of the addition of an acid to a base, in which case the curves will be the inverse.

Be sure to check which way round the data are given when answering questions on this topic.

Guiding Question revisited

What happens when protons are transferred?

In this chapter we have seen:

- A proton is $[H^+]$.
- Brønsted–Lowry acids are proton donors and Brønsted–Lowry bases are proton acceptors.
- A conjugate acid–base pair differs by a single proton.
- Amphiprotic species can act as both Brønsted–Lowry acids and bases.
- The pH scale is the negative base 10 logarithm of $[H^+]$.
- The ion product constant of water, K_w, is temperature dependent. It shows the inverse relationship between $[H^+]$ and $[OH^-]$.
- Strong acids and bases are fully ionized in solution, weak acids and bases are only partially ionized.
- Acids and bases react together in neutralization reactions to form salts. The overall reaction is $H^+(aq) + OH^-(aq) \rightarrow H_2O(l)$ which is exothermic.
- The equivalence point in a titration is where the acid and base have exactly neutralized each other in a stoichiometric ratio.
- pH curves show the change in pH as an acid and base react together in titration.
- Neutralization involving a strong acid and a strong base gives a characteristic pH curve. The pH at the equivalence point is 7.

Practice questions

1. Which is a conjugate Brønsted–Lowry acid–base pair?

$$CH_3COOH(aq) + H_2O(l) \rightarrow CH_3COO^-(aq) + H_3O^+(aq)$$

   A CH_3COO^- / H_3O^+ **B** H_2O / CH_3COO^-

   C H_2O / H_3O^+ **D** CH_3COOH / H_2O

2. All of the following would be described as typical of an acid except:

   A Produces bubbles when it reacts with calcium carbonate

   B Conducts electricity

   C Releases heat when it reacts with sodium hydroxide

   D Has a soapy feel

3. What is the conjugate base of HSO_3^-?

   A H_2SO_3

   B SO_3^{2-}

   C H_2O

   D $H_2SO_3^-$

4. Which of the following can be described as amphiprotic?

   A OH^-

   B NH_4^+

   C $BeCl_2$

   D SO_4^{2-}

5. When equal concentrations of a strong and a weak acid are compared, which of the following measurements will be the same?

   I.   The volume of base used to neutralize the acid

   II. The electrical conductivity

   III. The rate of the reaction with calcium carbonate

   A I only

   B I and II only

   C I, II, and III

   D None of I, II or III

6. CH_3OH is a very weak Brønsted–Lowry acid. What is true of the ion CH_3O^-?

   A It is a good proton donor.

   B It will be present in high concentration in CH_3OH at equilibrium.

   C It is a strong Brønsted–Lowry base.

   D It is a stronger Brønsted–Lowry acid than CH_3OH.

7. Two acids, X and Y, both of concentration $0.20\,mol\,dm^{-3}$ have pH values of 2 and 4 respectively. Which of the following statements is true?

   A X is a stronger acid than Y.

   B Y is more dilute than X.

   C X contains more undissociated molecules than Y.

   D Y contains a higher concentration of H^+ ions than X.

8. Which of the following statements is the defining property of a Brønsted–Lowry base?

   A donates protons

   B produces OH^- ions

   C conducts electricity

   D accepts H^+ ions

9. In the reaction $CH_3OH + NH_2^- \rightleftharpoons CH_3O^- + NH_3$ the two Brønsted-Lowry acids are:

   A NH_2^- and CH_3OH

   B NH_2^- and CH_3O^-

   C CH_3O^- and NH_3

   D CH_3OH and NH_3

10. Which of the following are conjugate acid-base pairs?

    I. HNO_3 / NO_3^-

    II. H_2O / OH^-

    III. H_2 / H^+

    A I only

    B I and II

    C I and III

    D I, II and III

11. Which of these reduces the production of acid rain?

I.   removing SO_2 from power station emissions

II. neutralizing acidified lakes with lime

III. switching to renewable energy sources

  A I and II   **B** I and III

  C II and III   **D** I, II and III

12. The pH scale can be used to distinguish between different acids and bases in the laboratory.

  (a) You are provided with five unlabelled bottles which are known to contain $0.10\,mol\,dm^{-3}$ solutions of each of the following:

      $HCOOH(aq)$, $KCl(aq)$, $HNO_3(aq)$, $NaOH(aq)$, $NH_3(aq)$

    (i) Describe **two** ways in which you could measure the pH of the solutions.   (2)

    (ii) Suggest expected pH values for each solution.   (3)

  (b) **(i)** In the laboratory, some solid sodium hydrogencarbonate was added to an aqueous solution of hydrochloric acid. Formulate an equation for this reaction including state symbols, and predict what you would expect to observe during the reaction.   (5)

    (ii) The hydrogencarbonate can be described as amphiprotic. Explain this statement with reference to the Brønsted–Lowry theory of acids and bases.   (4)

  (c) Conductivity can be used to distinguish between strong and weak acids of the same concentration. Justify the claim that the use of a conductivity meter is a more scientifically acceptable approach than the use of an indicator.   (2)

                                                        (Total 16 marks)

13. (a) Define the terms *acid* and *base* according to the Brønsted–Lowry theory. Distinguish between a weak base and a strong base. State **one** example of a weak base.   (3)

  (b) Weak acids in the environment may cause damage. Identify a weak acid in the environment *and* outline **one** of its effects.   (2)

                                                         (Total 5 marks)

14. Water is an important substance that is abundant on the Earth's surface.

  (a) State the expression for the ionic product constant of water, K_w.   (1)

  (b) Explain why even a very acidic aqueous solution still has some OH^- ions present in it.   (1)

  (c) State and explain the effect of increasing temperature on the value of K_w given that the ionization of water is an endothermic process.   (3)

  (d) State and explain the effect of increasing temperature on the pH of water.   (2)

                                                         (Total 7 marks)

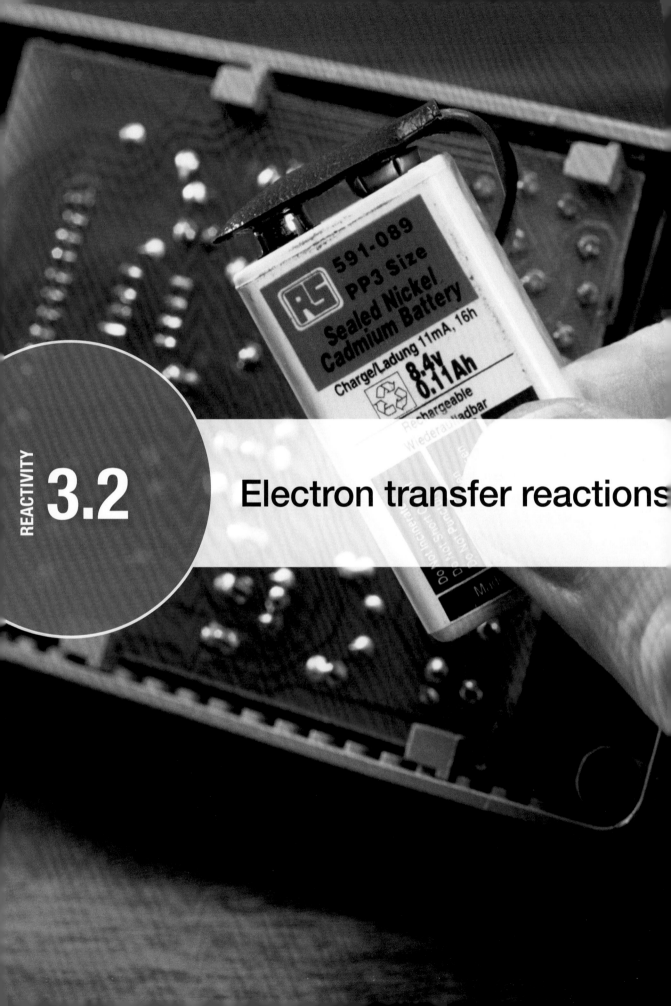

Electron transfer reactions

A nickel–cadmium battery. This battery can be recharged by reversing a spontaneous reaction of electron transfer.

Guiding Question

What happens when electrons are transferred?

Oxygen makes up only about 20% by volume of the air, yet it is the essential component for many reactions. Without it, fuels would not burn, iron would not rust, and we would be unable to obtain energy from our food molecules through respiration. Indeed, animal life on the planet did not evolve until a certain concentration of oxygen had built up in the atmosphere over 600 million years ago. The term oxidation has been in use for a long time to describe these and other reactions where oxygen is involved in chemical change. Closer inspection of these reactions shows that species generally lose electrons when they react with oxygen and so the term oxidation has been broadened to include any reaction which involves the loss of electrons. Oxidation, though, can only be half of the story, as the electrons which are lost cannot disappear. They must be taken away by another species which we say is **reduced**. The transfer of electrons from one species to another is the basis for all **redox** reactions.

If a redox reaction is set up such that the electron transfer between the reactants can only occur through an external circuit, the electrons produce an electric current. A simple **voltaic cell** or **battery** works in this way and converts chemical energy to electrical energy. This energy change can be reversed if electrical energy is used to drive electrons to force redox reactions to take place against their spontaneous flow. This process of **electrolysis** can be used to break down stable compounds into their elements. These applications of redox reactions are collectively known as **electrochemical cells**.

Redox reactions have truly revolutionized our world. They form the basis of great industries, such as steel-making and aluminium extraction, and it would be hard to imagine life without the many battery-powered mobile devices we use every day. An understanding of oxidation and reduction is therefore at the heart of the study of a large branch of chemistry both in the laboratory and beyond.

Reactivity 3.2.1 – Redox reactions

Reactivity 3.2.1 – Oxidation and reduction can be described in terms of electron transfer, change in oxidation state, oxygen gain/loss or hydrogen loss/gain.	
Deduce oxidation states of an atom in a compound or an ion.	
Identify the oxidized and reduced species and the oxidizing and reducing agents in a chemical reaction.	
Include examples to illustrate the variable oxidation states of transition element ions and of most main group non-metals.	Structure 3.1 – What are the advantages and limitations of using oxidation states to track redox changes?
Include the use of oxidation numbers in the naming of compounds.	Structure 2.3 – The surface oxidation of metals is often known as corrosion. What are some of the consequences of this process?

Different definitions of oxidation and reduction

Oxidation is the loss of electrons

Oxygen is the most reactive gas in air and an essential component of many reactions. The term oxidation has been in use for a long time to describe reactions where oxygen is involved in chemical change. For example, magnesium metal combines with oxygen to form magnesium oxide which, as discussed in Structure 2.1, is an ionic solid consisting of Mg^{2+} ions and O^{2-} ions. The energy released in the reaction is emitted as light and heat.

$$2Mg(s) + O_2(g) \rightarrow 2MgO(s)$$

In this oxidation reaction each magnesium atom has lost two electrons and formed a doubly charged Mg^{2+} ion. Magnesium undergoes the same electron loss when it reacts with other non-metals such as chlorine:

$$Mg(s) + Cl_2(g) \rightarrow MgCl_2(s)$$

The product, magnesium chloride, consists of Mg^{2+} ions and Cl^- ions. Each magnesium atom has again lost two electrons to become an Mg^{2+} ion. Although no oxygen is involved in the second reaction the magnesium has changed in the same way. This suggests that the meaning of the term oxidation can be extended to any reactions which involve *the loss of electrons.*

Reduction is the gain of electrons

Reduction traditionally referred to the extraction of a metal from its ore: the ore was *reduced* to the metal. Copper oxide can be reduced to copper by heating it with a more reactive metal.

Magnesium ribbon burning in air. The magnesium metal is oxidized to form magnesium oxide (MgO).

Magnesium reacting with chlorine.

Flames resulting from a reaction between magnesium metal and copper oxide. The more reactive magnesium reacts with the copper oxide to form magnesium oxide and copper metal.

$$Mg(s) + CuO(g) \rightarrow MgO(s) + Cu(s)$$

The magnesium has again been oxidized and lost electrons. The copper ions have been reduced and gained electrons to form Cu atoms. This suggests that reduction involves the gain of electrons. Electrons have been transferred from the Mg to the Cu^{2+} ions. More generally we can see that in all electron transfer reactions one species will lose electrons and another species will gain them. Oxidation and reduction must both occur together, so these reactions are also redox reactions.

Oxidation is hydrogen loss and reduction is hydrogen gain

Hydrogen sulfide solution can react with oxygen to give different products.

$$H_2S(aq) + 2O_2(g) \rightarrow H_2SO_4(aq)$$

$$2H_2S(aq) + O_2(g) \rightarrow 2S(s) + H_2O(l)$$

It seems reasonable to say that the hydrogen sulfide has been oxidized in both cases, although it has lost hydrogen rather than gained oxygen in the second reaction. As the hydrogen is often removed to form water from compounds containing hydrogen, the term oxidation can also be applied to the removal of hydrogen.

Conversely, reduction refers to the addition of hydrogen. For example, chlorine is reduced when it reacts with hydrogen:

$$H_2(g) + Cl_2(g) \rightarrow 2HCl(g)$$

Oxidation is an increase in the oxidation state and reduction is a decrease in the oxidation state

In reactions involving ions it is easy to identify which elements have been oxidized and which reduced. In the earlier example, magnesium was oxidized as it changed from atoms to ions with a positive charge, and copper ions were reduced as their charge was decreased from 2+ to 0. This suggests that oxidation involves an increase in charge and reduction involves a decrease in charge.

But for some reactions, such as the reaction between hydrogen and chlorine, it is quite difficult to identify electron loss as the reactions do not involve ions. This is the idea behind the oxidation state introduced in Structure 3.1. The oxidation state is the charge an atom would have if the compound were composed of ions. Considering again the reaction between chlorine and hydrogen, both elements have an oxidation state of 0 as the bonds in H_2 and Cl_2 are both non-polar. The product HCl, by contrast, is polar, with the H possessing a partial positive charge and the chlorine a partial negative charge.

$$H^0-H^0 + Cl^0-Cl^0 \rightarrow 2H^{\delta+}-Cl^{\delta-}$$

Given the partial charges of the molecule, we assign H an oxidation state of +1 and chlorine an oxidation state of −1.

Chlorine is reduced as its oxidation state has decreased. Hydrogen has been oxidized as its oxidation state has increased. It is now recognized that oxidation and reduction occur during chemical change whenever there is a shift in electron density from one atom to another. Oxidation states are a very sensitive measure of electron distribution and so are an efficient tool to analyze all redox reactions.

Oxidation is the loss of electrons. Reduction is the gain of electrons.

It may help you to remember OILRIG: Oxidation Is Loss (of electrons), Reduction Is Gain (of electrons).

Oxidation is the removal of hydrogen. Reduction is the addition of hydrogen.

Note that the charge on an ion X is written with the number first then the charge (as a superscript): e.g. X^{2+}. The oxidation state is written with the charge first then the number: e.g. +2.

Oxidation occurs when there is an increase in the oxidation state of an element. Reduction occurs when there is a decrease in the oxidation state of an element.

Do we justify our knowledge of oxidation states and ionic charges in the same way?

TOK

Returning to the second reaction between hydrogen sulfide and oxygen:

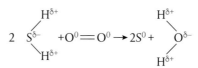

Given the partial charges of the hydrogen sulfide and water molecule, we assign H an oxidation state of +1 and sulfur and oxygen an oxidation state of −2.

Sulfur is oxidized as its oxidation state has increased and oxygen is reduced as its oxidation state has increased.

$$H_2S + O_2 \rightarrow S + H_2O$$

| +1 | 0 | 0 + 1 | oxidation state |

−2   −2

Oxidation and reduction are summarized below.

Oxidation is:	Reduction is:
• the addition of oxygen • the removal of hydrogen • electron loss • an increase in oxidation state	• the removal of oxygen • the addition of hydrogen • electron gain • a decrease in oxidation state

Nature of Science

The concepts of oxidation and reduction have evolved as they have been extended from their original meaning. A large number of what were thought to be diverse chemical reactions are now understood to be similar. This unification of diverse experience is a common feature of science and leads to a deeper understanding of nature.

Oxidation states enable us to analyze redox reactions

We have seen that the oxidation state of an atom in a compound is a measure of the electron possession it has relative to the atom in the pure element. Oxidation states allow us to keep track of the relative electron density of atoms in different compounds and how these change during a reaction.

So far in this chapter we have assigned oxidation states of elements in a compound as the charges an atom would have if the compound was ionic. We saw in Structure 3.1 that these charges could be formalized into a set of rules.

Worked example

Assign oxidation states to the metal ion in **(a)** $[Cu(NH_3)_6]^{2+}$ and **(b)** $[Cr(NH_3)_4Cl_2]^+$.

Solution

(a) NH_3 is a neutral ligand, so the charge on the complex is the same as the charge on the metal ion: ∴ Cu = +2

(b) Cl has a 1− charge   ∴ Cr + (2 × −1) = +1   ∴ Cr = +3

We can use these rules to deduce if an electron transfer has occurred during a reaction and to identify which species has been oxidized and which species has been reduced.

A redox reaction is a chemical reaction where changes in the oxidation states occur.

Worked example

Use oxidation states to deduce which species is oxidized and which is reduced in the following reactions:

(a) $Ca(s) + Sn^{2+}(aq) \rightarrow Ca^{2+}(aq) + Sn(s)$

(b) $4NH_3(g) + 5O_2(g) \rightarrow 4NO(g) + 6H_2O(l)$

(c) $2MnO_4^-(aq) + 6H^+(aq) + 5SO_3^{2-}(aq) \rightarrow 2Mn^{2+}(aq) + 3H_2O(l) + 5SO_4^{2-}(aq)$

(d) $4Fe(s) + 3O_2(g) + 6H_2O(l) \rightarrow 4Fe(OH)_3(s)$

(e) $Cl_2(g) + H_2O(l) \rightarrow HCl(aq) + HOCl(aq)$

TOK

Are oxidation states 'real' or are they artificial constructs invented by the chemist?

Solution

Assign oxidation states to the elements in the reactants and products:

(a)	$Ca(s) + Sn^{2+}(aq) \rightarrow Ca^{2+}(aq) + Sn(s)$			
Oxidation states	0	+2	+2	0

Ca is oxidized because its oxidation state increases from 0 to +2; Sn^{2+} is reduced because its oxidation state decreases from +2 to 0.

(b)	$4NH_3(g) + 5O_2(g) \rightarrow 4NO(g) + 6H_2O(l)$			
Oxidation states	+1(H)	0	−2(O)	+1(H)
	−3(N)		+2(N)	−2(O)

NH_3 is oxidized because the oxidation state of N increases from −3 to +2; O_2 is reduced because the oxidation state of O decreases from 0 to −2.

(c)	$2MnO_4^-(aq) + 6H^+(aq) + 5SO_3^{2-}(aq) \rightarrow 2Mn^{2+}(aq) + 3H_2O(l) + 5SO_4^{2-}(aq)$					
Oxidation states	−2(O)	+1(H)	−2(O)	+2(Mn)	+1(H)	+6(S)
	+7(Mn)		+4(S)		−2(O)	−2(O)

$MnO_4^-(aq)$ is reduced because the oxidation state of Mn decreases from +7 to +2; $SO_3^{2-}(aq)$ is oxidized because the oxidation state of S increases from +4 to +6.

(d)	$4Fe(s) + 3O_2(g) + 6H_2O(l) \rightarrow 4Fe(OH)_3(s)$			
Oxidation states	0	0	+1(H)	+3(Fe) +1(H)
			−2(O)	−2(O)

$Fe(s)$ is oxidized because the oxidation state of Fe increases from 0 to +3; $O_2(g)$ is reduced because the oxidation state of O decreases from 0 to −2.

(e)	$Cl_2(g) + H_2O(l) \rightarrow HCl(aq) + HOCl(aq)$			
Oxidation states	0	+1(H)	+1(H)	+1(H) +1(Cl)
		−2(O)	−1(Cl)	−2(O)

Cl_2 is oxidized because the oxidation state of one Cl increases from 0 to +1; but it is also reduced because the oxidation state of the other Cl increases from 0 to −1.

A **disproportionation reaction** occurs when the oxidation state of the same element both increases and decreases.

Corrosion, or rust, on a painted cast iron figure. Rust forms when iron spontaneously oxidizes by water and the oxygen of the air, forming iron oxide. Rust flakes off the surface of the iron, causing it to be degraded and weakened. Several studies over the past 30 years have shown that the annual direct cost of corrosion to an industrial economy is approximately 3.1% of Gross National Product (GNP).

Structure 2.3 – The surface oxidation of metals is often known as corrosion. What are some of the consequences of this process?

Structure 3.1 – What are the advantages and limitations of using oxidation states to track redox changes?

The limitations of the oxidation concept

We have seen that oxidation states allow us to analyze redox reactions by identifying which atoms are involved in the electron transfer. Oxidation states do, however, have some limitations as, unlike ionic charge, they have little structural or physical significance. The oxidation number of carbon in carbon dioxide is +4 but the carbon is a covalently bonded atom. It should also be noted that the oxidation states calculated in molecules with more than one atom of an element is an average value and can be a non-integer.

Worked example

Assign oxidation states to carbon in C_3H_8 and comment on your answer.

Solution

We assign H = +1

Therefore, $8 \times (+1) + 3\,C = 0 \therefore C = -\frac{8}{3} = -2\frac{2}{3}$

This is a fractional value because it is an average for three non-equivalent carbon atoms.

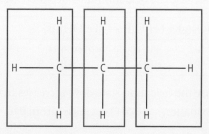

We can divide the molecule into three regions, each with a zero charge.

For the carbon atoms in the red regions $C + 3(+1) = 0 \therefore C = -3$

For the carbon atom in the blue region $C + 2(+1) = 0 \therefore C = -2$

The average oxidation state $= \dfrac{(2(-3) + -2)}{3} = -2\frac{2}{3}$ as expected.

Challenge yourself

1. Assign oxidation states to the overall reactants and products of photosynthesis to show that it is a redox reaction. The reactants are CO_2 and H_2O. The products are $C_6H_{12}O_6$, and O_2.

Oxidizing and reducing agents

We have seen that redox reactions *always* involve the simultaneous oxidation of one reactant with the reduction of another as electrons are transferred between them. The reactant that accepts electrons is called the **oxidizing agent** because it brings about oxidation of the other reactant. In the process, it becomes reduced. Likewise, the reactant that supplies the electrons is known as the **reducing agent** because it brings about reduction and so is oxidized.

For example, in the reaction where iron (Fe) is extracted from its ore (Fe_2O_3):

$$\begin{array}{cccccc}
\text{oxidizing} & & \text{reducing} & & & \\
\text{agent} & & \text{agent} & & & \\
Fe_2O_3(s) & + & 3C(s) & \rightarrow & 2Fe(s) & + & 3CO(g) \\
+3 & & 0 & & 0 & & +2
\end{array}$$

(The oxidation state of O is not shown as it does not change during the reaction.)

The reducing agent C brings about the reduction of Fe ($+3 \rightarrow 0$), while C itself is oxidized to CO ($0 \rightarrow +2$). The oxidizing agent Fe_2O_3 brings about the oxidation of C while itself being reduced to Fe.

> In a redox equation the substance that is reduced is the oxidizing agent, the substance that is oxidized is the reducing agent.

Worked example

When an acidic solution of potassium dichromate(VI), $K_2Cr_2O_7(aq)$, is mixed with iron(II) chloride, $FeCl_2(aq)$, $Fe^{3+}(aq)$ and $Cr^{3+}(aq)$ are produced. Identify the oxidizing and reducing agents.

Solution

Oxidation of Fe: $+2 \rightarrow +3$ so $FeCl_2(aq)$ is oxidized. It is the reducing agent.

Reduction of Cr: $+6 \rightarrow +3$ so $K_2Cr_2O_7(aq)$ is reduced. It is the oxidizing agent.

Some examples of useful oxidizing and reducing agents are given below:

- oxidizing agent: O_2, O_3, H^+/MnO_4^-, $H^+/Cr_2O_7^{2-}$, F_2, Cl_2, conc. HNO_3, H_2O_2
- reducing agent: H_2, C, CO, SO_2, reactive metals

- Note that some substances can act as an oxidizing agent in some reactions and as a reducing agent in others. Water, for example, can act as an oxidizing agent and be reduced to hydrogen with sodium, or act as a reducing agent and be oxidized to oxygen with fluorine.

- H_2O acting as an oxidizing agent:

$$2H_2O(l) + 2Na(s) \rightarrow 2NaOH(aq) + H_2(g)$$
$$+1 0$$

- H_2O acting as a reducing agent:

$$2H_2O(l) + 2F_2(g) \rightarrow 4HF(aq) + O_2(g)$$
$$-2 0$$

The ability of water to act as both an oxidizing and reducing agent makes it a useful solvent for many redox reactions.

Challenge yourself

2. Hydrogen peroxide, H_2O_2, can act as an oxidizing agent and as a reducing agent. With reference to the oxidation states of its atoms, suggest why this is so and which behaviour is more likely.

Systematic names of compounds use oxidation numbers

The use of oxidation numbers in the IUPAC nomenclature for naming compounds unambiguously is discussed in Structure 3.1. A Roman numeral, corresponding to the oxidation state, is inserted in brackets after the name of the relevant element.

The table shows some common examples.

Oxidation states are shown with a '+' or '−' sign and an Arabic numeral, e.g. +2. Oxidation numbers are shown by inserting a Roman numeral in brackets after the name or symbol of the element.

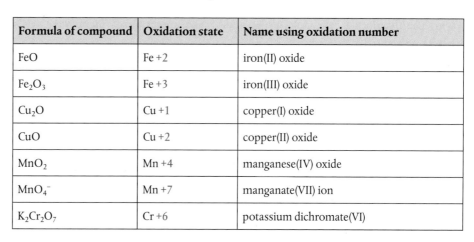

Formula of compound	Oxidation state	Name using oxidation number
FeO	Fe +2	iron(II) oxide
Fe_2O_3	Fe +3	iron(III) oxide
Cu_2O	Cu +1	copper(I) oxide
CuO	Cu +2	copper(II) oxide
MnO_2	Mn +4	manganese(IV) oxide
MnO_4^-	Mn +7	manganate(VII) ion
$K_2Cr_2O_7$	Cr +6	potassium dichromate(VI)

Worked example

Deduce the name of the following compounds using oxidation numbers.

(a) PbO_2 (b) $Cr(OH)_3$ (c) $KMnO_4$

Solution

1. First deduce the oxidation state:

   (a) PbO_2 (b) $Cr(OH)_3$ (c) $KMnO_4$

       $O = -2$ $OH = -1$ $K = +1$ and $O = -2$

       $Pb = +4$ $Cr = +3$ $Mn = +7$

2. The corresponding Roman numeral is inserted after the name of the element. There is no space between the name and the number, and the number is placed in brackets.

   (a) PbO_2 (b) $Cr(OH)_3$ (c) $KMnO_4$

       lead(IV) oxide   chromium(III)   potassium

                     hydroxide   manganate(VII)

The IUPAC system aims to help chemists communicate more easily in all languages by introducing systematic names for compounds. However, its success in achieving this will be determined by how readily it is adopted. What do you think might prevent chemists using exclusively the 'new names'?

Exercise

Q1. Which equation represents a redox reaction?

  A $NaOH(aq) + HNO_3(aq) \rightarrow NaNO_3(aq) + H_2O(l)$

  B $Zn(s) + 2HCl(aq) \rightarrow ZnCl_2(aq) + H_2(g)$

  C $CuO(s) + 2HCl(aq) \rightarrow CuCl_2(aq) + H_2O(l)$

  D $MgCO_3(s) + 2HNO_3(aq) \rightarrow Mg(NO_3)_2(aq) + H_2O(l) + CO_2(g)$

Q2. The oxidation state of iron is the same in all the following compounds except which one?

  A $FeCl_3$ **B** Fe_2O_3

  C $FeCO_3$ **D** $Fe_2(SO_4)_3$

Q3. Identify the correct descriptions of the redox reaction between manganese(IV) oxide and the chloride ions:

$$MnO_2(s) + 2Cl^-(aq) + 4H^+(aq) \rightarrow Mn^{2+}(aq) + Cl_2(g) + 2H_2O(l)$$

  I. H^+ is the reducing agent

  II. MnO_2 undergoes reduction

  III. $Cl^-(aq)$ is the reducing agent

  A I and II only   **B** I and III only

  C II and III only   **D** I, II and III

Q4. Assign oxidation states to all elements in the following compounds.

(a) CH_4 (b) $CuSO_4$ (c) C_2H_4 (d) CO

(e) $K_2Cr_2O_7$ (f) K_2CrO_4 (g) H_2O_2 (h) C_4H_{10}

Q5. Use oxidation states to deduce which species is oxidized and which is reduced in the following reactions.

(a) $Sn^{2+}(aq) + 2Fe^{3+}(aq) \rightarrow Sn^{4+}(aq) + 2Fe^{2+}(aq)$

(b) $2FeCl_2(aq) + Cl_2(aq) \rightarrow 2FeCl_3(aq)$

(c) $2H^+(aq) + S_2O_3{}^{2-}(aq) \rightarrow S(s) + SO_2(g) + H_2O(l)$

(d) $2H_2O_2(aq) \rightarrow 2H_2O(l) + O_2(g)$

(e) $I_2(aq) + SO_3{}^{2-}(aq) + H_2O(l) \rightarrow 2I^-(aq) + SO_4{}^{2-}(aq) + 2H^+(aq)$

Q6. The combustion of ethanol is a redox reaction.

$$C_2H_5OH(g) + 3O_2(g) \rightarrow 2CO_2(g) + 3H_2O(l)$$

(a) Assign oxidation states to carbon in C_2H_5OH and comment on your answer.

(b) Discuss the application of the different definitions of oxidation and reduction in analyzing the reaction.

Q7. Identify the oxidizing agents and the reducing agents in the following reactions.

(a) $H_2(g) + Cl_2(g) \rightarrow 2HCl(g)$

(b) $2Al(s) + 3PbCl_2(s) \rightarrow 2AlCl_3(s) + 3Pb(s)$

(c) $Cl_2(aq) + 2KI(aq) \rightarrow 2KCl(aq) + I_2(aq)$

(d) $CH_4(g) + 2O_2(g) \rightarrow CO_2(g) + 2H_2O(l)$

Reactivity 3.2.2 – Half-equations

Reactivity 3.2.2 – Half-equations separate the processes of oxidation and reduction, showing the loss or gain of electrons.

Deduce redox half-equations and equations in acidic or neutral solutions.

	Tool 1, Inquiry 2 – Why are some redox titrations described as 'self-indicating'?

Writing half-equations for redox equations

Although oxidation and reduction always occur together, it can be useful to separate out the two processes in a redox equation and write separate equations for the oxidation and reduction processes. These **half-equations** emphasize the electron transfers that occur during the reaction. Electrons are included in both half-equations to balance the charges.

Worked example

Deduce the two half-equations for the following reaction:

$$Zn(s) + Cu^{2+}(aq) \rightarrow Zn^{2+}(aq) + Cu(s)$$

Solution

Assign oxidation states so you can see what is being oxidized and what is reduced.

$$Zn(s) + Cu^{2+}(aq) \rightarrow Zn^{2+}(aq) + Cu(s)$$
$$\quad 0 \qquad +2 \qquad\quad +2 \qquad\quad 0$$

The Zn is oxidized and Cu^{2+} is reduced.

We write separate equations for oxidation and reduction, adding electrons to balance the charges.

oxidation: $Zn(s) \rightarrow Zn^{2+}(aq) + 2e^-$ electrons are lost by Zn(s)

reduction: $Cu^{2+}(aq) + 2e^- \rightarrow Cu(s)$ electrons are gained $Cu^{2+}(aq)$

Note there are equal numbers of electrons in the two half-equations, so that when they are added together the electrons cancel out.

Sometimes we may know the species involved in a redox reaction, but not the overall equation. We can work this out by making sure it is balanced for both atoms and charges. A good way to do this is to write half-equations for the oxidation and reduction processes separately and then add these two together to give the overall reaction. Many of these reactions take place in acidified solutions and we therefore use H_2O and/or H^+ ions to balance the half-equations. The process is best broken down into a series of steps, as shown in the example below.

Worked example

Deduce an equation for the reaction in which NO_3^- and Cu react together in acidic solution to produce NO and Cu^{2+}.

Solution

1. Write an unbalanced equation with the given reactants and products and assign oxidation states to determine which elements are being oxidized and which are being reduced:

$$NO_3^-(aq) + Cu(s) \rightarrow NO(g) + Cu^{2+}(aq)$$
$$+5\ -2 \qquad\ 0 \qquad +2\ -2 \quad +2$$

In this example Cu is oxidized ($0 \rightarrow +2$) and N is reduced ($+5 \rightarrow +2$).

2. Write separate half-equations for oxidation and reduction including the electrons needed to account for the changes in oxidation state. Balance the atoms other than H and O. In this example the Cu and N are already balanced.

oxidation: $Cu(s) \rightarrow Cu^{2+}(aq) + 2e^-$ Cu: $0 \rightarrow +2$, 2 electrons produced

reduction: $NO_3^-(aq) + 3e^- \rightarrow NO(g)$ N: $+5 \rightarrow +2$, 3 electrons added

3. Balance each half-equation for O by adding H_2O as needed.

In this example the reduction equation needs two more O atoms on the right-hand side. These are added as water molecules:

reduction: $NO_3^-(aq) + 3e^- \rightarrow NO(g) + 2H_2O(l)$

4. Balance each half-equation for H by adding H^+ as needed.

Here the reduction equation needs four H atoms on the left-hand side:

reduction: $NO_3^-(aq) + 4H^+(aq) + 3e^- \rightarrow NO(g) + 2H_2O(l)$

5. Check that each half-equation is balanced for atoms and for charge.

6. Equalize the number of electrons in the two half-equations by finding a common factor and multiplying each appropriately.

Here the common factor is 6. The equation for oxidation must be multiplied by 3, and the equation for reduction by 2. This gives six electrons in both equations:

oxidation: $3Cu(s) \rightarrow 3Cu^{2+}(aq) + 6e^-$

reduction: $2NO_3^-(aq) + 8H^+(aq) + 6e^- \rightarrow 2NO(g) + 4H_2O(l)$

7. Add the two half-equations together, cancelling out anything that is the same on both sides, which includes the electrons.

$$3Cu(s) + 2NO_3^-(aq) + 8H^+(aq) \rightarrow 3Cu^{2+}(aq) + 2NO(g) + 4H_2O(l)$$

Note the final equation is balanced for atoms and charge and has no electrons.

Summary of steps in writing redox equations.
1. Assign oxidation states to determine which elements are oxidized and which are reduced.
2. Write separate half-equations for oxidation and reduction. Balance the atoms other than H and O and add electrons to account for the changes in oxidation states.
3. Balance each half-equation for O by adding H_2O as needed.
4. Balance each half-equation for H by adding H^+ as needed.
5. Check that each half-equation is balanced for atoms and for charge.
6. Equalize the number of electrons in the two half-equations by multiplying each appropriately.
7. Add the two half-equations together, cancelling out anything that is the same on both sides.

Challenge yourself

3. Reactions in which the same element is simultaneously oxidized and reduced are known as **disproportionation reactions**. Deduce the equation for the reaction when $I_2(aq)$ reacts with KOH(aq) to produce KI, KIO_3, and H_2O, and show that this is a disproportionation reaction.

Redox titrations are used to determine the unknown concentrations in redox reactions

As discussed in Structure 1.3, titrations are used to determine the unknown concentration of a substance in solution. In redox titrations the technique is based on a redox reaction between the two reactants and finds the equivalence point where they have reacted stoichiometrically by transferring electrons. Redox titrations are carried out in much the same way as acid–base titrations, using a burette and pipette to measure volumes accurately, and a standard solution of one reactant. An indicator is often used to signal the equivalence point, although some redox changes are accompanied by a colour change and may not need an external indicator. From the volume of the solution added from the burette to reach equivalence, which is known as the **titre**, the concentration of the other reactant can be determined. The calculations are based on the equation for the redox reaction, which can be deduced from the half-equations.

Examples of redox titrations

Iodine–thiosulfate reaction

Several different redox titrations use an oxidizing agent to convert excess iodide ions to iodine.

$$2I^-(aq) + \text{oxidizing agent} \rightarrow I_2(aq) + \text{other products}$$

Examples of oxidizing agents used in this way include $KMnO_4(aq)$, $KIO_3(aq)$, $K_2Cr_2O_7(aq)$, and $NaOCl(aq)$.

The liberated iodine, $I_2(aq)$, is then titrated with sodium thiosulfate, $Na_2S_2O_3(aq)$.

Redox equations:

oxidation: $2S_2O_3^{2-}(aq) \rightarrow S_4O_6^{2-}(aq) + 2e^-$

reduction: $I_2(aq) + 2e^- \rightarrow 2I^-(aq)$

Adding the half-reactions:

$$2S_2O_3^{2-}(aq) + I_2(aq) \rightarrow 2I^-(aq) + S_4O_6^{2-}(aq)$$
$$\text{deep blue in}$$
$$\text{presence of starch}$$

Starch can be added as an indicator just before the equivalence point is reached. It forms a deep blue colour by forming a complex with free I_2. As the I_2 is reduced to I^- during the reaction, the blue colour disappears, making the equivalence point very distinct.

▲
At the start of an iodine titration the solution is brown in colour. A standard sodium thiosulfate solution of known concentration is added to the iodine solution until it becomes a pale yellow. A few drops of starch solution are added to turn it a very dark blue to sharpen the visibility of the end point. The end point occurs when the last trace of blue disappears and the final solution is colourless.

'Iodine solution' is made by dissolving I_2 in a solution of iodide ions. A complex ion, $I_3^-(aq)$, is formed which is more soluble in water than iodine, I_2, which has a very low solubility.

It is valid to use I_2 in equations, as the $I^-(aq)$ ions are spectator ions.

Worked example

Household bleach is an oxidizing agent that contains sodium hypochlorite, $NaOCl(aq)$, as the active ingredient. It reacts with iodide ions in acidic solution as follows:

$$OCl^-(aq) + 2I^-(aq) + 2H^+(aq) \rightarrow I_2(aq) + Cl^-(aq) + H_2O(l)$$

A 10.00 cm^3 sample of bleach was reacted with a solution of excess iodide ions, and the liberated iodine was then titrated with $Na_2S_2O_3$. The titration required 38.65 cm^3 of $0.0200 \text{ mol dm}^{-3}$ $Na_2S_2O_3(aq)$. Determine the concentration of $OCl^-(aq)$ in the bleach.

Solution

First we need the balanced equation for the titration, which can be solved by the half-equation method above.

Next we calculate the number of moles of $Na_2S_2O_3$ as we know both its concentration and its volume.

$$n = cV$$

$$n(S_2O_3^{2-}) = 0.0200 \text{ mol dm}^{-3} \times \frac{38.65}{1000} \text{ dm}^3 = 7.73 \times 10^{-4} \text{ mol } S_2O_3^{2-}$$

From the equation on the previous page we know the reacting ratio $S_2O_3^{2-} : I_2 = 2 : 1$

$$\therefore n(I_2) = 7.73 \times 10^{-4} \times 0.5 = 3.865 \times 10^{-4} \text{ mol}$$

This is a back-titration as the I_2 was liberated by the reaction of I^- with OCl^-, as given in the question. The reacting ratio is $OCl^-(aq) : I_2(aq) = 1 : 1$

$$\therefore n(OCl^-) = n(I_2) = 3.865 \times 10^{-4} \text{ mol}$$

The concentration of the OCl^- in the bleach can now be calculated.

$$n(OCl^-) = cV$$

$$\therefore 3.865 \times 10^{-4} \text{ mol} = c \times \frac{10}{1000} \text{ dm}^3$$

$$[OCl^-] = 0.0387 \text{ mol dm}^{-3}$$

Analysis of iron with manganate(VII)

SKILLS

Determination of the level of rust in a rusted nail. Full details of how to carry out this experiment with a worksheet are available in the eBook.

This redox titration uses $KMnO_4$ in an acidic solution to oxidize Fe^{2+} ions to Fe^{3+}. During the reaction MnO_4^- is reduced to Mn^{2+}. This is accompanied by a colour change from deep purple to colourless, so the reaction mixture acts as its own indicator, and the equivalence point is determined.

Redox equations:

oxidation: $Fe^{2+}(aq) \rightarrow Fe^{3+}(aq) + e^-$

reduction: $MnO_4^-(aq) + 8H^+(aq) + 5e^- \rightarrow Mn^{2+}(aq) + 4H_2O(l)$

So the overall equation is

$$5Fe^{2+}(aq) + MnO_4^-(aq) + 8H^+(aq) \rightarrow 5Fe^{3+}(aq) + Mn^{2+}(aq) + 4H_2O(l)$$

Worked example

All the iron in a 2.000 g tablet was dissolved in an acidic solution. The resulting solution of Fe^{2+}(aq) ions was then titrated with $KMnO_4$. The titration required 27.50 cm^3 of 0.100 mol dm^{-3} $KMnO_4$(aq). Calculate the percentage by mass of iron in the tablet. Describe what would be observed during the reaction and how the equivalence point can be detected.

Solution

The balanced equation for the reaction can be deduced by the half-equation method above.

The O and H are balanced by adding H_2O as a product and H^+(aq) as a reactant.

$$5Fe^{2+}(aq) + MnO_4^-(aq) + 8H^+(aq) \rightarrow 5Fe^{3+}(aq) + Mn^{2+}(aq) + 4H_2O(l)$$

Next we need to know the amounts of reactants used to reach equivalence. We start with $KMnO_4$(aq) as we know both its concentration and its volume.

$$n = cV$$

$$n(MnO_4^-) = 0.100 \text{ mol dm}^{-3} \times \frac{27.50}{1000} \text{ dm}^3 = 0.00275 \text{ mol } MnO_4^-$$

From the equation for the reaction, we know the reacting ratio:

$MnO_4^- : Fe^{2+} = 1 : 5$

$\therefore\ n(Fe^{2+}) = n(MnO_4^-) \times 5 = 0.00275 \text{ mol} \times 5 = 0.01375 \text{ mol}$

$M(Fe) = 55.85 \text{ g mol}^{-1}$

$$n = \frac{m}{M}$$

$\therefore\ m(Fe) = n \times M = 0.01375 \text{ mol} \times 55.85 \text{ g mol}^{-1} = 0.7679 \text{ g}$

$$\% \text{ Fe in tablet} = \frac{0.768 \text{ g}}{2.000 \text{ g}} \times 100 = 38.39\%$$

MnO_4^- in the burette is purple, but a nearly colourless solution in the flask is formed as it reacts to form Mn^{2+}. When the reducing agent Fe^{2+} in the flask has been used up at equivalence, MnO_4^- ions will not react when added after the equivalence point, and the purple colour persists.

◀ The browning of fruit when it is exposed to the air is an oxidation reaction. It occurs when phenols released in damaged cells are oxidized with molecular oxygen at alkaline pH. The brown products are tannins.

Tool 1, Inquiry 2 – Why are some redox titrations described as 'self-indicating'?

SKILLS

Redox titrations can be used to analyze many household chemicals such as food and drinks. Wines, for example, can be analyzed for the presence of sulfur dioxide, and foods for vitamin C.

In titration calculations, always look first for the reactant where you know both its reacting volume *and* its concentration. Start by using this data to calculate the number of moles. Then use the reacting ratio in the equation to deduce the moles of the other reactant.

Exercise

Q8. Consider the following unbalanced equation:

$$MnO_4^-(aq) + SO_3^{2-}(aq) + H^+(aq) \rightarrow Mn^{2+}(aq) + SO_4^{2-}(aq) + H_2O(l)$$

Deduce the amount of MnO_4^- which reacts with 1 mol of SO_3^{2-} ions.

A 0.40 mol   **B** 3 mol   **C** 1.50 mol   **D** 7 moles

Q9. Consider the following unbalanced equation:

$$MnO_4^-(aq) + H_2O_2(aq) + H^+(aq) \rightarrow Mn^{2+}(aq) + H_2O(l) + O_2(g)$$

Deduce the amount of $H_2O_2(aq)$ which reacts with 1 mol of $MnO_4^-(aq)$.

A 0.40 mol   **B** 1.25 mol   **C** 2.50 mol   **D** 5 moles

Q10. Deduce the half-equations of oxidation and reduction for the following reactions.

(a) $Ca(s) + 2H^+(aq) \rightarrow Ca^{2+}(aq) + H_2(g)$

(b) $2Fe^{2+}(aq) + Cl_2(aq) \rightarrow 2Fe^{3+}(aq) + 2Cl^-(aq)$

(c) $Sn^{2+}(aq) + 2Fe^{3+}(aq) \rightarrow Sn^{4+}(aq) + Fe^{2+}(aq)$

(d) $Cl_2(aq) + 2Br^-(aq) \rightarrow 2Cl^-(aq) + Br_2(aq)$

Q11. Write balanced equations for the following reactions that occur in acidic solutions.

(a) $I^-(aq) + HSO_4^-(aq) \rightarrow I_2(aq) + SO_2(g)$

(b) $I_2(aq) + OCl^-(aq) \rightarrow IO_3^-(aq) + Cl^-(aq)$

(c) $MnO_4^-(aq) + H_2SO_3(aq) \rightarrow SO_4^{2-}(aq) + Mn^{2+}(aq)$

Q12. A bag of 'road salt', used to melt ice and snow from roads, contains a mixture of calcium chloride, $CaCl_2$, and sodium chloride, NaCl. A 2.765 g sample of the mixture was analyzed by first converting all the calcium into calcium oxalate, CaC_2O_4. This was then dissolved in H_2SO_4 and titrated with 0.100 mol dm^{-3} $KMnO_4$ solution. The titration required 24.65 cm^3 of $KMnO_4(aq)$ and produced $Mn^{2+}(aq)$, $CO_2(g)$, and $H_2O(l)$.

(a) What would be observed at the equivalence point of the titration?

(b) Write the half-equation for the oxidation reaction, starting with $C_2O_4^{2-}$.

(c) Write the half-equation for the reduction reaction, starting with MnO_4^-.

(d) Write the overall equation for the redox reaction.

(e) Determine the number of moles of $C_2O_4^{2-}$.

(f) Deduce the number of moles of Ca^{2+} in the original sample.

(g) What was the percentage by mass of $CaCl_2$ in the road salt?

Q13. Alcohol levels in blood can be determined by a redox titration with potassium dichromate, $K_2Cr_2O_7$, according to the following equation:

$$C_2H_5OH(aq) + 2Cr_2O_7^{2-}(aq) + 16H^+(aq) \rightarrow 2CO_2(g) + 4Cr^{3+}(aq) + 11H_2O(l)$$

(a) Determine the alcohol percentage in the blood by mass if a 10.000 g sample of blood requires 9.25 cm^3 of 0.0550 mol dm^{-3} $K_2Cr_2O_7$ solution to reach equivalence.

(b) Describe the change in colour that would be observed during the titration.

Reactivity 3.2.3 – Trends in ease of oxidation and reduction of elements

Reactivity 3.2.3 – The relative ease of oxidation and reduction of an element in a group can be predicted from its position in the periodic table.

The reactions between metals and aqueous metal ions demonstrate the relative ease of oxidation of different metals.

Predict the relative ease of oxidation of metals.

Predict the relative ease of reduction of halogens.

Interpret data regarding metal and metal ion reactions.

The relative reactivity of metals observed in metal/metal ion displacement reactions does not need to be learned; appropriate data will be supplied in examination questions.	Structure 3.1 – Why does metal reactivity increase, and non-metal reactivity decrease, down the main groups of the periodic table?
	Tool 1, Inquiry 2 – What observations can be made when metals are mixed with aqueous metal ions and solutions of halogens are mixed with aqueous halide ions?

More reactive non-metals are stronger oxidizing agents

Some oxidizing and reducing agents are stronger than others, depending on their relative tendencies to lose or gain electrons. We saw in Structure 3.1 that the halogens (group 17 elements) react by gaining electrons and forming negative ions, so act as oxidizing agents by removing electrons from other substances. We also learned that this tendency to gain electrons decreases down the group.

This gives the following trend:

F_2 strongest oxidizing agent, most readily becomes reduced

Cl_2

Br_2

I_2 weakest oxidizing agent, least readily becomes reduced

For example, adding a more reactive halogen to a solution containing the ions of a less reactive halogen (known as halide ions) leads to a reaction. For example:

$$Cl_2(aq) + 2KI(aq) \rightarrow 2KCl(aq) + I_2(aq)$$

Species which do not change during a reaction, like $K^+(aq)$ here, are called spectator ions and can be cancelled out in the equation. This gives the ionic equation:

$$Cl_2(aq) + 2I^-(aq) \rightarrow 2Cl^-(aq) + I_2(aq)$$

As discussed in Structure 3.1, a reaction occurs because Cl atoms attract electrons more strongly than I atoms do due to the smaller atomic radius of Cl. You can think of it as a competition for electrons where the stronger oxidizing agent, in this case chlorine, 'wins'.

Chlorine gas bubbling through a clear solution of potassium iodide KI(aq). The solution turns brown due to the formation of iodine in solution, as chlorine oxidizes the iodide ions and forms chloride ions.

Adding a less reactive halogen to the ions of a more reactive halogen results in no reaction.

$$I_2(aq) + 2Cl^-(aq) \rightarrow \text{no reaction}$$

As non-metals generally react by gaining electrons this pattern can be summarized: more reactive non-metals are stronger oxidizing agents than less reactive non-metals.

More reactive non-metals oxidize the ions of a less reactive non-metal.

More reactive metals are stronger reducing agents

The different strengths of metals as reducing agents can be compared in a similar way. We saw in Structure 2.3 that metals have a tendency to lose electrons and form positive ions. They act as reducing agents – they give their electrons to another substance. More reactive metals lose their electrons more readily and are stronger reducing agents than less reactive metals. We saw in Structure 3.1 that the reactivity of the group 1 metals increases down the group, so potassium is a stronger reducing agent than sodium. More generally we can compare the reactivity of the metals in displacements similar to those discussed earlier between the halogens and the halide ions.

Structure 3.1 – Why does metal reactivity increase, and non-metal reactivity decrease, down the main groups of the periodic table?

If we add zinc metal, for example, to an aqueous solution of copper sulfate, we see a reaction (Figure 1). The blue colour of the solution fades, the pink/brown colour of the copper metal appears, and there is a rise in temperature. The $Cu^{2+}(aq)$ ions are displaced from solution as they are reduced by the more reactive $Zn(s)$, which simultaneously dissolves as it is oxidized to $Zn^{2+}(aq)$.

$$Zn(s) + CuSO_4(aq) \rightarrow ZnSO_4(aq) + Cu(s)$$

A strip of zinc metal half submerged in a solution of $CuSO_4(aq)$. Solid copper that appears as a pink/brown solid is precipitated and the blue colour of the solution fades as the copper ions are reduced by the zinc. Zinc is a stronger reducing agent than copper.

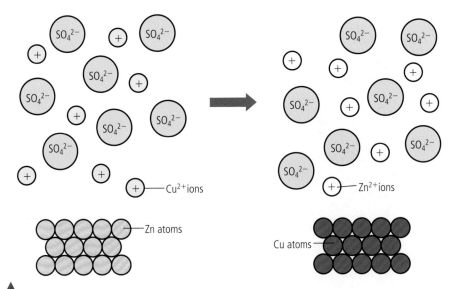

▲ **R3.2 Figure 1** The reaction of zinc with copper(II) sulfate solution.

Again we can write this as an ionic equation. The sulfate ions are not included because they act as spectator ions and are not changed during the reaction.

More reactive metals are stronger reducing agents than less reactive metals. A more reactive metal is able to reduce the ions of a less reactive metal.

Zinc is acting as the reducing agent – it is therefore the more reactive metal. We can think of it as having the reducing strength to 'force' copper ions to accept the electrons. This can be confirmed by trying the reaction the other way round by adding copper metal to a solution of zinc ions. There is no reaction because copper is not a strong enough reducing agent to force electrons to reduce Zn^{2+}(aq) ions. This is another way of saying that Cu(s) is a less reactive metal, less able to push the electrons onto Zn^{2+}(aq).

By repeating displacement reactions like these between different combinations of metals and their ions, we can build up a list of relative strengths of the metals as reducing agents. This is called the **activity series** and it enables us to predict whether a particular redox reaction between a metal and the ions of another metal will be feasible. Later in this chapter we will learn how to quantify these differences in metal reactivity, which is of great importance in many industrial processes. For example, the extraction of a metal from its ore often involves choosing a suitable reducing agent by reference to this data.

Here is a small part of the activity series of metals:

Mg — strongest reducing agent, most readily becomes oxidized
Al
Zn
Fe
Pb
Cu
Ag — weakest reducing agent, least readily becomes oxidized

The activity series is not something that needs to be learned. The key point is that you are able to interpret it to deduce the order of reactivity from given data.

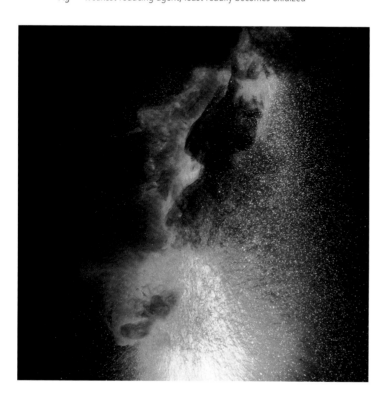

The thermite reaction between powdered aluminum and iron oxide to produce aluminum oxide and iron. Aluminum is a stronger reducing agent than iron.

Worked example

Refer to the activity series given on the previous page to predict whether the following reactions will occur:

(a) $ZnCl_2(aq) + 2Ag(s) \rightarrow 2AgCl(s) + Zn(s)$

(b) $2FeCl_3(aq) + 3Mg(s) \rightarrow 3MgCl_2(aq) + 2Fe(s)$

Solution

(a) This reaction would involve Ag(s) reducing $Zn^{2+}(aq)$ in $ZnCl_2(aq)$. Ag(s) is a weaker reducing agent than Zn, so this will not occur.

(b) This reaction involves Mg(s) reducing $Fe^{3+}(aq)$ in $FeCl_3(aq)$. Mg(s) is a stronger reducing agent than Fe, so this will occur.

We can investigate how some non-metals such as carbon and hydrogen would fit into this activity series of metals by similar types of displacement reactions. Carbon is able to reduce the oxides of iron and metals below it in the series, which provides one of the most effective means for the extraction of these metals. The position of hydrogen relative to the metals is discussed on page 514.

Tool 1, Inquiry 2 – What observations can be made when metals are mixed with aqueous metal ions, and solutions of halogens are mixed with aqueous halide ions?

Investigating the activity series. Full details of how to carry out this experiment with a worksheet are available in the eBook.

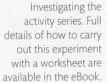

Challenge yourself

4. Carbon can be used as the reducing agent in the extraction of iron and zinc. Lead salts are produced when lead is added to dilute acids. Suggest where you place the non-metals carbon and hydrogen in the activity series on page 511.

Exercise

Q14. Use the activity series on the previous page to predict what happens when zinc powder is added to an aqueous solution of sodium chloride.

    A $Cl_2(g)$ is produced   **B** Na(s) is produced

    C $ZnCl_2(aq)$ is produced   **D** No reaction

Q15. Consider the following reactions of three unidentified metals P, Q and R.

    1. $PCl_2(aq) + 2Q(s) \rightarrow P(s) + 2QCl(aq)$

    2. $RCl_2(aq) + 2Q(s) \rightarrow R(s) + 2QCl(aq)$

    3. $RCl_2(aq) + 2P(s) \rightarrow$ no reaction

    Deduce the order of **increasing** reactivity of the metals.

    A $P < Q < R$ **B** $P < R < Q$

    C $R < P < Q$ **D** $Q < R < P$

Q16. Use the activity series given to predict if a reaction occurs between the following reactants, and write equations where relevant.

    (a) $CuCl_2 + Ag$ **(b)** $Fe(NO_3)_2 + Al$

    (c) $NaI + Br_2$ **(d)** $KCl + I_2$

Q17. An activity series can be deduced from observing displacement reactions.

(a) Use the following reactions to deduce the order of reactivity of the elements W, X, Y and Z, putting the most reactive first.

$W + X^+ \rightarrow W^+ + X$ $Y^+ + Z \rightarrow$ no reaction

$X + Z^+ \rightarrow X^+ + Z$ $X + Y^+ \rightarrow X^+ + Y$

(b) Which of the following reactions would you expect to occur according to the activity series you established in part (a)?

(i) $W^+ + Y \rightarrow W + Y^+$ (ii) $W^+ + Z \rightarrow W + Z^+$

Reactivity 3.2.4 – Oxidation of metals by acids

Reactivity 3.2.4 – Acids react with reactive metals to release hydrogen.

Deduce equations for reactions of reactive metals with dilute HCl and H_2SO_4.

Acids react with metals to form salts and release hydrogen

As discussed in Reactivity 3.1, the term **salt** refers to an ionic compound formed when the hydrogen of an acid is replaced by a metal or another positive ion.

For example:

$$2HCl(aq) + Zn(s) \rightarrow ZnCl_2(aq) + H_2(g)$$

$$H_2SO_4(aq) + Fe(s) \rightarrow FeSO_4(aq) + H_2(g)$$

$$2CH_3COOH(aq) + Mg(s) \rightarrow Mg(CH_3COO)_2(aq) + H_2(g)$$

$ZnCl_2$, $FeSO_4$, and $Mg(CH_3COO)_2$ are all salts.

Magnesium reacting with dilute hydrochloric acid.
$Mg(s) + 2HCl(aq) \rightarrow MgCl_2(aq) + H_2(g)$
The tiny bubbles are hydrogen gas being produced.

We can also write these as ionic equations, showing only the species that change during the reaction. For example:

$$2H^+(aq) + Zn(s) \rightarrow Zn^{2+}(aq) + H_2(g)$$

As zinc is oxidized from 0 to +2 during the reaction it is acting as the reducing agent. We can think of it as having the reducing strength to 'force' hydrogen ions to accept the electrons. As there is no reaction between copper and dilute acids we can deduce that hydrogen is more reactive than copper. These results show that hydrogen is below zinc and above copper in the activity series. You can demonstrate the release of hydrogen from acids in simple experiments by adding a small piece of metal to a dilute solution of the acid. There is a big range, however, in the reactivity of metals in these reactions. More reactive metals such as sodium and potassium in group 1 would react so violently that it would not be safe to perform the experiment, while copper and other less reactive metals such as silver and gold will usually not react at all. These results show that we can insert hydrogen between copper and lead in the activity series.

These reactions of metals with acids illustrate why acids are so corrosive and why, for example, it is important to keep car battery acid well away from the metal bodywork of a car. This also explains why the less reactive metals such as silver and gold are so valuable – they are resistant to corrosion.

Exercise

Q18. Identify the metals which will react with 0.1 mol dm^{-3} hydrochloric acid.

I. Ag   II. Fe   III. Zn

A I and II only   **B** I and III only

C II and III only   **D** I, II and III

Q19. Identify the products formed when a metal reacts with a dilute acid.

I. a salt   II. water   III. hydrogen gas

A I and II   **B** I and III

C II and III   **D** I, II and III

Q20. Write complete and ionic equations for the following reactions.

(a) hydrochloric acid reacting with aluminium

(b) sulfuric acid reacting with magnesium

(c) ethanoic acid reacting with zinc

Q21. Write balanced equations for the following reactions that occur in acidic solutions.

(a) $Zn(s) + SO_4^{2-}(conc) \rightarrow Zn^{2+}(aq) + SO_2(g)$

(b) $NO_3^-(conc) + Zn(s) \rightarrow NH_4^+(aq) + Zn^{2+}(aq)$

(c) $NO_3^-(aq) + Cu(s) \rightarrow NO(g) + Cu^{2+}(aq)$

(d) $NO_3^-(conc) + Cu(s) \rightarrow NO_2(g) + Cu^{2+}(aq)$

Reactivity 3.2.5 – Comparing voltaic and electrochemical cells

Reactivity 3.2.5 – Oxidation occurs at the anode and reduction occurs at the cathode in electrochemical cells.

Label electrodes as anode and cathode, and identify their signs/polarities in voltaic cells and electrolytic cells, based on the type of reaction occurring at the electrode.

Voltaic and electrolytic cells are collectively known as electrochemical cells.

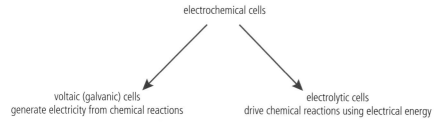

electrochemical cells

voltaic (galvanic) cells
generate electricity from chemical reactions

electrolytic cells
drive chemical reactions using electrical energy

A voltaic cell converts chemical energy to electrical energy: a spontaneous chemical reaction drives electrons around a circuit. An electrolytic cell converts electrical energy to chemical energy. An electric current reverses the normal directions of chemical change, and non-spontaneous reactions occur. Stable compounds like sodium chloride can be broken down into their reactive constituent elements.

The anode and cathode in electrochemical cells are defined by the chemical changes that take place. Oxidation occurs at the anode and reduction occurs at the cathode in both types of cell. The polarity of the cell, however, changes. The anode is the negative terminal in a voltaic cell but the positive terminal in an electrolytic cell. The cathode is the positive terminal in a voltaic cell but the negative in an electrolytic cell.

'An Ox and a Red Cat' may help you remember the reaction at the electrodes of an electrochemical cell (**ox**idation occurs at the **an**ode, **red**uction occurs at the **cat**hode).

Type of reaction	Type of cell	
	Voltaic	**Electrolytic**
Type of reaction	spontaneous	non-spontaneous
Electrode of oxidation	anode	anode
Electrode of reduction	cathode	cathode
Polarity of anode	negative	positive
Polarity of cathode	positive	negative

Voltaic and electrolytic cells are explored in more detail in the sections that follow.

Exercise

Q22. Identify the processes that occur at the positive electrode in voltaic and electrolytic cells.

	Electrolytic cell	Voltaic cell
A	oxidation	reduction
B	reduction	oxidation
C	oxidation	oxidation
D	reduction	reduction

Q23. Identify the best description of the reaction that occurs at the negative electrode of an electrolytic cell.

	Type of reaction	Voltaic cell
A	spontaneous	reduction
B	not spontaneous	reduction
C	spontaneous	oxidation
D	not spontaneous	oxidation

Reactivity 3.2.6 – Primary (voltaic) cells

Reactivity 3.2.6 – A primary (voltaic) cell is an electrochemical cell that converts energy from spontaneous redox reactions to electrical energy.

Explain the direction of electron flow from anode to cathode in the external circuit, and ion movement across the salt bridge.

Construction of primary cells should include: half-cells containing metal/metal ion, anode, cathode, electric circuit, salt bridge.	Reactivity 1.3 – Electrical energy can be derived from the combustion of fossil fuels or from electrochemical reactions. What are the similarities and differences in these reactions?

Voltaic cells generate electricity from spontaneous redox reactions

In oxidation reactions electrons are lost and in reduction reactions they are gained. If the half-reactions occur at spatially separated electrodes, the electrons need to pass through an external circuit between the electrodes and an electric current is produced. Let us consider again the reaction we discussed on page 510 in which zinc reduced copper ions. Remember that here zinc was the reducing agent and became oxidized while copper ions were reduced. When the reaction is carried out in a single test tube, as shown in the photo on page 510, the electrons flow spontaneously from the zinc atoms to the copper ions in the solution and, as we noted, energy is released in the form of heat. It is an exothermic reaction. Alternatively, we can separate the two half-reactions:

oxidation: $Zn(s) \rightarrow Zn^{2+}(aq) + 2e^-$

reduction: $Cu^{2+}(aq) + 2e^- \rightarrow Cu(s)$

Electrons pass from the zinc to the copper ions through the external circuit.

This is known as a voltaic or a **galvanic cell**, and we will see how it is constructed in the next section. The half-cells connected in this way are often called **electrodes.** As the chemicals are not renewed during the process it is also known as a **primary cell.** A battery normally consists of several cells connected together.

(a)  **(b)**

▲
Electrical energy can be produced from chemical energy **(a)** indirectly by the combustion of fuels or **(b)** directly in a battery. Both involve redox reactions. Electrical energy is produced directly from the oxidation of a fuel in a fuel cell. These are discussed on page 526.

Reactivity 1.3 – Electrical energy can be derived from the combustion of fossil fuels or from electrochemical reactions.
What are the similarities and differences in these reactions?

Half-cells generate electrode potentials

There are many types of half-cell but probably the simplest is made by putting a strip of metal into a solution of its ions (Figure 2).

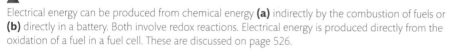

▲
R3.2 Figure 2 Copper and zinc half-cells.

In the zinc half-cell, zinc atoms form ions by releasing electrons that will make the surface of the metal negatively charged with respect to the solution (Figure 3).

$$Zn(s) \rightarrow Zn^{2+}(aq) + 2e^- \text{ (electrons give the electrode a negative charge)}$$

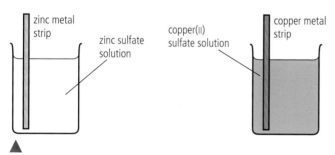

▲
A copper half-cell consisting of a piece of copper metal dipped into a solution of a copper salt. An equilibrium is set up between the Cu metal and its ions:

$$Cu^{2+}(aq) + 2e^- \rightleftharpoons Cu(s)$$

◀ **R3.2 Figure 3** Zinc atoms form zinc ions by releasing electrons. An equilibrium is set up between the metal and its solution of ions.

The negative charge on the electrode attracts zinc ions which can gain electrons in the reverse process:

$$Zn(s) \leftarrow Zn^{2+}(aq) + 2e^-$$

At a particular charge separation the rates of both processes will be the same and equilibrium is reached:

$$Zn(s) \rightleftharpoons Zn^{2+}(aq) + 2e^-$$

The position of this equilibrium depends on the reactivity of the metal. The more reactive the metal, the more the equilibrium favours the formation of ions, and the greater the negative charge on the electrode. As copper is a less reactive metal, the equilibrium position favours the formation of metal in its half-cell:

$$Cu^{2+}(aq) + 2e^- \rightleftharpoons Cu(s)$$

The copper metal strip will have fewer electrons and be less negative than the zinc metal strip.

When these two half-cells are connected by an external wire and a salt bridge, electrons will have a tendency to flow spontaneously through the external circuit from the more negative zinc half-cell to the less negative copper half-cell. The tendency for the electrons to flow can be quantified by measuring the potential difference between the two electrodes using a voltmeter.

Investigation of voltaic cells. Full details of how to carry out this experiment with a worksheet are available in the eBook.

SKILLS

Zinc–copper voltaic cell showing a copper half-cell and a zinc half-cell connected by a salt bridge. Electrons flow from the zinc electrode to the copper electrode through the electrical wires, while ions flow through the salt bridge to complete the circuit. The voltmeter is showing 1.10 V, the potential difference of this cell.

The electrode where oxidation occurs is called the **anode**, in this case it is the zinc electrode, and it has a negative charge:

$$Zn(s) \rightarrow Zn^{2+}(aq) + 2e^-$$

Oxidation always occurs at the anode; reduction always occurs at the cathode. In the voltaic cell, the anode has a negative charge and the cathode has a positive charge.

The electrode where reduction occurs is called the **cathode**, in this case it is the copper electrode and it has a positive charge:

$$Cu^{2+}(aq) + 2e^- \rightarrow Cu(s)$$

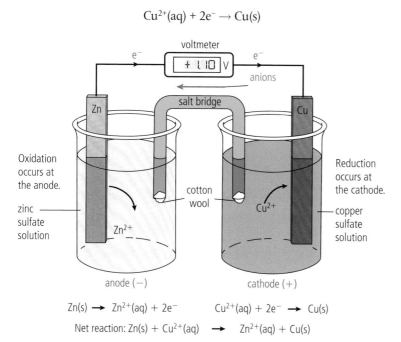

The voltaic cell (Figure 4) must therefore have the following connections between the half-cells:

SKILLS

Building voltaic cells with different combinations of metals/metal ion electrodes will help in the understanding of this topic.

- An external electronic circuit, connected to the metal electrode in each half-cell. A voltmeter can also be attached to this external circuit to record the voltage generated. Electrons will flow from the negative terminal, the anode, to the positive terminal, the cathode, through the wire.
- A salt bridge completes the circuit. The salt bridge is a glass tube or strip of absorptive paper that contains an aqueous solution of ions. Movement of these ions neutralizes any build up of charge and maintains the potential difference. Anions move in the salt bridge from the cathode to the anode. Anions move in the opposite direction to the electrons in the external circuit. Cations move in the opposite direction to anions in the salt bridge from the anode to the cathode. The solution chosen is often aqueous $NaNO_3$ or KNO_3 as these ions will not interfere with the reactions at the electrodes. Without a salt bridge, the circuit is incomplete and no voltage is generated.

In general, the more reactive the metal, the greater the build-up of electrons and the more negative the electrode potential in its half-cell.

Cell diagram convention

We can summarize the arrangements of electrodes in a voltaic cell in a **cell diagram.** This has the following features:

- A single vertical line represents a phase boundary such as that between a solid electrode and an aqueous solution within a half-cell.
- A double vertical line represents the salt bridge.
- The aqueous solutions of each electrode are placed next to the salt bridge.

- The anode is generally put on the left and the cathode on the right, so electrons flow from left to right.
- Spectator ions are usually omitted from the diagram.
- If a half-cell includes two ions, they are separated by a comma because they are in the same phase.

The cell diagram for the copper–zinc cell discussed earlier is:

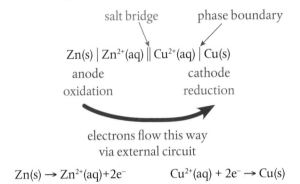

Note that the cell diagram is set out to show the direction of the two half-reactions.

Anode	Cathode
$Zn(s) \mid Zn^{2+}(aq) : Zn(s) \rightarrow Zn^{2+}(aq)$	$Cu^{2+}(aq) \mid Cu(s) : Cu^{2+}(aq) \rightarrow Cu(s)$

Different half-cells make voltaic cells with different voltages

Any two half-cells can be connected together to make a voltaic cell. For any such cell, the direction of electron flow and the voltage generated will be determined by the *difference* in reducing strength of the two metals. In most cases this can be judged by the relative position of the metals in the activity series. For example, if we connect a zinc half-cell to a silver half-cell the corresponding cell reaction is:

$$Zn(s) + 2Ag^+(aq) \rightarrow Zn^{2+}(aq) + 2Ag(s)$$

A larger voltage would be produced than with the copper half-cell discussed earlier because the difference in electrode potentials of zinc and silver is greater than that between zinc and copper.

The two half-reactions are:

$$Zn(s) \rightarrow Zn^{2+}(aq) + 2e^-$$

$$2Ag^{2+}(aq) + 2e^- \rightarrow 2Ag(s)$$

Electrons would flow from zinc (anode), where electrons are produced, to silver (cathode), where they are accepted, as shown in Figure 5.

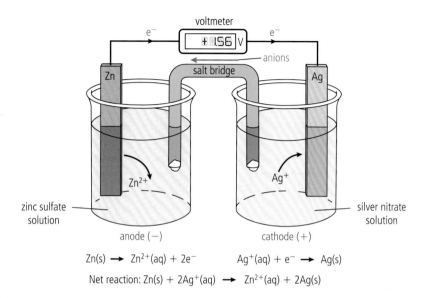

R3.2 Figure 5 A silver–zinc voltaic cell.

If we now make a voltaic cell with one copper electrode and one silver electrode (Figure 6), the cell reaction is:

$$Cu(s) + 2Ag^+(aq) \rightarrow Cu^{2+}(aq) + 2Ag(s)$$

and the half-reactions are:

$$Cu(s) \rightarrow Cu^{2+}(aq) + 2e^-$$

$$2Ag^+(aq) + 2e^- \rightarrow 2Ag(s)$$

The direction of electron flow is now *away* from copper and towards silver due to the greater reducing power of copper. The copper electrode is the negative electrode of the cell as it has the lower electrode potential.

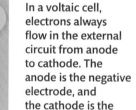

In a voltaic cell, electrons always flow in the external circuit from anode to cathode. The anode is the negative electrode, and the cathode is the positive electrode.

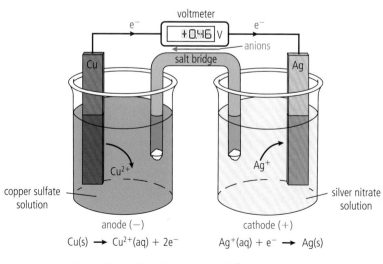

R3.2 Figure 6 A silver–copper voltaic cell.

Batteries are an application of a voltaic cell, making electrical energy available as a source of power. Our reliance on batteries increases as global demand for mobile electronic devices such as smartphones, laptops and tablets continues to grow. While demand for batteries looks set to continue, there is concern over toxicity and environmental damage from battery disposal. Batteries should be recycled as they contain metals such as zinc, cobalt, manganese, copper, cadmium and iron.

We can now summarize the parts of a voltaic cell and the direction of movement of electrons and ions (Figure 7):

- The anode is the negative electrode and the cathode is the positive electrode.
- Electrons flow from anode to cathode through the external circuit.
- Anions migrate from cathode to anode through the salt bridge.
- Cations migrate from anode to cathode through the salt bridge.

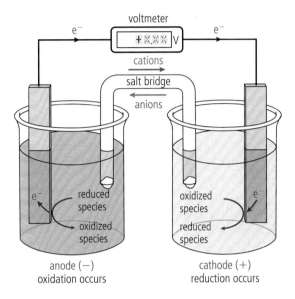

▲
R3.2 Figure 7 Summary of the components of a voltaic cell showing the ion and electron movements.

Nature of Science

The Italian scientist Luigi Galvani noticed the leg muscles of a dead frog twitched when they were in contact with two different metals. After many experiments, he concluded that the contractions were caused by an electrical fluid that was carried to the muscles in the nerves. His contemporary and fellow countryman Alessandro Volta had an intuition that the source of the electricity was not biological but originated instead from the interaction between the two metals. He confirmed his hypothesis by building a 'voltaic pile' from pairs of metals separated by a conducting solution and demonstrated that this generated an electric current. The terms galvanic cell and voltaic cell are used in recognition of both men's contribution to their discovery. The role of chance observations or serendipity is often a feature of scientific discovery. Scientific progress is not, however, down to luck, as the significance of these observations needs to be recognized. As Louis Pasteur said' 'Chance favours only the prepared mind'.

▲
Italian physicist Count Alessandro Volta (1745–1827) demonstrates his newly invented battery or 'voltaic pile' to Napoleon Bonaparte in 1801. Constructed from alternating discs of zinc and copper with pieces of cardboard soaked in brine between the metals, his voltaic pile was the first battery that produced a reliable, steady current of electricity.

Exercise

Q24. Use the metal activity series on page 514 to predict which electrode will be the anode and which will be the cathode when the following half-cells are connected. Write half-equations for the reactions occurring at each electrode.

(a) Zn/Zn^{2+} and Fe/Fe^{2+}

(b) Fe/Fe^{2+} and Mg/Mg^{2+}

(c) Mg/Mg^{2+} and Cu/Cu^{2+}

Q25. A voltaic cell can be constructed with one half-cell consisting of Mg and a solution of Mg^{2+} ions and the other half-cell consisting of Zn and a solution of Zn^{2+} ions.

(a) Draw a diagram of the cell and label the electrodes with name and polarity, and the direction of electron and ion movement, and write equations for the reactions occurring at each electrode.

(b) Write the cell diagram to represent the above voltaic cell.

Q26. Predict what would be observed if an iron spatula was left in a solution of copper sulfate overnight.

Reactivity 3.2.7 – Secondary (rechargeable) cells

> **Reactivity 3.2.7 – Secondary (rechargeable) cells involve redox reactions that can be reversed using electrical energy.**
>
> Deduce the reactions of the charging process from given electrode reactions for discharge, and vice versa.
>
> | Include discussion of advantages and disadvantages of fuel cells, primary cells and secondary cells. | Reactivity 2.3 – Secondary cells rely on electrode reactions that are reversible. What are the common features of these reactions? |
> | --- | --- |

Secondary (rechargeable) cells involve redox reactions that can be reversed

A primary cell or battery is one that cannot be recharged. A secondary cell or battery is one that can be recharged.

We have so far discussed primary cells where the electrochemical reaction is irreversible. The cell reactions involve the production of aqueous ions or gas molecules which become dispersed during the reaction. In this section we will discuss **secondary cells** which involve redox reactions that can be reversed using electrical energy. As they can be recharged, they have a longer life than primary cells. Rechargeable batteries, which contain many rechargeable cells connected together, are more expensive to buy but they become more economical with use.

The lithium-ion battery

One of the most promising reusable batteries is the lithium-ion battery, which benefits from lithium's low density and high reactivity (Figure 8). It can store a lot of electrical energy per unit mass. As lithium metal is reactive, steps must be taken to prevent it from forming an oxide layer which would decrease contact with the electrolyte. The lithium cathode is placed in the lattice of a metal oxide (MnO_2) and the lithium anode is mixed with graphite. A non-aqueous polymer-based electrolyte is used. As discussed in Reactivity 1.8, an electrolyte is a substance that does not conduct electricity in the solid state, but does conduct electricity when molten or in aqueous solution as it has ions which are free to move.

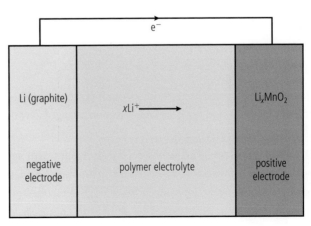

▲ **R3.2 Figure 8** The lithium-ion battery.

Discharging a lithium-ion battery

At the negative electrode, lithium is oxidized:

$$Li(polymer) \rightarrow Li^+(polymer) + e^-$$

At the positive electrode, the Li^+ ion and MnO_2 are reduced:

$$Li^+(polymer) + MnO_2(s) + e^- \rightarrow LiMnO_2(s)$$

The two half-reactions are reversed when the battery is recharged.

Lithium-ion batteries are used in cell (mobile) phones, laptops, computers and cameras.

Lithium-ion battery. This is a rechargeable battery often used in electrical equipment. The energy is stored in the battery through the movement of lithium ions. The first commercial lithium-ion battery was produced in 1991.

Worked example

Rechargeable nickel–cadmium batteries are used in electronics and toys. When the battery is discharged, the positive electrode is nickel hydroxide and the negative electrode is cadmium hydroxide. The electrolyte is aqueous potassium hydroxide.

During the charging process a cadmium(II) hydroxide electrode is reduced to the element and a nickel(II) hydroxide electrode is oxidized to NiO(OH).

(a) Deduce the reduction half-reaction during charging.

(b) Deduce the oxidation half-reaction during charging.

(c) Deduce the chemical equation for the charging process.

(d) State the chemical equation for the discharging process and explain why the process can be reversed.

Solution

(a) $Cd(OH)_2(s) + 2e^- \rightarrow Cd(s)$: 2 electrons are needed as Cd is reduced from $+2 \rightarrow 0$

$Cd(OH)_2(s) + 2e^- \rightarrow Cd(s) + 2OH^-(aq)$: 2 $OH^-(aq)$ are needed to balance the charges and the atoms

(b) $Ni(OH)_2(s) \rightarrow NiO(OH)(s) + e^-$: 1 electron is produced as Ni is oxidized from $+2 \rightarrow +3$

$Ni(OH)_2(s) + OH^-(aq) \rightarrow NiO(OH)(s) + e^-$: 1 $OH^-(aq)$ is needed to balance the charges

$Ni(OH)_2(s) + OH^-(aq) \rightarrow NiO(OH)(s) + H_2O(l) + e^-$: 1 $H_2O(aq)$ is added to balance the atoms

(c) State the half-reactions for 2 electrons:

$Cd(OH)_2(s) + 2e^- \rightarrow Cd(s) + 2OH^-(aq)$

$2Ni(OH)_2(s) + 2OH^-(aq) \rightarrow 2NiO(OH)(s) + 2H_2O(l) + 2e^-$

Add the half-reactions:

$2Ni(OH)_2(s) + \cancel{2OH^-(aq)} + Cd(OH)_2(s) + \cancel{2e^-}$
$\rightarrow 2NiO(OH)(s) + 2H_2O(l) + \cancel{2e^-} + Cd(s) + \cancel{2OH^-(aq)}$

which gives

$2Ni(OH)_2(s) + Cd(OH)_2(s) \rightarrow 2NiO(OH)(s) + 2H_2O(l) + Cd(s)$

Reactivity 2.3 – Secondary cells rely on electrode reactions that are reversible. What are the common features of these reactions?

(d) $2NiO(OH)(s) + Cd(s) + 2H_2O(l) \rightarrow 2Ni(OH)_2(s) + Cd(OH)_2(s)$

This reaction can be reversed as both metal hydroxides are insoluble and so cannot be dispersed. A common feature of secondary cells is that the products cannot be dispersed away from the electrodes.

A fuel cell converts chemical energy directly to electrical energy

As combustion reactions are redox reactions, they can be used to produce an electric current if the reactants are physically separated. This is the basis of a fuel cell which, unlike primary voltaic cells, requires a constant supply of fuel and oxygen to the electrodes to produce electricity continuously as they are consumed in the process.

The hydrogen fuel cell

The use of the hydrogen fuel cell could reduce our dependence on fossil fuels. One mole of hydrogen can release 286 kJ of heat energy when it combines directly with oxygen:

Most of this energy can be converted to electrical energy in a fuel cell.

The hydrogen fuel cell is a clean and efficient power source. Hydrogen (red) is pumped into the cell, where protons (the hydrogen nuclei) are separated from their electrons (yellow). The electrons flow around a conducting loop as an electric current, which can be harnessed. The hydrogen nuclei (H^+) pass through a membrane and then combine with the electrons and oxygen from the air to form steam, and water is the only waste product.

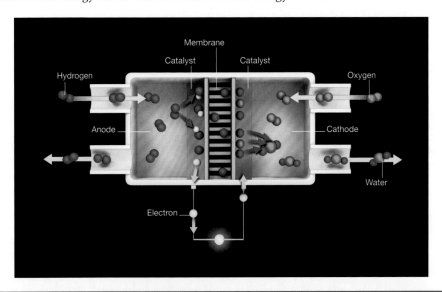

Worked example

Deduce the two half-reactions that occur at the anode and the cathode in the hydrogen–oxygen fuel cell in acid conditions.

Solution

At the negative electrode (anode), $H_2(g)$ is oxidized:

$$2H_2(g) \rightarrow 4H^+(aq) + 4e^-$$

At the positive electrode (cathode), $O_2(g)$ is reduced:

$$4H^+(aq) + O_2(g) + 4e^- \rightarrow 4H_2O(l)$$

Challenge yourself

5. Deduce the two half-reactions that occur at the anode and the cathode in the hydrogen–oxygen fuel cell with aqueous sodium hydroxide as the electrolyte.

Advantages and disadvantages of fuel cells, primary cells and secondary cells

Some advantages and disadvantages of the different systems are summarized below.

Type of cell/battery		Advantages	Disadvantages
Primary	general	• inexpensive • convenient • lightweight • good shelf life	• can only be used once • leads to a large amount of waste as batteries need to be recycled • batteries put into landfill sites have severe environmental impact • can only deliver small currents
	fuel cell	• more efficient than direct combustion as more chemical energy is converted to useful electrical energy • no pollution if hydrogen is the fuel • low density of hydrogen	• hydrogen is a potentially explosive gas • hydrogen must be stored and transported in large/heavy containers • very expensive • technical problems due to catalytic failures, leaks and corrosion • can deliver only small currents • low reaction rates and therefore low power output
Rechargeable / secondary	general	• materials can be regenerated • can deliver high current	
	lead–acid	• can deliver large amounts of energy over short periods	• heavy mass • lead and sulfuric acid could cause pollution
	cadmium–nickel	• longer life than lead–acid batteries	• cadmium is very toxic • produces a low voltage • very expensive
	lithium ion	• low density of lithium • high voltage • does not contain a toxic heavy metal	• expensive • limited life span

Exercise

Q27. The reaction taking place when a lead–acid storage battery discharges is:
$$Pb(s) + PbO_2(s) + 2H_2SO_4(aq) \rightarrow 2PbSO_4(s) + 2H_2O(l)$$

(a) Use oxidation numbers to explain what happens to the Pb(s) in terms of oxidation and reduction during this reaction.

(b) Write a balanced half-equation for the reactions taking place at the negative terminal during this discharge process.

(c) Identify the property of $PbSO_4$ which allows this process to be reversed.

(d) State one advantage and one disadvantage of using a lead–acid battery.

Q28. One type of fuel cell uses methanol as the fuel with an acid electrolyte. The overall equation for the process is:
$$2CH_3OH\ (aq) + 3O_2\ (g) \rightarrow 2CO_2\ (g) + 4H_2O\ (l)$$

(a) State the half-equations for the reactions at the negative electrode.

(b) State the half-equations for the reactions at the positive electrode.

(c) Suggest **one** advantage and **one** disadvantage of a methanol fuel cell over a lead–acid battery as an energy source in a car.

Q29. One type of fuel cell uses hydrogen as the fuel with hot aqueous potassium hydroxide as the electrolyte. The overall equation for the process is:
$$2H_2(g) + O_2(g) \rightarrow 2H_2O(l)$$

(a) State the two half-equations for the reactions involving each reactant.

(b) Hydrogen is a more expensive source of chemical energy than gasoline (petrol). Explain why fuel cells are considered to be more economical than gasoline engines.

(c) State one common feature of primary, fuel and rechargeable cells.

Reactivity 3.2.8 – Electrolytic cells

Reactivity 3.2.8 – An electrolytic cell is an electrochemical cell that converts electrical energy to chemical energy by bringing about non-spontaneous reactions.	
Explain how current is conducted in an electrolytic cell.	
Deduce the products of the electrolysis of a molten salt.	
Construction of electrolytic cells to include DC power source connected to anode and cathode, electrolyte.	Structure 2.1 – Under what conditions can ionic compounds act as electrolytes?

An external source of electricity can drive non-spontaneous redox reactions

As we have seen, a voltaic cell takes the energy of a spontaneous redox reaction and harnesses it to generate electrical energy. An **electrolytic cell** reverses this process: it uses an external source of electrical energy to bring about a non-spontaneous redox reaction that would otherwise not take place. You can think of it in terms of an external power supply pumping electrons into the electrolytic cell, driving reactions of oxidation and reduction. As the word *electro-lysis* suggests, it is the process where electricity is used to bring about chemical reactions which break down substances.

The reactant in the process of electrolysis must contain mobile ions which allow the current to pass between the electrodes. It is an **electrolyte**, which is a liquid, usually a molten ionic compound or an aqueous solution of an ionic compound. As the electric current passes through the electrolyte, redox reactions occur at the electrodes, removing the charges on the ions and forming products that are electrically neutral. The ions are therefore said to be **discharged** during this process.

An electrolyte is a substance which does not conduct electricity in the solid state but does conduct electricity when molten or in aqueous solution as it has ions which are free to move.

An electrolytic cell

The components of an electrolytic cell are shown in Figure 9 and described below.

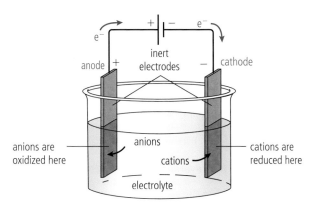

◄ **R3.2 Figure 9** Components of an electrolytic cell.

 SKILLS

Investigating electrolytic cells. Full details of how to carry out this experiment with a worksheet are available in the eBook.

- The source of electric energy is a battery or a DC power source. This is shown in the diagram as ┃┃ where the longer line represents the positive terminal and the shorter line the negative terminal.
- The electrodes are placed in the electrolyte and connected to the power supply. Electrodes are made from a conducting substance – generally a metal or graphite. They are described as **inert** when they do not take part in the redox reactions. They must not touch each other as this would 'short' the circuit, and the current would bypass the electrolytic cell.
- Electric wires connect the electrodes to the power supply.

Electricity can reverse the natural direction of change

In Structure 2.1 we discussed the spontaneous reaction between sodium and chlorine to produce sodium chloride:

$$Na(s) + \tfrac{1}{2}Cl_2(g) \rightarrow NaCl(s)$$

Electrons are transferred from the sodium to the chlorine. Sodium and chloride ions are formed which then bond to form a sodium chloride lattice.

This process can be reversed if solid sodium chloride is first converted into the molten state:

$$NaCl(s) \rightarrow Na^+(l) + Cl^-(l)$$

The ions are now mobile so the molten state conducts electricity and the ions act as an electrolyte if placed in an electrochemical cell (Figure 10).

R3.2 Figure 10 Electrolysis of molten sodium chloride. Chloride ions are oxidized at the anode and sodium ions are reduced at the cathode.

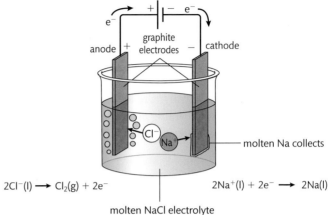

$$2Cl^-(l) \rightarrow Cl_2(g) + 2e^-$$

$$2Na^+(l) + 2e^- \rightarrow 2Na(l)$$

molten Na collects

molten NaCl electrolyte

Reactive metals, including aluminum, lithium, magnesium, sodium and potassium, are found naturally in compounds such as Al_2O_3 and NaCl where they exist as positive ions. Extraction of the metal therefore involves reduction of these ions. This is best done using electrolysis.

The battery or DC power source pushes electrons towards the negative electrode, or cathode, where they are accepted as Na^+ ions are reduced in the electrolyte.

$$Na^+(l) + e^- \rightarrow Na(l)$$

Electrons are released at the positive terminal, the anode, as chloride ions are oxidized to chlorine atoms which then form chlorine molecules.

$$2Cl^-(l) \rightarrow Cl_2(g) + 2e^-$$

The released electrons then move towards the positive terminal of the power source.

Balancing the two half-reactions at the electrodes we have the overall reaction:

$$2Na^+(l) + 2e^- + 2Cl^-(l) \rightarrow 2Na(l) + 2e^- + Cl_2(g)$$

$$2Na^+(l) + 2Cl^-(l) \rightarrow 2Na(l) + Cl_2(g)$$

Electrodes are electrically conducting, electrolytes are ionically conducting. Ionic compounds can act as electrolytes in the molten state or in solution.

The chemical reactions occurring at each electrode remove the ions from the electrolyte. The compound has been split into its constituent elements and electrolysis has occurred.

Note the current is passed through the electrolyte not by electrons but by the mobile ions as they migrate to the electrodes.

Structure 2.1 – Under what conditions can ionic compounds act as electrolytes?

Ionic compounds can also conduct electricity in aqueous solution. This is discussed later in the chapter.

The electrolysis of molten salts

We can generalize the results for the electrolysis of sodium chloride to other molten compounds. When the electrolyte is a molten salt, the only ions present are those from the compound itself, so one ion migrates to each electrode. Positive ions are reduced and gain electrons at the cathode – the negative electrode – and negative ions are oxidized and lose electrons at the anode – the positive electrode.

Worked example

Describe the reactions that occur at the two electrodes during the electrolysis of molten lead(II) bromide. Write an equation for the overall reaction and comment on any likely changes that would be observed.

Solution

1. Identify the ions present in the electrolyte and to which electrode they will be attracted:

$$PbBr_2(l) \rightarrow Pb^{2+}(l) + 2Br^-(l)$$

to cathode   to anode

2. Deduce the half-reactions at each electrode:

anode: Br^- is oxidized: $2Br^-(l) \rightarrow Br_2(l) + 2e^-$

cathode: Pb^{2+} is reduced: $Pb^{2+}(l) + 2e^- \rightarrow Pb(l)$

3. Add the half-equations to obtain the overall reaction:

$$Pb^{2+}(l) + 2Br^-(l) \rightarrow Pb(l) + Br_2(l)$$

4. $Br_2(g)$, a brown vapour with a strong smell, is produced at the anode and $Pb(l)$, a molten grey metal, is produced at the cathode.

This can be set out in a table.

	Cathode	Anode
Ions present	$Pb^{2+}(l)$	$2Br^-(g)$
Reaction at electrode	$Pb^{2+}(l) + 2e^- \rightarrow Pb(l)$ lead is a liquid at the temperatures used	$2Br^-(l) \rightarrow Br_2(g) + 2e^-$ bromine is a gas at the temperatures used
Observations	a molten grey metal is produced	a brown vapour with a strong smell is produced

Lead(II) bromide has a relatively low melting point of 373 °C, but most other ionic compounds have very high melting points, so electrolysis of molten compounds generally involves working at high temperatures.

The electrolytic extraction of aluminum was developed almost simultaneously by Charles Martin Hall and Paul Héroult, who worked independently on different sides of the Atlantic. They both discovered the process in the same year, 1886. Both were born in the same year (1863) and died in the same year (1914).

Worked example

Molten aluminium oxide is electrolyzed in the cell shown in the figure below. The graphite-lined steel on the outside acts as the cathode and graphite anodes are dipped into the molten electrolyte.

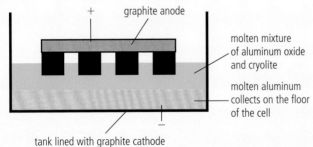

Describe and explain what happens at the cathode and anode. Explain your answer in terms of the redox reactions occurring at the electrodes, including equations in your answer.

Worker siphoning off molten aluminium in an aluminium processing plant. Electrolysis of a solution of alumina (Al_2O_3) dissolved in cryolite is used to reduce the Al^{3+} ions at the cathode. The molten aluminium is siphoned off and cooled before any further processing.

Solution

Aluminium ions are attracted towards the negative electrode where they are reduced to aluminium atoms:

$$Al^{3+}(l) + 3e^- \rightarrow Al(l)$$

Oxygen is produced at the anode from the oxide ions:

$$2O^{2-}(l) \rightarrow O_2(g) + 4e^-$$

Challenge yourself

6. Explain with chemical equations why the carbon anodes used in the electrolysis of aluminium oxide need to be replaced at regular intervals.

7. Cryolite is added to molten aluminium oxide to increase its conductivity. Explain the low conductivity of aluminium oxide.

Exercise

Q30. Identify the electrode to which potassium ions are attracted during the electrolysis of molten potassium bromide.

   A   The anode which has a positive charge.

   B   The anode which has a negative charge.

   C   The cathode which has a positive charge.

   D   The cathode which has a negative charge.

Q31. Identify the processes that occur in an electrical circuit with an electrolytic cell with a lithium iodide electrolyte.

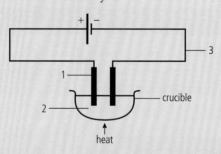

	1	2	3
A	electrode where reduction occurs	ions migrating	electrons moving anti-clockwise
B	electrode where oxidation occurs	ions migrating	electrons moving clockwise
C	electrode where reduction occurs	electrons moving anti-clockwise	ions migrating
D	electrode where oxidation occurs	electrons moving clockwise	ions migrating

Q32. Which statement is **not** correct for the electrolysis of copper(II) chloride?

    A   Copper ions and chloride ions move through the electrolyte.

    B   Oxidation of chloride takes place at the anode.

    C   Electrons move through the external circuit.

    D   Oxidation of chlorine takes place at the cathode.

Q33. Write half-equations for the electrode reactions occurring during the electrolysis of the following molten salts using graphite electrodes.

    (a) KBr

    (b) MgF_2

    (c) ZnS

    (d) $CaCl_2$

Q34. Magnesium metal is produced by the electrolysis of molten magnesium chloride using inert electrodes.

    (a) Draw a fully labelled diagram of the electrolytic cell, including the polarity on the electrodes and the direction of electron and ion migration.

    (b) Write equations for the reactions occurring at each electrode and the overall cell reaction.

In industrial processes. another compound is often added to lower the melting point of the compound to be electrolyzed. This makes the process more economical. Molten cryolite – a mineral form of Na_3AlF_6 – is added to reduce the melting point of aluminium oxide and so reduces the energy requirements of the process. Pure aluminium oxide would not be a suitable electrolyte because it has a very high melting point, and it is a poor electrical conductor even when molten.

Reactivity 3.2.9 – Oxidation of functional groups in organic compounds

Reactivity 3.2.9 – Functional groups in organic compounds may undergo oxidation.

Deduce equations to show changes in the functional groups during oxidation of primary and secondary alcohols, including the two-step reaction in the oxidation of primary alcohols.

Include explanation of the experimental set-up for distillation and reflux.	Structure 3.2 – How does the nature of the functional group in a molecule affect its physical properties, such as boiling point?
Include the fact that tertiary alcohols are not oxidized under similar conditions.	Reactivity 1.3 – What is the difference between combustion and oxidation of an alcohol?
Names and formulas of specific oxidizing agents and the mechanism will not be assessed.	

Oxidation of alcohols

As discussed in Reactivity 1.3, the combustion of organic compounds breaks the carbon chain.

Worked example

Use oxidation numbers to show the difference in the extent of oxidation of complete and incomplete combustion of propane.

Solution

In C_3H_8 the average oxidation state of C is $-\frac{8}{3}$

Complete combustion produces CO_2 in which the carbon has an oxidation state of +4.

Incomplete oxidation produces CO, in which the carbon has an oxidation state of +2, and C, in which the carbon has an oxidation state of 0.

Challenge yourself

8. Use oxidation numbers to show the differences in the extent of oxidation in the complete combustion of propane, $CH_3CH_2CH_3$, and propene, CH_3CHCH_2.

More subtle oxidation can occur when an oxidizing agent selectively oxidizes only the carbon atom of a functional group. Alcohols can be oxidized into other important organic compounds when oxidizing agents selectively oxidize the carbon atom attached to the $-OH$ group, keeping the carbon skeleton of the molecule intact. The products of these reactions are determined by whether the alcohol is a primary, secondary or tertiary alcohol. The distinction between the different alcohols is described in Structure 3.2.

Various oxidizing agents can be used for these reactions, but commonly the laboratory process uses acidified potassium dichromate(VI). This is a bright orange solution, which is reduced to green $Cr^{3+}(aq)$ as the alcohol is oxidized on heating. The colour of a transition metal depends on the oxidation state of the metal.

When writing equations for these reactions it is sometimes convenient to show the oxidizing agent simply as '+ [O]'.

The degree of oxidation can be related to the number of H atoms attached to the carbon with the hydroxyl group. Primary alcohols have two H atoms attached to the carbon with the hydroxyl group and can be oxidized in a two-step reaction. The removal of the first H atom leads to the formation of the aldehyde. The second H atom is removed under prolonged reaction, and a carboxylic acid is formed.

For example: ethanol is oxidized to ethanal in one step and to ethanoic acid in two steps:

And acidified dichromate $Cr_2O_7^{2-}$ is reduced to $Cr^{3+}(aq)$:

$$Cr_2O_7^{2-} \xrightarrow{\text{reduction}} Cr^{3+}(aq)$$

It is possible to remove the aldehyde product as it is formed by distillation because it has a lower boiling point than the primary alcohol or the carboxylic acid. Aldehydes do not have hydrogen bonding between their molecules, unlike the alcohol or the acid.

When a bottle of wine is exposed to the air, bacteria slowly oxidize the ethanol to ethanoic acid, giving the smell of vinegar – 'vin aigre' means 'bitter wine' in French.

Distillation is discussed in more detail in Structure 1.1.

 Distillation apparatus is used to separate liquids with different boiling points. In the oxidation of primary alcohols, the aldehyde has a lower boiling point than the alcohol or the acid, and so is collected as a gas and passes into the condensing tube, which is surrounded by cold flowing water. The gas condenses back to a liquid which is collected in the beaker at the bottom.

If we want to produce the carboxylic acid, the aldehyde needs to react with the oxidizing agent for a prolonged period of time. The reaction mixture is heated under **reflux** which means the aldehyde condenses back into the reaction mixture when it evaporates. The apparatus for reflux is shown below.

Structure 3.2 – How does the nature of the functional group in a molecule affect its physical properties, such as boiling point?

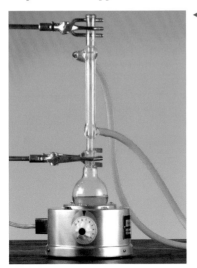

 Reflux heating using an electric mantle. The heat boils the liquid, allowing the reaction to proceed at the boiling point of the solvent. The tubes bring cooling water to the condenser. Any vapour condenses back into the reaction vessel. This arrangement is used to oxidize primary alcohols to carboxylic acids.

 When you are asked for the conditions for the oxidation reactions for primary alcohols, make sure that you specify the aldehyde is produced by distillation and the carboxylic acid by reflux.

Secondary alcohols are oxidized to **ketones** by a similar process of oxidation. As there is only one H attached to the carbon bonded to the hydroxyl group, this is a one-step reaction.

 Reactions of alcohols. Full details of how to carry out this experiment with a worksheet are available in the eBook.

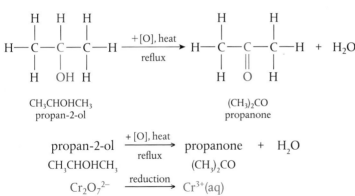

$CH_3CHOHCH_3$
propan-2-ol

$(CH_3)_2CO$
propanone

$$\text{propan-2-ol} \xrightarrow[\text{reflux}]{+[O],\ \text{heat}} \text{propanone} + H_2O$$
$$CH_3CHOHCH_3 \qquad\qquad (CH_3)_2CO$$
$$Cr_2O_7^{2-} \xrightarrow{\text{reduction}} Cr^{3+}(aq)$$

535

Tertiary alcohols have no H atoms attached to the carbon bonded to the hydroxyl group and so are not readily oxidized without breaking the carbon chain. The carbon skeleton of the molecule is broken in combustion.

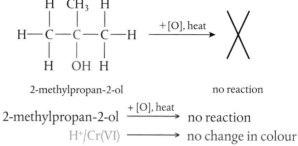

2-methylpropan-2-ol

no reaction

$$\text{2-methylpropan-2-ol} \xrightarrow[\text{H}^+/\text{Cr(VI)}]{+ \text{[O], heat}} \begin{array}{l} \text{no reaction} \\ \text{no change in colour} \end{array}$$

The colour change observed with potassium dichromate(VI), when it acts as an oxidizing agent from orange to green, is used to distinguish primary and secondary alcohols from tertiary alcohols. As there is no reaction with the tertiary alcohols, there is no colour change.

Reactivity 1.3 – What is the difference between combustion and oxidation of an alcohol?

The colour of the alcohol solutions after oxidation with potassium dichromate(VI) solution, $K_2Cr_2O_7$, which is orange. Primary and secondary alcohols are oxidized, to carboxylic acids and ketones respectively, and the $Cr_2O_7^{2-}$ is reduced to green Cr^{3+}. Tertiary alcohols are not oxidized.

Alcoholic drinks are widely consumed globally. However, as blood alcohol levels rise, a person's balance and judgement could be impaired. Most countries have set legal limits of alcohol consumption for activities such as driving. Analysis of ethanol concentration is based on samples of breath, blood or urine. Common techniques use the oxidation reactions of ethanol in fuel cells or photocells (in which orange $Cr_2O_7^{2-}$ is reduced to green Cr^{3+}). Alcohol abuse can be a major contributor to social and economic problems in society.

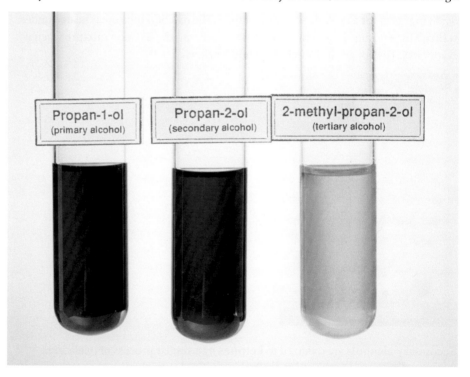

Alcohol	Oxidation product	Colour change with acidified $K_2Cr_2O_7$(aq)
primary alcohol	aldehyde → carboxylic acid	orange → green
secondary alcohol	ketone	orange → green
tertiary alcohol	not oxidized	no colour change

Worked example

(a) Deduce the half-equations of oxidation and reduction for the reaction between ethanol and acidified potassium dichromate(VI) to produce ethanoic acid.

(b) Deduce the overall reaction.

Solution

(a) Consider first the oxidation: ethanol → ethanoic acid

We need to calculate the average oxidation state of carbon in each molecule.

The average oxidation state of C in ethanol $= \dfrac{(-6+2)}{2} = -2$

The average oxidation state of C in ethanoic acid $= \dfrac{(-4+4)}{2} = 0$

As both carbon atoms increase their oxidation state by +2, 4 electrons are released.

$$C_2H_5OH(aq) \rightarrow CH_3COOH(aq) + 4e^-$$

We add $4H^+$ to balance the charges:

$$C_2H_5OH(aq) \rightarrow CH_3COOH(aq) + 4e^- + 4H^+(aq)$$

We add H_2O to balance the H and the O atoms:

$$C_2H_5OH(aq) + H_2O(l) \rightarrow CH_3COOH(aq) + 4e^- + 4H^+(aq)$$

Now consider the reduction: $Cr_2O_7^{2-}(aq) \rightarrow 2Cr^{3+}(aq)$

The oxidation state of Cr in $Cr_2O_7^{2-} = +6$ and the oxidation state of Cr in $Cr^{3+} = +3$

As both chromium atoms decrease their oxidation state by 3, 6 electrons are released.

$$Cr_2O_7^{2-}(aq) + 6e^- \rightarrow 2Cr^{3+}(aq)$$

Add $14H^+(aq)$ to balance the charges:

$$Cr_2O_7^{2-}(aq) + 6e^- + 14H^+(aq) \rightarrow 2Cr^{3+}(aq)$$

Add $7H_2O(l)$ to balance the H and O atoms:

$$Cr_2O_7^{2-}(aq) + 6e^- + 14H^+(aq) \rightarrow 2Cr^{3+}(aq) + 7H_2O(l)$$

(b) To deduce the overall reaction, balance the number of electrons in the half-reactions:

$$Cr_2O_7^{2-}(aq) + 6e^- + 14H^+(aq) \rightarrow 2Cr^{3+}(aq) + 7H_2O(l)$$

$$C_2H_5OH(aq) + H_2O(l) \rightarrow CH_3COOH(aq) + 4e^- + 4H^+$$

12 is a common multiple

$$2Cr_2O_7^{2-}(aq) + 12e^- + 28H^+(aq) \rightarrow 4Cr^{3+}(aq) + 14H_2O(l)$$

$$3C_2H_5OH(aq) + 3H_2O(l) \rightarrow 3CH_3COOH(aq) + 12e^- + 12H^+(aq)$$

Adding the half-reactions and simplifying:

$$2Cr_2O_7^{2-}(aq) + \cancel{12e^-} + 28H^+(aq) + 3C_2H_5OH(aq) + 3H_2O(l)$$
$$\rightarrow 4Cr^{3+}(aq) + 14H_2O(l) + 3CH_3COOH(aq) + \cancel{12e^-} + 12H^+(aq)$$

$$2Cr_2O_7^{2-}(aq) + 16H^+(aq) + 3C_2H_5OH(l) \rightarrow 4Cr^{3+}(aq) + 11H_2O(l) + 3CH_3COOH(aq)$$

Challenge yourself

9. Deduce the half-equations of oxidation and reduction for oxidation of ethanol to produce ethanal with acidified potassium dichromate(VI) and state the overall reaction.

10. Deduce the half-equations for the oxidation of ethanol focusing on the change of oxidation state of the carbon in the hydroxyl group.

Exercise

Q35. Identify the product when $CH_3CH_2CH(OH)CH_3$ is reacted with acidified potassium dichromate(VI).

    A $CH_3CH_2COOCH_3$

    B $CH_3CH_2CH_2CHO$

    C $CH_3CH_2COCH_3$

    D $CH_3CH_2CH_2COOH$

Q36. Identify the compound produced when propan-1-ol is oxidized using acidified potassium dichromate(VI), $K_2Cr_2O_7(aq)$, under two different conditions.

	Reaction mixture heated under reflux	**Product distilled as the oxidizing agent is added**
A	propanone	propanal
B	propanone	propanoic acid
C	propanoic acid	propanal
D	propanal	propanoic acid

Q37. Identify the compound with the highest boiling point.

    A $CH_3CH_2\,CH_2CH_3$

    B $CH_3CH_2CH_2OH$

    C $CH_3OCH_2CH_3$

    D CH_3CH_2CHO

Q38. Write equations for the complete combustion of butanol and pentanol.

Q39. Predict the products of heating the following alcohols with acidified potassium dichromate(VI) solution, and the colour changes that would be observed in the reaction mixture.

    (a) butan-2-ol

    (b) methanol (product collected by distillation)

    (c) 2-methylbutan-2-ol

Reactivity 3.2.10 – Reduction of functional groups in organic compounds

Reactivity 3.2.10 – Functional groups in organic compounds may undergo reduction.	
Deduce equations to show reduction of carboxylic acids to primary alcohols via the aldehyde, and reduction of ketones to secondary alcohols.	
Include the role of hydride ions in the reduction reaction. Names and formulas of specific reducing agents and the mechanisms will not be assessed.	Structure 3.1 – How can oxidation states be used to show that the following molecules are given in increasing order of oxidation: CH_4, CH_3OH, $HCHO$, $HCOOH$, CO_2?

Reduction of carbonyl compounds

It is often convenient in organic chemistry to analyze redox reactions in terms of gain and loss of oxygen or hydrogen, rather than in terms of electrons. As we saw in the last section, the oxidation of secondary alcohols to ketones involves the removal of two H atoms. We can reverse this process: ketones can be reduced to secondary alcohols by adding two H atoms. The reducing agent used provides the hydrogens in the form of H^- and H^+ ions. The carbonyl group is polar with a partial positive charge on the carbon atom and so can be attacked by species with a lone pair of electrons. As discussed in Reactivity 3.4, the hydride ion, H^- is acting as a **nucleophile** on the electron-deficient carbonyl carbon.

Aldehydes can be reduced to primary alcohols in a similar reaction.

Carboxylic acids can also be reduced to aldehydes. We can show these reactions using '[+ H]' to represent reduction, in the same way as we used '[+ O]' to represent oxidation of alcohols. Some examples are given below.

$$CH_3COOH \xrightarrow{[+H]} CH_3CHO \xrightarrow{[+H]} CH_3CH_2OH$$

ethanoic acid — ethanal — ethanol
carboxylic acid — aldehyde — primary alcohol

Conditions: heat with $LiAlH_4$ in dry ether. The reaction cannot be stopped at the aldehyde as it reacts too readily with $LiAlH_4$.

$$(CH_3)_2CO \xrightarrow{[+H]} (CH_3)_2CHOH$$

propanone — propan-2-ol
ketone — secondary alcohol

The reduction reactions are summarized in the table.

Attacking species	Reactive site	Functional group attacked	Functional group formed
H^- (hydride ion)	$R-C\overset{\delta^+}{\underset{}{}}\overset{O}{\diagup}$	carboxylic acid aldehyde	primary alcohol $R-\overset{H}{\underset{H}{C}}-OH$
		ketone	secondary alcohol $R-\overset{R}{\underset{H}{C}}-OH$

Worked example

(a) Predict the product when pentan-2-one is reduced.

(b) Predict the product when butanoic acid is reduced.

Solution

(a) Draw the condensed formula with carbonyl group: $$\begin{array}{c} O \\ \parallel \\ CH_3CH_2CH_2CCH_3 \end{array}$$ pentan-2-one	Add H atoms to the carbonyl group: $$\begin{array}{c} OH \\ \mid \\ CH_3CH_2CH_2CHCH_3 \end{array}$$ pentan-2-ol
(b) Draw the condensed formula with carbonyl group: $$\begin{array}{c} O \\ \parallel \\ CH_3CH_2CH_2COH \end{array}$$ butanoic acid	Replace OH group by H: $$\begin{array}{c} O \\ \parallel \\ CH_3CH_2CH_2CH \end{array}$$ butanal
Aldehydes are further reduced by $LiAlH_4$ with H atoms added to carbonyl group.	$$\begin{array}{c} OH \\ \mid \\ CH_3CH_2CH_2CH_2 \end{array}$$ butan-1-ol

Worked example

Deduce the half-equations for the reduction of methanoic acid to methanol.

Solution

$$HCOOH \rightarrow CH_3OH$$

The oxidation state of C in methanoic acid = $-2 + 4 = +2$

The oxidation state of C in methanol = $-4 + 2 = -2$

The carbon atom decreases its oxidation state by 4, so 4 electrons are needed:

$$HCOOH + 4e^- \rightarrow CH_3OH$$

Add $4H^+$ to balance the charges:

$$HCOOH + 4H^+ + 4e^- \rightarrow CH_3OH$$

Add H_2O to balance the H and the O atoms:

$$HCOOH + +4e^- + 4H^+ \rightarrow CH_3OH + H_2O$$

Challenge yourself

11. Deduce the half-equations for the reduction of methanal.

In the previous sections we have seen how alcohols, aldehydes, ketones and carboxylic acids can be interconverted by redox reactions. The degree of oxidation can be indicated by the oxidation state of the carbon atom as shown in the previous Worked example.

Structure 3.1 – How can oxidation states be used to show that the following molecules are given in increasing order of oxidation: CH_4, CH_3OH, $HCHO$, $HCOOH$, CO_2?

Exercise

Q40. Identify the compound which can be reduced to an aldehyde.

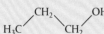

Q41. Identify the class of compounds which are produced when RCOR' are reduced.

    A ROR'   **B** RCOOR'   **C** RCH(OH)R' **D** RCOOH

Q42. Explain how the following reduction reactions are carried out in the laboratory:

    (a) propanoic acid to propanol

    (b) ethanal to ethanol

Q43. Deduce the half-equations for the reduction of ethanal with sodium borohydride.

Reactivity 3.2.11 – Reduction of unsaturated compounds

Reactivity 3.2.11 – Reduction of unsaturated compounds by the addition of hydrogen lowers the degree of unsaturation.

Deduce the products of the reactions of hydrogen with alkenes and alkynes.

	Reactivity 3.4 – Why are some reactions of alkenes classified as reduction reactions while others are classified as electrophilic addition reactions?

Alkenes and alkynes are unsaturated hydrocarbons and undergo **addition reactions** to form a range of different saturated products. The alkenes generally react by electrophilic addition as discussed in Reactivity 3.4.

Alkenes can be reduced to alkanes by reacting with hydrogen:

$$C_nH_{2n} + H_2 \rightarrow C_nH_{2n+2}$$

Propene, for example, reacts with hydrogen in the presence of a nickel catalyst at about 150 °C:

$$H-\underset{\underset{H}{|}}{\overset{\overset{H}{|}}{C}}-\overset{\overset{H}{|}}{C}=\overset{\overset{H}{|}}{C}-H + H_2 \longrightarrow H-\underset{\underset{H}{|}}{\overset{\overset{H}{|}}{C}}-\underset{\underset{H}{|}}{\overset{\overset{H}{|}}{C}}-\underset{\underset{H}{|}}{\overset{\overset{H}{|}}{C}}-H$$

$$CH_3CHCH_2 + H_2 \xrightarrow[150\,°C]{Ni\ catalyst} CH_3CH_2CH_3$$

propene   propane

The nickel catalyst provides a surface for the reactant molecules to come together with the correct orientation. The hydrogen atoms add to the two carbon atoms simultaneously and the products leave the surface once the reaction is complete.

The hydrogenation of ethenes should be contrasted with halogenation which is an electrophilic addition where the chlorine atoms are added in two separate steps.

Reactivity 3.4 – Why are some reactions of alkenes classified as reduction reactions while others are classified as electrophilic addition reactions?

Unsaturated fatty acids have one or more carbon–carbon double bonds. As double bonds cannot rotate, the unsaturated carbon chains have kinks which make it more difficult for the molecules to pack closely together. The intermolecular forces are weaker, and the melting points are lower. Unsaturated fats are liquid oils under normal conditions. Hydrogenation is used in the margarine industry to convert oils containing many unsaturated hydrocarbon chains into more saturated compounds which have higher melting points. This is done so that margarine will be a solid at room temperature. However, there are widespread concerns about the health effects of so-called *trans* fats, produced by partial hydrogenation.

Similarly, alkynes can be reduced to alkenes and alkanes.

$$H-C\equiv C-H + H_2 \longrightarrow \underset{\underset{H}{|}\ \ \underset{H}{|}}{\overset{\overset{H}{|}\ \ \overset{H}{|}}{C=C}}$$

$$H-C\equiv C-H + 2H_2 \longrightarrow H-\underset{\underset{H}{|}}{\overset{\overset{H}{|}}{C}}-\underset{\underset{H}{|}}{\overset{\overset{H}{|}}{C}}-H$$

Be careful not to confuse the terms *hydrogenation* (addition of hydrogen) with *hydration* (addition of water).

The number of molecules of H_2 needed to convert the organic molecule to the corresponding saturated molecule is related to the number of multiple bonds in the molecule. The alkenes have the general molecular formula C_nH_{2n} so one molecule of hydrogen reacts with the carbon–carbon double bond of the alkenes to form alkanes with the general formula C_nH_{2n+2}. Similarly, alkynes with a triple bond have the general molecular formula C_nH_{2n-2} and react with two molecules of hydrogen to form the saturated alkanes.

Exercise

Q44. Outline how vegetable oils are converted to margarine and explain the difference in their melting points.

Q45. Deduce the molecular structure of the reduction products of the following.

     (a) CH_3COCH_3 **(b)** CH_3CO_2H **(c)** CH_2CHCl **(d)** C_3H_4

What happens when electrons are transferred?

In this chapter we have explored electron transfer or redox reactions:

- Oxidation is electron loss and reduction is electron gain. The two processes can be shown in terms of half-reactions.

- Oxidation and reduction can be identified by changes in oxidation state. A species with an element whose oxidation state is increased is oxidized. A species with an element with a decreased oxidation state is reduced.

- The oxidizing agent is the species that is reduced, and the reducing agent is the species that is oxidized.

- When the electron transfer in a spontaneous reaction occurs via an outside circuit, it can be used to produce an electric current. Electrons flow from anode to cathode in the external circuit and ions move between the electrodes in the electrolyte.

- Only reactive metals with negative electrode potentials produce hydrogen gas when added to dilute acids.

- A primary cell produces electricity from reagents which are used up in an irreversible chemical reaction. A secondary cell can be recharged as the cell reaction is reversible. In a fuel cell the reagents are supplied as it operates.

- Electricity can be used to drive non-spontaneous reactions.

- Electrolysis can be used to break down stable compounds in the molten state. It is used to extract reactive metals.

- Reduction occurs at the cathode and oxidation occurs at the anode in both voltaic and electrolytic cells.

- Oxidation reactions in organic chemistry involve the addition of oxygen or the removal of hydrogen. Alcohols and aldehydes are oxidized by oxidizing agents. When oxidizing agents contain a transition metal, there is a colour change as the oxidation state changes.

- Reduction reactions in organic chemistry involve the addition of hydrogen or the removal of oxygen. The alkenes and alkynes are reduced by H_2 adding across a C=C bond. The aldehydes, ketones and carboxylic acids are reduced by reducing agents with the H^- ion, which react with the partially positive charge of the carbon atom in the carbonyl group.

Practice questions

1. Which conditions are required to obtain a good yield of a carboxylic acid when ethanol is oxidized using potassium dichromate(VI), $K_2Cr_2O_7(aq)$?

    I.   add sulfuric acid

    II. heat the reaction mixture under reflux

    III. distil the product as the oxidizing agent is added

    A I and II only

    B I and III only

    C II and III only

    D I, II and III

2. Which species could be reduced to form NO_2?

    A N_2O

    B NO_3^-

    C HNO_2

    D NO

3. Which statements are correct for electrolysis?

    I.   An exothermic reaction occurs.

    II. Oxidation occurs at the anode (positive electrode).

    III. The reaction is non-spontaneous.

    A I and II only

    B I and III only

    C II and III only

    D I, II and III

4. A student performed displacement reactions using metals W and X and solutions of salts of metals W, X, Y and Z. The results are summarized in the table.

		Salt solution			
		W^{2+}	X^{2+}	Y^{2+}	Z^{2+}
Metal	W		Reaction	No reaction	No reaction
	X	Reaction		Reaction	No reaction

    Which of the four metals is most reactive?

    A W

    B X

    C Y

    D Z

5. Consider the following reaction.

$$MnO_4^-(aq) + 8H^+(aq) + 5Fe^{2+}(aq) \rightarrow Mn^{2+}(aq) + 5Fe^{3+}(aq) + 4H_2O(l)$$

Which statement is correct?

A MnO_4^- is the oxidizing agent and it loses electrons.

B MnO_4^- is the reducing agent and it loses electrons.

C MnO_4^- is the oxidizing agent and it gains electrons.

D MnO_4^- is the reducing agent and it gains electrons.

6. Which coefficients correctly balance this redox equation?

(a) $Fe^{2+}(aq) + MnO_4^-(aq) +$

(b) $H^+(aq) \rightarrow$

(c) $Fe^{3+}(aq) + Mn^{2+}(aq) +$

(d) $H_2O(l)$

	a	b	c	d
A	1	8	1	4
B	5	4	5	2
C	3	4	3	2
D	5	8	5	4

7. What is the reducing agent in the reaction below?

$$2MnO_4^-(aq) + Br^-(aq) + H_2O(l) \rightarrow 2MnO_2(s) + BrO_3^-(aq) + 2OH^-(aq)$$

A Br^-

B BrO_3^-

C MnO_4^-

D MnO_2

8. Which changes could take place at the positive electrode (cathode) in a voltaic cell?

I. $Zn^{2+}(aq)$ to $Zn(s)$

II. $Cl_2(g)$ to $Cl^-(aq)$

III. $Mg(s)$ to $Mg^{2+}(aq)$

A I and II only

B I and III only

C II and III only

D I, II and III

9. What happens at the negative electrode in a voltaic cell and in an electrolytic cell?

	Voltaic cell	Electrolytic cell
A	oxidation	reduction
B	reduction	oxidation
C	oxidation	oxidation
D	reduction	reduction

10. (a) Magnetite, Fe_3O_4, is an ore of iron that contains both Fe^{2+} and Fe^{3+}.

    Deduce the ratio of $Fe^{2+} : Fe^{3+}$ in Fe_3O_4. (1)

    (b) In acidic solution, hydrogen peroxide, H_2O_2, will oxidize Fe^{2+}.

    $$Fe^{2+}(aq) \rightarrow Fe^{3+}(aq) + e^-$$

    Write the half-equation for the reduction of hydrogen peroxide to water in acidic solution. (1)

    (c) Deduce a balanced equation for the oxidation of Fe^{2+} by acidified hydrogen peroxide. (1)

    (Total 3 marks)

11. (a) The diagram shows an incomplete voltaic cell with a light bulb in the circuit.

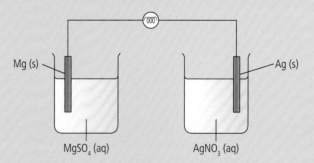

    Identify the missing component of the cell and its function. (2)

    (b) Deduce the half-equations for the reaction at each electrode when current flows. (2)

    (c) Annotate the diagram with the location and direction of electron movement when current flows. (1)

    (Total 5 marks)

12. (a) Outline two differences between an electrolytic cell and a voltaic cell.   (2)

(b) Explain why solid sodium chloride does not conduct electricity but molten sodium chloride does.   (2)

(c) Molten sodium chloride undergoes electrolysis in an electrolytic cell. For each electrode deduce the half-equation and state whether oxidation or reduction takes place. Deduce the equation of the overall cell reaction including state symbols.   (5)

(d) Electrolysis has made it possible to obtain reactive metals such as aluminium from their ores, which has resulted in significant developments in engineering and technology. State one reason why aluminium is preferred to iron in many uses.   (1)

(Total 10 marks)

13. Fuel cells may be twice as efficient as the internal combustion engine. Although fuel cells are not yet in widespread use, NASA has used a basic hydrogen–oxygen fuel cell as the energy source for space vehicles.

(a) State the half-equations occurring at each electrode in the hydrogen-oxygen fuel cell in an alkaline medium.   (2)

(b) Describe the composition of the electrodes and state the overall cell equation of a nickel–cadmium battery.   (3)

(Total 5 marks)

14. The high activity of lithium metal leads to the formation of an oxide layer on the metal which decreases the contact with the electrolyte in a battery.

(a) Suggest how this is overcome in the lithium-ion battery.   (2)

(b) Describe the migration of ions taking place at the two electrodes in the lithium-ion battery when it produces electricity.   (2)

(c) Discuss one similarity and one difference between fuel cells and rechargeable batteries.   (2)

(Total 6 marks)

15. Aluminium and its alloys are widely used in industry.

    (a) Aluminium metal is obtained by the electrolysis of alumina dissolved in molten cryolite.

        (i) Suggest the function of the molten cryolite. (1)

        (ii) State the half-equations for the reactions that take place at each electrode. (2)

    (b) Outline **two** different ways that carbon dioxide may be produced during the production of aluminium. (2)

    (Total 5 marks)

16. (a) Explain why iron is obtained from its ores using chemical reducing agents but aluminium is obtained from its ores using electrolysis. (2)

    (b) Both carbon monoxide and hydrogen can be used to reduce iron ores. State the equations for the reduction of magnetite, Fe_3O_4, with

        (i) carbon monoxide

        (ii) hydrogen. (2)

    (Total 4 marks)

17. Aluminium is the most abundant metal in the Earth's crust.

    (a) State the materials used for the positive and negative electrodes in the production of aluminium by electrolysis. (2)

    (b) Discuss why it is important to recycle aluminium. (2)

    (Total 4 marks)

18. Brass is a copper-containing alloy with many uses. An analysis is carried out to determine the percentage of copper present in three identical samples of brass. The reactions involved in this analysis are shown below.

    Step 1: $Cu(s) + 2HNO_3(aq) + 2H^+(aq) \rightarrow Cu^{2+}(aq) + 2NO_2(g) + 2H_2O(l)$

    Step 2: $4I^-(aq) + 2Cu^{2+}(aq) \rightarrow 2CuI(s) + I_2(aq)$

    Step 3: $I_2(aq) + 2S_2O_3^{2-}(aq) \rightarrow 2I^-(aq) + S_4O_6^{2-}(aq)$

    (a) (i) Deduce the change in the oxidation numbers of copper and nitrogen in step 1. (2)

        (ii) Identify the oxidizing agent in step 1. (1)

    (b) A student carried out this experiment three times with three identical small brass nails and obtained the following results.

    mass of brass = 0.456 g ± 0.001 g

Titre	1	2	3
Initial volume of 0.100 mol dm^{-3} S$_2$O$_3^{2-}$ (±0.05 cm³)	0.00	0.00	0.00
Final volume of 0.100 mol dm^{-3} S$_2$O$_3^{2-}$ (±0.05 cm³)	28.50	28.60	28.40
Volume added of 0.100 mol dm^{-3} S$_2$O$_3^{2-}$ (±0.10 cm³)	28.50	28.60	28.40
Average volume added of 0.100 mol dm^{-3} S$_2$O$_3^{2-}$ (±0.10 cm³)	28.50		

(i) Calculate the average amount, in mol, of $S_2O_3^{2-}$ added in step 3.   (2)

(ii) Calculate the amount, in mol, of copper present in the brass.   (1)

(iii) Calculate the mass of copper in the brass.   (1)

(iv) Calculate the percentage by mass of copper in the brass.   (1)

(v) The manufacturers claim that the sample of brass contains 44.2% copper by mass. Determine the percentage error in the result.   (1)

(Total 9 marks)

Electron sharing reactions

◀ The kayaker's clothing and equipment, including the kayak, paddle and personal flotation device, are made almost entirely of organic compounds that have been artificially synthesized. Organic chemists use knowledge of the functional groups in organic molecules to design compounds with the required properties such as waterproofing, breathability, strength, density, flexibility and insulation. Addition polymers such as polyethene, which the kayak is made of, are synthesized using electron sharing reactions involving reactive species called radicals.

Guiding Question

What happens when a species possesses an unpaired electron?

Radicals are highly reactive species that contain an unpaired electron. This unpaired electron makes radicals energetically unstable so they readily react with other species to form products that are energetically more stable. These reactions can involve radicals breaking bonds in other compounds or reacting with each other by combining their unpaired electrons and forming a covalent bond.

The high reactivity of radicals means they are capable of breaking down compounds which would otherwise be stable. This can be beneficial when used in applications such as wastewater treatment and the breakdown of persistent pollutants, as well as in the industrial production of additional polymers. However the high reactivity of radicals can also result in environmentally harmful processes such as smog formation and atmospheric ozone depletion. Beneficial reactions involving radicals occur within our bodies, including various enzyme reactions and the destruction of pathogens. However, radicals can also cause cell damage and have been associated with various illnesses such as cancers, strokes and heart disease as well as being identified as a possible cause of the ageing process.

Radical reactions occur via a mechanism involving three stages: initiation, propagation and termination. In this chapter we will use the substitution of alkanes by halogens as a case study to illustrate how stable compounds can react with radicals through this mechanism.

Nature of Science

The development of scientific knowledge relies on the ability of scientists to obtain empirical evidence that either supports or discredits proposed theories. Early chemists, relying on personal observations and/or simple measurements, were often only able to postulate or make inferences about the nature of reactions being studied and the presence of reactive intermediates. Progressive developments in technologies and instrumentation have allowed chemists to obtain increasingly detailed empirical evidence, including the detection and measurement of previously undetectable species such as radicals. This has led to many breakthroughs in our understanding of chemical reactivity.

Reactivity 3.3.1 – Radicals

Reactivity 3.3.1 – A radical is a chemical entity that has an unpaired electron. Radicals are highly reactive.

Identify and represent radicals, e.g. $\bullet CH_3$ and $Cl\bullet$.

	Structure 2.1 – How is it possible for a radical to be an atom, a molecule, a cation or an anion? Consider examples of each type.

Radicals have unpaired electrons and are highly reactive

A **radical** is a chemical species that contains an unpaired electron. The presence of an unpaired electron results in radicals having a high enthalpy. This makes radicals very reactive as it is energetically favourable for them to take an electron from other species or to combine with other radicals to form a covalent bond, as both can result in products with a lower enthalpy.

> The free-radical theory of ageing suggests that the physiological changes associated with ageing are the result of oxidative reactions in cells causing damage to membranes and large molecules such as DNA. These changes accumulate with time and may explain the increase in degenerative diseases, such as cancer, with age. The theory suggests that supplying cells with antioxidants will help to slow down the damaging oxidative reactions. Antioxidants are particularly abundant in fresh fruit and vegetables, as well as in tea and cocoa. Although there is strong evidence that antioxidant supplementation may help protect against certain diseases, it has not yet been shown to produce a demonstrated increase in the human life span.

Because they are highly reactive, radicals tend to have short lifetimes and do not exist for very long. One notable exception is when radicals are formed in the upper atmosphere where they can persist for a significant length of time as there is a low probability of them colliding and reacting with other chemical species.

▲ A researcher using electron resonance spectroscopy to examine the role of radicals in anticancer drug activity.

The only formal requirement of a radical is the presence of an unpaired electron. It is therefore possible for radicals to be atoms, molecules or ions. Examples of the different types of radicals include:

- atomic: chlorine ($Cl•$), bromine ($Br•$), hydrogen ($H•$)
- molecular: neutral molecules with an odd number of electrons will be radicals, e.g. nitric oxide ($NO•$), hydroxyl radical ($OH•$), methyl radical ($•CH_3$)
- anionic: molecules that gain electrons can become radical anions, e.g. superoxide radical ($O_2•^-$), benzene radical anion ($C_6H_6•^-$)
- cationic: molecules that lose electrons can become radical cations, e.g. propane cation ($C_3H_8•^+$), ethanol cation ($C_2H_5OH•^+$).

▲ Fresh fruits and vegetables are good sources of antioxidants, which may help prevent damaging oxidative reactions from occurring in cells.

TOK

The free-radical theory of ageing is a hypothesis that is not universally supported. Will the explanation of certain natural processes ever go beyond hypotheses?

The presence of radicals in specific chemical reactions was first postulated by chemists in the 1900s but these radicals could not be isolated or observed directly. The development of Electron Spin Resonance (ESR) spectroscopy in the 1940s enabled radicals to be detected due to the unique magnetic properties associated with unpaired electrons. Further advancements in ESR and other spectroscopic techniques combined with increased computer processing power, have demonstrated the importance of radical reactions in many areas of chemistry.

The different species given on the previous page are all identified as radicals by including a dot (•) in their chemical formulas. This dot represents an unpaired electron and is usually aligned vertically with the middle of the chemical formula.

Nature of Science

As chemistry has progressed, the term radical has been applied in different contexts. In earlier definitions a radical was described as a group of atoms that could exist on its own, or as part of a larger molecule. If the group was to exist *free* of the other atoms in the molecule, this would require the homolytic breaking of a covalent bond so a *free* radical would have an unpaired electron.

In modern contexts, the term functional group (see Structure 3.2) is now used instead of radical to describe a specific group of atoms within a molecule, with 'radical' specifically referring to a chemical species that contains an unpaired electron. However the term 'free radical' is still commonly used.

Nature of Science

Classification of reactions can be approached in many different ways. For example, combustion reactions can be defined as a redox reaction, because carbon atoms in the fuel are oxidized to give CO_2, or as an electron sharing (radical) reaction, because they proceed via a chain reaction involving various radical species.

It is estimated that around 50% of the world's polymers are addition polymers which have been made via radical reactions. Two examples of addition polymers are polypropene and acrylonitrile butadiene styrene (ABS), which can be used in 3D printers to create custom shapes and designs. At addition polymers are covered in more detail in Structure 2.4.

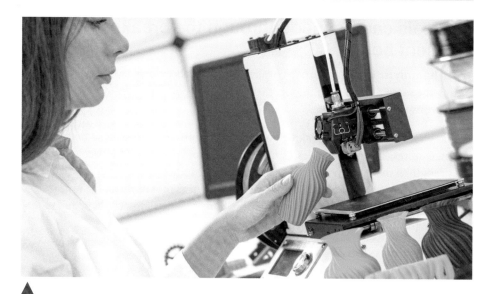
3D printing can use addition polymers to make custom shapes and designs.

Structure 2.1 – How is it possible for a radical to be an atom, a molecule, a cation or an anion? Consider examples of each type.

An unpaired electron is represented in a chemical formula by a dot (•).

Make sure that you understand the difference between a radical and an ion. A radical has an unpaired electron but no net charge; an ion carries a charge. Species that are radicals *and* carry a charge are known as radical cations or radical anions.

A beach campfire at sunset. Combustion reactions, such as the burning of wood, occur via chain reactions of many steps, with radical species progressively breaking the fuel down into smaller compounds, eventually forming carbon dioxide, water and other simple oxides.

Exercise

Q1. Which of the following species is a radical?

    A   F   **B** Cl^- **C**   Ne   **D** Li^+

Q2. Which of the following molecules is a radical?

    A CO_2 **B** NO_2 **C** CH_4 **D** NH_3

Q3. Which of the following ions is a radical?

    A Cl^- **B** NO_3^- **C** Mg^{2+} **D** CH_4^+

Reactivity 3.3.2 – Homolytic fission

Reactivity 3.3.2 – Radicals are produced by homolytic fission, e.g. of halogens, in the presence of ultraviolet (UV) light or heat.

Explain, including with equations, the homolytic fission of halogens, known as the initiation step in a chain reaction.

The use of a single-barbed arrow (fish hook) to show the movement of a single electron should be covered.	Reactivity 1.2 – Why do chlorofluorocarbons (CFCs) in the atmosphere break down to release chlorine radicals but not fluorine radicals?
	Structure 2.2 – What is the reverse process of homolytic fission?
	Structure 2.2 – Chlorine radicals released from CFCs are able to break down ozone, O_3, but not oxygen, O_2, in the stratosphere. What does this suggest about the relative bond strengths of bonds in the two allotropes?

Radicals can be made through the breaking of covalent bonds

When **homolytic fission** occurs, a covalent bond breaks to form two radicals. 'Homo' means 'the same' and refers to the fact that the two products have an equal assignment of electrons from the bond.

As shown in Reactivity 1.2, breaking bonds is an endothermic process, so homolytic fission requires energy for it to happen. The amount of energy required depends on the strength of the covalent bond. For weaker bonds, sufficient energy can come from heating the compound and the process is known as **thermolytic fission**. The breaking of stronger bonds requires the absorption of high-energy UV light and is known as **photolytic fission**.

Thermolytic fission

$$X \!:\! X \xrightarrow{\text{Heat}} X\bullet + X\bullet$$

Photolytic fission

$$X \!:\! X \xrightarrow{\text{UV}} X\bullet + X\bullet$$

The Cl–Cl bond in chlorine, Cl_2, can be broken through photolytic fission. The molecule splits into two chlorine atoms, each has an unpaired electron and both atoms are therefore radicals.

$$Cl_2 \xrightarrow{\text{UV light}} 2Cl\bullet \text{ radicals}$$

2 chlorine radicals

This process can also be shown using single-headed curly arrows, sometimes known as 'fish hooks', to show the movement of a single electron.

As we will see later in this chapter, termination reactions occur when two radicals combine, sharing their unpaired electrons and forming a covalent bond. This is effectively the reverse process of homolytic fission and can also be represented using single-headed curly arrows. For example, the termination reaction between $CH_3\bullet$ and $Cl\bullet$ radicals to form CH_3Cl can be represented by:

$$H_3C\bullet + \bullet Cl \xrightarrow{\text{termination}} C_3H-Cl$$

Chlorine radicals can cause ozone depletion

Chlorofluorocarbons (CFCs) were widely used in aerosols, refrigerants, solvents and plastics due to their low reactivity and low toxicity in the troposphere. But when CFCs enter the stratosphere, the higher energy UV radiation breaks them down, releasing chlorine atoms, which are reactive radicals.

In 1985, measurements showed unusually low levels of atmospheric ozone over Antarctica and the Southern Ocean. The main culprit for ozone depletion, chlorofluorocarbons (CFCs), had been identified as far back as 1974, but efforts to ban CFCs had proved controversial due to economic concerns and difficulties in finding alternative substances. The levels of depletion observed in 1985 led to urgent action with the Montreal Protocol, which banned ozone depleting substances, being adopted in 1987. Global cooperation in adhering to this Protocol has stopped the decline in atmospheric ozone levels, with measurements in recent years showing signs of a recovery.

The level observed in 2000 was one of the lowest measured, as shown by the large purple region in the figure on the left below. The greater amount of blue present in 2020 provides evidence that atmospheric ozone levels are now recovering.

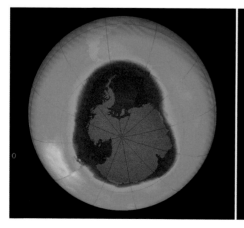
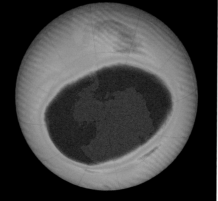

Homolytic fission occurs when a bond breaks by splitting the shared pair of electrons between the two products. It produces two radicals, each with an unpaired electron:

$$X:X \rightarrow X\bullet + X\bullet$$

Structure 2.2 – What is the reverse process of homolytic fission?

▲ Used spray paint cans. Before being banned under the Montreal Protocol, chlorofluorocarbons (CFCs) were used as refrigerants and propellants for spray paints, deodorant and hair sprays, along with other applications.

Minimum annual ozone levels measured over Antarctica in September 2000 (left) and October 2020 (right). The colours represent ozone concentrations and follow the progression: purple (lowest), blue, green yellow (highest).

Reactivity 1.2 – Why do chlorofluorocarbons (CFCs) in the atmosphere break down to release chlorine radicals but not fluorine radicals?

The success of the Montreal Protocol in banning CFCs in order to avoid serious environmental consequences is often compared to the current global heating crisis. The scientific evidence is now unequivocal that the Earth is warming due to human activities, and that this will have major environmental and societal impacts. However, this is proving to be a much more controversial issue due to the significant political, economic, technological and social challenges that governments, industry and individuals all face in making the significant reductions in greenhouse gas emissions required. How might developments in scientific knowledge trigger political controversies or controversies in other areas of knowledge?

TOK

Although CFCs contain both C–Cl and C–F bonds, the C–Cl bond has a much lower bond enthalpy (324 kJ mol⁻¹) compared to the C–F bond (492 kJ mol⁻¹). When UV radiation is absorbed, the weaker C–Cl bond breaks in preference to the C–F bond.

For example, the CFC dichlorodifluoromethane, commonly known as Freon-12, undergoes photochemical decomposition as follows:

$$CCl_2F_2(g) \xrightarrow{\text{UV}} CClF_2\bullet(g) + Cl\bullet(g)$$

$$CCl_2F_2(g) \overset{\text{UV}}{\cancel{\longrightarrow}} CCl_2F\bullet(g) + F\bullet(g)$$

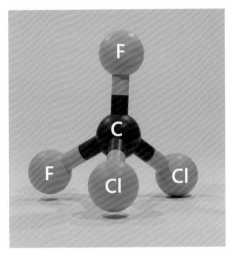

▲ A molecular model of dichlorodifluoromethane (Freon-12), a chemical that was commonly used as a refrigerant and aerosol propellant before being banned under the Montreal Protocol.

Challenge yourself

1. The average bond enthalpies for the C–Cl and C–F bonds are 324 kJ mol⁻¹ and 492 kJ mol⁻¹ respectively. Calculate the wavelengths of UV radiation required to break C–Cl and C–F bonds.

The chlorine radicals released then catalyze the decomposition of ozone, O_3.

$$Cl\bullet(g) + O_3(g) \rightarrow O_2(g) + ClO\bullet(g)$$

$$ClO\bullet(g) + O\bullet(g) \rightarrow O_2(g) + Cl\bullet(g)$$

Here Cl•(g) has acted as a catalyst and the net reaction is:

$$O_3(g) + O\bullet(g) \rightarrow 2O_2(g)$$

Because it acts as a catalyst and is regenerated in the reaction, one chlorine radical is capable of catalyzing the decomposition of many thousands of ozone molecules.

Although they can break down ozone, the chlorine radicals do not readily react with oxygen, O_2, present in the stratosphere. The O=O double bond in oxygen is stronger than the bonds present in ozone, and this makes O_2 more resistant to reactions with chlorine radicals.

Structure 2.2 – Chlorine radicals released from CFCs are able to break down ozone, O_3, but not oxygen, O_2, in the stratosphere. What does this suggest about the relative bond strengths of bonds in the two allotropes?

Nature of Science

Sometimes the unintended consequences of scientific innovation are not known until problems have developed on a large scale and are difficult to reverse. Many ozone-depleting chemicals are expected to persist in the atmosphere for a long time. Other chemicals such as hydrofluorocarbons (HFCs), which replaced CFCs as refrigerants and have high global warming potentials, are now being detected in increasing amounts in the atmosphere. As a result, the use of HFCs is now also being phased out under the Kigali amendment to the Montreal Protocol.

Exercise

Q4. Which of the following reactions involves homolytic fission?

    A $CH_4 + Br_2 \rightarrow CH_3Br + HBr$

    B $Br_2 \rightarrow Br^+ + Br^-$

    C $C + O_2 \rightarrow CO_2$

    D $HBr \rightarrow H + Br$

Q5. The homolytic fission of a Cl–Cl bond:

    A   is exothermic

    B   forms two ions

    C   always requires heat

    D   is endothermic.

Q6. Use curly arrows to illustrate the homolytic fission of the covalent bond in iodine, I_2.

Q7. Explain the difference between homolytic fission that occurs thermolytically or photolytically.

Reactivity 3.3.3 – Radical substitution reactions of alkanes

Reactivity 3.3.3 – Radicals take part in substitution reactions with alkanes, producing a mixture of products.

Explain, using equations, the propagation and termination steps in the reactions between alkanes and halogens.

Reference should be made to the stability of alkanes due to the strengths of the C–C and C–H bonds and their essentially non-polar nature.	Reactivity 2.2 – Why are alkanes described as kinetically stable but thermodynamically unstable?

Substitution reactions of alkanes with halogens occur via a radical mechanism

Alkanes contain strong C–C and C–H bonds which take a lot of energy to break. Because alkanes are non-polar compounds, they do not attract reactive species such as electrophiles and nucleophiles. These two factors contribute to alkanes being relatively unreactive compounds.

The low reactivity of alkanes is also explained by the high activation energies that are associated with their reactions. At regular temperatures, the reactant molecules lack sufficient kinetic energy to overcome the activation energy, and alkanes are therefore regarded as being **kinetically stable**.

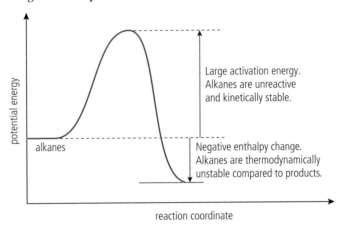

However, as covered in Reactivity 1.2, alkanes do undergo highly exothermic combustion reactions. Because the reaction is exothermic, this means that the alkane reactants have a higher enthalpy and are **thermodynamically unstable** compared to the products.

Radical substitution reactions occur via three stages

As alkanes are saturated molecules, the main type of reaction that they can undergo is **substitution**. This occurs when another reactant, such as a halogen, takes the place of a hydrogen atom in the alkane. The products of a substitution reaction between an alkane and a halogen will be a halogenoalkane and the hydrogen halide. For example, methane, CH_4, reacts with chlorine, Cl_2, to produce chloromethane, CH_3Cl, and hydrogen chloride, HCl.

$$CH_4(g) + Cl_2(g) \xrightarrow{\text{UV light}} CH_3Cl(g) + HCl(g)$$

The reaction does not take place in the dark as the energy of UV light is necessary to break the covalent bond in the chlorine molecule and form chlorine radicals.

Once formed, these radicals will start a **chain reaction** in which a mixture of products including the halogenoalkane is formed. We can describe the reaction as a sequence of steps, known as the **reaction mechanism**, and it occurs via three stages; initiation, propagation and termination.

Initiation

The initiation stage involves the formation of the radical species, which can then react with the alkane. For the substitution reaction of methane with chlorine, the initiation step is the photolytic fission of chlorine:

$$Cl_2 \xrightarrow{\text{UV light}} 2\,Cl\bullet$$

The high kinetic stability of alkanes is one reason that they have become the world's predominant fuel source as they can be safely compressed, transported and stored before use.

Reactivity 2.2 – Why are alkanes described as kinetically stable but thermodynamically unstable?

A Liquefied Natural Gas (LNG) tanker at sea. LNG is natural gas (mainly methane) that has been purified and cooled to below –162 °C, which condenses it into a liquid. In liquefied form it only requires 1/600th the volume of the gas so larger amounts can be transported in special, highly-insulated tanks.

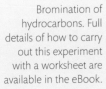

Bromination of hydrocarbons. Full details of how to carry out this experiment with a worksheet are available in the eBook.

SKILLS

Propagation

In the propagation stage, the radicals formed in the initiation stage react with other species present to form new radicals. There are many possible propagation steps, which all allow the reaction to continue. This is why this type of reaction is often called a chain reaction.

For example:

$$Cl\bullet + CH_4 \rightarrow CH_3\bullet + HCl$$

$$CH_3\bullet + Cl_2 \rightarrow CH_3Cl + Cl\bullet$$

$$CH_3Cl + Cl\bullet \rightarrow CH_2Cl\bullet + HCl$$

$$CH_2Cl\bullet + Cl_2 \rightarrow CH_2Cl_2 + Cl\bullet$$

Termination

Termination reactions remove radicals from the mixture when they react with each other and pair up their electrons to form a covalent bond. Again there are many possible termination steps.

For example:

$$Cl\bullet + Cl\bullet \rightarrow Cl_2$$

$$CH_3\bullet + Cl\bullet \rightarrow CH_3Cl$$

$$CH_3\bullet + CH_3\bullet \rightarrow C_2H_6$$

$$CH_2Cl\bullet + Cl\bullet \rightarrow CH_2Cl_2$$

So the reaction mixture may contain mono- and disubstituted halogenoalkanes, as well as HCl and larger alkanes.

As a termination step involves two unpaired electrons combining to form a covalent bond, this is, in effect, the reverse process of homolytic fission where a bond is broken to give two radicals.

A similar reaction to the example given above, for methane and chlorine, also occurs with other alkanes and bromine, Br_2. The change in colour from brown to colourless of bromine water when it reacts with an alkane occurs only in the presence of UV light, as shown in Figure 1. This reaction is sometimes used to distinguish between alkanes and alkenes, as will be discussed in more detail in the following chapter.

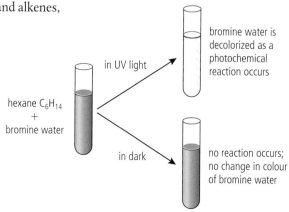

hexane C_6H_{14}
+
bromine water

in UV light → bromine water is decolorized as a photochemical reaction occurs

in dark → no reaction occurs; no change in colour of bromine water

The reaction of methane, CH_4, with chlorine, Cl_2, via a free radical mechanism involves a number of different possible propagation and termination steps resulting in a mixture of products. The primary product of interest is chloromethane, CH_3Cl. Focusing on the steps that result in the formation of this product gives a simplified version of the mechanism:

Initiation:
$$Cl_2 \xrightarrow{\;\;UV\;\;} 2Cl\bullet$$

Propagation:
$$Cl\bullet + CH_4 \rightarrow CH_3\bullet + HCl$$

Termination:
$$CH_3\bullet + Cl\bullet \rightarrow CH_3Cl$$

Overall:
$$CH_4(g) + Cl_2(g) \xrightarrow{\;\;UV\;\;} CH_3Cl(g) + HCl(g)$$

R3.3 Figure 1 The reaction between hexane, C_6H_{14}, and bromine water takes place only in the presence of UV light.

Alkanes are saturated hydrocarbons and undergo substitution reactions.

Exercise

Q8. What are the three stages of a radical mechanism? Use the reaction of methane, CH_4, with bromine, Br_2, to explain your answer, giving relevant equations for each stage.

Q9. Alkanes react with halogens in the presence of UV light to form halogenoalkanes.

    (a) Write equations showing the possible steps leading to the formation of bromoethane, C_2H_5Br, when bromine, Br_2, and ethane, C_2H_6, react together in UV light.

    (b) Explain, using equations, how this reaction also leads to the formation of trace amounts of butane, C_4H_{10}.

    (c) If the reaction is allowed to proceed for a sufficient length of time, large amounts of dibromoethane products can be formed. Draw and name the structures of possible dibromoethane products in this reaction.

Q10. State and explain what would be observed during the following experiments carried out at room temperature.

    (a) Bromine water is added to hexane in a test tube exposed to UV light.

    (b) Bromine water is added to hexane in a test tube that is covered in foil.

Q11. Chlorofluorocarbons (CFCs) such as trichlorofluoromethane, $CFCl_3$, are ozone-depleting substances.

    (a) Explain, using curly arrows, how trichlorofluoromethane reacts with UV light to produce chlorine radicals.

    (b) Provide chemical equations showing how chlorine radicals catalyze the decomposition of ozone, O_3.

Guiding Question revisited

What happens when a species possesses an unpaired electron?

In this chapter we have learnt that:

- Radicals are chemical species that contain an unpaired electron and are highly reactive.
- Radicals are formed through homolytic fission of covalent bonds.
- Alkanes are kinetically stable but react with halogens, in the presence of UV light, through a radical mechanism.
- Radical mechanisms have three stages: initiation, propagation and termination.
- Ozone depletion due to chlorofluorocarbons (CFCs) also occurs through a radical mechanism.

Practice questions

1. Which reaction occurs via a radical mechanism?

   A $C_2H_6 + Br_2 \rightarrow C_2H_5Br + HBr$

   B $C_2H_4 + Br_2 \rightarrow C_2H_4Br_2$

   C $C_4H_9I + OH^- \rightarrow C_4H_9OH + I^-$

   D $(CH_3)_3CI + H_2O \rightarrow (CH_3)_3COH + HI$

2. Which statement is true about radical mechanisms?

   A Radical mechanisms involve four stages.

   B Radicals form spontaneously in the initiation stage.

   C Radicals react with non-radical species in the termination stage.

   D Radicals react with other radicals in the termination stage.

3. Under the correct conditions, pentane reacts with bromine to form a number of different products. The reaction can be represented by the equation

   $$C_5H_{12} + Br_2 \rightarrow C_5H_{11}Br + Y$$

   (a) Identify the product Y. (1)

   (b) State the essential condition for this reaction. (1)

   (c) The reaction takes place through a stepwise mechanism with three stages. Give the names of these stages along with one equation for each stage. (3)

   (d) The reaction can form three different products with the formula $C_5H_{11}Br$.

       (i) State the relationship between these products. (1)

       (ii) Draw and name the structures of the three products. (3)

   (Total 9 marks)

4. Propane, C_3H_8, reacts with chlorine, Cl_2, in the presence of UV light.

   (a) State the type of reaction that occurs. (1)

   (b) Give equations to explain how this reaction occurs through a stepwise mechanism. (3)

   (c) Draw and name two possible monosubstituted products of the reaction. (2)

   (d) Explain why propane is regarded as kinetically stable but thermodynamically unstable. (2)

   (Total 8 marks)

5. **(a)** Explain, using equations, how ethane can be converted to chloroethane. (4)

   (b) Identify a possible alkane, other than ethane, that can be found in the product mixture. Give an equation that explains the formation of this alkane. (2)

   (Total 6 marks)

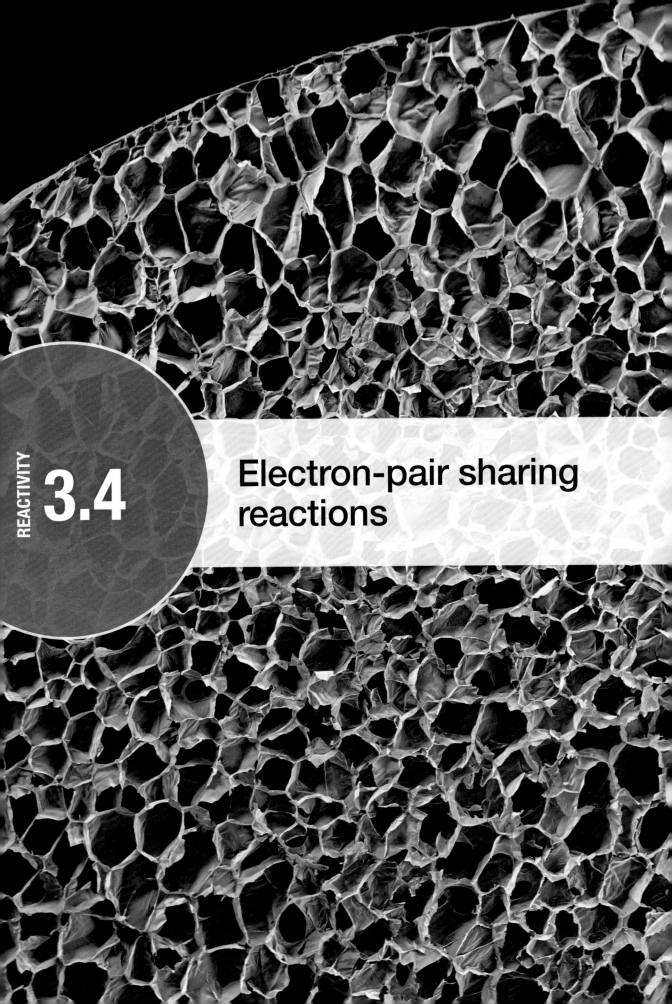

3.4

Electron-pair sharing reactions

A scanning electron microscope (SEM) image of polystyrene foam, commonly used for packaging or insulation. We owe the enormous variety and impact of plastics in our world to synthesis reactions that convert simple molecules such as alkenes into more complex structures. As our understanding of reaction mechanisms has increased, so has our ability to control the products and create materials with highly specialized properties. Polystyrene is the common name for poly(1-phenylethene) which can be made into foam by adding highly volatile compounds such as pentane. When heated, the pentane becomes gaseous and expands, forming air spaces, clearly visible in the image above, that make up 95% of polystyrene foam. It is the presence of these air spaces that gives polystyrene foam its insulating properties. Its low density also makes it a useful material for applications such as flotation devices.

Guiding Question

What happens when reactants share their electron pairs with others?

When one reactant shares an electron pair with another reactant, this results in the formation of a coordination bond. This is a common theme in many areas of chemistry, including the reactions of acids and bases, transition metal ions and many organic compounds. In this chapter, we will focus on two specific examples of electron-pair sharing reactions involving organic compounds:

- nucleophilic substitution of halogenoalkanes and
- electrophilic addition of alkenes.

Reactivity 3.4.1 – Nucleophiles

Reactivity 3.4.1 – A nucleophile is a reactant that forms a bond to its reaction partner (the electrophile) by donating both electrons.	
Recognize nucleophiles in chemical reactions.	
Both neutral and negatively charged species should be included.	

Nucleophiles form bonds by donating electron pairs

A **nucleophile** is a reactant that forms a covalent bond to its reaction partner (the **electrophile**) by donating both bonding electrons. Nucleophiles themselves tend to be electron rich, having one or more lone pairs of electrons and they may also carry a negative charge. Examples of nucleophiles include;

Neutral: H_2O, NH_3, ROH (alcohols), RNH_2 (amines)

Anions: OH^-, F^-, Cl^-, Br^-, I^-, CN^-, R^- (carbanions)

The names nucleophile and electrophile reflect earlier definitions for these species that were based on their attraction to charges on other species. The suffix *-phile* implies 'a strong liking for' and refers to the fact that:

• *nucleophiles*, typically being electron rich, are attracted to positive charges such as that of a *nucleus*

• *electrophiles*, typically being electron deficient, are attracted to negative charges such as that of an *electron*.

Nucleophiles donate electron pairs to form coordination bonds

When nucleophiles donate a pair of electrons to other species (electrophiles), a new covalent bond is formed. This is illustrated in the diagram below, with the green colouring of the new bond formed in the product showing that both bonding electrons came from the nucleophile.

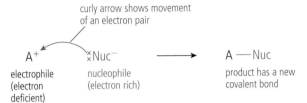

Because the nucleophile donated both electrons in forming the new covalent bond, this is a **coordination bond** which is introduced in Structure 2.2.

Exercise

Q1. What is the essential feature of a nucleophile?

    A a lone pair of electrons

    B a negative charge

    C a pi bond

    D a positive charge

Q2. Which of the following species can act as a nucleophile?

    I. NH_3 II. CH_4 III. H_2O

    A I, II and III

    B II and III

    C I and III

    D I and II

Reactivity 3.4.2 – Nucleophilic substitution reactions

Reactivity 3.4.2 – In a nucleophilic substitution reaction, a nucleophile donates an electron pair to form a new bond, as another bond breaks, producing a leaving group.

Deduce equations with descriptions and explanations of the movement of electron pairs in nucleophilic substitution reactions.

$$Nu\overset{-}{:} + R\!-\!\ddot{x}\!: \longrightarrow R\!-\!Nu + :\ddot{x}\!:^{-}$$

substrate

Nu = nucleophile
R = electrophile
x = leaving group

Further details of the mechanisms are not required at SL.	

Halogenoalkanes react via nucleophilic substitution

Halogenoalkanes are polar compounds due to the fact that the halogen atom is more electronegative than carbon and so exerts a stronger pull on the shared electrons in the carbon–halogen bond. As a result, the halogen gains a partial negative charge ($\delta-$), and the carbon gains a partial positive charge ($\delta+$) and is said to be **electron deficient**.

$$-\overset{|}{\underset{|}{C}}\overset{\delta+}{\longrightarrow}\overset{\delta-}{Cl}$$

Nucleophiles, such as OH^-, are attracted to the electron-deficient carbon in a halogenoalkane. This leads to reactions in which substitution of the halogen occurs, known as **nucleophilic substitution** reactions.

$$CH_3Cl \;+\; OH^- \;\rightarrow\; CH_3OH \;+\; Cl^-$$
chloromethane \qquad\qquad methanol

$$C_3H_7Br \;+\; OH^- \;\rightarrow\; C_3H_7OH \;+\; Br^-$$
1-bromopropane \qquad\qquad propan-1-ol

The halogen atom can be substituted by a variety of nucleophiles, and this means that halogenoalkanes can be converted into many different classes of compounds. This makes them very useful intermediates in organic synthesis pathways.

The overall reaction that occurs between a nucleophile and an electron-deficient substrate molecule such as a halogenoalkane, can be illustrated using curly arrows:

$$Nu:^- + R-\overset{..}{\underset{..}{x}}: \longrightarrow R-Nu \;+\; :\overset{..}{\underset{..}{x}}:^-$$
substrate

Nu = nucleophile
R = electrophile
x = leaving group

For the nucleophile to donate an electron pair to the substrate molecule and form a new bond, another bond must break. This causes a species, known as a **leaving group,** to break away from the substrate molecule. The leaving group is represented by an x in the figure above.

The curly arrows shown in the mechanism represent the movement of electron pairs and this is explained further on page 567. You should be able to explain how these arrows relate to the formation and breaking of covalent bonds in nucleophilic substitution reactions.

Halogens, in general, make good leaving groups as they form relatively weak bonds with carbon. Their higher electronegativity also means the bonded electrons are drawn towards the halogen atom, making the carbon atom electron deficient and susceptible to nucleophilic attack.

The nucleophilic substitution reactions of hydroxide ions with chloromethane and 1-bromopropane are illustrated below using curly arrows. Note that the bond that forms and the bond that breaks, must both involve the carbon atom that is bonded to the leaving group:

$$HO\overset{..}{\underset{..}{x}}^- \quad H_3C-\overset{\delta+}{\underset{\times\times}{C}}\overset{\delta-}{\underset{\times\times}{l\times}} \longrightarrow H_3C-OH \;+\; \times Cl^-$$
chloromethane \qquad methanol

$$HO\overset{..}{\underset{..}{x}}^- \quad C_2H_5-CH_2-\overset{\delta+}{\underset{\times\times}{}}\overset{\delta-}{\underset{\times\times}{Br\times}} \longrightarrow CH_3H_7OH \;+\; \times Br^-$$
1-bromopropane \qquad\qquad propan-1-ol

If the nucleophile is a neutral species, such as water, then the initial product formed from the substrate molecule will be positively charged. However, this subsequently deprotonates and loses an H^+ to give a neutral product. If the leaving group was a halide ion, this could combine with the H^+ ion to form a hydrogen halide.

Overall Equation: $H_2O + RX \rightarrow ROH + HX$

Exercise

Q3. In the reaction $R–X + OH^- \rightarrow R\text{-}OH + X^-$

X^- acts as a:

A nucleophile

B electrophile

C reducing agent

D leaving group

Q4. Explain what is meant by nucleophilic substitution, using the reaction between NaOH and chloroethane to illustrate your answer.

Q5. Use curly arrows to show how hydroxide ions react with bromomethane to form methanol.

Q6. Use Section 12 in the data booklet to predict which halide ion will be the best leaving group in a nucleophilic substitution reaction.

Reactivity 3.4.3 – Heterolytic fission

Reactivity 3.4.3 – Heterolytic fission is the breakage of a covalent bond when both bonding electrons remain with one of the two fragments formed.

Explain, with equations, the formation of ions by heterolytic fission.

Curly arrows should be used to show the movement of electron pairs during reactions.	Reactivity 3.3 – What is the difference between the bond-breaking that forms a radical and the bond-breaking that occurs in nucleophilic substitution reactions?

Heterolytic fission creates oppositely charged ions

The bond-breaking that occurs in nucleophilic substitution reactions to generate the leaving group involves **heterolytic fission.** This occurs when a covalent bond breaks such that one of the products gains both of the shared electrons. *Hetero-* means 'different' and refers to the fact that the two products have an unequal assignment of electrons from the bond.

$$\times \!:\! \times \rightarrow \times \!:\!^- + \times^+$$

In Reactivity 3.3 we learnt that homolytic fission produces two radicals as products. Here we see that heterolytic fission results in the formation of oppositely charged ions. When the bond contains two different atoms, the more electronegative atom will gain the negative charge.

Examples of heterolytic fission:

$$Cl–Cl \rightarrow Cl^+ + Cl^-$$

$$H–Cl \rightarrow H^+ + Cl^-$$

Reactivity 3.3 – What is the difference between the bond-breaking that forms a radical and the bond-breaking that occurs in nucleophilic substitution reactions?

In Reactivity 3.3 we also learnt that the movement of single electrons in homolytic fission could be represented by single-headed curly arrows. Heterolytic fission can also be illustrated by curly arrows but it is important to distinguish the important fact that this type of bond breakage occurs through the movement of an electron *pair*. The distinction is made by using a *double-headed* arrow ($\frown$) to show the movement of an electron pair. The tail shows where the electron pair comes from and the head of the arrow shows where it is moving to.

For example, the heterolytic fission of a H–Cl bond would be represented as:

$$H — \overset{\times\times}{\underset{\times\times}{Cl}}\times \xrightarrow{\text{heterolytic fission}} H^+ + \overset{\times\times}{\underset{\times\times}{\times Cl}}\times^-$$

The opposite process to heterolytic fission occurs when a nucleophile donates a pair of electrons to an electrophile, forming a coordination bond. This bond formation can also be represented using curly arrows:

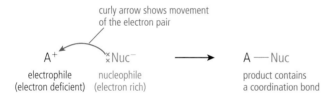

curly arrow shows movement of the electron pair

$$A^+ \qquad {}^\times_\times Nuc^- \longrightarrow A — Nuc$$

electrophile   nucleophile   product contains
(electron deficient) (electron rich)   a coordination bond

Exercise

Q7. Which of the following reactions best represents heterolytic fission?

    A $C + O_2 \rightarrow CO_2$

    B $HF \rightarrow H^+ + F^-$

    C $BF_3 + F^- \rightarrow BF_4^-$

    D $Br_2 \rightarrow Br\bullet + Br\bullet$

Q8. Use curly arrows to explain the difference between the homolytic and heterolytic fission of a H–Br bond.

Reactivity 3.4.4 – Electrophiles

Reactivity 3.4.4 – An electrophile is a reactant that forms a bond with its reaction partner (the nucleophile) by accepting both bonding electrons from that reaction partner.

Recognize electrophiles in chemical reactions.

Both neutral and positively-charged species should be included.	

Electrophiles form bonds by accepting electron pairs

An electrophile is a reactant that forms a covalent bond to its reaction partner (the nucleophile) by accepting both electrons. Electrophiles tend to be electron deficient, or contain an electron-deficient region, so they often have a positive charge or partial positive charge. Examples of electrophiles include:

Neutral: HX (hydrogen halides), X_2 (halogens), H_2O, halogenoalkanes (RX)

Cations: H^+, NO_2^+, NO^+, CH_3^+, R^+ (carbocations)

Nature of Science

Definitions in science need to be broad enough to be inclusive. For example, electrophiles are often described as 'electron deficient', but this definition would exclude molecules such as bromine, Br_2, that are not electron deficient but can act as electrophiles in addition reactions.

Exercise

Q9. What is the essential feature of an electrophile?

    A   It has a lone pair of electrons.

    B   It can accept a pair of electrons.

    C   It has a negative charge.

    D   It has a positive charge.

Q10. Which of the following species can act as an electrophile?

    I. HBr   II. Br_2 III. NH_3

    A   I, II and III

    B   II and III

    C   I and III

    D   I and II

Reactivity 3.4.5 – Electrophilic addition of alkenes

> **Reactivity 3.4.5 – Alkenes are susceptible to electrophilic attack because of the high electron density of the carbon–carbon double bond. These reactions lead to electrophilic addition.**
>
> Deduce equations for the reactions of alkenes with water, halogens and hydrogen halides.
>
> | The mechanisms of these reactions will not be assessed at SL. | Reactivity 3.3 – Why is bromine water decolorized in the dark by alkenes but not by alkanes? |
> | --- | --- |
> | | Structure 2.4 – Why are alkenes sometimes known as 'starting molecules' in industry? |

Addition reactions of alkenes give saturated products

Alkenes have the general formula C_nH_{2n} and are **unsaturated hydrocarbons** containing a carbon–carbon double bond.

$$
\begin{array}{c}
H \diagdown \qquad \diagup H \\
C = C \\
H \diagup \qquad \diagdown H
\end{array}
$$

Alkenes are more reactive than alkanes because the double bond has a high electron density which makes it attractive to electrophiles. The reaction occurs at the site of the double bond, as one of these bonds is relatively easily broken, which creates two new bonding positions on the carbon atoms. This enables alkenes to undergo **addition reactions** with electrophiles and form a range of different saturated products.

Alkenes are unsaturated hydrocarbons and undergo addition reactions.

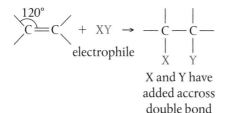

X and Y have added accross double bond

Common addition reactions of alkenes are described below.

Addition of water

The reaction with water is known as **hydration**, and converts the alkene into an alcohol. Water is actually a poor electrophile and this addition reaction requires the use of a strong acid catalyst. In the laboratory, it can be achieved using concentrated sulfuric acid as a catalyst.

The overall equation for the hydration of ethene is:

$$
CH_2CH_2 \xrightarrow[\text{H}_2\text{O}]{\text{H}_2\text{SO}_4\text{(conc.)}} CH_3CH_2OH
$$

ethene ethanol

The hydration reaction of ethene is of industrial significance because ethanol is a very important solvent and is manufactured on a large scale.

The modern industrial process for ethanol synthesis is by direct catalytic hydration of ethane over a phosphoric acid catalyst absorbed onto silica (SiO_2).

▲

Test tubes containing solutions of chlorine, Cl_2 (left), bromine, Br_2 (middle) and iodine, I_2 (right) in water (lower layer) and cyclohexane (top layer). Being non-polar, the halogens are more soluble in cyclohexane where they produce vivid colours.

Addition of halogens

Halogens react with alkenes to produce dihalogeno compounds. These reactions happen quickly at room temperature and are accompanied by the loss of colour of the reacting halogen. Note that the name and structure of the product indicate that a halogen atom attaches to each of the two carbon atoms of the double bond.

Solutions of halogens have distinct colours: Cl_2 is green, Br_2 is orange/brown and I_2 is purple. When dissolved in organic solvents, these colours are particularly vivid.

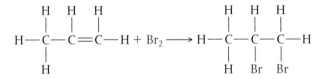

propene   1,2-dibromopropane

▲

Ethanol is an important component of many hand sanitizers. Health agencies recommend a minimum concentration of 60% ethanol to ensure it is effective in killing bacteria and viruses.

Reactivity 3.3 – Why is bromine water decolorized in the dark by alkenes but not by alkanes?

As mentioned on page 559, bromine water is often used to distinguish between alkanes and alkenes. Alkanes react with bromine water through a radical mechanism which requires the presence of UV light to initiate the reaction. However, alkenes readily react with Br_2 and do not require ultraviolet (UV) light, so this reaction will happen in the dark.

Nature of Science

A simple chemical test to distinguish between alkanes and alkenes using bromine water has been outlined in this section.

Alkenes decolorize bromine water rapidly at room temperature, whereas alkanes will only do this in the presence of UV light.

This test might be considered as an appropriate way to distinguish between the two classes of compound, but it relies on qualitative data and is subjective, based on the eye of the observer. Wherever possible, experiments are designed to produce quantitative data that is more objective and able to be analyzed. Consider what instrumentation and sensors could be used to develop these observations into more rigorous tests to distinguish between saturated and unsaturated hydrocarbons.

▲
Use of bromine water to distinguish between an alkane (hexane) and an alkene (hex-1-ene). Without UV light, the brown colour is decolorized by the alkene but not by the alkane. This is due to the addition reaction that occurs with the unsaturated alkene but does not occur with the saturated alkane.

Addition of halogen halides

Hydrogen halides, such as HCl and HBr, react with alkenes to produce halogenoalkanes. These reactions take place rapidly in solution at room temperature.

$$H-\underset{\underset{H}{|}}{\overset{\overset{H}{|}}{C}}=\underset{\underset{H}{|}}{\overset{\overset{H}{|}}{C}}-H + HCl \longrightarrow H-\underset{\underset{H}{|}}{\overset{\overset{H}{|}}{C}}-\underset{\underset{Cl}{|}}{\overset{\overset{H}{|}}{C}}-H$$

$$CH_2CH_2 + HCl \longrightarrow CH_3CH_2Cl$$
ethene   chloroethane

All the hydrogen halides are able to react in this way, but the reactivity is in the order HI > HBr > HCl due to the decreasing strength of the hydrogen halide bond down group 17. So HI, with the weakest bond, reacts the most readily.

The examples of addition reactions provided above show that alkenes can react with specific electrophiles to make different classes of organic compounds such as alcohols, halogenoalkanes and dihalogenoalkanes. These products can be further converted into many other classes of compounds so this makes alkenes very useful starting molecules in organic synthetic pathways.

Structure 2.4 – Why are alkenes sometimes known as 'starting molecules' in industry?

The following diagram summarizes the important addition reactions covered in this chapter and shows the different classes of compounds formed along with a specific example of the reactants required for each conversion.

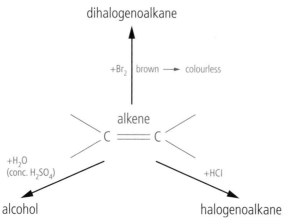

dihalogenoalkane

+Br$_2$ brown ⟶ colourless

alkene

$$C = C$$

+H$_2$O
(conc. H$_2$SO$_4$)

+HCl

alcohol   halogenoalkane

Nature of Science

It is hard to assess the vastness of the scale of the impact that alkanes and alkenes have had on society. As fuels, alkanes have made energy widely available, and as chemical feedstock, alkenes have led to the synthesis of diverse plastics. Yet these innovations have come at significant cost to the environment, and the long-term consequences are still being determined. While scientific progress is responsible for both the intended and unintended consequences, it is scientists who must respond to the problems. The issues of climate change, air pollution, and plastic disposal are currently pressing areas of research and development.

Exercise

Q11. In the reaction between ethene and hydrogen chloride, $CH_2CH_2 + HCl$, the product is:

    A CH_2ClCH_2Cl **B** CH_3CH_3

    C   no reaction occurs   **D** CH_3CH_2Cl

Q12. Give the condensed formulas and names of the products for the following reactions:

    (a) $CH_3CH=CHCH_3 + HBr$ **(b)** $CH_2=CH_2 + $ conc. H_2SO_4

    (c) $CH_3CH=CH_2 + Br_2$

Q13. Explain how bromine water, $Br_2(aq)$, can be used to distinguish between an alkane and an alkene.

Guiding Question revisited

What happens when reactants share their electron pairs with others?

In this chapter we have seen that the donation of an electron pair from an electron-rich species to an electron-deficient species results in the formation of a coordination bond. Curly arrows are used to show the movement of electron pairs in these electron sharing reactions.

- A nucleophile forms a bond by donating a pair of electrons to its reaction partner.
- In a nucleophilic substitition reaction, a leaving group forms as another bond breaks.
- Heterolytic fission of a covalent bond produces two oppositely charged ions
- An electrophile forms a bond by accepting a pair of electrons from its reaction partner.
- Alkenes undergo electrophilic addition reactions in which the carbon-carbon double bond breaks.

Electron-pair sharing reactions between nucleophiles and electrophiles were then examined in the context of two important reactions of organic compounds:

- Nucleophilic substitution reactions, with the specific example of halogenoalkanes reacting with nucleophiles.
- Electrophilic addition reactions, with the specific example of alkenes reacting with electrophiles.

The following tables should prove a useful summary of the various conventions and definitions that have been introduced for these reactions, along with the radical substitution mechanism introduced in Reactivity 3.3.

Types of reactant

Saturated	Unsaturated
• compounds which contain only single bonds • for example: alkanes	• compounds which contain double or triple bonds • for example: alkenes, arenes

Electrophile (electron-seeking)	Nucleophile (nucleus-seeking)
• a reactant that forms a covalent bond by accepting both bonding electrons from another reactant • electrophiles are typically positive ions or have a partial positive charge and are therefore attracted to parts of molecules which are electron rich • for example: NO_2^+, H^+, Br^+	• a reactant that forms a covalent bond by donating both bonding electrons to another reactant • nucleophiles have a lone pair of electrons and may also have a negative charge; they are therefore electron rich and are attracted to parts of molecules which are electron deficient • for example: Cl^-, OH^-, NH_3

Types of reaction

Addition	• occurs when two reactants combine to form a single product • characteristic of unsaturated compounds • for example: $C_2H_4 + Br_2 \rightarrow C_2H_4Br_2$
Substitution	• occurs when one atom or group of atoms in a compound is replaced by a different atom or group • characteristic of saturated compounds and aromatic compounds • for example: $$CH_4 + Cl_2 \rightarrow CH_3Cl + HCl$$

Types of bond-breaking (bond fission)

Homolytic fission	Heterolytic fission
• is when a covalent bond breaks by splitting the shared pair of electrons between the two products • produces two radicals, each with an unpaired electron	• is when a covalent bond breaks with both the shared electrons going to one of the products • produces two oppositely charged ions

Convention for depicting organic reaction mechanisms

Describing organic reaction mechanisms often involves showing the movement of electrons within bonds and between reactants. The convention adopted for this is a **curly arrow**, drawn from the site of electron availability, such as a pair of non-bonding electrons, to the site of electron deficiency, such as an atom with a partial positive charge.

For example:

X—Y	represents the electron pair being pulled towards Y so Y becomes $\delta-$ and X becomes $\delta+$
X: ⟶ C $\delta+$	the nucleophile X: is attracted to the electron-deficient C ($\delta+$)

A 'normal' double-barbed arrow ( ) represents the motion of an electron pair (as above). Often the mechanism involves several steps. The electrons are transferred ultimately to an atom or group of atoms that then detaches itself and is known as the leaving group. We used blue throughout this chapter to show curly arrows and the pull of electrons.

Note that a single-barbed arrow ( ), known as a **fish hook**, represents the movement of a single electron. These single arrows are often used in reactions involving radicals (Reactivity 3.3).

Practice questions

1. Nucleophiles are species that can:

   A undergo substitution reactions with alkenes

   B accept an electron pair from other species

   C undergo addition reactions with alkenes

   D donate an electron pair to other species

2. Which of the following species can act as an electrophile?

   I. HI

   II. I_2

   III. H_2O

   A I, II and III   **B** II and III

   C I and III   **D** I and II

3. Which reaction type is typical for halogenoalkanes?

   A electrophilic substitution   **B** electrophilic addition

   C nucleophilic substitution   **D** nucleophilic addition

4. Which compound will decolorize bromine water in the dark?

   A pentane   **B** cyclopentane

   C pent-1-ene   **D** methane

5. What is the product in the reaction of ethene with bromine, Br_2?

   A bromoethane   **B** 1,2-dibromoethane

   C 1,1-dibromoethane   **D** ethane

6. What is the product in the reaction of ethene with HI?

   A 1,2-diiodoethane   **B** 1,1-diiodoethane

   C iodoethane   **D** ethane

7. The conversion of C_4H_9Cl to C_4H_9OH involves what type of reaction?

   A electrophilic addition   **B** radical substitution

   C electrophilic substitution   **D** nucleophilic substitution

8. **(a)** Explain, in terms of electron pairs, the difference between a nucleophile and an electrophile. (2)

   (b) State the type of bond that is formed when a nucleophile reacts with an electrophile. (1)

   (Total 3 marks)

9. Ethene, C_2H_4, can be converted into ethanol, C_2H_5OH, via a two-step reaction:

$$C_2H_4 \xrightarrow[\text{HBr (g)}]{\text{I}} X \xrightarrow[\text{NaOH(aq)}]{\text{II}} C_2H_5OH$$

   (a) State the names of the reactions occurring in steps I and II. (2)

   (b) Draw the full structure and name the compound formed in step I of the reaction. (2)

   (Total 4 marks)

10. Alkenes are very useful starting materials in organic synthesis reactions as they can be converted to many other classes of compounds. Some reactions of but-2-ene, C_4H_8, are outlined in the diagram below.

$$C_4H_8Br_2 \longleftarrow C_4H_8 \longrightarrow C_4H_9OH$$
$$\text{2,3-dibromobutane} \quad \downarrow \text{HBr} \quad \text{butan-2-ol}$$
$$X$$

   (a) Draw the full structure of but-2-ene. (1)

   (b) State the reagents needed to convert but-2-ene to butan-2-ol. (1)

   (c) State what changes would be observed in the conversion of but-2-ene to 2,3-dibromobutane. (1)

   (d) Draw the full structure and give the name of the product, X, formed in the reaction of but-2-ene with HBr. (2)

   (Total 5 marks)

11. 1-Iodopropane, C_3H_7I, reacts with aqueous sodium hydroxide, NaOH(aq), via a nucleophilic substitution reaction to give a product with the formula C_3H_8O.

   (a) Give a balanced equation for this reaction. (1)

   (b) Draw the full structure and give the name of the product formed in this reaction. (1)

   (c) Identify the species acting as the nucleophile and as the leaving group in this reaction. (2)

   (d) Show, using curly arrows, how the reaction occurs through the breaking and forming of covalent bonds. (3)

   (e) State the type of bond breaking that occurs in this reaction. (1)

   (Total 8 marks)

Green Chemistry

'Green chemistry is the utilization of a set of principles that reduces or eliminates the use or generation of hazardous substances in the design, manufacture and application of chemical products.'

Green Chemistry: Theory and Practice, P. T. Anastas and J. C. Warner,
Oxford University Press, Oxford, 1998

Green Chemistry is also known as sustainable chemistry and applies across the life cycle of a chemical product, from design to ultimate disposal.

The 12 principles of Green Chemistry are:

1. Pollution prevention

It is better to prevent waste than to treat or clean up waste after it has been created.

2. Atom economy

Synthetic methods should be designed to maximize the incorporation of all materials used in the process into the final product.

3. Less hazardous chemical synthesis

Whenever practicable, synthetic methods should be designed to use and generate substances that possess little or no toxicity to people or the environment.

4. Design safer chemicals

Chemical products should be designed to effect their desired function while minimizing their toxicity.

5. Safer solvents and auxiliaries

The use of auxiliary substances (e.g. solvents or separation agents) should be made unnecessary whenever possible and not harmful when used.

6. Design for energy efficiency

Energy requirements of chemical processes should be recognized for their environmental and economic impacts and should be minimized. If possible, synthetic methods should be conducted at ambient temperature and pressure.

7. Use of renewable feedstocks

A raw material or feedstock should be renewable rather than being used up whenever technically and economically practicable.

8. Reduce derivatives

Unnecessary derivatization (use of blocking groups, protection/de-protection, and temporary modification of physical/chemical processes) should be minimized or avoided if possible because such steps require additional reagents and can generate waste.

9. Catalysis

Catalytic reagents (as selective as possible) are superior to stoichiometric reagents.

Molecular graphic of the structure of a synthetic zeolite catalyst showing its microchannel architecture. Zeolites, often called molecular sieves, contain a silica–alumina crystalline structure and are used as catalysts in the petroleum industry. They are becoming increasingly important in processes such as biomass conversion, fuel cells and air and water purification.

10. Design for degradation

Chemical products should be designed so that at the end of their function they break down into harmless degradation products and do not remain in the environment.

11. Real-time analysis for pollution prevention

Analytical methods need to be further developed to allow for real-time, in-process monitoring and control prior to the formation of hazardous substances.

12. Inherently safer chemistry for accident prevention

Substances and the form of a substance used in a chemical process should be chosen to minimize the potential for chemical accidents, including releases, explosions and fires.

The 12 principles of Green Chemistry.

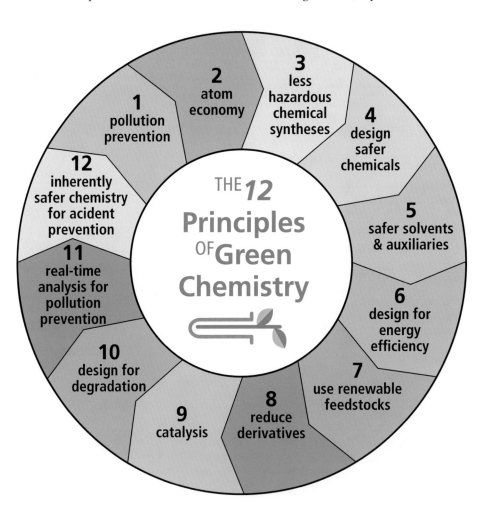

Application of these principles in research and industry is beneficial to human health and the environment in many ways. With an emphasis on *prevention* of waste and pollution rather than on treatment of the problems, Green Chemistry helps reduce both the negative impacts and the cost of chemical products and processes.

Applications of Green Chemistry in IB chemistry

During the IB chemistry course, there are many ways in which the principles of Green Chemistry can be considered and applied. For example:

- In all laboratory work, consideration can be given to reducing the volumes and/or concentrations of reagents used.

- Micro-scale experimentation can be an effective alternative to full-scale laboratory work to reduce waste.

- Disposal of chemicals should be carried out on a case-by-case basis, and where possible, excess reactants and products should be reused.

- The choice of solvent for a reaction should be carefully considered, and where possible aqueous solutions or simple alcohols or alkanes should be used.

- In the research design phase of the investigation, knowledge and assessment of risk should be made before developing a chemical procedure.

- In Reactivity 2.2, the study of catalysts can include consideration of how waste is minimized and energy costs of a process reduced.

- In Reactivity 2.1 and Structure 2.4, determination of the atom economy of a reaction helps to assess the efficiency of the process in terms of the incorporation of reactants into the final product.

Green Chemistry is a rapidly growing and interdisciplinary area of study. You can find out more about this field on the two websites below:

Theory of Knowledge in chemistry

Chemistry $\rightleftharpoons$ TOK

Passengers are warned to mind the gap as they leave an underground train in London. We often make a 'jump' when we accept knowledge claims and should always be aware of the gap that we have crossed. When we see bubbles in a test tube and infer that hydrogen is produced, or draw a best fit line through a series of points, we are 'leaping' across such a gap. When we go from the specific to the general we cross a precarious gap, as discussed later in the chapter.

Introduction

In theory of knowledge (TOK) you are encouraged to think critically about the production and the application of knowledge in the subjects that you study. You are asked to reflect on the nature of the different subjects, looking for similarities and differences. You are also asked to reflect on the impact of the learning experience on you as a learner who must effectively navigate and make sense of a world with many uncertainties. Chemistry is a significant part of your Diploma experience and can add to and enrich your TOK experience, particularly as you discuss the natural sciences area of knowledge. This is a reversible process: the critical thinking encouraged in TOK will deepen your understanding of what you learn in chemistry.

Does chemistry give you a 'true' picture of reality? How does the knowledge you gain in your chemistry class differ from that gained in other subjects? What are the ethical implications of developments in the subject? How have technological developments impacted our knowledge of chemistry? Chemistry has been hugely successful in giving us explanations of the material world and has also made a significant contribution to improving our quality of life. But how reliable is the knowledge that it offers?

Knowledge questions are questions about knowledge – about how it is produced, acquired, shared and used; what it is and what it is not; who has it and who does not; and who decides the answers to these questions.

Knowledge questions and the knowledge framework

Most of the questions in this book are about the chemistry of the material world and not about the nature of chemical knowledge. TOK, by contrast, focuses on **knowledge questions**. These are contestable questions about knowledge that draw on TOK concepts. The table below shows how TOK questions explore the fundamental nature of the knowledge we seek in chemistry.

Chemistry questions	Knowledge questions
What is the balanced equation for the reaction?	Can all knowledge be expressed in words or symbols?
What is the electron configuration of the sodium atom?	Are the models and theories that scientists use merely pragmatic instruments, or do they truthfully describe the natural world?
What are the acid and basic properties of the period 3 oxides?	To what extent do the classification systems we use limit the pursuit of knowledge?
Which molecule has a structure consistent with the spectral data?	How do the tools we use shape the knowledge that we produce?
Which species is acting as a Lewis acid?	How significant have notable individuals been in shaping the development of the different areas of knowledge?
Why is fluorine more reactive than chlorine?	What counts as a good explanation?

Knowledge questions draw on the 12 concepts below.

evidence	certainty	truth	interpretation	power	justification
explanation	objectivity	perspective	culture	values	responsibility

Challenge yourself

1. Identify four of the concepts above which you think are most related to chemical knowledge and explain your choices.

Reference to these concepts is illustrated later in the text by references in **bold**.

The different areas of knowledge can be characterized by a knowledge framework. This provides a structure for considering knowledge questions throughout the TOK course. The framework is made up of four elements: scope, perspective, methods and tools, and ethics.

The knowledge framework consists of four common elements: scope, perspectives, methods and tools, and ethics.

Knowledge and the knower

What shapes your perspective as a knower? What does Figure 1 below represent?

◀ **TOK Figure 1** A circle in a hexagon?

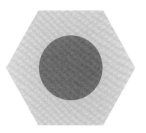

▲
TOK Figure 1 A circle in a hexagon?

To many people this is simply a circle and a hexagon, but to an IB Diploma Programme chemistry student it might have more significance. It can be **interpreted** as a benzene ring or the IB Diploma Programme with six subjects as sides and a core of TOK, CAS and the Extended Essay at its centre. How you choose to interpret the figure will depend on the context in which it is presented and your prior knowledge. This illustrates the importance of perspectives, one of the elements in the TOK knowledge framework and also one of the 12 TOK concepts.

What shapes your perspective, as a knower? Why did you choose to study chemistry? How does your chemistry class differ from your other classes? How much of our knowledge depends on our interactions with other knowers? How does the knowledge that you study in class have an impact on the world outside your classes?

Chemistry is a study of the material world. How do your interactions with the material world shape your chemical knowledge? Is your motivation for studying chemistry to have a better understanding of the properties of the materials around you? What criteria can we use to distinguish between knowledge and belief? Do you know or believe that salt is made of ions? What could you do to improve your knowledge?

In a 2011 speech, Sir Harold Kroto, the 1996 Nobel Prize Laureate in Chemistry, talked about the importance of science, which, in his view, is the academic pursuit best positioned to develop the **truth** about the natural world. He also challenges us all to take **responsibility** for what we accept as knowledge.

> 'Common sense says the Sun goes around the Earth. Who agrees with me? ... It's uncommon sense that was needed to recognize that the Earth was turning on its axis... Let me just check – how many of you know the **evidence** for Galileo to say that the Earth was going around the Sun? Put your hand up. You've accepted it. Almost nobody's put their hand up. It's incredible. Look at yourself, you've accepted this. You've accepted a lot of things without evidence.'

Do you agree with Kroto or do you think you can become too skeptical?

Chemical knowledge

The subject of chemistry has a body of highly structured and systematic knowledge. It is the work of many individuals and is in a sense anonymous – although there would have been no Hamlet without Shakespeare, atomic theory would still have been developed without Dalton. While individuals can and do contribute to this body of knowledge, their work is subject to peer review and their experimental results need to be replicated by others if they are going to be accepted by the scientific community. We all, however, have different experiences of chemistry, and this affects our knowledge of the chemical content and our procedural knowledge of how to do something such as a titration.

Challenge yourself

2. 'One aim of the physical sciences has been to give an exact picture of the material world. One achievement... has been to prove that this aim is unattainable.'

J. Bronowski

What are the implications of this claim for the aspirations of science?

The scope of chemical knowledge

Chemistry is about understanding the nature of matter and its interactions. It helps us understand interactions between different materials and predict if and how they will react. This gives us some control of the material world and the potential to improve human life and our environment.

Chemistry has applications in the biological, material and environmental sciences. The scientific approach assumes that the material world behaves in a coherent way and is rationally comprehensible.

> 'The most incomprehensible thing about the world is that it is at all comprehensible.'

Albert Einstein

The application of chemical knowledge can have ethical, environmental, economic, cultural and social impacts.

The scope of the subject is limited by the dangerous nature of some substances to the individual and the environment.

The periodic table is the map of chemistry

Maps are often used as a metaphor for knowledge in TOK and the periodic table can be thought of as the 'map' of chemistry; it suggests new avenues of research for the professional chemist and is a guide for students – it disentangles a mass of observations and reveals hidden order. Chemistry is not the study of a random collection of elements, but of the trends and patterns in their chemical and physical properties, and we should not forget that 'the map is not the territory' (Alfred Korzybski)! The periodic table is a human construct that helps us describe the elements, but it is an incomplete representation as it does not incorporate all their properties.

Perspectives in chemical knowledge

Chemistry and alchemy

It is worth reflecting that many of the experimental techniques of chemistry have their origins in the pseudoscience of alchemy, a medieval forerunner of chemistry which attempted to purify and transform matter.

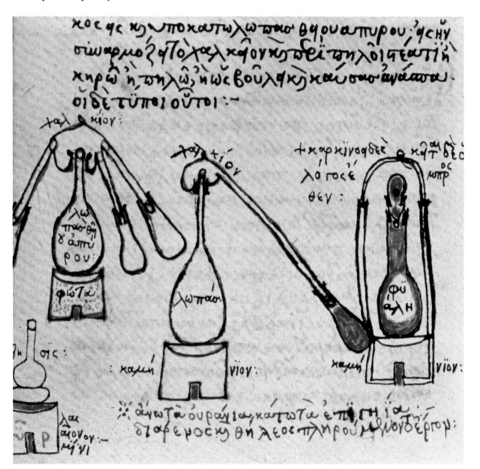

A page from a treatise on alchemy written by Zosimus of Panapolitus (4th century). Some of the equipment drawn is still found in a modern chemistry laboratory.

The alchemist's hunt for the Philosopher's Stone, which was believed to give eternal life and could turn base metals into gold, seems very naïve to us now. How do you think our current chemical knowledge will be viewed in 500 years?

Ernest Rutherford, the father of nuclear physics, described himself as an alchemist as he was able to change one element into another by nuclear reactions.

A web and hierarchy of disciplines

Is it possible to place the different scientific disciplines in a hierarchy, and which criteria would you use for your choice (Figure 2)? Chemical theories can help our understanding of biology: for example, hydrogen bonding explains the double helix structure of DNA. But much of chemistry relies on physics: for example, hydrogen bonding is explained in terms of electrostatic attraction.

Is this direction of **explanation** ever reversed? The view that one subject can be explained in terms of the components of another is called **reductionism**. Is physics in some way 'better' than the other sciences as it explains more? Where in this hierarchy would you place mathematics, which has been described as both the queen and the servant of the sciences? Some would argue that biology cannot be reduced to physics as it involves levels of organization that are not captured by simple concepts of physics.

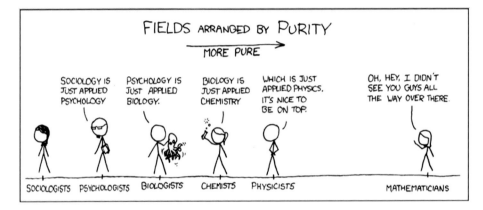

TOK Figure 2 Chemistry, the central science?

Is chemistry in some way the most complete science? Our knowledge of the periodic table and atomic structure suggests that there are limits to the number of elements in nature. Chemical knowledge is **justified** by empirical evidence which is acquired by the senses and enhanced, if necessary, by technology. Which of the natural sciences is most clearly based on direct observation?

Same data, different hypothesis: Occam's razor

It is always possible to think of a range of hypotheses that are consistent with a given set of data. For the same reason, an infinite number of patterns can be found to fit the same experimental data.

TOK Figure 3 The curve and the straight line in the graph both fit the experimental data.

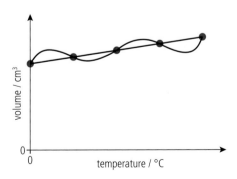

You may argue that the straight line in Figure 3 is a more suitable fit to the data as it is simpler, but on what grounds do we base the assumption that simplicity is a criterion for truth? The idea that the simplest explanations are the best is inspired by the principle – named after the medieval philosopher William of Occam – known as Occam's razor that a theory should be as simple as possible while maximizing explanatory power.

> *'Explanations must be as simple as possible – and no simpler.'*
>
> Albert Einstein

Do the natural sciences rely on any assumptions that are themselves unprovable by science?

Rejecting anomalous results: confirmation bias

We often dismiss results which do not fit the expected pattern as being due to experimental error. In Figure 4, is it reasonable to reject the point that is not on the line?

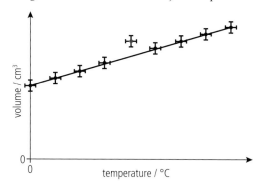

TOK Figure 4 When are you justified in dismissing a data point which does not fit the general pattern?

How does chemical knowledge change with time?

Lavoisier attached numbers to substances by measuring their masses and helped chemistry develop as a physical science. Mendeleev's periodic table systematized a mass of chemical information, and atomic theory explained the structure of the periodic table and the nature of chemical interactions. It was originally believed that a 'vital force' was needed to explain organic chemistry, but we now realize that it fits in with the rest of our chemical understanding.

Chemical knowledge has grown due to the contributions of individuals such as Boyle, Avogadro, Mendeleev, Dalton, Bohr, Lewis, Gibbs, Brønsted, Lowry and Kekulé. Their contributions have made the subject more systematic and coherent.

The above discussion suggests that there has been steady progress in chemical knowledge and our knowledge of the material world is increasing and improving in accuracy in the same way a map's description of a territory might improve with time. We will see when we discuss the ideas of Popper and the provisional nature of scientific knowledge that this view is problematic and an alternative was offered by Thomas Kuhn (1922–1996). Kuhn suggested that science does not develop by the orderly accumulation of facts and theories, but by dramatic revolutions which he called **paradigm shifts**. In this context, a paradigm can be thought of as a model or world view accepted by the scientific community. Kuhn distinguished between periods of **normal science** in which new discoveries are placed within the current paradigm and **extraordinary science** which produces results which do not fit the current paradigm. Isaac Newton, John Dalton, Charles Darwin, and Albert Einstein are all revolutionary scientists who changed the way we look at the world by proposing new paradigms.

A paradigm is a system of concepts, language, assumptions, methods, values, and interests that define research in an area of knowledge. It is related to the knowledge framework in TOK.

The idea of a paradigm shift can be illustrated by considering the picture in Figure 5.

TOK Figure 5 In one paradigm, the picture can be interpreted as a duck; in another paradigm, it can be interpreted as a rabbit. Turn the page through 90° and experience a paradigm shift.

'What were ducks in the scientists' world before the revolution, are rabbits afterwards…'

Thomas Kuhn

Atomic theory is one of the most important paradigms of chemistry. Dalton's model of the atom as indivisible spheres collapsed with the discovery of the proton, neutron and electron. And this model has developed further with advances in particle physics and the discovery of quarks.

Paradigm shifts: phlogiston theory and the discovery of oxygen

When a solid such as magnesium burns, it crumbles into ash and becomes smaller. It seems quite natural to assume that the metal is giving something off as it burns. It was originally believed that all flammable materials contain phlogiston (a word derived from the Greek for flame), which was given off and absorbed by the air as substances burn. In this theory, substances stop burning when all the phlogiston has been released or when the air is saturated with the phlogiston released. The crisis for the paradigm occurred when careful measurements showed that a mass increase occurred during combustion.

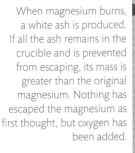

When magnesium burns, a white ash is produced. If all the ash remains in the crucible and is prevented from escaping, its mass is greater than the original magnesium. Nothing has escaped the magnesium as first thought, but oxygen has been added.

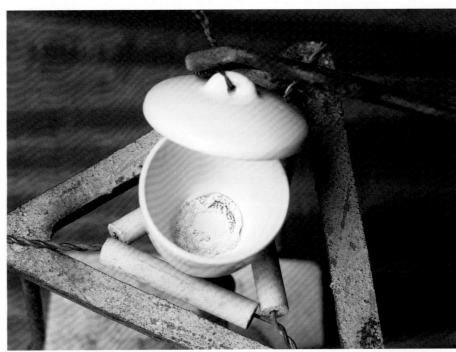

Some explained this result within the phlogiston paradigm by suggesting that phlogiston could have a negative mass, but that explanation was rejected in favour of the modern oxygen theory of combustion.

As substances burn more brightly in oxygen, the gas was originally called 'dephlogisticated air' by Joseph Priestley, one of the scientists credited with its discovery. However, the discovery of oxygen made the term 'phlogiston' meaningless.

Priority disputes

There have been some priority disputes over who discovered oxygen and arguments can be made for Scheele, Priestley and Lavoisier. Their research techniques were very different – while Priestley and Scheele heated and sniffed, Lavoisier heated, measured, and calculated. Perhaps then, it was Lavoisier who changed the biggest paradigm in the history of chemistry as he explained the mass increase during combustion as due to the combination with oxygen. Does competition between scientists help or hinder the production of knowledge?

Science revolutionaries

A widely held stereotype of scientific progress is that of an idealistic young innovator challenging the ideas of the establishment. This is rationalized in Kuhn's model of science because established individual scientists are often reluctant to make such leaps. The need to be a risk-taker, one of the attributes of the IB learner profile, and thinking outside the box is emphasized by Richard Feynman:

> 'One of the ways of stopping science would be to only experiment in the region where you know the law.'

This emphasizes the role of creative thinking in scientific progress.

'Air and Fire' (1782) by Scheele. Scheele discovered several new chemical elements and compounds, but was unlucky in that his contemporaries often published their findings before him. Scheele published in German.

Which Einstein discovered the Theory of Relativity?

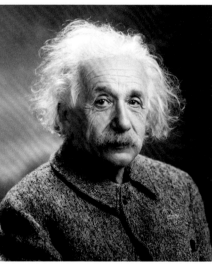

Max Planck emphasized the societal element of progress in science.

> 'A new scientific theory does not triumph by convincing its opponents and making them see the light, but rather because its opponents eventually die out and a new generation grows up that is familiar with it.'

587

How does the social context of scientific work affect the methods and findings of science?

Methodology and tools in chemical knowledge

Chemists follow a scientific methodology with an emphasis on experimental work. Results are explained using theories and models and evaluated by peer review. Intuition and creativity are needed to produce hypotheses, but reason must govern coherent explanations. Knowledge is shared via agreed-upon systems of language and convention including scientific terminology, chemical symbols and mathematics. Observations and measurements are made using sense perception, and emotion can provide the personal motivation for a scientist to focus on a particular area of research.

Experimental results and well-established theories count as facts and explanations which reduce chemical change and structure to simple, well-understood concepts such as mass, charge, and attractive and repulsive electrostatic forces. Models can be used to simplify a problem to simple cause and effect relationships.

What to observe?

Look at the picture (right) and make a list of observations.

Did you notice that the flame has gone orange where it heats the test tube? Many people miss this, even when they are asked to heat an empty test tube, but to the chemist, it is a significant observation. It shows that there is sodium in the glass.

'In the field of observation, chance favours only the prepared mind.'

Louis Pasteur

The scientific method

Should the natural sciences be regarded more as a method or more as a system or body of knowledge?

Richard Feynman (1918–1988), one of the great physicists of the 20th century, gave the following description of the scientific method:

'In general we look for a new law by the following process. First, we guess it. Then we compute the consequences of the guesses to see what would be implied if the law was right. If it disagrees with experiment, it is wrong. In that simple statement is the key to science... It does not make any difference how smart you are, or what is your name – if it disagrees with experiment, it is wrong... It is true that one has to check a little to check that one is wrong.'

During your IB chemistry course, you are continually challenged to think about the Nature of Science. To what extent is Feynman's view of the scientific method still a valid description today? Does a single method adequately describe scientific studies in all disciplines from astronomy to geology? Could you use your experience in the chemistry laboratory or on the collaborative sciences project to write your own definition of the scientific method?

Zinc (Zn) and black copper(II) oxide (CuO) powders are mixed in a stoichiometric ratio. The mixture is strongly heated in a flame of a propane Bunsen burner.

Measurement: the observer effect

Measurement has allowed the chemist to attach numbers to the properties of materials, but the act of measurement can change the property being measured. Adding a thermometer to a hot beaker of water, for example, will cause the temperature to decrease slightly, and adding an acid–base indicator, which is itself a weak acid or base, will slightly change the pH of the solution.

Generally, such effects can be ignored, as measures are taken to minimize them. Only a few drops of indicator are used, for example, and thermometers are designed to have a low heat capacity. The **observer effect** can cause significant problems at the atomic scale, however, which led the physicist Werner Heisenberg (1901–1976) to comment:

'What we observe is not nature itself, but nature exposed to our mode of questioning.'

The observer effect is also significant in the human sciences. Can your school director observe a 'normal' chemistry class?

What is the role of inductive reasoning?

How do individual observations lead to theories and scientific laws of nature? Imagine an experiment in which you test the pH of aqueous solutions of some different oxides. The results are shown in the table.

Methyl orange is a weak acid. The addition of the indicator will change the pH of the solution it is measuring.

Oxide	Acid/Alkali
$Na_2O(s)$	alkali
$MgO(s)$	alkali
$CO_2(g)$	acid
$SO_2(g)$	acid
$CaO(s)$	alkali
$N_2O_5(g)$	acid

Two possible conclusions which fit the pattern are:

- All solid oxides are alkalis. All gaseous oxides are acidic.
- All metal oxides are alkalis. All non-metal oxides are acidic.

Challenge yourself

3. Suggest another conclusion which fits the results.

We have used **induction** to draw these conclusions. Inductive logic allows us to move from specific instances to a general conclusion. Although it appeals to common sense, it is logically flawed. Both conclusions above are equally valid, based on the evidence, but both could be wrong. Just because something has happened many times in the past does not prove that it will happen in the future. This is the **problem of induction**. The philosopher Bertrand Russell illustrated the danger of generalization by considering the case of the philosophical turkey. The bird reasoned that since he had been fed by the farmer every morning, he always would be. Sadly, this turkey discovered the problem with induction on Thanksgiving Day! Are you acting in the same way as the turkey when you draw conclusions in your experimental work?

Induction means inferring general conclusions from particular examples.

How do we justify the use of induction? Consider the following form of reasoning:

- On Monday I used induction and it worked.
- On Tuesday I used induction and it worked.
- On Wednesday I used induction and it worked.
- On Thursday I used induction and it worked.
- On Friday I used induction and it worked.
- Therefore I know that induction works.

> ### Challenge yourself
>
> **4.** What does this illustrate about the problem of induction?

Falsification?

Karl Popper (1902–94) realized that scientific verification does not actually prove anything and decided that science finds theories, not by verifying statements, but by falsifying them. Popper believed that even when a scientific theory had been successfully and repeatedly tested, it was not necessarily true. Instead, it had simply not yet been proved false. Observing a million white swans does not prove that all swans are white, but the first time we see a black swan, we can firmly disprove the theory.

What is the role of deductive reasoning?

Deductive logic allows us to move from general statements to specific examples. The conclusion of deductive logic must be true if the general statements on which it is based are correct. Deductive reasoning is the foundation of mathematics. We can also use this reasoning to test our scientific hypotheses. Consider again the pH of the oxides tabulated earlier. The two competing hypotheses could be distinguished by considering the pH of a non-metal oxide such as phosphorus oxide, which is a solid at room temperature. If we use the two hypotheses as starting points (premises), they lead to two different conclusions using deductive reasoning:

- All solid oxides are alkalis.
- Phosphorus oxide is solid.

Therefore: phosphorus oxide is an alkali.

OR

- All non-metal oxides are acidic.
- Phosphorus oxide is a non-metal oxide.

Therefore: phosphorus oxide is an acid.

When the pH of phosphorus oxide solution is tested and shown to be an acid, we can be certain that the first hypothesis is false, but we cannot be sure that the second is definitely true. If it survives repeated tests, it may, however, become accepted as scientific truth, but does not mean that it is certain. No matter how many tests a hypothesis survives, we will never prove that it is true in the same way as a mathematical proof is true. All scientific knowledge is **provisional**.

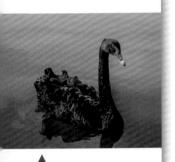

▲
No matter how many times we record in our notebooks the observation of a white swan, we get no closer to proving the universal statement that all swans are white. Black swans inhabit lakes, rivers and coastal areas in Australia.

There are still problems, however, with this view of science. When your results do not match up with the expected values in a chemistry investigation, do you abandon the accepted theories, or do you explain the differences as being due to experimental error, faulty instruments, or contaminated chemicals?

Are the models and theories that scientists use merely pragmatic instruments or do they actually describe the natural world?

We started the chapter with the metaphor of a map for knowledge. Models play a similar role to maps. Scientists use models to clarify certain features and relationships. They can become more and more sophisticated, but they can never become the real thing. Models are just a representation of reality. For example, in pre-IBDP courses, the sodium atom can be represented by the diagram in Figure 6 with a nucleus at the centre of the atom and electrons orbiting the nucleus like planets orbiting the Sun. This model helps us understand the reactivity of the element. Sodium reacts by losing its outer electron to adopt the electron arrangement of neon with only complete energy levels.

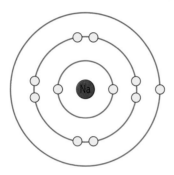

TOK Figure 6 A model you may have used to explain the chemistry of sodium.

The model is not complete, however, as it fails to explain the electron arrangement of atoms after calcium in the periodic table. To explain the chemistry of all the elements, we need a more sophisticated model, that of electrons occupying orbitals. Can an inaccurate model give us knowledge?

Models are not the territory

The methane molecule can be represented in a number of ways (Figure 7).

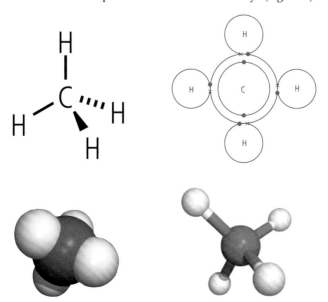

TOK Figure 7 Different models of the methane molecule.

Since a model is a simplified representation of the world, it tends to focus on features that are directly relevant. This highlights the need to consider the relationship between the model and the context in which it has been applied.

What is the role of imagination?

The evidence for the wave nature of electrons comes from the diffraction pattern observed when a beam of electrons is passed through a metal structure, which is similar to that observed when light is passed through a diffraction grating. (see below).

(a)   **(b)**

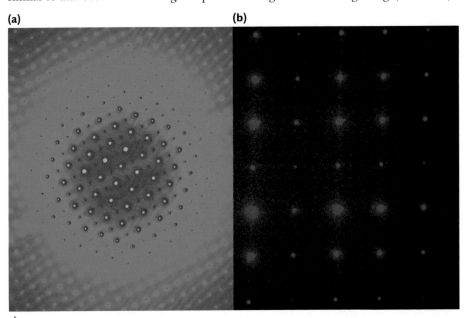

▲
(a) The electron diffraction pattern of molybdenum trioxide (MoO_3). **(b)** Diffraction grating pattern formed by laser light (red).

We now recognize that the 'true' nature of the electron, or indeed of anything, cannot be completely described by a single model. Both the wave and particle models are needed. This should not surprise us as models are often based on our everyday experience which does not apply to the small scale of the atom. To understand nature at this level, scientists must go beyond their experience and use their imagination. Scientists need imagination to form a mental representation of something when it is beyond sense experience. Imagination is also associated with creativity, problem-solving and originality, all key skills of scientists.

The wave–particle duality of the electron illustrates a general knowledge issue. Two sources of knowledge can often give conflicting results and it is for us to see if the conflict is real or a limitation of human knowledge.

E.J. Corey, who was awarded the Nobel Prize for his work on retrosynthetic analysis, also stressed the role of imagination in synthetic chemistry. The retrosynthetic approach starts with the target molecule and works backwards by focusing on the functional group to identify possible precursors leading to a possible starting material.

'The synthetic chemist is more than a logician and strategist; he is an explorer strongly influenced to speculate, to imagine and even to create.'

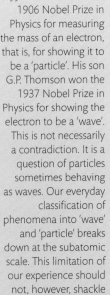

J.J. Thomson won the 1906 Nobel Prize in Physics for measuring the mass of an electron, that is, for showing it to be a 'particle'. His son G.P. Thomson won the 1937 Nobel Prize in Physics for showing the electron to be a 'wave'. This is not necessarily a contradiction. It is a question of particles sometimes behaving as waves. Our everyday classification of phenomena into 'wave' and 'particle' breaks down at the subatomic scale. This limitation of our experience should not, however, shackle our understanding of the subatomic world.

Our imagination can help us to solve problems and develop original ideas. We may use imagination to make connections between separate ideas, to build models or to create theories. But to what extent should we trust our imagination, given that it is derived in the mind rather than through sense perception? Can imagination reveal truths that reality hides? What other examples of discoveries in chemistry illustrate the importance of imagination?

Science and pseudoscience: homeopathy

If you have ever dropped a Mentos candy into a bottle of a carbonated drink you will have noticed that it fizzes with an eruption of foam. Can you explain this? If someone suggested that it was because of an 'evil demon' hidden in the drink responding angrily to being assaulted by a solid lump, you would probably reject it as a non-scientific explanation. But what makes this non-scientific? A scientific theory must be testable. It is difficult to imagine an experiment which could be designed to prove the 'evil demon' theory *false*. A theory that can be used to explain everything explains nothing.

There are a number of activities that are claimed by those who practice them to be scientific, such as astrology, homeopathy, and crystology (crystal therapy). The label 'science' is used to add authority to the claims, but how do we distinguish a genuine science from a fake or **pseudoscience**? It is the methods used in the natural sciences that are the key distinguishing factor. What is it about these methods which makes the knowledge they generate reliable?

Homeopathy is an alternative therapy that aims to treat diseases by giving extremely dilute doses of compounds that cause the same symptoms as the disease. The more dilute the dosage, the more powerful the remedy. Some doses are so dilute that it is likely they do not contain a single molecule of the active substance. Practitioners attribute this to the 'memory' of the water in which they have been diluted. Conventional science has subjected these claims to intense scrutiny, but firm evidence of anything other than the placebo effect has not yet been found.

Ethics

Progress in science and technology can profoundly affect our lives. Often these effects are positive – medicines have made our lives safer, and the development of new materials has made our lives more comfortable – but technological developments can also bring suffering and injustice. Industry and technology have had a negative impact on the environment. How do we decide what is the right and wrong use of science? These are ethical questions. There is often disagreement about what is 'right', and ideas of what is considered to be acceptable often change over time. Ethical issues raise difficult questions about risk versus benefits. In many cases, the science is so new that judging long-term benefits and risks is difficult. Developments in chemistry can create new ethical issues. There are also more direct concerns about how scientists should conduct their work. Scientists generally work in communities, not in isolation, and are expected to report their results honestly and openly in an atmosphere of trust. How would you respond if a student you were working with wanted to 'make up' their experimental results? What are your responsibilities in such a situation? The IB's mission statement and academic honesty policy is clear about the active role students have on their own learning. Your work should be authentic and genuine.

▲
Homeopath preparing herbal remedies.

Science is, however, a human endeavour and scientists, like the rest of humanity, can be motivated by envy, vanity and ambition.

Is science, or should it be, **value**-free? Some would argue that it is the aim of science to describe the world as it is and not as it should be.

Dioxin pollution can be produced by the incomplete combustion of plastics. Dioxin toxicity research has involved tests on rats. Are such tests ethically justified?

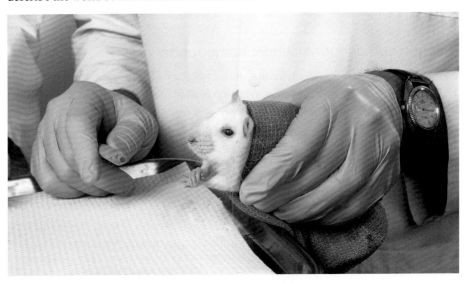

Some people do, however, fear the consequence of scientists 'interfering with nature'. Who decides what is and what is not a legitimate area of scientific research? The scientific community or society at large?

What are the responsibilities of the scientific community in controlling the quality of knowledge? It was originally believed, for example, that the medicine thalidomide was so safe that it was prescribed to pregnant women to control morning sickness, but this led to tragic consequences as the drug can cause many malformations in the fetus. Who is responsible when things go wrong?

Individual chemists have to take some responsibility. They must use their knowledge responsibly and consider the full implications of their work. There are ethical considerations that limit the scope of inquiry.

Chemistry and the optional themes

The TOK optional themes are focused on issues beyond the school experience that have a huge impact on the world today and a particularly key role in shaping your perspectives and identity. Some possible links can be made between the optional themes and what you study in chemistry. We outline some examples here.

Knowledge and technology

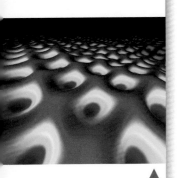

This picture is obtained by firing a beam of polarized ^{3}He atoms towards a metal surface. The metal can be seen as a lattice of positive metal ions (blue) in a sea of electrons (red).

Chemistry deals with what is acquired by the senses, and enhanced, if necessary, by technology. We are now able to see things which are beyond the direct limits of our senses. Look at the photograph (right). Is this what a metal surface really looks like?

Does technology merely extend the range of the human senses, or does it give a new way of seeing the world?

Technology has also had a big impact on experimental work. You will have used data loggers to collect and analyze experimental data and spreadsheets to analyze and present results. You may have also made measurements with new instruments that would not have been possible a few years ago. You will have used the internet to collaborate with other students, possibly from different schools in different parts of the world. How are the questions you explore limited by the technology that is available to you?

How have these technologies shaped your knowledge? Will the chemical knowledge of future IB Diploma Programme students progress as technologies continue to develop?

Robots are already used to perform some experimental procedures. Can a machine have procedural knowledge in the same way as an experienced laboratory chemist? Could these practical skills be one day redundant and not worth learning?

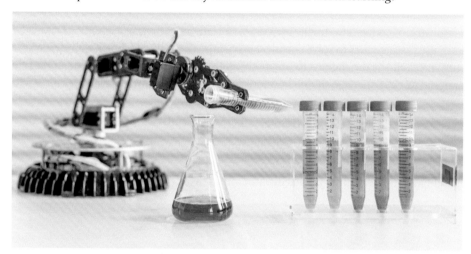

Robotic arm pouring a chemical into a flask.

Part of the final assessment of your chemical knowledge includes a multiple-choice paper where you are asked to choose the best answer from a selection of four answers. This is marked by a computer which has the answers stored as data. The computer does not know the answer in the same way that you hopefully will! In this assessment, your knowledge is demonstrated by your response to a list of questions.
With developments in Artificial Intelligence, it may not be too long before computers can be programmed to perform equally well in the different assessment tasks. Would this make the knowledge of the computer in some way equivalent to yours? In what sense, if any, can a machine be said to 'know' something?

Knowledge and language

The language of alchemy was cryptic and secretive because the knowledge it communicated was thought to be too powerful to share with the general public. The names of many chemicals were derived from their natural origin and were not related to their composition. The language of modern chemistry, by contrast, is precise, and is an effective tool for thought. When asked to draw the different isomers of C_7H_{16} it is very easy to draw the same structure twice by mistake. The IUPAC nomenclature, however, allows you to distinguish between the different isomers. They have different structures if they have different names.

The nine isomers of C_7H_{16}. The names help you to distinguish the different structures.

heptane	2-methylhexane	3-methylhexane
2,2-dimethylpentane	2,3-dimethylpentane	2,4-dimethylpentane
3,3-dimethylpentane	3-ethylpentane	2,2,3-trimethylbutane

The use of oxidation states has also allowed us to develop a systematic nomenclature for naming inorganic substances. The IUPAC aims to provide scientists with a common language as it saves time and resources. How does language allow humans to pool resources and share knowledge?

Today if a scientist is going to coin a new term, or publish a new discovery, it is most probably in English. Back in the late 19th century, however, the dominant language in chemistry was German and learning the language was often required to study chemistry at university. The change from German to English was influenced by the politics of the times. What role does language play in allowing knowledge to be shared with future generations?

Chemistry, of course, also has its own universal language. Balanced equations allow us to use mathematics to solve chemical problems. The ability to attach numbers to substances allows the chemist to use mathematics as a precise tool to investigate the material world. Language should be a tool and not an obstacle to knowledge.

Knowledge and politics

Chemistry has applications in many important areas of our lives such as food, energy security, environment, defence and health, and so can have an impact on politics. Developments in scientific knowledge can often trigger political controversies.

Cans of genetically modified tomato puree. Different countries have adopted very different approaches to the marketing of GM foods. In the European Union (EU), strict rules apply to the labelling of GM foods, while in the USA and Canada there is no mandatory labelling of GM content. It has been said that the EU is adopting the 'precautionary principle' in this regard, and this raises several concerns for international trade.

Knowledge and religion

There have been many historical conflicts between science and religion but are they necessarily irreconcilable? The chemist Peter Atkins argues that there is no evidence and no reason to have religious belief. He argues that science and religion are incompatible, and science can answer the questions that religion has traditionally answered.

To others, however, science and religion are complementary areas of knowledge which are concerned with different spheres of experience, asking different questions, and justified in different ways. According to Einstein: science is concerned with *what is* whereas religion with *what should be*.

> '*Science without religion is lame, religion without science is blind.*'
>
> Albert Einstein

To what extent do scientific developments have the power to influence thinking about religion?

Knowledge and indigenous societies

Many modern medicines have primitive origins based on folklore. In 400 BCE Hippocrates recommended a solution derived from a willow leaf as a treatment for the pain of childbirth, and the bark of the willow tree was later used as a treatment for fever. As chemical techniques advanced, the active chemical salicylic acid was extracted; its structure was modified, and the analgesic aspirin synthesized.

The traditional knowledge of many indigenous societies is rooted in their long-standing and close relationships with the natural world. It is more dependent on the direct use of the senses than modern science. The French anthropologist Claude Lévi-Strauss, in his essay on The Science of the Concrete, contrasts indigenous knowledge with chemistry by giving an example of how similar discoveries can be made but in different ways. The fragrances of lavender, wintergreen and bananas have a common feature identified by indigenous societies and so they group them together. Analytical chemistry related this feature to them all being esters. Similarly, wild cherries, cinnamon, vanilla and sherry, which are grouped together by indigenous societies by their fragrances, are all identified as aldehydes by the analytical chemist.

In this case the indigenous knowledge of the plants is related to a sense of smell and a basic process of categorization which, according to Lévi-Strauss, is more 'concrete' in nature than chemistry. This suggests that indigenous knowledge is very highly empirical and pragmatic in nature.

Chemistry and TOK assessment

TOK in Chemistry assessment

In your chemistry exams the questions use different command terms which indicate the depth of treatment required. 'Describe' and 'outline' questions require less depth of knowledge than 'explain' and 'predict' questions.

Is there a categorical distinction between a description and an explanation?

Chemistry in TOK assessment

The TOK exhibition

The TOK exhibition explores how TOK manifests itself in the world around you. You are required to create an exhibition of three objects that connect to **one** of the 35 'IA prompts' in the TOK guide. You are encouraged to choose objects of personal interest that you have come across in your school studies and/or your lives beyond the classroom.

An extremely wide variety of objects are suitable for use in a TOK exhibition. For example, the pictures of experimental equipment, molecular models or the periodic table discussed in this chapter, as well as many of the illustrations in the text or the textbook itself, are all objects of interest. It is, however, strongly recommended that you base your exhibition on one of the TOK themes, either the core theme or one of the optional themes and not areas of knowledge such as the natural sciences.

The TOK essay

In your TOK essay, you are expected to make connections between the knowledge question and your own experiences as a learner. It is helpful to support your argument with examples drawn from your IB diploma courses as well as from other sources. This includes your chemistry class.

Some examples of prescribed essay titles for you to consider.

- Do good explanations have to be true? (May 2019)
- To produce knowledge just observe and then write down what you observe. Discuss the effectiveness of this strategy in two areas of knowledge. (Nov 2019)
- Is there a trade-off between skepticism and successful production of knowledge? (Nov 2019)
- 'There is a sharp line between describing something and offering an explanation of it.' To what extent do you agree with this claim? (May 2020)
- 'Present knowledge is wholly dependent on past knowledge.' Discuss this claim with reference to two areas of knowledge. (May 2020)
- 'Reliable knowledge can lack certainty.' Explore this claim with reference to two areas of knowledge. (Nov 2020)
- 'If all knowledge is provisional, when can we have confidence in what we claim to know?' Answer with reference to two areas of knowledge. (Nov 2021)
- 'We are rarely completely certain, but we are frequently certain enough.' Discuss this statement with reference to two areas of knowledge. (Nov 2021)

Internal Assessment

Purpose

During your two-year IB Diploma Programme, you are expected to carry out a **scientific investigation.** This must be written up as a full report and is an integral part of your final assessment for the course.

Your investigation will be based on a topic of your own interest, and have a focused research question, which is rationalized through appropriate scientific context. Your topic may be selected individually or through collaboration within a group of no more than three students (yourself included). Your approach and methodology can rely on the collection and analysis of primary data through experimental work, or the analysis of secondary data from a database or from the use of an open-ended simulation. Your investigation may also use a mix of approaches and data sources.

In all cases, the investigation will be assessed according to the same **four** assessment criteria which are summarized below.

Note the following general points:

- The scientific investigation may be completed individually or collaboratively. Details on group work are presented later in this chapter.
- The scientific investigation contributes 20% towards your final chemistry grade.
- Your written report will be marked internally by your teacher who will allocate a predicted grade.
- A sample of the reports from your class will be submitted to the IB for moderation and standardization to ensure fairness in applying the standards across all candidates. The mark awarded upon moderation may be different from the mark predicted by your teacher.
- The scientific investigation should take approximately 10 hours to complete including any preparatory work, lab work and write up.
- The report has an upper word-limit of 3000 words.
- The following words are not included in the word count:
  - charts and diagrams
  - data tables
  - equations, formulae and calculations
  - citations/references
  - bibliography
  - headers
- The topic selected for the scientific investigation can be based on the course content, or it can be on extension material beyond the specifications of the chemistry subject guide.
- It is recommended that you are proactive and initiate a discussion with your teacher early on regarding your scientific investigation ideas, and always seek to obtain advice and information from teachers and peers.

Academic honesty

There is a strong expectation from IB candidates to show the highest level of academic integrity and academic honesty. Under no circumstances is plagiarism and/or collusion in your scientific investigation or any other piece of work submitted to the IB for assessment acceptable. It is encouraged that an appropriate web-based plagiarism detection engine is used prior to submitting work.

Appropriate referencing to sourced information used in the scientific investigation report is *required*. You can follow any of the accepted referencing styles.

Remember that, as stated in the IB chemistry subject guide, *omitted or improper referencing will be considered as academic malpractice* and flagged for investigation as a potential breach of regulations.

The assessment criteria

The scientific investigation will be marked using a best-fit approach to match the level of work with the descriptors given at each mark band. Each criterion is marked separately and contributes to the final mark of 24 (later scaled to 20% of overall grade) as shown below:

Criterion	Maximum number of marks available	Weighting (%)
Research design	6	25
Data analysis	6	25
Conclusion	6	25
Evaluation	6	25
Total	**24**	**100**

The descriptors of each criterion are presented below. Guidance *in addition to* what is already presented in the chemistry subject guide is given. You are strongly recommended to focus on the highest mark-band level descriptors each time – there is no point in aiming for less!

Research design

Marks	Level descriptor
0	The report does not reach the standard described by the descriptors below.
1–2	• The research question is stated without context. • Methodological considerations associated with collecting data relevant to the research question are stated. • The description of the methodology for collecting or selecting data lacks the detail to allow for the investigation to be reproduced.
3–4	• The research question is outlined within a broad context. • Methodological considerations associated with collecting relevant and sufficient data to answer the research question are described. • The description of the methodology for collecting or selecting data allows for the investigation to be reproduced with few ambiguities or omissions.
5–6	• The research question is described within a specific and appropriate context. • Methodological considerations associated with collecting relevant and sufficient data to answer the research question are explained. • The description of the methodology for collecting or selecting data allows for the investigation to be reproduced.

Checklist

A downloadable copy of this checklist is available in the eBook.

✓ My research question investigates the relationship between two correlated variables.

✓ I have provided a justification for the relationship between the dependent and independent variables hypothesized using appropriate scientific context.

✓ I have provided additional background which is focused and deepens the understanding of the investigated research question.

✓ I have clearly stated and defined the variables (dependent/independent/controlled) involved in my investigation.

✓ My choice of investigation methodology is justified. I have explained why the methodology chosen is appropriate to answer the research question within specific parameters.

✓ I have highlighted and explained any decisions taken in the design of a new (or the modification of an existing) methodology.

✓ I have explained and justified key steps of the methodology and I have ensured that my report is clear on why specific conditions are used in the methodology and what the importance of these conditions are.

✓ I have assessed and reported any associated Health and Safety considerations including Ethical and Environmental risks, which may not be immediately obvious.

✓ My methodology is clearly stated. Another student may potentially repeat the work by following the methodology that I have described.

✓ I have reviewed the methodology and any unnecessary steps or repetition of steps have been removed.

✓ My methodology allows for the collection of enough data.

✓ My presented methodology is an honest representation of the work undertaken in the lab, stating the procedure I used and not the one that I should have used in retrospect.

✓ My methodology is internally consistent with the work presented in the report.

Data analysis

Marks	Level descriptor
0	The report does not reach the standard described by the descriptors below.
1–2	• The recording and processing of the data is communicated but is neither clear nor precise. • The recording and processing of data shows limited evidence of the consideration of uncertainty. • Some processing of data relevant to addressing the research question is carried out but with major omissions, inaccuracies or inconsistencies.
3–4	• The communication of the recording and processing of the data is either clear or precise. • The recording and processing of data shows evidence of a consideration of uncertainties but with some significant omissions or inaccuracies. • The processing of data relevant to addressing the research question is carried out but with some significant omissions, inaccuracies or inconsistencies.
5–6	• The communication of the recording and processing of the data is both clear and precise. • The recording and processing of data shows evidence of an appropriate consideration of uncertainties. • The processing of data relevant to addressing the research question is carried out appropriately and accurately.

Checklist

A downloadable copy of this checklist is available in the eBook.

✓ My qualitative and quantitative data are presented appropriately.

✓ The origin of the presented data is clear.

✓ Tables, graphs, figures and schemes are clearly labelled, and in-text citations are used when appropriate to help the reviewer follow my work and thinking process.

✓ My data are presented in a consistent manner following SI conventions.

✓ My data-analysis protocol is clearly shown. I have provided an example of my calculations.

✓ I have taken into consideration the uncertainties in my measurements and the error propagation in the various mathematical operations used.

✓ My data analysis protocol is relevant, and the produced results are appropriate to address the research question.

✓ I have presented/visualized my results with appropriate graphs that enhance the reader's understanding of my work and are relevant in addressing my research question.

Conclusion

Marks	Level descriptor
0	The report does not reach the standard described by the descriptors below.
1–2	• A conclusion is stated that is relevant to the research question but is not supported by the analysis presented. • The conclusion makes superficial comparison to the accepted scientific context.
3–4	• A conclusion is described that is relevant to the research question but is not fully consistent with the analysis presented. • A conclusion is described and makes some relevant comparison to the accepted scientific context.
5–6	• A conclusion is justified that is relevant to the research question and fully consistent with the analysis presented. • A conclusion is justified through relevant comparison to the accepted scientific context.

Checklist

A downloadable copy of this checklist is available in the eBook.

✓ My conclusion directly addresses the research question and is justified by the data and the data-analysis I have presented in the report.

✓ I have correctly interpreted my processed data, considering the associated uncertainties and their impact on the conclusion.

✓ I have compared my conclusion to the established scientific context, such as journal articles, science textbooks, encyclopedia articles or other peer-reviewed sources. I have considered the extent of my agreement or disagreement with the established scientific understanding and I have explored the reasons that may have contributed to my conclusion.

This point is critical as it will enable you to appropriately evaluate your work!

Evaluation

Marks	Level descriptor
0	• The report does not reach the standard described by the descriptors below.
1–2	• The report states generic methodological weaknesses or limitations. • Realistic improvements to the investigation are stated.
3–4	• The report describes specific methodological weaknesses or limitations. • Realistic improvements to the investigation, that are relevant to the identified weaknesses or limitations, are described.
5–6	• The report explains the relative impact of specific methodological weaknesses or limitations. • Realistic improvements to the investigation, that are relevant to the identified weaknesses or limitations, are explained.

Checklist

A downloadable copy of this checklist is available in the eBook.

✓ I have considered the **weaknesses** related to the control of variables, the choice of equipment, the choice of methodology and any choices regarding the data-analysis processes used. I have considered how these choices affected my conclusion. I have considered whether I would have made different choices given the opportunity.

✓ I have considered the **limitations** of my investigation. I took into consideration any assumptions made in the data-analysis, the range of the collected data, and the boundaries of the system investigated. I have considered the extent to which the obtained results are applicable.

✓ Considering the above, I have proposed *realistic* changes that could lead to more accurate results and/or extend the range and/or scope of the investigation.

Group work

The table below distinguishes between individual and collaborative work with regards to the scientific investigation.

Individual work	**Student works individually**	Students formulate and investigate their individual research question.
		Students develop their own methodology to answer their research question.
		A student cannot present the same raw dataset as another student.
Collaborative work	**Small group (no more than 3 students in a group)**	Students collaborate in selecting a topic of investigation.
		Each student formulates and investigates their individual research question by choosing:
		A A different independent variable than the other members of the group -or-
		B The same independent variable but with a dependent variable different than the one considered by the rest of the group -or-
		C A different part of the dataset to that used by the other members of the group from within a larger data set that the group has collected together.
		Each student submits their own individual report, clearly indicating that the investigation was developed through collaboration in a group.
		A student cannot present the same raw dataset as another student within the group.
	Class collaboration	A school or a class may participate or undertake a large-scale activity collecting data to generate a database.
		A student may participate in adding data to the database by using standardized protocols that they did not develop themselves.
		The student may then use the database to generate data for a scientific investigation that is treated as a scientific investigation that utilized secondary data from an external source.
		A student cannot present the same raw dataset as another student.

If you decide to work in a group, you still have to develop your own research question within the group-selected topic (see table above). While working in a group you may collaborate in developing a methodology or a standardized protocol that will allow for the collection of different sets of raw data that can address the needs of each group member's research question.

You may help, and receive help from, other group members in collecting the data and running the experiment(s), but in the end, you need to produce and submit your own individual report according to the criteria discussed on the previous page.

Types of investigations

Below you are presented with some examples of types of investigations and some important factors you need to consider when planning your scientific investigation.

Hands-on practical laboratory work

Hands-on laboratory work is the most traditional type of investigation. When planning for hands-on laboratory work, you need to ensure that your school:

- has the chemical materials needed
- has the required equipment available and in good shape
- and that the physical space and time needed will be available to you.

Time management is essential as practical work can be delayed by unexpected experimental procedure failures, for example, broken equipment, shipping delays in materials or other procedural challenges.

Fieldwork

Fieldwork has similar constraints as hands-on laboratory work with the additional requirement of collecting data in the field prior to analyzing/processing it in a lab. You need to ensure that you have collected more than enough data in the field before returning to the lab, as multiple visits to perform fieldwork may be unfeasible and, in some cases, reduce the validity of your work, as environmental conditions will vary in each visit.

Use of spreadsheet for analysis and modelling

The use of spreadsheets for analysis and modelling is becoming more and more popular and allows for faster processing of extended datasets. You need to collect your own primary data and then process them with the use of a spreadsheet or application.

Therefore, one of the challenges is that you need to be, or become, familiar with using a spreadsheet program such as Microsoft Excel® and its required syntax. Furthermore, when analyzing large datasets, you must be very careful in considering your personal bias in selecting the range and type of data to analyze.

Extraction and analysis of data from a database

The use of databases is very similar to the use of spreadsheets for analysis; the difference is that instead of generating your own primary data to use in the spreadsheet, you make use of secondary data found in publicly available databases. There are numerous databases available online that you can refer to, and examples are provided later in this chapter.

Use of a simulation

Simulations have also increased in popularity. As such, any open-ended simulation can be used to produce primary data that can be then analyzed as part of your scientific investigation. The challenge here is to find a good open-ended simulation that will enable you to design your own experiment rather than follow a set of instructions that will lead to a pre-defined result.

Developing a research question

All journeys start with a first step. In your scientific investigation journey, your first step is developing your own research question. This can be challenging at the beginning and your ideas may be rejected by your teacher, either because they are not open-ended enough, they have been overdone in the past or, simply, your school does not have the equipment or resources necessary for the work you are proposing.

Remember that your work does not need to be groundbreaking or world-changing, even though it could be. You also need to consider that you only have 10 hours to research, experiment and write up your report for your scientific investigation, so be realistic in your goal setting.

In general, your scientific investigation should be looking to investigate the relationship between two correlated variables within a system. Your research question can be simply stated as:

How does the change of an <independent variable> in <specific system> affect a <dependent variable>?

Hulley *et al.* (2007) have suggested a set of criteria (FINER) that should be considered when developing your research question. These are presented below, adapted to the requirements of the scientific investigation in IB chemistry.

Feasible	You can address the question within the allowed time.
	The required equipment and materials are available.
	You have, or you will be able to develop, the required technical skills to be successful in researching your question.
Interesting	Your research question is open-ended, and it is not a simple yes or no question.
	The research question is of personal interest and will keep you, as well as your peers and eventually the moderators, intrigued and interested.
Novel	Your research question confirms, refutes, or extends previous findings.
Ethical	You have considered significant ethical, environmental and health concerns related to the research question posed.
Relevant	Your research question is relevant to chemistry.

Depending on the type of your investigation, your research question may be developed in slightly different ways.

Manipulation of an independent variable

Consider the following:

You have a well-defined chemical system where a causal relationship between two variables is obvious.

You develop a protocol that can test and establish the mathematical relationship between the dependent and independent variables.

Example research question:

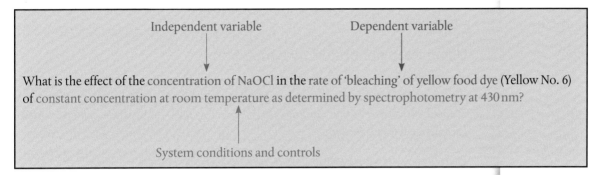

When reviewing the above example research question, the dependent and independent variables are immediately obvious. The system investigated, as well as the specific conditions that focus the scope of the investigation, are also stated.

Selecting variables through fieldwork

Consider the following:

You want to investigate how a dependent variable is affected by manipulating an independent variable on samples collected in the field under specific conditions.

Example research question:

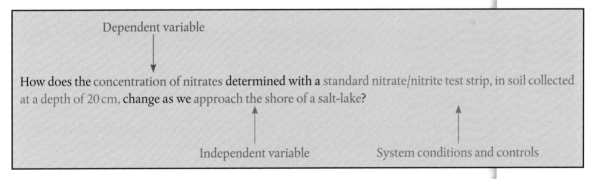

In the example above, the variable is the concentration of nitrates in a solid sample that is collected at specific distances from the shore of a salt lake. The system of investigation and the range of the scope of the investigation are also clearly stated.

Selecting data from databases

Analyzing topics from databases such as (but not limited to) those from EPA, European Environment Agency, United Nations Environment Programme, Database of Environmental Education, NETROnline, World Health Organization, EMA, CDC, ECDC, NASA, PubMed, NIST, ChemExpert etc., can allow for various research questions to be developed.

See a comprehensive list of databases by following the links on the right.

As you are using secondary data and you are not generating your own through experimentation, these scientific investigations are expected to have a large dataset and rely heavily on the mathematical analysis of said dataset.

The use of spreadsheets and/or models in the analysis of these data sets is highly encouraged.

Using data generated by simulations

The use of simulations to generate data is approached in a similar fashion to databases. Predictive software such as MOPAC, WebMO, GAMESS, SAPT, Spartan Student, Molden or Tinker (among others) as well as PhET simulations can be adapted to generate data for this use.

Example research question:

> What are the trends observed in the dissociation enthalpy of the O–H bond of straight-chain alcohols with no side-chain substituents in respect to their increasing mass, as reported by the National Standard Reference Data Series, National Bureau of Standards, no. 31, Washington, 1970?

In the example above, the variable is the dissociation enthalpy of the O–H bond in a well-focused area of investigation, that of straight-chain alcohols with no side-chain substituents. Notice how the scope and the range of the investigation are very clearly stated. The question is also very clear that it is a data-based investigation, as the database of reference is stated at the end.

When utilizing databases, the sheer amount of available data can obscure the focus and the scope of your investigation and easily take you on a tangent from what you are investigating. It is, therefore, extremely important to limit the range of your investigation and set a very clear boundary on the extent of your investigation. By doing so, you will allow yourself to run a robust analysis of the dataset that can lead to a valid conclusion relevant to your research question.

If you expand the boundary of your investigation too much, then you may fall prey to several *unknown unknowns* – or variables whose influence on your hypothesis you are unaware of – that can and will affect the validity of your conclusion.

Data paralysis: in the words of Winston Churchill: 'too much analysis can lead to paralysis'. It is of high importance to narrow the scope and range of your scientific investigation, no matter the kind of investigation you have decided to pursue. It can be very tempting to do more experiments, collect additional data or explore further avenues of investigation; but there is simply not enough time to do so well. It will eventually become difficult to separate meaningful data from irrelevant data that 'muddy' your analysis, prohibiting you from reaching a meaningful conclusion to your question. Focus on the question at hand instead, and any ideas or observations that you may have had during the investigation that could have been worthy of their own investigation can be included at the end of your work under 'Further Investigation'. It may also help your teacher in designing new experiments or providing ideas to new students the year after.

Bibliography

Hulley, S., Cummings, S., Browner, W., Grady, D. and Newman, T. (2007). *Designing clinical research, 3rd ed*. Philadelphia (PA): Lippincott Williams and Wilkins.

Skills in the study of chemistry

Central to learning chemistry is the application of techniques and gaining experimentation skills. This course is designed to provide a number of activities to enrich your understanding of chemical concepts through experimentation and inquiry. These skills are integrated into the learning throughout the course, as illustrated in the tables that follow.

Tools

Tool 1: Experimental techniques

Chemistry is an experimental science. Through the ages, chemistry emerged from the domain of craftspeople and artisans such as smelters, metallurgists, painters, potters and bakers who were the initial lore-masters of chemistry. Understandings in the fundamentals of chemistry were developed from their experimental knowledge and observations. Technological advancement has allowed the development of increasingly sophisticated instrumentation that enables scientists to collect ever more accurate data and make highly detailed observations to further refine our chemistry models and understandings.

Skill	Description	Section references
Addressing safety of self, others and the environment In the duration of the course, you will be involved in many hands-on and other experimental activities. When in the lab, you must always be aware of the potential risks and hazards associated with the work at hand and always comply with related safety regulations. These risks may be **chemical** due to the nature of the materials used, or **environmental** due to the potential harm that the direct disposal of the materials used, as well as the products formed, can have on the environment and on health. Besides these obvious risks and hazards, an investigation may have ethical implications which may not be as obvious but still need to be taken into consideration.	*Recognize and address relevant safety, ethical or environmental issues in an investigation.*	S1.1.1 S1.4.4 S1.4.5 S1.5.4 S2.1.3 S2.2.9 S2.2.10 S2.3.1 R1.1.2 R1.1.4 R2.1.3 R2.2.1 R2.2.3 R3.1.4 R3.1.6 R3.1.7 R3.1.8 R3.2.2 R3.2.3

You should consider the environmental effects of your work in the laboratory and learn safe protocols for the storage, use and disposal of chemicals. Micro-scale chemistry and Green Chemistry can be considered here.

The best sources of information for carrying out risk assessments for the use of chemicals are material safety data sheets (MSDS), which should be available whenever you are handling chemicals. The Laboratory Safety Institute (LSI) has developed useful guidelines on laboratory practice.

SKILLS

Measuring variables	*Understand how to accurately measure with an appropriate level of precision:*	
Measurement is an important part of chemistry. In the laboratory, you will use different measuring equipment and there will be times when you will need to select the instrument that is most appropriate for your task from a range of possibilities. The conclusions reached from a scientific investigation are only as good as the data collected and the quality of the analysis that follows. For these reasons it is imperative to understand how to accurately measure variables to ensure high quality data collection and analysis.	• mass	S1.4.4 S1.4.5 R1.1.4 R2.1.3 R2.2.1 R2.2.3
	• volume	S1.4.5 S1.5.4 R2.1.3 R2.2.1 R2.2.3
	• time	R2.2.1 R2.2.3
	• temperature	S1.5.4 R1.1.2 R1.1.4 R2.2.1 R2.2.3
	• length	S2.2.2
	• pH of solution	R2.2.1 R2.2.3 R3.1.4 R3.1.6 R3.1.8
	• electric current	S2.1.3 S2.2.9 S2.3.1 R2.2.1 R3.1.6
	• electric potential difference	R2.2.3 R3.2.3
Related activities	• Reading an SDS • Common laboratory equipment • Treatment of waste • Measuring mass and volume • Use of a pH meter and universal indicator to measure pH	

Applying techniques	Show awareness of the purpose and practice of:	
There are some techniques that are considered fundamental for any chemistry student. Extended practice and familiarization with these techniques are essential for students in chemistry designing their scientific investigation.	• preparing a standard solution	S1.4.5 S2.2.9 R2.2.1
	• carrying out dilutions	S1.4.5
	• drying to constant mass	S1.4.4
	• distillation and reflux	S1.1.1 R3.1.7 S2.2.9
	• paper or thin layer chromatography	S1.1.1 S2.2.10
	• separation of mixtures	S1.1.1 R3.1.7
	• calorimetry	R1.1.2 R1.1.4 R2.2.1
	• acid–base and redox titration	R2.2.1 R2.2.3 R3.1.4 R3.1.6 R3.1.8 R3.2.2 R3.2.3
	• electrochemical cells	R3.2.3
	• colorimetry or spectrophotometry	R2.2.3
	• physical and digital molecular modelling	S1.5.4 S2.1.3 S2.2.9
	• recrystallization	S.1.1.1
	• melting point determination	S1.1.1 S2.2.9

It is not expected that you will have individual access to all of the equipment specified. During the course you should gain experience of carrying out these processes in small groups, or of using simulations, or observations of demonstrations as needed.

Related activities	
	• filtering and separating heterogeneous mixtures
	• preparing solutions
	• preparing a calibration curve for spectrophotometric determinations
	• thin-layer chromatography (TLC) in a sample of paracetamol
	• setting up a reflux and distillation apparatus
	• preparation of AgI(s) precipitate from aqueous solutions
	• sublimation of iron and iodine
	• the heat curve of water
	• preparation and standardization of aqueous solutions
	• racing compounds on TLC

Tool 2: Technology

Chemistry does not only happen on the bench by heating and mixing solutions. A lot of modern chemistry, especially research related to the evaluation of theoretical models, identification of trends in large data sets and/or predictions from large data sets, takes place behind a computer screen. Computational chemistry is a well-established field that finds applications in both theoretical and applied chemistry.

In general, when we want to generate data from models and simulations, the logical path employed is that shown in Figure 1.

Skill	Description	Section references
Applying digital technology to collect data Early analysts relied on their senses and their skills in collecting data but continuous improvement in instrumentation and technology has not only allowed us to explore the world beyond the constraints of our senses but also to collect large datasets in short time with relative ease. Modelling and manipulating data Molecular modelling and prediction of properties	Use sensors.	S1.5.3 R1.1.2 R3.1.4 R3.1.8
	Identify and extract data from databases.	S1.5.4 S2.1.3 S3.1.4
	Generate data from models and simulations.	S1.5.4 S3.1.4 S3.2.3

Applying technology to process data	Use spreadsheets to manipulate data.	S1.5.3
Digital technology can be used not only to generate but also to process data. Whether data are generated/collected digitally or in traditional investigations, they can be analyzed with the help of computers through spreadsheets or more specialized analysis/modelling software.	Represent data in a graphical form.	S1.5.3 S3.1.4 R3.1.4
	Use computer modelling.	S1.5.3

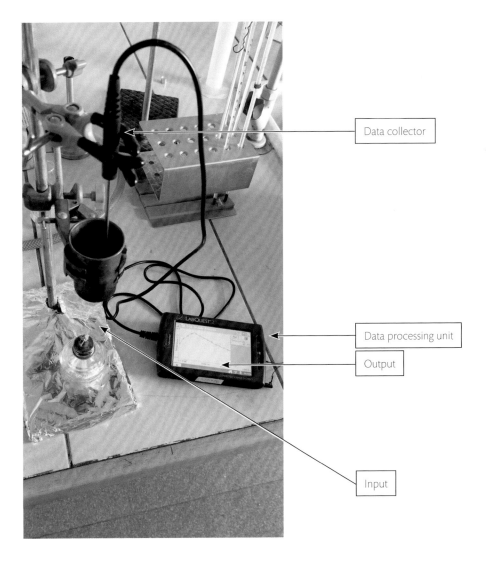

Figure 1 Experimental set-up for calorimetry using a temperature probe for data collection.

Tool 3: Mathematics

It has been stated that a chemist is *someone who can solve a problem you did not know you have.* Chemistry problems are usually of low mathematical complexity in the sense that you do not need advanced mathematics or calculus to solve them, but can rely on simple mathematical operations as itemized below.

The challenging part in chemistry problem solving is, more often than not, understanding what the question is asking and having a good grasp of the underlying concepts at play. Problem-solving strategies will be presented in each topic to help you master this skill.

Skill	Description	Section references
Applying general mathematics Data is the lifeblood of scientists. It may be qualitative, obtained from observations, or quantitative, collected from measurements. Quantitative data are generally more reliable as they can be analyzed mathematically but nonetheless will still suffer from unavoidable uncertainties. This analysis helps identify significant relationships and eliminate outliers. Scientists look for relationships between the key factors in their investigations. The errors and uncertainties in the data must be considered when assessing the reliability of the data. Errors and uncertainties in chemistry	Use basic arithmetic and algebraic calculations to solve problems.	S1.1.1 S1.4.4 S1.4.5 S1.5.3 R1.1.4 R2.2.1 R2.2.3 R3.1.4 R3.1.8
	Carry out calculations involving decimals, fractions, percentages, ratios, reciprocals, exponents.	S1.4.4
	Carry out calculations involving logarithmic functions.	R3.1.4
	Determine rates of change from tabulated data.	R2.2.1
	Calculate mean and range.	Errors and uncertainties
	Use and interpret scientific notation (e.g. 3.5×10^6).	S1.3.6 R3.1.4
	Use approximation and estimation.	S1.4.4
	Appreciate when some effects can be ignored and why this is useful.	TOK – What are the implications of our experimental choices?
	Compare and quote values to the nearest order of magnitude.	Errors and uncertainties
	Understand direct and inverse proportionality, as well as positive and negative correlations between variables.	S1.5.3
	Calculate and interpret percentage change and percentage difference.	Errors and uncertainties
	Calculate and interpret percentage error and percentage uncertainty.	Errors and uncertainties
	Distinguish between continuous and discrete variables.	R2.2.1

Using units, symbols and numerical values	Apply and use SI prefixes and units.	S1.5.3 R1.1.4
The measurement of a physical property is always expressed as:	Identify and use symbols stated in the guide and the data booklet.	Errors and uncertainties
physical property = numerical value × unit	Express quantities and uncertainties to an appropriate number of significant figures or decimal places.	S1.3.6
Units are treated in the same way that any algebraic symbol is used in calculations.		
SI stands for *Système International d'Unités*, an internationally agreed system of units established by the International Bureau of Weights and Measures and endorsed by IUPAC.		
The system, which is metric in nature, has 7 base units from which all other units are derived.		
<table><tr><td>metre</td><td>m</td><td>length</td></tr><tr><td>kilogram</td><td>kg</td><td>mass</td></tr><tr><td>second</td><td>s</td><td>time</td></tr><tr><td>ampere</td><td>A</td><td>electric current</td></tr><tr><td>kelvin</td><td>K</td><td>temperature</td></tr><tr><td>mole</td><td>mol</td><td>amount of substance</td></tr><tr><td>candela</td><td>cd</td><td>luminous intensity</td></tr></table>		
This book contains some updates to the definitions of units agreed by IUPAC, now based on established and timeless universal constants.		
Processing uncertainties	Understand the significance of uncertainties in raw and processed data.	Errors and uncertainties
An uncertainty range applies to any experimental value. Some pieces of apparatus state the degree of uncertainty; in other cases, you will have to make a judgement. Appropriate consideration of the impact of measurement uncertainty on the analysis is a requirement in any experimental process.	Record uncertainties in measurements as a range (±) to an appropriate precision.	
	Propagate uncertainties in processed data, in calculations involving addition, subtraction, multiplication and division.	
The experimental error in a result is the difference between the recorded value and the generally accepted or literature value. Errors can be categorized as **random** or **systematic**.	Express measurement and processed uncertainties – absolute, fractional (relative), percentage – to an appropriate number of significant figures or level of precision.	
	Apply the coefficient of determination (R^2) to evaluate the fit of a trend line or curve.	

Graphical representations of data are widely used in areas such as finance and climate modelling, and can be the basis of predicted trends.

Graphing

Data collected from investigations are often presented in graphical form. This provides a pictorial representation of how one quantity is related to another. A graph is also a useful tool to assess errors as it identifies data points which do not fit the general trend and so gives another measure of the reliability of the data.

Sketch graphs, with labelled but unscaled axes, to qualitatively describe trends.	R2.2.1 R2.2.10	
Construct tables, charts and graphs for raw and processed data including bar charts, histograms, scatter graphs and line and curve graphs.	R2.2.3	
Plot linear and non-linear graphs showing the relationship between two variables with appropriate scales and axes.	S1.4.5 R2.2.1 R2.2.3	
Draw lines or curves of best fit.	R2.2.3	
Interpret features of graphs including gradient, changes in gradient, intercepts, maxima and minima, and areas.	S1.4.5 R2.2.1 R2.2.3 R3.1.8	
Draw and interpret uncertainty bars.	Errors and uncertainties	
Extrapolate and interpolate graphs.	S1.4.5 R2.2.1 R2.2.3 R3.1.8	

Related activities

Coffee-cup calorimeter

Comparing the enthalpy of isopropanol and ethanol

Finding the enthalpy change of hydration

Comparing the enthalpy of pentane and hexane.

Synthesis and purification of Aspirin

Acid–base titrations using a pH sensor – identifying the equivalence point

Determination of the level of rust in a rusted nail

Redox titrations

Inquiry process

The components of the Inquiry cycle and their links in the book are presented below, and they are individually discussed in more detail in the Internal Assessment chapter of this book.

Inquiry 1: Exploring and designing

Skill	Description	Section references
Exploring	Demonstrate independent thinking, initiative and insight.	R1.1.4 R2.1.3
	Consult a variety of sources.	
	Select sufficient and relevant sources of information.	
	Formulate research questions and hypotheses.	
	State and explain predictions using scientific understanding.	
Designing	Demonstrate creativity in the designing, implementation and presentation of the investigation.	
	Develop investigations that involve hands-on laboratory experiments, databases, simulations, modelling.	
	Identify and justify the choice of dependent, independent and control variables.	
	Justify the range, interval and quantity of measurements.	
	Design and explain a valid methodology.	
	Pilot methodologies.	
Controlling variables	*Appreciate when and how to:*	
	calibrate measuring apparatus including sensors	
	maintain constant environmental conditions of reaction systems	
	insulate against heat loss/gain	

 SKILLS

Laboratory work usually involves 'hands-on' experimental work, which leads to the collection of primary data. These are data obtained directly, such as by measuring changes in a variable during an experiment or by collecting information through surveys or observations. Secondary data refer to data obtained from another source, such as via reference material or third-party results. The source of secondary data must always be cited in a report. Analysis of secondary data can also form an important part of experimental work.

Inquiry 2: Collecting and processing data

Skill	Description	Section references
Collecting data	Identify and record relevant qualitative observations.	S1.3.2 S1.4.5
	Collect and record sufficient relevant quantitative data.	S1.5.4 S2.1.3
	Identify and address issues that arise during data collection.	S2.2.9 S2.3.1
Processing data	Carry out sufficient, relevant and accurate data processing.	S3.1.4 R1.1.2
Interpreting results	Interpret qualitative and quantitative data.	R1.1.4 R1.3.2
	Interpret diagrams, graphs, and charts.	R2.1.3 R2.2.1
	Identify, describe, and explain patterns, trends and relationships.	R2.2.10 R3.1.6
	Identify and justify the removal or inclusion of outliers in data (no mathematical treatment required).	R3.2.2 R3.2.3
	Assess accuracy, precision, reliability and validity.	

Inquiry 3: Concluding and evaluating

Skill	Description	Section references
Concluding	Interpret processed data and analysis to draw and justify conclusions.	R1.1.4 R2.1.3 R2.2.3
	Compare the outcome of an investigation to the accepted scientific context.	
	Relate the outcome of an investigation to the stated research question or hypothesis.	
	Discuss the impact of uncertainties on the conclusions.	
Evaluating	Evaluate hypotheses.	
	Identify and discuss sources and impacts of random and systematic errors.	
	Evaluate the implications of methodological weaknesses, limitations, and assumptions on conclusion.	
	Explain realistic and relevant improvements to an investigation.	

Strategies for success in IB chemistry

During the course

Take responsibility for your own learning. As you finish your study of a topic, it is a really good time to check back through the numbered Understandings for each chapter and make sure that you are comfortable with all the expectations.

Each Understanding is shown as a three-part box, for example:

Content statement →

Outcomes of learning and teaching →

Guidance on the coverage expected →

Structure 1.4.2 – Masses of atoms are compared on a scale relative to ^{12}C and are expressed as relative atomic mass (A_r) and relative formula mass (M_r)

Determine relative formula masses M_r from relative atomic masses A_r.

Relative atomic mass and relative formula mass have no units. The values of relative atomic masses given to two decimal places in the data booklet should be used in calculations.	Structure 3.1 – Atoms increase in mass as their position descends in the periodic table. What properties might be related to this trend?

Linking Questions

Spend extra time on parts where you are less confident of your knowledge and understanding. Use additional sources of information such as online simulations, videos, other books and journals to help spark your curiosity and grasp the wider contexts of the topic. The more you do, the more you will enjoy the course and the more successful you will be.

As you learn more about each topic, you should be able to answer the Guiding Question at the start of each chapter with increasing depth. For example from Structure 1.1:

Guiding Question

How can we model the particulate nature of matter?

Use the Guiding Question revisited at the end of each chapter as a checklist summary of what has been covered and a prompt for more work if anything is not clear. For example, from Structure 1.2:

Guiding Question revisited

How do the nuclei of atoms differ?

Structure 2.4 — Why are some substances solid while others are fluid under standard conditions?

The Linking Questions help you to find interconnections between different parts of the course, linking the concepts in different topics. Brief answers to these questions are given in the text, and you can also try to make up your own questions to explore the connections between different parts of the course.

Note that laboratory work and other skills are an integral part of the course to support your learning of the topics. These skills are assessed in both the Internal Assessment and parts of the External Assessment.

Make sure that you have completed all the exercises at the end of each Understanding and checked your answers in the eBook. The Practice questions at the end of each chapter are mostly questions from previous years' IB exam papers, so they are a very good way of testing yourself at the end of each topic. The answers and guidance used by examiners in marking the papers are also given in the eBook.

SKILLS

The halogen–halide reactions can be investigated in the lab. Full details of how to carry out this experiment with a worksheet are available in the eBook.

Exercise

Q1. What increases **in equal steps of one** from left to right in the periodic table for the elements sodium to argon?

    A   the relative atomic mass

    B   the number of occupied electron energy levels

    C   the number of neutrons in the most common isotope

    D   the number of electrons in the atom

Practice questions

1. Which is a conjugate Brønsted–Lowry acid–base pair?

$$CH_3COOH(aq) + H_2O(l) \rightarrow CH_3COO^-(aq) + H_3O^+(aq)$$

    A CH_3COO^- / H_3O^+ **B** H_2O / CH_3COO^-

    C H_2O / H_3O^+ **D** CH_3COOH / H_2O

Preparing for the examination

Organize your time for review well ahead of the exam date on a topic-by-topic basis. While you are studying make sure that you test yourself as you go – being able to recognize the content on the page is very different from being able to produce it yourself on blank paper. Effective revision generally involves using lots of scrap paper for testing your knowledge and understanding. Practise writing balanced equations, and drawing diagrams, structural formulas and so on.

Remember that you need to cover the work from the entire course. So try to make your studying cumulative; this means building up a connected picture of knowledge and understanding by recognizing how the topics are inter-related and reinforce the same concepts. It can feel a bit like doing a jigsaw, where the more you have done, the easier it gets to add in the new pieces.

There is no choice of questions in the exam so make sure that you do not miss anything out. When you have finished your review of a particular topic, it is a good idea to test yourself with IB questions and time yourself according to how much time you are given for each type of question.

In the examination

The external assessment accounts for 80% of the final grade awarded for Standard Level Chemistry. The assessment consists of two exam papers, which are scheduled separately, and total 3 hours.

	Duration / hours	Total marks	Description of examination	% of total mark
Paper 1 Paper 1A Paper 1B	1.5	55	Paper 1 is presented as two separate booklets, 1A and 1B, which are both to be completed within the allocated time, with no break between. • 30 multiple-choice questions • data-based questions • questions on experimental work	36
Paper 2	1.5	50	• short-answer and extended-response questions	44

In both Paper 1 and Paper 2:

- calculators may be used
- a clean copy of the data booklet will be provided
- all questions are compulsory
- questions may test a general understanding and application of the Nature of Science (NOS)
- you are given 5 minutes reading time at the start of the exam when you can look at the questions but are not allowed to start writing.

Guidance for the exam papers

At the start of the exam, use the 5 minutes reading time to turn through all the pages of the exam booklets, to make sure you are aware of the full extent of the paper.

In Paper 1 you are responsible for the division of time between 1A and 1B. As the two sections are worth similar numbers of marks, you should aim to spend approximately 1 hour on each part.

Multiple choice questions (Paper 1A)

- There is no penalty for wrong answers in multiple-choice questions, so make sure that you do not leave any blanks.
- Read *all* the given answers A–D for each question. It is likely that more than one answer is close to being correct but you must choose the best answer available.

Written response questions (Paper 1B and Paper 2)

- Note the number of marks given in parentheses for each part of a question and use this to guide you in the detail required. In general, one mark represents one specific fact or answer.

- Take note of the command terms used in the questions as these also give an important clue about exactly what is required. It is a good idea to underline the command terms on the question paper to help you focus your answer.

- Sometimes questions include several different instructions, and it can be easy to miss a part of the question; avoid this by crossing off the parts of the question on the paper as you go so you can see what you might have missed.

- Remember these questions are testing your knowledge and understanding of *chemistry*, so be sure to give as much relevant detail as you can, with equations and specific examples where possible. You can only be given credit for what you write down so do not assume anything – show off!

- In calculations, it is essential that you show all your workings very clearly. Also pay attention to significant figures and include units always. When a question has several parts which all follow on from each other, you will not be penalized more than once for the same mistake. So, for example, if you make a mistake in part (a) of a question, but then use that wrong answer correctly in part (b), you will still get full marks for part (b) – *provided your method is clear*. So never give up!

- Be careful with your writing of formulas and structures. Notation in chemistry is very picky, and you need to be precise with upper vs lower case letters, the exact positions of bonds between atoms, etc. For example, Co means something completely different from CO; a hydroxyl group is —OH, not OH—.

▲
Understanding of concepts in chemistry, such as hybridization, is often helped by the use of models and visual analogies. Can you identify the molecules and explain the bond angles of the models in the foreground?

Advice on the Extended Essay

The Extended Essay is a compulsory part of the IB Diploma Programme. It is an independent 40-hour research project in an IBDP subject of your choice. The final essay, of up to 4000 words of formally presented, structured writing, is the longest assignment of the two-year programme. Although this may sound daunting, it is a great opportunity to investigate a topic of particular interest and produce knowledge that is new to you. It shares with the theory of knowledge (TOK) course a focus on interpreting and evaluating evidence and constructing reasoned arguments. The marks awarded for the Extended Essay and TOK are combined to give a maximum of three bonus points.

An Extended Essay in chemistry must have a clear chemical emphasis and deal with chemical reactions, chemical bonding or chemical analysis. Sometimes you need to be careful that it is not more closely related to another subject. For example, a chemistry Extended Essay in an area such as biochemistry will be assessed on its chemical and not its biological content. It should include chemical principles and theory.

Some advice

Why do an Extended Essay in chemistry?

If chemistry is your favourite subject and you want to study it at university, an Extended Essay in chemistry can support your university application as it shows that you have a genuine interest in the subject. Most chemistry essays are from Higher Level students, but this is not a requirement, and many Standard Level students have submitted excellent essays. This is something you can discuss with your supervisor.

Possible approaches

There are a number of possible approaches to an Extended Essay in chemistry.

Experimental	Design and implementation of an experimental procedure and the collection and analysis of the data.
Data-based	Location and extraction of raw or processed secondary data which are analyzed and evaluated.
Theoretical or computer simulation	Development of a quantitative or semi-quantitative model. Perhaps best done in combination with another approach.
Literature or survey	The essay must have a firm chemical basis. Essays written at the level of a newspaper or news magazine article are unlikely to achieve a high mark.
Combination	Some combination of the approaches listed above.

Although it is not a requirement, the best chemistry Extended Essays tend to be those based on experiments performed in a school laboratory. It is easier to have personal input if you are able to plan and modify your experimental procedures and use familiar equipment. If the investigation is undertaken in an external laboratory, you must show clearly that you understand the methods and materials, and describe your role in collecting the data.

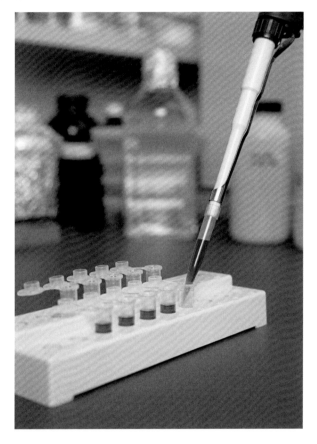

The chemistry Extended Essay and the Internal Assessment

An Extended Essay (EE) is different from the Internal Assessment (IA) and is not a lab report. The IA is more likely to be directly related to syllabus content, whereas the EE could explore aspects of chemistry not covered in the syllabus. An EE must construct a theoretical framework for the underlying chemistry of the chosen topic, whereas the IA task focuses on the application of the scientific method to a problem of interest and will only include some background information. The EE explicitly assesses your ability to analyze and evaluate scientific arguments. The assessment criteria for the two tasks are distinct. The IA is discussed in the Internal Assessment chapter.

Before you start

Read a copy of the subject-specific details of an Extended Essay in chemistry, including the assessment criteria. Read some previous essays and try to identify their strengths and weaknesses. Draw up a list of possible research questions including the techniques you would use to address them. Many of the best essays involve investigations of relatively simple phenomena using apparatus and materials found in most school laboratories.

You may find it useful to consider some of the following techniques when planning your research; it is often appropriate to use a combination of two or more of these approaches:

- titration: acid–base or redox
- chromatography: paper, partition, thin-layer, column
- electrophoresis
- spectrophotometry
- measuring mass or volume changes
- calorimetry
- qualitative and quantitative analysis
- separation and purification techniques in organic chemistry
- use of data-logging probes for some of the above.

The research question

Spend time working out your research question. This is the key to a successful Extended Essay as the overall quality of an essay often depends on the quality of the research question. You should choose a topic that interests you as you will be spending 40 hours on it! Do not choose anything that is too complicated or difficult. Your question must be sharply focused and capable of being addressed in 40 hours and 4000 words. For example, *The ratio of oxygen and chlorine produced at the anode during the electrolysis of different concentrated solutions of aqueous sodium chloride solution* is better than *The electrolysis of salt*. It is likely that your research question will be modified during the process as you complete your research. The crucial thing is that your essay is a true analysis of the research question. You need to directly address the research question in your conclusion.

The research process

- Safety is a priority. Do not do anything in the laboratory without first checking with your supervisor.
- If your approach is experimental make sure that you can produce results that you can analyze and evaluate in detail.
- Use a range of resources to find out what others have done in the area. Textbooks should never be the only source of information. Do not spend all your time online! Use your school and other local libraries if possible. Peer-reviewed articles in scientific journals are the most reliable sources.
- Keep written records of everything that you do and make a note of all references, including the date when internet sites were accessed so that you can build up your footnotes and bibliography as you go along.
- Record all experimental data, including the dates when the experiments were performed and any uncertainties in your measurements. In your preliminary investigations, write down any problems and challenges you encounter and record any modifications. Use your imagination to design new equipment if necessary.

Writing the essay

- Make sure that you address your research question and the Extended Essay assessment criteria.
- Include explanations of any theory not covered in the IB subject guide, including the chemistry of any specialized techniques you have used.
- Use the appropriate chemical language and make sure that all chemical equations are balanced.
- Include sufficient details of any experimental procedures to allow others to repeat the work.
- Check any calculations and make sure that all experimental data are presented correctly including units and significant figures.
- Discuss the limitations of the experimental method and any systematic errors.
- Consider any questions that are unresolved at the end of your research and suggest new questions and areas for possible further investigation.
- Let your enthusiasm and interest in the topic show and emphasize clearly your own personal input.
- Ensure that your word count is close to 4000. You will often find you can cut quite a number of words as you polish your essay at the end.

After completing the essay

Check and proofread the final version carefully.

Use the assessment criteria to grade your essay. Are you satisfied with the grade you award yourself?

Assessment

An excellent Extended Essay

The grade descriptor for an excellent Extended Essay is as follows:

Demonstrates effective research skills resulting in a well-focused and appropriate research question that can be explored within the scope of the chosen topic; effective engagement with relevant research areas, methods and sources; excellent knowledge and understanding of the topic in the wider context of the relevant discipline; the effective application of source material and correct use of subject-specific terminology and/or concepts further supporting this; consistent and relevant conclusions that are proficiently analyzed; sustained reasoned argumentation supported effectively by evidence; critically evaluated research; excellent presentation of the essay, whereby coherence and consistency further support the reading of the essay; and present and correctly applied structural and layout elements. Engagement with the process is conceptual and personal, key decision making during the research process is documented, and personal reflections are evidenced, including those that are forward-thinking.

The assessment criteria

All Extended Essays are assessed according to five criteria, A–E. Each criterion is made up of a series of strands that act as the core focus for that criterion. The tables in the following section give a checklist to help you achieve the highest level for each criterion. The notes in italics refer specifically to essays in chemistry.

A. Focus and method

Strand	Checklist
Topic	Is the topic communicated accurately and effectively?
	Is the research topic clearly identified and explained?
	Is the purpose and focus of the research clear and appropriate?
	The topic for study may be generated from the chemistry course or may relate to a subject area beyond the syllabus content.
Research question	Is the research question focused and clearly stated in the introduction?
	Does the research question address an issue appropriately connected to the discussion in the essay?
Methodology	Have an appropriate range of relevant source(s) and/or method(s) been selected?
	Is there evidence of an effective and informed selection of sources and/or methods?
	You must research the existing literature on the topic and choose an appropriate methodology to pursue the investigation. If practical work is undertaken, the rationale for choosing the procedure should be clearly explained.

Do	Don't
✓ Cite the source of the method used.	✗ State multiple research questions.
✓ Discuss possible alternative methods (experimental or otherwise) and justify your choice.	✗ Make statements that are not justified including unsupported generalizations.
✓ Discuss any modifications you have made to a standard procedure.	✗ Include superficial, irrelevant or unresearched background information.
✓ Avoid repetition.	

B. Knowledge and understanding

Strand	Checklist
Context	Are the source materials clearly relevant and appropriate to the research question?
	Is the knowledge of the topic clearly demonstrated and coherent?
	Are the sources used effectively with understanding?
	Literature cited should generally come from acknowledged scientific sources.
Communication	Are subject-specific terminology and concepts used accurately and consistently?
	Does the essay demonstrate effective knowledge and understanding?
	Chemical nomenclature and terminology should be used consistently and effectively. This includes relevant chemical and structural formulas, balanced equations with state symbols, reaction mechanisms, significant figures and SI units.

Do	Don't
✓ Maintain a consistent linguistic style throughout the essay. ✓ Select a range of sources. ✓ Explain the chemistry behind the techniques used. ✓ Process data correctly and show sample calculations.	✗ Give unnecessary explanations for basic chemistry.

C. Critical thinking

Strand	Checklist
Research	Is the research appropriate to the research question? Is the research consistently relevant? Does the research support the argument?
Analysis	Is the research analyzed effectively and clearly focused on the research question? Are conclusions to individual points of analysis effectively supported by the evidence? *Analysis often includes mathematical transformations, tables of processed data and graphs. Do not use statistical analysis unless it adds to the chemistry aspect of the question. The analysis should support and clarify the argument leading to the conclusion.*
Discussion and evaluation	Is an effective and focused reasoned argument developed from the research? Is the conclusion reflective of the evidence presented? Is the reasoned argument well-structured and coherent? Is the research critically evaluated? *Unresolved issues should be identified, and suggestions made as to how these could be explored further. Any weaknesses in experimental design or systematic errors should be identified. The quality, balance and quantity of sources should be evaluated. The validity and reliability of any data, including measurement uncertainties, should be evaluated and discussed.*

Do	Don't
✓ Consider different approaches or counterarguments. ✓ Question the reliability of any sources used. ✓ Include any experimental uncertainties. ✓ Discuss any assumptions made in your arguments. ✓ Address the research question in the conclusion.	✗ Write a descriptive or narrative account. ✗ Introduce new material into the conclusion that has not been referred to earlier. ✗ Suggest superficial or unrealistic areas for further investigation.

D. Presentation

Strand	Checklist
Structure	Is the structure of the essay appropriate in terms of the expected conventions for chemistry, the topic, and the argument? *Large tables of raw data are best included in an appendix, where they should be carefully labelled. Tables of processed data should be designed to clearly display the information in the most appropriate form. Graphs or charts drawn from the analyzed data should be selected to highlight only the most pertinent aspects related to the argument. Too many graphs, charts and tables will distract from the overall quality of the communication.*
Layout	Do the structure and layout support the reading, understanding and evaluation of the essay? *If an experimental method is long and complex, you may include a summary of the methods in the body of the essay but be careful to ensure that the summary contains all elements that contribute to the quality of the investigation.*

Do	Don't
✓ Acknowledge any unoriginal material. ✓ Label diagrams clearly. ✓ Use the appendix to show materials, unsuccessful trials, risk assessment, preparation of solutions, etc.	✗ Use superficial diagrams or images.

E. Engagement

Strand	Checklist
Research focus	Are your reflections on decision-making and planning evaluative? Do your reflections communicate a high degree of engagement? Do your reflections demonstrate initiative and/or creativity?
Planning and process	Do you discuss the challenges you experienced in the research process and include any actions taken or new ideas generated? *Reflections include the relative success of approach and strategies chosen, and the approaches to learning skills you developed during the process.*

Do	Don't
✓ Discuss the challenges you faced and details of how you overcame them. ✓ Include questions that emerged as a result of the research. ✓ Discuss how your understandings developed or changed during the process.	✗ Give a simple descriptive account of what you have done.

Bibliography and references

It is required that you acknowledge all sources of information and ideas in an approved academic manner. Your supervisor or school librarian will be able to give you advice on which format to follow.

World studies Extended Essays

As you think about topics for an Extended Essay in chemistry, it may be that you find an area of current global interest which overlaps with another subject. In this case, it may be appropriate to submit the essay as world studies, which is a separate IB Extended Essay subject. An essay in world studies must have the following characteristics:

- It must focus on an issue of contemporary global significance.
- It must involve an in-depth interdisciplinary study, making reference to at least two IB subjects.

The most successful topics reveal connections between specific or local places, people, phenomena or experiences and the larger global framework in which they take place.

Remember to use a range of resources including the internet and any libraries available.